The Evolutionary Biology
of Human Female Sexuality

The Evolutionary Biology of Human Female Sexuality

Randy Thornhill and Steven W. Gangestad

OXFORD
UNIVERSITY PRESS
2008

Oxford University Press, Inc., publishes works that further
Oxford University's objective of excellence
in research, scholarship, and education

Oxford   New York
Auckland   Cape Town   Dar es Salaam   Hong Kong   Karachi
Kuala Lumpur   Madrid   Melbourne   Mexico City   Nairobi
New Delhi   Shanghai   Taipei   Toronto

With offices in
Argentina   Austria   Brazil   Chile   Czech Republic   France   Greece
Guatemala   Hungary   Italy   Japan   Poland   Portugal   Singapore
South Korea   Switzerland   Thailand   Turkey   Ukraine   Vietnam

Copyright © 2008 by Oxford University Press, Inc.

Published by Oxford University Press, Inc.
198 Madison Avenue, New York, New York 10016

www.oup.com

Oxford is a registered trademark of Oxford University Press

Library of Congress Cataloging-in-Publication Data
Thornhill, Randy.
The evolutionary biology of human female sexuality / Randy Thornhill and
Steven W. Gangestad.
      p. cm.
Includes bibliographical references and index.
ISBN 978-0-19-534098-3; 978-0-19-534099-0 (pbk.)
1. Women—Sexual behavior.   2. Evolution (Biology)   3. Sex (Biology)
I. Gangestad, Steven W.   II. Title.
[DNLM: 1. Evolution. 2. Sexual Behavior—physiology. 3. Fertility—physiology.
4. Menstrual Cycle—physiology. 5. Sex Characteristics. HQ 29 T512e 2008]
HQ29.T46 2008
306.7082—dc22   2008002779

9 8 7 6 5 4 3 2 1

Printed in the United States of America
on acid-free paper

This book is dedicated with perpetual love to our children
Aubri, Margo, Patrick, Reed, and Sophie Thornhill
and
Max Gangestad

PREFACE

This book grew out of our synergistic and complementary research collaboration over many years. The writing here, too, represents synergy. Every paragraph—indeed, virtually every sentence—reflects the thinking and writing of both of us. The order of authorship could have been determined by a coin flip.

Our research collaboration took root in 1991, growing out of a shared interest in evolutionary processes that have created sexual and related features, especially in humans, and out of the convenience of our academic residences in neighboring buildings, albeit different academic departments. We first investigated the role of developmental instability in the human mating system. Our attempts to understand the sexual selection processes responsible for this role led us to explore ways in which women's sexuality changes across the menstrual cycle. Our findings raised questions of how men respond to these changes, as well as how male partners' features moderated changes in women's sexuality, which we also jointly studied.

After a decade of collaboration, we came to realize that our findings and related work by others did not merely add incrementally to an empirical literature; they pointed to the need for a new interpretation of human evolutionary history, an orientation contrasting with traditional ones. The new data suggested that women have estrus, that their estrus is possibly concealed, and that women's sexuality functions differently during and outside estrus. Furthermore, we found that traditional accounts often did not fully integrate some important perspectives—notably, comparative reproductive biology of vertebrates in general and salient components of evolutionary theories pertaining to sexual selection, signaling, and life history.

This book offers our new interpretation of the evolutionary history of women's sexuality. As women's and men's sexualities have coevolved, our book also casts men's history in a different light. We emphasize that our book is not intended to be a comprehensive survey text on our topic (though we do broadly consider all of the primary components of female sexuality highlighted in the literature). Its purpose is to offer new perspectives on human reproductive biology and thereby, we hope, stimulate new lines of research and thinking.

Although *Homo sapiens* is our primary topic, the book discusses how recent knowledge of human sexuality may importantly illuminate major topics concerning nonhuman primate sexuality and vertebrate sexuality in general. As well, it illustrates the reverse: that human sexuality is informed significantly by comparative sexuality across the vertebrates. We trace the phylogenetic origin of estrus to the first vertebrate and propose that it has a common evolved function throughout vertebrates, including women. Complete understanding of the evolutionary history of human sexuality requires elucidation of both of its distinct, causal contexts: the relevant traits' phylogenetic origins and their evolutionary maintenance after origin events.

We both draw on and speak to many literatures. Our grandest hope, perhaps, is that this book will prove valuable to a wide range of scholars—researchers in human reproduction, anthropologists, psychologists, primatologists, mammalogists, ornithologists, herpetologists, and ichthyologists. In drawing on diverse literatures, we hope to have represented them fairly. Naturally, we hope, too, that we have treated fairly authors whose publications we cite. Many scholars are responsible for ideas we present here, and we hope that we have properly credited them. We appreciate, too, that even ideas we criticize have importantly contributed to a dialogue among scientists interested in reproductive biology, one that has shaped our own views. We hope that a few of the ideas we ourselves offer prove to be right. But we fully expect that, in places, we have erred in our own thinking and conjecture, and we wish that our mistakes may help others find truth. By our own standards, we will have succeeded if we at least convince readers that many conventional interpretations that are widely accepted are problematic and if our attempts to reorient thinking stimulate productive scholarly activity.

We acknowledge colleagues who read and critically evaluated portions of or our entire manuscript at various stages of its development: Rosalind Arden, David Buss, Chris Eppig, Peter Ellison, Christine Garver-Apgar, Melissa Heap, Chris Jenkins, Astrid Kodric-Brown, Jane Lancaster, Kenneth Letendre, Ilanit Tal, Josh Tybur, and Paul Watson, as well as four anonymous reviewers. Corey Fincher generously provided input on two drafts of the entire manuscript. Ric Charnov did, as well, and offered steadfast encouragement throughout the project. Anders Møller provided valuable insights on the entire penultimate draft. We are grateful to all. We benefited greatly not only from these explicit critiques of our writing but also from informal discussions with these scholars and many others. We thank, too, Anne Rice, who assisted with word processing, and Amanda Humphrey and Phuong-Dung Le for various forms of assistance. We thank Elsevier Press for permission to use in chapter 2 the quotations from Joan Silk (2001), *Evolution and Human Behavior* 20: 443–448. Finally, for support of some of our own research, we thank the National Science Foundation (Grant Award 0136023) and the Sense of Smell Institute (formally the Olfactory Research Fund). (Any opinions, findings, and conclusions or recommendations we express are our own and do not necessarily reflect the views of the National Science Foundation.)

CONTENTS

The Evolutionary Biology
of Human Female Sexuality

1 Background and Overview of the Book

An Abridged History of the Study of Woman's Sexuality

"I should say that the majority of women (happily for them) are not very much troubled with sexual feeling of any kind." So wrote William Acton in 1857. Acton was an acclaimed expert on the topic—a physician and the leading sexologist of Victorian England. His book, *Functions and Disorders of the Reproductive Organs in Youth, in Adult Age, and in Advanced Life*, appeared in eight editions in the nineteenth century. He may well have expressed not only his own views but also those of generations of men who read his work.

Just two years after the first edition of Acton's book was published, Darwin's (1859) *Origin of Species* appeared, followed a dozen years later by *The Descent of Man* (1871). The latter book, in particular, elaborated Darwin's theory of evolution driven by female choice, as pushed along by female sexual preferences, views that met widespread resistance. A main problem, skeptics surmised, was that the theory presumed that, over many generations, females agree about which male features they find sexually attractive (Bajema, 1984). Yet conventional wisdom viewed women, and by extrapolation other females, as erratic, fickle, undecided, or absent in their sexual interests. Darwin himself held fast to his ideas to his death; his last public defense of his theories, read at a meeting of the Zoological Society of London just hours before he died, emphasized the important role of female choice in evolution (Bajema, 1984, p. 150). Yet more than a century would pass before Darwin's claim that female choice is a salient evolutionary force was vindicated and became widely accepted in evolutionary biology—this did not occur until the past few decades, supported by literally hundreds of empirical

3

documentations of female choice and male adaptations to impress choosy females in many animal taxa (see Andersson, 1994, for a review of empirical studies through the early 1990s).

When women's sexual feelings were not entirely absent, prominent nineteenth-century Victorians thought they caused problems. *Hysteria* (or "wandering womb"; the Greek root is shared by *hysterectomy*) was the name for an alleged neurotic disorder thought to be responsible for physical and psychological maladies. A common cure was to masturbate the female sufferer to "hysterical paroxysm"—orgasm—as performed by a physician, not by the woman's lover or husband. The heavy demand on physicians to perform this therapy ultimately led to the invention of the vibrator, which allowed them to treat patients quickly and reliably (Maines, 1999).

Coexisting with the view that most women lack sexual passions (or possess patho-logical ones) was a common belief that sexual passions in women, when they do occur, should be feared. The original text of Acton's observation was that women lacked sex-ual desire, "happily for them." In later editions, the text was altered to say that women lacked sexual desire, "happily *for society*" (emphasis added). Three centuries earlier, Montaigne wrote that husbands should approach their wives "prudently and severely," for fear of awakening in them lasciviousness (as cited by Ellis, 1922), which could under-mine the certainty of husbands' paternity. As it made women unfit for marriage, then, wantonness in women was to be harshly discouraged during girls' upbringing. Girls were instructed to be modest in interactions with men and boys and sexually reserved with husbands (Shalit, 1999). As R. J. Culverwell, in his 1844 *Porneiopathology*, put it, "Continence in females is...the brightest ornament a woman possesses" (as quoted in Maines, 1999, p. 36). Though it appeared in dramatic form in nineteenth-century Victorian England, demand for female sexual modesty has arisen frequently in cultures, in some cases leading to practices designed to do more than cultivate sexual modesty, cruelly preventing forms of female sexual pleasure (e.g., claustration and genital modi-fication; Betzig, 1986; Dickemann, 1981; Gaulin & Boster, 1990; Hrdy, 1997; Lancaster, 1997; N. W. Thornhill & Thornhill, 1987; for cross-cultural patterns in girls' training in sexual modesty, see Low, 1989).

Amidst a chorus advocating harsh sexual restraint, voices of dissent spoke at the dawn of the twentieth century. Elizabeth Blackwell (1902, p. 58), the first woman to obtain a medical degree in the United States, concluded, "it must be distinctly recog-nised that the assertion that sexual passion commands more of the vital force of men than of women is a false assertion, based upon a perverted or superficial view of the facts of human nature." Havelock Ellis (1903, 1922) wrote of ways that prevailing Western views deprived women of "erotic rights." George Bernard Shaw and other prominent intellectuals of the time expressed their disdain for the sexual control of women, even for marriage as an institution.

Around the same time, serious scientific investigation of women's sexuality was seeded and took root. From the outset, scientific inquiry posed questions of function: What is the nature of women's sexuality (or, as we shall see, sexualities) and what are its adaptive functions? The twentieth century and the beginning of the twenty-first century witnessed three major overlapping waves of such adaptation-minded research.

Wave One: The Physiology of Women's Fertility and Investigation Into Estrus

A first set of questions that researchers tackled concerned physiology more than psychology: What is the nature of women's fertility across the menstrual cycle, and what mechanisms are responsible for it? Researchers pursued answers in earnest from 1890 onward (Corner, 1942). By 1930, a consensus emerged: Ovulation occurs at mid-cycle, not during menses, as formerly thought. Research during this period lacked sophisticated selectionist thinking; rather, researchers embraced a comparative perspective, which assumed that humans, having evolved from "lower animals," differed in degree, not kind, from other species. Domesticated mammals were observed to experience estrus, or "heat," at ovulation. Women, then, were hypothesized to similarly experience heat at ovulation. Researchers furthermore proposed that ovarian hormones, most notably estrogen, play a role in women's estrus, just as demonstrated in nonhuman female mammals.

In the 1930s, these hypotheses guided psychological investigations, the first of their kind, of patterns of women's sexual behavior across the menstrual cycle (Corner, 1942; Wallen, 2000). Prevailing ideas about estrus and its nature presented two predictions to researchers. First, women will experience a mid-cycle increase in eroticism, sexual motivation, and sexual behavior. Second, women's male partners will find women more sexually attractive when they are mid-cycle. Over a period of decades, research that tested these predictions yielded mixed results. Some studies reported enhanced sexual interest at mid-cycle, others detected it at infertile cycle phases, and still others found no change in female sexuality across women's cycle (for summaries and reviews, see Brewis & Meyer, 2005; Hill, 1988; Steklis & Whiteman, 1989). Similarly, evidence failed to convincingly demonstrate predicted changes in male sexual interest as a function of women's fertility (see Hill, 1988; Steklis & Whiteman, 1989, for reviews and summaries). By the 1960s, with no clear-cut evidence demonstrating estrus, a prevailing view emerged and solidified: In the evolution of humans, women lost estrus (Etkin, 1964; Jolly, 1972). This finding purportedly established, researchers' and theorists' next question naturally followed: *Why* did women lose estrus?

Many perspectives on this question that took shape (see "Wave Two," below) were influenced implicitly by a prevailing notion about the function of estrus in those species possessing it (including species ancestral to modern humans): Ewes, mares, sows, and dogs in heat are highly motivated to mate because, as only at this time can they conceive, they are motivated to get sperm. Estrus fulfills the function of sperm acquisition (Corner, 1942; Wallen, 2000). This notion that the heightened sexual motivation of estrous females is the indiscriminate pursuit of any sperm remains widely held in mammalian reproductive biology (Nelson, 2000). As women lack estrus, women's sexuality, according to this way of thinking, is not merely about getting sperm.

Wave Two: Women's Means of Obtaining Nongenetic Material Benefits Through Their Sexuality

Robert Trivers's (1972) classic treatment of sexual selection and parental investment ended the century-long neglect of Darwin's seminal ideas about female choice and its influence on the evolution of male features. Though exceptions (e.g., Fisher, 1930; Williams, 1966) are important and noteworthy, Trivers's paper catapulted female choice to its lofty position within current selectionist thinking.

Donald Symons's (1979) book, *The Evolution of Human Sexuality*, appeared shortly thereafter, a landmark in the study of human sexuality. It was the first serious effort to investigate and inquire into the nature of human sexuality, guided by both the sophisticated, adaptationist thinking offered by Trivers and the more general framework for thinking about adaptations, by-products, and how to infer the forces of historical selection explicated by George Williams (1966) and followers. Symons analyzed various data sources—homosexual behavior, pornography, classical literature, cross-cultural patterns in the anthropological record, and so on—to test hypotheses about the functional design of men's and women's sexualities. He argued that men's sexuality includes adaptation designed for pursuit of many sex partners without continued investment in them and for pursuit of young adult sex partners. Of women, he argued that evidence revealed design for caution and discrimination surrounding mating decisions. Symons focused on women's preference for male traits related to status and resource holdings. His research strongly supported an evolved sexuality of women: They have sexual adaptations for mate choice based on the quantity and quality of nongenetic material goods and services possessed by males.

This view that women have sexual adaptations to obtain material goods and services offered by males shaped one view of the reason women lost estrus. Loss of estrus during this period was typically referred to as "concealed ovulation" (though, as we forcefully argue, there are very important distinctions between these phenomena), as reflected in the absence of marked changes in female sexual behavior or female attractiveness to males across the menstrual cycle and of purported signals of ovulation (e.g., sexual swellings). (Researchers often included the permanence of female sexual traits such as breasts as still other manifestations of concealed ovulation.) As one adaptive explanation, Symons (1979) proposed that, through concealed ovulation, women could obtain material benefits from male partners throughout the cycle, not merely when fertile, for men are motivated to obtain sexual access to partners (and will not be shaped by selection to lose sexual interest in women when they are not fertile if they cannot discriminate women's fertile state from infertile states). Women's sex, Symons argued, is not merely about getting sperm; women's sexuality has been designed primarily to obtain material benefits for herself and her offspring, with concealed ovulation playing a central role in this endeavor. Alexander and Noonan (1979) presented a related hypothesis.

(Alternative explanations for women's concealed ovulation, which we discuss later, were offered this same year; Benshoof & Thornhill, 1979; Burley, 1979; Symons, 1979.)

In the ensuing quarter century, a number of researchers, with David Buss at the vanguard, investigated women's criteria of mate choice and aspects of their sexual behavior, with an eye toward comparing them with male attributes and expecting sex differences along the lines that Symons outlined. In many respects, Symons's ideas and Buss's expansions of them have received support (for summaries, see Buss, 1994, 2003b; Geary, 1998; Townsend, 1998). Among these respects is the idea that woman's sexuality includes what we refer to as *extended sexuality*, sexual adaptation that functions to gain access to nongenetic material benefits from males through its expression when women are not fertile within their menstrual cycles.

While Symons focused his attention on human sexuality, another important researcher of the era, Sarah Hrdy, theorized about female primate sexuality more generally. Hrdy studied langur monkeys in the field. She showed that female mating with multiple males occurs in langurs and other nonhuman primate taxa and argued that this behavior was ancestral among primates in general, including the hominin descent lineage (Hrdy, 1979, 1981; Hrdy & Whitten, 1987). (Hominins are the genera [e.g., *Ardipithecus*, *Australopithecus*, *Homo*] comprising the tribe Hominini of the subfamily Homininae of the family Hominidae. Hence, modern humans are hominids [family], hominines [subfamily], and hominins [tribe]; e.g., Wood & Constantino, 2004.) Hrdy's findings appeared to be contrary to arguments by Trivers, following from Bateman (1948), that female reproductive success does not increase as a function of number of sexual partners. If ancestral female primates did not benefit from mating with multiple males, the fact that female primates do so now is puzzling. Hrdy argued that primate female sexuality (and multiple mating) cannot be understood solely in terms of its direct reproductive effects on number of conceptions and thereby offspring number; rather, its effects on female reproductive success must be partly understood in terms of its effects on the survival and well-being of individual offspring (1981). Specifically, she proposed that langur females and females of many other species of nonhuman primates have adaptation that motivates them to mate with multiple males to confuse paternity and thereby prevent the males from maltreating females' future offspring. In so arguing, Hrdy claimed that female primates exhibit situation-dependent sexual receptivity and motivation, which ultimately led to a paradigm shift in how female, nonhuman primates are viewed by primatologists: as conditional sexual tacticians, adaptively modifying their sexual behavior to meet changing ecological and social conditions (Shahnoor & Jones, 2003).

Symons and Hrdy emphasized contrasting, perhaps even contradictory, aspects of female sexuality—Symons stressing female sexual modesty and caution, Hrdy highlighting female promiscuity. The spirit of their work nonetheless is kindred in one very important respect: Both argued that the nature of the function of female sexuality must be understood not only by its direct effects on conceptions but also by its effects on other ways by which females enhance reproductive success. According to Symons, female sexual modesty and receptivity throughout the cycle functions to obtain material benefits for offspring, just as sexual promiscuity and paternity confusion do according to Hrdy. By 1980, the study of female sexuality was unshackled from the mistaken idea that sexuality is merely about getting conceptions, and the manifestations were multiple.

Wave Three: Good Genes and Intersexual Conflict

So female sexuality, and women's sexuality in particular, is clearly not merely about getting sperm. Even getting sperm, however, is not *merely* about getting sperm. Sperm contain packages of DNA, which combine with packages of DNA that females produce—eggs—to conceive offspring. Not all packages of DNA, however, are identical; indeed, not all contribute equally well to developmental processes that affect the adaptedness of offspring. To the extent that some males produce sperm with packages of DNA that enhance offspring fitness better than others, selection will favor females who choose those males. Robert Trivers (1972) wrote of female choice for male attributes associated with the acquisition of genetic benefits to be passed onto offspring and, in so doing, introduced the term *good genes*. At least another two decades would pass, however, before researchers seriously examined women's sexuality for adaptation that functions to secure paternal genetic benefits for offspring that enhance their reproductive capacity (health and general condition, survival, and/or mating success). Following suggestions by Trivers (1972; see also Williams, 1966), Benshoof and Thornhill (1979) and Symons (1979, in part) proposed that concealed ovulation in women was positively selected because crypsis allowed females to copulate outside their pair bonds and with males possessing genes that were superior to those of their primary pair bond partners without those partners' knowledge. As Buss (2003b, p. 225) remarked about the earlier period, however, on the whole, "the theory downplayed the role of 'genetic quality' in mate selection."

Researchers neglected female choice for good genes partly because theories that female choice can function to secure superior genes for offspring quickly became controversial in evolutionary biology. Following Trivers's seminal article, Amotz Zahavi introduced in 1975 (and expanded in 1977) a theory of how sexual selection exerted by choice for genetic quality may shape characteristics of the chosen sex. In particular, he argued that signals of genetic quality must be "honest" and thereby will be costly—too costly for relatively unfit individuals to bear (see review in Zahavi & Zahavi, 1997). But important assumptions were left unaddressed. Notably, selection removes "bad genes." Does selection leave sufficient heritable genetic variation between potential sires to make it worthwhile for females to choose for good genes? What prevents all meaningful genetic differences between sires from being eliminated by selection? In addition, Zahavi's theory was a verbal argument. Could it be quantitatively modeled into a theory whose assumptions and derivations are precise? As we later review, by the early 1990s, many biologists were finally convinced that choice for good genes could evolve. Only then, however, did researchers seriously investigate sexual selection for good genes in any species, and eventually in humans.

Another possible impediment to appreciating female selection for good genes was a tendency by evolutionary biologists to focus on male features that regulate competition for mates and hence failing to recognize ways by which females actively regulate this competition or choose mates despite male competition. This bias can be seen in the study of sperm competition dating to the 1970s. Geoff Parker announced discovery of sperm competition in his 1970 classic, "Sperm competition and its evolutionary consequences in the insects." This competition between the sperm of different males for the egg(s) of a single female was a previously unrecognized type of sexual selection. Soon after Parker's discovery, biologists widely reinterpreted male traits in animals as having evolved due to sperm competition. In many cases, however, sperm competition was the only hypothesis considered. In "Alternative hypotheses for traits believed to have evolved by sperm competition," Thornhill (1984a) stressed two points: First, sperm competition should be studied against other reasonable, alternative, ultimate hypotheses in investigations of the design of male reproductive behavior, morphology, and physiology; second, females are important players in intersexual conflict games pertaining to control of fertilization, as Parker himself had emphasized earlier. In reference to sperm competition adaptation, Parker (1970, p. 551) stated, "the female cannot be regarded as an inert environment around which this form of adaptation evolves." In the late 1970s and in the '80s and '90s, some researchers stressed the value of theory that included male and female conflicting reproductive interests and females as active evolutionary players in sperm competition games among males (see especially Charnov, 1979; Møller's, 2001, review of bird studies in the 1980s; Parker, 1979a; also Birkhead & Møller, 1992; Thornhill, 1983, 1984a, 1984b; Thornhill & Alcock, 1983).

Some biologists, however, viewed females as mere vessels in which sperm of different males compete, rather than players with reproductive interests of their own that often differ from those of males. As Eberhard commented in *Female Control: Sexual Selection By Cryptic Female Choice* (1996, p. 420), "Abandoning the idea that females are morphologically and behaviorally passive and inflexible in male–female interactions promises to give a more complete understanding of sexual selection." The point is not that sperm competition is an unimportant cause of evolution; it is a salient sexual selective force that has designed male traits in all major taxa investigated (for humans, see review in Shackelford & Pound, 2006; for other animal taxa, see reviews in Birkhead & Møller, 1992, 1998; Simmons, 2001). Rather, female reproductive interests must always be considered, too. Intersexual conflict, whether involving conflicting male and female interests in control of fertilization or in other contexts, gives rise to evolutionary arms races between the sexes (Parker, 1979a; Rice, 1996; reviewed in Arnqvist & Rowe, 2005). Buss's (1989a) work on what he referred to as "strategic interference" properly casts both sexes of humans in an evolutionary race to achieve their sexually distinct optima against the interests of the opposite sex. In coevolutionary races, neither party may actually achieve an optimum; selection on one sex favoring an optimum is opposed by selection on the other sex. Though each sex may not be fully adapted to the other sex, however, each sex's features resulting from antagonistic coevolution must be understood in strategic terms.

Research in the past decade, then, advanced understanding of women's sexuality in two major ways. First, a strong case for the role of good-genes sexual selection was established. Female choice has partly been designed as a result of it. But male choice for females has been similarly shaped; certain traits in women, such as estrogen-facilitated features of women's faces and bodies (which we later treat in detail), appear to be designed by sexual selection to signal personal quality pertaining to future or residual reproductive value. Second, researchers became increasingly aware of the ways by which intersexual conflict fuels antagonistic coevolution of the sexes. In so doing, they more fully incorporated Hrdy's view of female sexuality being strategic and situation dependent.

Estrus Redux

The scientific investigation of women's sexuality has a noble history. The dedicated pursuit of the truth of women's sexual design and the phylogenetic origin of the design is apparent throughout it. As in all histories of scientific discovery, errors have been made. Many have been corrected by subsequent research. Girdwood (1842) dissected cadaver ovaries and found a tight correlation between the number of ovarian scars (due to ovulation) and the calculated number of menses. On this basis, he reasoned that women ovulate during menses. His interpretation was reasonable and was taken as "truth" by a generation of researchers. As established by later research, it was not truth. Scientific bodies of knowledge not uncommonly move in this way—not as slow, steady accumulations, but rather as moves down false pathways, only to have the scientific community realize its error and suddenly backtrack. Researchers make observations and offer interpretations. The interpretations become "fact." Subsequent researchers come to see the observations in light of a very different interpretation and show that many "facts" are wrong.

Recent research has, we show, gone beyond previous views by, for instance, leading to the recognition that female choice for good genes, as well as for nongenetic material benefits, has played an important role in human sexual evolution. We argue, however, that this recent research does not merely add to the accumulation of "facts" contributed by earlier research. Rather, reflection on recent research ultimately has led us to question previous interpretations of women's sexuality, some of which have come to be taken as "fact."

As we detail in later chapters, much of the most compelling evidence for female choice for good genes in humans comes from research examining changes in female sexuality across the menstrual cycle. The features that women are most attracted to when fertile in their cycles are not precisely the features they are most attracted to when nonfertile. Many features most sexually attractive to fertile women, we argue, are markers of good genes (or were ancestrally). These changes furthermore have real consequences for patterns of women's sexual attraction to men across the cycle—despite small and uneven overall changes in female "sexual desire" across the cycle. Previous research on cyclic changes in women's sexuality did not find consistent patterns because it looked in the wrong places; recent research clearly establishes these changes.

Do women, then, have "estrus" after all? In spite of generations of researchers' knowing the "fact" that women *lost* estrus, we argue that the answer is yes. We furthermore argue that this terminology is appropriate not merely in a metaphorical sense. Women's estrus, we propose, is in key respects homologous with the estrus of other mammalian species—indeed, homologous with what we refer to as estrus in vertebrates in general.

That is, women have estrus because, through deep historical time, they and their ancestors never lost estrus.

The nature of estrus in women, however, has even more profound implications for an understanding of female sexuality in general. The function of estrus in women is not to obtain sperm per se—to merely increase the chances of conception. It is to enhance the probability that sires of offspring have good genes for offspring. The function of estrus in women has not changed in this regard, however; rather, careful reconsideration of estrus in mammalian females in general (indeed, vertebrate females even more broadly) suggests that it is very unlikely that the function of estrus has *ever* been to merely facilitate access to sperm and enhance the chances of conception, contrary to enduring belief. Rather, just as with estrus in women, estrus in all vertebrates likely has functioned for several hundred million years to obtain good genes.

Further reflection on recent research exposes other misconceptions. "Concealed ovulation" is not the loss of estrus. Women have not lost estrus. We argue, however, that selection has operated to "conceal" ovulation in women to others, most notably men. Despite this selection, men are not completely insensitive to changes in female fertility status across the cycle. Female sexual receptivity across the cycle is similarly not loss of estrus. "Extended sexuality"—sexual receptivity during nonfertile periods—despite estrus in women reveals dual sexuality, two distinct sexualities that have different, even opposing, functions. Finally, permanent sexual ornaments, such as breasts, do not reflect loss of estrus, deceptive extended estrus, or concealed ovulation, but rather reflect other adaptations.

We need not expand on these points here; in many ways, the goal of this entire book is to do so. For precisely the reason that, in many ways, the purpose of this book is to expand on these themes, however, we do mention these themes here. Naturally, a good deal of what we offer describes details of the studies and key findings of the past decade or two. This book pulls together and concisely summarizes bodies of work that we find exciting. We try to go well beyond what can be found in the literature, however. The major focus of this book is on more general theses that, together, constitute a framework for thinking about women's sexuality. Several of these theses contradict received wisdom and, in that sense, are bold, perhaps provocative. At the same time, we argue, they are grounded firmly in well-founded theory and empirical findings and, in that sense, are cautiously constructed.

Disagreement with received wisdom calls for humility in a couple of respects. Numerous researchers, including some who have pioneered the investigation of the evolution of women, have made significant errors in theorizing and interpreting women's continuous sexuality across the menstrual cycle, loss of estrus, concealed ovulation, and sexual ornamentation. We critique these ideas but applaud the noble efforts of those whose ideas we criticize; our observations necessarily build on those efforts, and it is senseless for us to think we tear them down. We are also keenly aware of the possibility that our own interpretations may subsequently be shown to be gravely in error— indeed, the likelihood that, in some respects at least, they are. We naturally hope that we get at least a few things right in the end. If we do not, however, we hope that our own efforts assist others who do.

Overview of the Book

We analyze four major features of woman's sexuality using evolutionary methodology. Chapter 2 discusses the methods we use, those of adaptationism and phylogenetics. Three of the four features we address—continuous copulability, permanent sexual ornamentation, and concealed ovulation—prominently figure in many theoretical treatments of the phylogenetic origin of human social behavior (e.g., see especially Alexander, 1979, 1990; Baker & Bellis, 1995; Benshoof & Thornhill, 1979; Cartwright, 2000; Fisher, 1982; Geary & Flinn, 2001; Gray & Wolfe, 1983; Hill, 1982; Hrdy, 1981, 1997; Lovejoy, 1981; Symons, 1979; for discussion of major human origin theories before 1979, see Alexander & Noonan, 1979; Benshoof & Thornhill, 1979). The fourth feature we treat— variation in women's sexuality across the menstrual cycle—has largely been denied until recently and speaks forcefully to the question of whether women experience mid-cycle estrus, or heat. (For reviews of relevant literature, see Dixson, 1998; Hrdy, 1997; Gray & Wolfe, 1983; Manson, 1986; Meuwissen & Over, 1992; Pawlowski, 1999a; Small, 1993; Tarín & Gómez-Piquer, 2002; Wallen, 2000.)

In many animal species, females are sexually active only or primarily at and/or near periods of high conception risk. Although many species deviate from this pattern, women represent an extreme: They show interest in copulating when adolescent (prior to achieving reliable conceptive cycles), across all days of the menstrual cycle, and during pregnancy and lactation. What selection pressure(s) is (are) responsible for women's continuous sexual activity or copulability outside of fertile periods? Recent game theoretical and other modeling and across-species comparative evidence support the hypothesis that extended female sexual activity is selected directly when males deliver nongenetic material benefits to females and/or their offspring and functions to increase access to these benefits by exchanging mating for them. In chapter 3, we review this modeling and comparative evidence and suggest that women's extended sexuality fits the pattern across species and is an adaptation to compete effectively for male material assistance. In chapter 4, we address the specific ways in which extended sexuality functions in humans and attempt to answer the question of why women represent an extreme.

We move next to consider the evolution of permanent facial and bodily sexual ornamentation of women that arises at puberty through adolescence (e.g., the estrogen-facilitated waist-to-hip ratio, velvety skin texture, gracile facial features, and copious fat in the breasts, thighs, and buttocks) but remains throughout the reproductive lives of women. Female sexual ornaments across species have recently received attention from evolutionary biologists after a lengthy period of almost exclusive focus on investigation of male sexual ornaments (see Amundsen, 2000; Andersson, 1994). Evidence now indicates that female ornaments in a number of species have evolved by sexual selection acting on females and function as honest signals of female phenotypic and possibly genetic quality to increase access to male-provided material benefits. In women, however, these features are not restricted to the time of a reproductive bout and/or periods of moderate to high conception probability, as is usual in ornamented females of other animal species. Moreover, women's ornamentation achieves full development at adolescence, well before peak age-related fertility. Expanding on Marlowe's (1998) hypothesis about the evolution of women's breasts, we hypothesize that women's ornaments function to signal individual quality pertaining to future reproductive potential, that is, residual reproductive value (Fisher, 1930). We review evidence supporting this hypothesis, including research on some female nonhuman adolescent primates, which also possess exaggerated ornamentation. Based on this comparative evidence and theory, we argue that permanence of female ornaments was sexually selected in the human evolutionary line because of the relatively long adult life with periodic reproductive episodes, highly dependent offspring in need of copious and extended parental care, and long-term pair bonding between male and female in ancestral hominins (chapters 5 and 6).

Women's extended sexuality and the specific nature of their sexual ornamental signals would not have evolved in the absence of substantial male parental investment. By contrast, primary elements of estrus, we argue, function to obtain good genes for women's offspring. In chapter 7, we review evidence for good-genes sexual selection in general. We also discuss men's sexual ornamentation, which we suggest functions to advertise honestly current reproductive capacity, especially the quality of genes affecting offspring reproductive value. In chapter 8, we turn to discussing the evolution of estrus. Though previous workers have argued that estrus functions to increase the probability of conception when females are fertile, we argue from theoretical bases that it is unlikely that females require any such adaptation, let alone estrus. Estrus in female vertebrates does not function to ensure insemination and conception. These are incidental effects of its functional design to obtain a sire for offspring of superior genetic quality.

In chapter 9, we review the evidence that women's sexuality changes across the menstrual cycle. Women appear to have specialized estrous sexuality, including preferences for male traits that probably connote superior genetic quality. Elaborating an earlier hypothesis (Gangestad & Thornhill, 1998; Penton-Voak et al., 1999; Thornhill & Gangestad, 1999a, 1999b), we argue that the motivational and behavioral manifestations of woman's estrus are designed by direct selection to achieve mating with a sire of superior genetic quality. Salient aspects of woman's estrus are comparable, almost certainly homologous, to estrus or heat shown by females of other mammalian species. Indeed, hormonal and neurobiological homologies and functional uniformity argue that *estrus* is a term that should be applied, regardless of vertebrate taxon, to the sexual motivation and related behavior of females that are fertile in their reproductive cycles. Chapter 10 places women's estrus in light of the mating system typical of human populations and pair bonds involving substantial male parental investment. Evidence suggests, we claim, that estrus functions in humans to obtain superior genes for offspring, often through extra-pair copulation (EPC).

Chapter 11 discusses concealed estrus. Concealed estrus—which is disguise of fertility cues that has actively been selected—must be distinguished from "undisclosed estrus," the lack of overt cues of fertility status resulting simply from lack of active selection for them. We propose, in an extension of the cuckoldry hypothesis of Benshoof and Thornhill (1979) and Symons (1979), that, in human evolutionary history, the importance to female reproductive success of securing a sire of high genetic quality by extra-pair mating generated direct selection for fertility disguise and produced adaptations of concealed menstrual-cycle fertility in women. These adaptations function to disguise physiological, emotional, and behavioral by-products of estrus, as well as the associated, contextually expressed sexual interest in an extra-pair mate(s). Women's eroticism when fertile is emotion designed to be manifested very selectively, that is, toward men with high genetic quality when benefits exceed costs and to be hidden otherwise. Chapter 11 also discusses the possibility that adaptations for concealed ovulation occur in certain nonhuman species.

Functional estrus is present in woman, but it is partially concealed as a result of direct selection for concealment. Hence, women have both estrous adaptation and concealed-estrus adaptation. Contrary to the traditional knowledge in human reproductive biology, loss of estrus is not equivalent to the evolution of concealed ovulation. We examine empirically all the major hypotheses in the literature that have been proposed to explain in woman what has been labeled in the literature as concealed ovulation or, erroneously, loss of estrus. The cuckoldry hypothesis, that the function of concealment of cycle-related fertility is extra-pair mating, receives considerable support, and the other hypotheses are hard to sustain based on current evidence of women's estrous sexuality and men's ability to detect women's peak fertility in the menstrual cycle. Contrary to conventional interpretation, recent evidence indicates that men have significant, but still quite incomplete, "knowledge" (ability to perceive and respond to relevant discriminative stimuli) of women's cycle phase of high conception probability and that such knowledge is used by pair-bonded men in ways that may increase paternity confidence.

Theory and evidence for a coevolutionary arms race in human evolutionary history between males' detection of peak cycle fertility and females' fertility disguised from the main partner are reviewed in chapter 12. Contrary to a widely held view in the literature of sexuality, however, we argue that females in the human evolutionary line never had adaptation (e.g., sexual skin/swelling, scent, or behavior) that functioned to signal cycle-related fertility. The evolution of woman's concealed fertility in the menstrual cycle did not involve the evolutionary loss of signals of cycle-related fertility. Rather, by-products of female peak fertility that males were selected to detect were disguised by selection on females during the evolution of woman's concealed fertility. We suggest that the slight sexual ornamentation that the earliest females in the human evolutionary line apparently possessed (Sillén-Tullberg & Møller, 1993) signaled individual quality, not cycle-related fertility, and subsequently was replaced by permanent ornaments with an individual-quality signaling function.

Summary

Traditionally in the West, women's sexual motivation has often been thought to be absent, erratic, or pathological. At the same time, a latent lasciviousness that, if not strongly discouraged during a girl's upbringing, will render a woman unfit for marriage has been feared. Evolution-minded research on women's sexuality began at the end of the nineteenth century and has continued to this day. Past research can be characterized in terms of three waves.

The first wave led to the discovery that women typically ovulate at mid-cycle and that ovulation is triggered by estrogen, as in other mammals. Accordingly, researchers looked for estrus in women. Mixed findings led to the conclusion, drawn about 1960, that estrus had been lost during recent human evolution. We argue that this conclusion is mistaken. Women have both estrus and concealed ovulation. Early researchers did not find human estrus because they assumed that estrus functions to obtain sperm. Instead, estrus is typically characterized by adaptations that function to obtain good genes for offspring.

The second wave emphasized that women's sexuality functions not merely for conception but also to obtain male-delivered material benefits. In particular, sexual motivation and mating outside of the fertile phase functions to obtain these benefits. Many researchers assumed that women's loss of estrus was linked to these functions. Once again, we argue that sex outside of the fertile phase coexists with estrus. Women's dual sexuality has been shaped by selection for dual functions.

Wave three brought to the forefront the roles of good-genes sexual selection and intersexual conflict. The existence of estrus in women and the importance of coevolutionary arms races between the sexes have been established.

Recent research on women's estrus holds major implications for understanding female sexuality in general, including the phylogenetic origin and function of estrus and the function of nonfertile sex. Other recent research has enlightened understanding of the function of women's permanent sexual ornaments.

2 Methodology

Adaptationism and Phylogenetics

Evolutionary biologists use a variety of methods to analyze the traits of organisms and understand how they have come about and been maintained. Our approach to understanding woman's sexuality applies two classes of broadly defined methods: adaptationism and phylogenetics. For several reasons, we explicitly review the rationale of these two approaches prior to discussing their application to women's sexuality. First, methodology in evolutionary biology has been controversial, and we wish to be clear about our own assumptions. Second, evolutionary biologists use analytic methods that fall outside of these two classes; we wish to state the reasons we implement our methods. Third, it is not always recognized that our two methods are not alternatives in the study of evolutionary history; they typically are used to address different ultimate causes that complement one another—indeed, both are needed to build a complete understanding of evolutionary history.

Adaptationism and phylogenetics both address *ultimate causation*, that is, causation that brought about effects in evolutionary history. As Tinbergen (1963) explained, the evolutionary history of any species' trait itself has two different aspects, phylogenetic origin and selective history (which fall under two of Tinbergen's famous four questions). Put otherwise, a biological trait is ultimately the result of *two distinct categories of causation*: First, phylogenetic origin, the cause(s) of the trait's *appearance* in evolutionary history; second, causes for *evolutionary maintenance* of the trait after its origin. Evolutionary maintenance of a newly arisen trait—its persistence and, in some instances, its spread through descendant phylogenetic branches (species that arise after the trait's appearance)—involves the evolutionary agent of either drift or selection. If the trait has been maintained by selection in the Tree of Life, the selection may be either direct (selection for the trait because of its fitness benefits) or indirect (selection for the trait because of its linkage with a trait that has been directly selected). Traits directly selected are referred to as *adaptations*. Traits maintained through indirect selection are referred to as *by-products*.

Adaptationism addresses causes of trait maintenance. Some critics (e.g., Gould & Lewontin, 1979) have suggested that the adaptationist program (or adaptationism) assumes what kind of traits exist in nature—specifically, that every trait is an adaptation. In fact, this view is badly mistaken. From its inception, adaptationism has been a *method*, one that explicitly strives to *distinguish* traits that have been selected from traits that have not and, of selected traits, to discriminate between those selected directly and those selected indirectly (Alcock, 2001; Andrews, Gangestad, & Matthews, 2003; Symons, 1992; Thornhill, 1990, 1997; Williams, 1966), without making any a priori assumptions about whether traits are adaptations. The goal of adaptationism, then, is to identify the mechanisms involved in a trait's evolutionary maintenance, including maintenance of evolutionary modifications of the trait, following its appearance on the Tree of Life.

Though a highly useful method with broad applicability, by no means can adaptationism answer all questions of evolutionary history. Indeed, it is simply impotent to explain a trait's origins. The reason is quite simple: Traits do not arise through processes whose signatures adaptationism is sensitive to. Selection or drift never explain how traits arise; they can possibly operate *only* after traits exist. As West-Eberhard (2003, p. 197) put it, "[r]esearch on selection and adaptation may tell why a trait persisted and spread, but it will not tell us where the trait came from" (see also Thornhill, 1990; Reeve & Sherman, 2001; Hauser, Tsao, Garcia, & Spelke, 2003). Instead, as we discuss in greater detail later, a trait's origin in evolutionary history is always caused by variation in developmental processes. Phylogenetic analysis can pinpoint the time and "location" of origin of a trait on the Tree of Life. Evolutionary biologists may also be able to discern how a trait originated in developmental processes.

Mammary Glands

We illustrate this crucial distinction between phylogenetic origin and evolutionary maintenance with a simple example, one highly relevant to women's reproduction: the evolutionary history of mammary glands. At some point in evolutionary history, a primordial mammary gland appeared. As all extant species within all three of the major mammalian taxa (monotremes, marsupials, and placental mammals) possess mammary glands, this primordial gland almost certainly debuted in the species that was ancestral to all three of these taxa (approximately 130 million years ago; on the origin of lactation, see Cowen, 1990, p. 278). This primordial gland not only appeared, however; it was maintained in all branches of the Tree of Life comprising Mammalia. Had mammary glands not appeared, current mammals would not now possess them. Had they not been maintained, however, current mammals likewise would not now possess them. The causes of both the origin and maintenance of mammary glands jointly explain their existence today. But these causes are themselves distinct.

That all mammals have mammary glands demonstrates that there was evolutionary maintenance of the glands during the evolutionary history of mammals. Some kind of evolutionary mechanism must account for the retention of the glands in the Tree of Life after their appearance. Were the glands selected directly, indirectly, or neither? If neither, their retention is explained by drift. Given the obvious functional design of the glands—their precision and efficiency in nourishing young mammals—their maintenance across the history of mammalian evolution is likely the result of direct selection for this function; mammary glands are almost certainly adaptations.

Selection, then, is very important for understanding the maintenance of mammary glands, and we can ask a variety of interesting questions about the responsible selection pressures. But selection cannot explain the origin of mammary glands, for it can only act on a trait that already exists. To identify the origins of mammary glands, we cannot look to selection. Instead, we look to phylogeny and developmental processes. In fact, we have some idea of what primordial mammary glands looked like based on comparison of features of mammary glands in mammals now possessing them. Interestingly, the mammary gland of monotremes, the earliest appearing group of mammals, looks much like a sweat gland. Furthermore, mammary glands within the two other major taxa are structurally similar to sweat glands and far more similar to those than they are to any other skin gland. It seems likely that the primordial mammary gland was a modification of a sweat gland, realized through the developmental process that typically gave rise to sweat glands, but that in some variation it gave rise to what would ultimately evolve to be mammary glands. Charles Darwin was, in fact, himself keen to know the origin of mammary glands and first hypothesized that sweat glands were precursors (Cowen, 1990).

All Traits Originate Phylogenetically Through an Ontogenetic Process

New traits are often thought to originate through chance mutation—which, if the manifested trait is selected, can be thought of as fortuitous mutation. In fact, however, this view is incomplete at best and, in worst cases, simply wrong. Mary Jane West-Eberhard (2003) argues convincingly that the origin of a phenotypic trait on the Tree of Life is caused ultimately by the incidental developmental transformation of an ancestral, pre-existing phenotype. Jablonka and Lamb (2005) offer a largely complementary general theory of phylogenetic origins of traits. That is, at their phylogenetic origin, all traits are novel phenotypes that incidentally arise as by-products of another trait or traits in conjunction with some developmental cause of the novelty itself. The novel by-product can only be understood in terms of some developmental process, which is hence ultimately responsible for its generation. In the ancestral species in which primordial mammary glands arose, sweat glands were the typical end results of a particular developmental process. This ontogenetic process was somehow transformed such that a novel phenotype, a primordial mammary gland, arose. Any complete understanding of how primordial mammary glands arose must recognize this developmental process as a causal condition.

What role, then, do mutations play? Mutations do often causally contribute to the transformation of developmental processes. As such, selection for the transformation and its phenotypic products (e.g., primordial mammary glands) may lead to the increase in frequency of a fortuitous mutation. Mutations hence do play important roles in origins. Ultimately, however, mutations have no effects in the absence of preexisting developmental processes, in which they act as partial causes to transform. Mutations are not an *alternative* to development as an explanation; indeed, the effects of mutations can be meaningfully interpreted *in the larger context* of a developmental system. Furthermore, mutations are *not necessary* for developmental processes to be transformed. Transformations and hence novel phenotypes can occur through alterations in nongenetic elements of the developmental system as well. If, given background genetic variation, individuals heritably differ in their tendency to encounter and experience these alterations, selection can operate on the novel phenotype to produce evolution (change in allele frequencies), even absent a novel mutation. (See West-Eberhard, 2003, on "genetic accommodation.")

Important lessons follow. Many biologists and evolution-minded psychologists treat development as only a proximate cause of traits—that is, causation acting during the individual's lifetime. They tend to see development solely as an outcome of past direct or indirect selection (or, in some cases, perhaps drift). This common view is plainly mistaken. Development is *both* proximate *and* ultimate, depending on its causal timing in giving rise to a trait. The ontogeny of the trait in an individual is proximate causation. But phylogenetic origin of a trait by developmental transformation is an ultimate cause of the trait (West-Eberhard, 2003).

Parsimony in Phylogenetic Inference

The inference that mammary glands arose in the species that was ancestral to all mammals applies the phylogenetic *principle of parsimony*. All mammals have mammary glands. Two explanations are possible. First, mammary glands may have existed in an ancestral species common to all and maintained in all. Second, mammary glands may have had multiple first origins in mammals. The principle of parsimony appropriately assumes that the evolutionary rise of novel traits is improbable relative to the evolutionary maintenance of a trait once it has arisen. This principle hence favors the evolutionary history that invokes the fewest number of independent origins for a trait—in this sense, the most parsimonious evolutionary explanation. That mammary glands arose once and then persisted in descendant lineages is more likely than multiple origins of mammary glands.

Homologies and Analogies

Homology and *analogy* define relations between traits in different species. Homologous traits are traits in different species that are similar because the species' common ancestral species also possessed the trait, which was then maintained, giving rise to similarity across species. Mammary glands are homologous across all mammals. Homologous traits need not be identical in form, for, as we discuss later, traits may evolve through "descent with modification." Analogy (or convergence or homoplasy) is similarity of traits in different species caused by independent evolution—that is, traits that had different points of origin in the Tree of Life and were then maintained (see Hall, 2003, for an excellent discussion of homologous and analogous traits). The milk-like regurgitant that pigeon mothers feed their nestlings is functionally analogous to mammalian milk. Though serving a similar function, the gland that produces milk in the pigeon (found in its crop) had an origin distinct from that of the mammary glands. Because analogous characters do not derive from phylogenetic relations, but homologous characters do, the fundamental task of phylogeneticists is to distinguish homology from analogy.

The Concept of Phylogenetic Inertia

Some biologists invoke "phylogenetic inertia" (Wilson, 1975; alternatively, just "phylogeny") to explain the evolutionary persistence of traits in the Tree of Life. The idea implies that, once a trait originates, it will persist through processes accounting for descent (inheritance) alone. If a mother has mammary glands, and if mammary glands are inherited, offspring too will have mammary glands. This sort of simple and seemingly straightforward logic is appealing. It is not only simple, however; it is simplistic and for that reason misleading. Descent through a phylogenetic tree does not occur with inheritance alone. Descent (and hence persistence) is not an automatic consequence of inheritance; it is affected by differential reproduction of individuals. Evolutionary agents are those that affect this differential reproduction. Differential reproduction can arise from chance alone (i.e., not as a result of differences in features systematically affecting reproduction), in which case it is caused by drift. It may also arise from systematic differences between individuals, in which case it is caused by selection. "Inertia" is not an evolutionary agent (see also Reeve & Sherman, 2001; West-Eberhard, 2003). If mammary glands were no longer directly or indirectly selected in descendants, they would likely degrade through mutation and drift if neutral.

Biologists sometimes also invoke genetic correlation to explain the maintenance of a trait in the Tree of Life. Just as inheritance cannot by itself explain persistence, neither can correlation. A trait that is genetically correlated with a directly selected trait (through pleiotropy or linkage disequilibrium) persists because indirect selection maintains it. The concept of "developmental constraint" is related. Nonadaptive outcomes may be carried along with directly selected traits because developmental systems do not readily transform (e.g., through mutation) into ones yielding directly selected traits in absence of the nonadaptive outcomes. In these instances, however, indirect selection is once again the agent that maintains the nonadaptive outcomes. Developmental "constraint" pertains to explanations of why a "better" variant has not *originated*, not why it has not been selected (see also Reeve & Sherman, 2001; West-Eberhard, 2003).

Feature-Specific Origins

Mammary glands, we argued earlier, had one common origin and were then maintained by direct selection throughout the mammalian branches of the Tree of Life. But not all mammary glands are alike. They differ in size. They differ in structure. They differ with respect to the ingredients of mother's milk they produce. What explains these differences across species?

The differences are captured by a process Darwin referred to as "descent with modification." The original gland was modified by the evolutionary process after its appearance, at least partly due to selection for new features that proficiently solved lineage-specific problems of nourishing offspring. The recipe of milk from which human infants highly benefit calls for ingredients different from the recipe that highly benefits seal infants. A proper understanding of this process that leads to modification and differences across species, however, demands that we appreciate the fact that the two different causes that apply to the evolution of a trait *also both apply to each and every instance of modification within a lineage*. That is, each change in a trait occurring along a branch within the Tree of Life required, first, an origin of that modification (a developmental novelty) and a process that maintained the newly arisen phenotype (in Darwin's concept of "descent with modification," often implied to be selection, though drift could also explain maintenance).

When we speak of the "origin" of a trait, then, we can avoid confusion only by defining the precise nature of the trait to which we refer. Mammary glands, writ large, had their origin in the species that was ancestral to all mammals. The precise design of *human-specific* mammary glands, fully specified, originated sometime during the hominin lineage. But a more complete analysis is possible. In theory, we could fully characterize the human mammary gland with a large catalog of features (many perhaps nonindependent). And in theory, each feature within this catalog has some point of origin (and subsequent maintenance) located somewhere along the branch in the Tree of Life originating with the ancestral mammal and terminating with modern humans.

We have focused on homology in mammary glands—that is, similarity from common ancestry. Similarity may also arise through convergence resulting from independent evolutionary origin and maintenance of mammary features and associated milk composition (analogy as opposed to homology). For example, slow-growing juveniles may typically benefit from chemistry of milk different from that which particularly benefits fast-growing juveniles, leading to convergence across distantly related mammalian taxa. In the study of the phylogyny of mammary features, as in all phylogenetic research, a fundamental task is to identify similarity arising from common descent and to distinguish it from analogous similarity. Figure 2.1 illustrates the principles of phylogenetic analysis that we have discussed in relation to the topic of estrus.

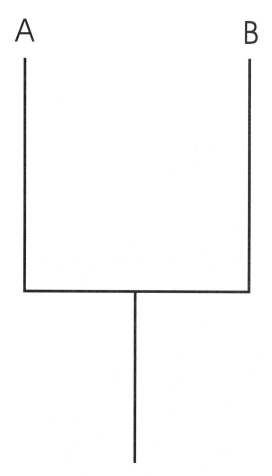

Figure 2.1 A and B are two hypothetical species; for example, they could be two species of primates or a species of fish and a species of primate. Suppose that both have homologous estrus. The estrus is homologous because the nature of the similarity of estrus across A and B requires that the ancestral species of A and B had estrus, too. The deduction that the ancestral species had estrus is based on parsimony: Trait perpetuation in combination with inheritance are inevitable, whereas the rise of novel traits is relatively rare. Although the estrus of A and of B could be independently evolved (i.e., be convergent), this is unlikely given the nature of the similarity in A's and B's estrus. Despite the similarity due to common descent (homology) of estrus between A and B, there are differences in estrus's design between A and B. These differences are due to lineage-specific direct selection for species-specific estrous adaptation.

Evolved Adaptations and Other Traits

Later, we present a specific hypothesis: Estrus possessed by a woman reflects special-purpose, evolved adaptation that functions to obtain superior genes for her offspring, just as do her specialized adaptations for seeing color, estimating object distance, digesting fat, responding to stress, and a multitude of other problems that gave rise to effective selection for functional traits in human evolutionary history. This hypothesis can be restated in a variety of ways: Human estrus is functionally designed or organized to obtain superior genes for offspring; human estrus enhanced individual female's reproductive success during human evolutionary history because estrous females produced genetically superior offspring compared with females without estrus; human female ancestors became ancestors (i.e., outreproduced females who failed to become ancestors) in part because estrus led to fitter offspring.

Each of these statements concerns *past* direct selection for a purpose or function; in the case of estrus, the hypothetical function or purpose is producing offspring of high genetic quality by mating with a high-quality male or males. In no way, however, do these statements imply that women's estrus is *currently adaptive* or that women's estrus *currently* enhances their reproductive success. An evolved adaptation may be currently nonadaptive or even maladaptive because the current ecological setting in which it occurs differs importantly from the evolutionary historical setting that selected it. Quite possibly, for instance, preferences for foods with high caloric content (fatty or sweet foods) are now maladaptive in certain segments of Western society, despite reflecting evolved adaptation. Reeve and Sherman (1993) define an adaptation as any trait that is currently adaptive. Clearly, their concept of adaptation is distinct from the concept of an *evolved adaptation* that we use. When the current ecological setting is the selection pressure that shaped it, an evolved adaptation will be adaptive currently, but an evolved adaptation need not exist within the ecological setting that shaped it to qualify as an evolved adaptation (Symons, 1992; Thornhill, 1990, 1997; Williams, 1966). Current adaptiveness is neither a sufficient nor necessary attribute of an evolved adaptation. The sole criterion used to identify evolved adaptation is functional design (Symons, 1992; Thornhill, 1990, 1997; Williams, 1966), a concept we discuss shortly.

The distinction between evolved adaptation and current adaptiveness is anything but subtle, but commentators still commonly confuse them. Setchell and Kappeler (2003), for instance, advised caution in interpreting research findings on human sexual behavior because no paternity analysis has been performed and evidence on lifetime reproductive success is lacking; it therefore does not "truly investigate sexual selection." These criticisms would be perfectly valid had researchers offered claims about current adaptiveness. The researchers in question, however, investigated whether aspects of men and women's sexuality qualified as evolved adaptations, to which paternity analysis and lifetime reproductive success—current adaptiveness—are irrelevant.

Other statements by Setchell and Kappeler (2003) illustrate more subtle but related confusions. They claim that research on the sexual behavior of largely nonreproducing individuals (university students) cannot yield "true evidence for sexual selection" because the behavior does not beget offspring. Yet human sexual adaptations evolved because they promoted reproductive success in human evolutionary history, not because they promote reproduction now. There is no reason to conclude, a priori, that university students lack evolved adaptations related to their sexual behavior simply because they do not currently reproduce (though, of course, one interested in human functional design should also want to study other human populations as well). Generalized, Setchell and Kappeler's view implies that true evidence for the nature of natural selection that crafted the eyes, livers, and immune systems of humans cannot be gleaned from investigations of these systems in university students because they are largely nonreproductive people. (For that matter, this view implies that study of most laboratory animals, few of which are permitted to reproduce and thereby pass on replicates of their genes, cannot shed any light on the functional design of adaptations.) For some reason, the error of this reasoning is apparent when the object of study is not sex itself, but all too often committed when the object of study *is* sex.

Yet another related but equally misplaced and common criticism of studies of evolved adaptation is that they do not establish whether there is any heritable variation in the adaptation. If a trait is under selection, evolution through natural selection can occur. Evolution occurs, however, only when there is some heritable variation in the trait. To establish evolution by natural selection at present, one must demonstrate heritable variation. Features that are evolved adaptations, however, need not currently be evolving; hence there is no need to establish that heritable variation affects them. Some evolutionary genetic studies have shown that specific alleles within the human genome (e.g., the seven-repeat allele of the dopamine receptor D4 gene locus; Ding et al., 2002) have been favored by direct selection in recent human history and hence contribute to adaptation. Such results, though fascinating, should not deceive us into thinking that claims of evolved adaptation require them. These sorts of observations are possible only when a favored allele has not yet been driven to fixation (or is maintained at an equilibrium frequency by negative frequency-dependent selection), whereas the alleles that support most adaptations are probably found in almost everyone (or vary only at neutral sites).

An evolved adaptation, then, is a feature that *now* exists in a species because selection favored it *historically*, not because it is favored now. Some commentators have defined adaptations as traits that have persisted in phylogenetic lines of species and higher taxonomic categories (e.g., Martins, 2000). Phylogenetic persistence by itself, however, is too broad a criterion. Not all traits that persist are adaptations. A trait that persists simply because it is linked to another trait that is favored directly by selection does not qualify as an adaptation; such a trait is a by-product of adaptation. Traits can be conserved phylogenetically by drift as well. But phylogenetic conservation is also too strict a criterion. Each species possesses some unique adaptation. When a species becomes extinct without descendant lineages, so do its unique adaptations.

All organisms are historical documents. Here, we simply mean that all organisms possess the features they do as a result of evolutionary processes that occurred in the past. Each organism's complex features are primarily of two types. Adaptations are one type—traits that are the product of direct selection for a function. By-products or incidental effects are a second type—traits that are the product of indirect selection and exist because direct selection favored a trait that had side or correlated effects. Aside from direct and indirect selection, two additional evolutionary agents account for features of organisms: mutation and drift. These agents, in all likelihood, do not commonly cause complex traits. Mutations are typically detrimental to performance and hence to reproductive success. Although all human individuals possess tens, if not hundreds, of them, maintained by the balance between recurrent mutations and selection against them, these mutations typically degrade complex organization, not promote it. Drift is differential reproductive success of individuals due to chance. Although drift may account for the perpetuation of simple features and their variants, complex traits are, in all likelihood, almost always maintained by selection, either direct or indirect, because some benefit must typically offset the costs of complex traits if they are to persist.

Incidental effects—by-products of adaptation—are anything but rare. In humans, consider just one: the distance from the top of a person's left kneecap to the proximal end of her appendix. Little imagination is needed to extrapolate from this one instance and see that incidental effects are far more common than adaptations in the totality of individual traits. An adaptation is a piece of the individual's phenotype that exhibits functional or purposeful design and that evolved by direct selection for the design because the design was a solution to an evolutionary historical problem that affected differential reproductive success of individuals. What distinguishes adaptations from incidental effects is functional design.

Functional Design: Fit of Phenotype to Problem

Williams (1966) first emphasized that adaptation is an onerous concept in biology. It should be inferred only when a trait exhibits evidence of *functional design*. Functional design, in turn, is demonstration of sufficient fit or coordination between a trait and a problem faced by an organism to rule out chance association that can arise from by-product, mutation or drift, thereby leaving only direct selection as an evolutionary force that shaped the trait. As we have already emphasized, selection cannot cause the initial appearance of novel, adaptive phenotypes—developmental transformations do. Selection is the separate causal process that arises after ontogenetic origin; it occurs when individuals with a particular trait outreproduce those lacking the trait because the trait has a beneficial effect. With accompanying heritability of trait variation, it may lead to an evolutionary response (change in allele frequency). Over deep evolutionary time, cumulative evolution by direct selection yields features that exhibit functional design—they have been selected to yield particular beneficial effects. A deeply rooted evolved adaptation, then, contains evidence of the kind of historical selection that shaped it. Adaptations, therefore, give the biologist information about the deep-time selection pressures that were actually effective in bringing about the evolutionary change that led to adaptation. Indeed, demonstration of the nature of functional design is the only means to unambiguously infer these selection pressures (Andrews et al., 2003; Symons, 1992; Thornhill, 1990, 1997; Williams, 1966)—precisely why the concept of evolved adaptation is so important and fundamental in biology.

Complexity Itself Does Not Demonstrate Adaptation

Any trait of the individual that has complex organization raises the *possibility* that the trait may be an adaptation. As noted previously, complexity is unlikely to be the result of either mutation alone or drift. But complexity can be an outcome of indirect selection and hence is by no means sufficient evidence for adaptation. Arguably, the human navel is complex. Yet it is an incidental effect of the umbilical cord, totally lacking in purpose.

The complex organization of the lateral line of certain fishes suggested that it may be an adaptation, but no function was immediately apparent. Eventually, biologists demonstrated its functional design for processing sound waves (Williams, 1966). Evidence for adaptation was not found in complexity itself; demonstration of functional design was the necessary and the sufficient evidence.

Camouflage

The concept of functional design is illustrated by adaptations for camouflage. Impressive examples of camouflage adaptations are seen in the common walkingstick insect in the southwestern United States, a master of disguise among the branches of its food source, the *Dalea* shrub. This walkingstick's cryptic color and behavior mimics a *Dalea* branch, its preferred habitat. Aspects of this animal's morphology (its color and shape), behavior (its movement patterns), and psychology (its habitat preference) are phenotypic solutions to the evolutionary historical problem of hiding from visual predators.

It is difficult to imagine that any evolutionary force other than direct selection for these solutions shaped these features. The fit between phenotype and environmental problem is remarkable. This coordination cannot be explained by chance arising from incidental matching, genetic drift, or mutation alone. Observers of the walkingstick on *Dalea* would come to this conclusion without use of any statistical tests; most people would agree that it is virtually impossible for the associations to be due to incidental chance alone. The functional design also is communicated easily to biologists and ordinary people. One need not have heard of Darwinism to appreciate the alignment between the traits considered and cryptic function, though Darwinism (specifically, evolution by selection) is an ultimate scientific explanation for coordinated functional design.

Functional design does not just demonstrate that features of the walkingstick are adaptations. It is also powerful evidence for actual historical selection pressures that effectively brought about phenotypic evolution in this organism. Specifically, the existence of camouflage adaptations places past selection for hiding from visual predators in this organism on biologists' list of realities of the deep-time history of life on Earth. (We note that the phylogenetic origin of crypsis in the walkingstick remains unknown; as emphasized earlier, a *complete* evolutionary historical account of camouflage in the walkingstick requires evidence of origin.)

In many adaptationist studies, tests of hypotheses about the coordination of a trait and a problem are conducted by experiments or in other controlled test settings, rather than by naturalistic observation. All findings, regardless of the method generating them, are evaluated in terms of whether the phenotype solves a problem.

The specialized habitat preference behavior of the walkingstick and the walkingstick's behavior of assuming a stick-mimicking posture reveal the presence of information-processing and decision-making neural adaptations functionally designed for crypsis in the habitat. The evidentiary basis of inferring psychological adaptation is identical to that with nonpsychological adaptation: fit between phenotype and problem. In the case of psychological adaptation, the problem domain is information processing, decision making, and motivation.

Walkingstick adaptations illustrate a fundamental property of adaptation: special-purpose organization. As Symons (1987) has argued in detail, specialized function is expected in adaptation because the problems that give rise to selection for their solutions are specific; thus specific solutions will best solve, or only solve, these problems. Adaptations are hence typically specific in function. The gross functional systems of the human body—digestion, excretion, skeletal support, immunity, growth, sexuality, reproduction, internal regulation, and so on—actually consist of multiple functionally specific adaptations (e.g., in the immune system, multiple kinds of cytokines, immunoglobulins, leukocytes, to name just a few). The visual processing system of the human brain similarly consists of multiple functionally specific adaptations. It seems highly likely that human decision making and motivational systems—for example, those regulating female sexual attraction and behavior—similarly consist of multiple functionally specific adaptations. Throughout this book, we detail evidence of functional design for some of these adaptations.

Two Approaches Within Adaptationism

Studies of walkingstick camouflage and the lateral line of fishes illustrate one approach to studying adaptation. The observation of a possible adaptation yields the question, Does the trait have a function? This question, in turn, leads to alternative hypotheses. Suspicion of adaptation (walkingstick behavior appears to be functionally designed) often simultaneously generates a functional hypothesis (it appears to be designed for camouflage), which then may lead to more specific questions about function and alternative hypotheses. Each hypothesis reflects an attempt to identify the type of past selection that may have built the focal trait; each, then, conjectures possible, deep-time historical causation. Each hypothesis entails consequences that should be observed if the hypothesis is true—predictions that follow from the hypothesis. These predictions concern functional design—the organization of a trait as a solution to a problem imposed by the environment. Verified predictions for which one has no alternative hypothesis ("strange coincidences" [Salmon, 1984] if the hypothesis is not correct) count in favor of the hypothesis. Naturally, predictions not supported count against the hypothesis, and very strongly so when the hypothesis cannot be rescued through reasonable and subsequently supported modifications. Over time, with sufficient information about the details of functional design gathered, researchers can offer cautiously constructed, well-founded conclusions about the historical selection responsible for a trait (obeying Williams's demand that adaptation be an onerous concept) and simultaneously conclude that it is an evolved adaptation. In sum, adaptationism involves standard scientific practice.

In a second approach, investigation does not begin with the observation of a possible adaptation but instead is motivated by observation of a significant ecological or social problem that the researcher suspects has been part of the study organism's environment for an evolutionarily relevant period of time. This approach yields the question, How has the organism solved the problem?, which, in turn, may suggest alternative hypotheses about solutions, whose predictions are then tested. This approach has characterized studies of optimal foraging (e.g., Stephens & Krebs, 1986), mate searching (e.g., Parker, 1984), sex allocation (Charnov, 1982), and human psychological adaptation (Thornhill, 1997).

In the study of women's sexuality, both approaches have been used. Various physical features of women (e.g., permanent breast tissue) are conspicuous. Researchers have asked whether they exhibit functional design. Other researchers have first identified a problem and then searched for a possible evolved solution. For instance, researchers asked whether women possess adaptations for seeking genetic benefits for their offspring from sires (e.g., Gangestad & Thornhill, 1998), which then led to the hypothesis that women's sexual preferences for men's traits that may connote superior genetic quality will vary across their menstrual cycle and be particularly pronounced on the fertile days of their cycles, when women could potentially conceive an offspring. (For a detailed discussion of the application of the second form of the adaptationist research program, see Thornhill & Gangestad, 2003a.)

Adaptationism Can Lead to Discovery of Incidental Effects

Adaptationism is not merely concerned with identifying adaptation. As we emphasized earlier, adaptationism makes no a priori assumption about whether traits are adaptations or incidental effects. And the methods of adaptationism can establish that a trait is an incidental effect just as it can establish that a trait is an adaptation. A classic example is the foveal blind spot. The wiring of the photocells of the vertebrate eye exit the cells on the inner face, not the exterior, of the retina. Where they converge and together exit the eye, they create a mass that prohibits any reception of light on the retinal surface: the blind spot. The blind spot has no functional design itself; indeed, it appears to be maladaptive itself. Given the functional design of the eye and many of its components that contribute to it being an effective optical device, we can be quite certain that the blind spot is an incidental effect of selection for overall functionality. (Had the wiring of the photocells of the primordial eye exited on the exterior of the retina, selection never would have maintained a blind spot as an incidental effect.) Precisely the same adaptationist logic that tells us that many components of the vertebrate eye are adaptations also tells us that the blind spot is clearly not.

It is worth emphasizing the reason that we can be so sure that a feature is an incidental effect in this instance: It is that we can be quite sure of the nature of functional design. Incidental effects arise through indirect selection that operates to directly favor some other feature. When we know the nature and function of the directly favored feature—the adaptation (in our example, the eye)—we can straightforwardly appreciate how other features were indirectly selected along with it—the incidental effects (in this example, the blind spot). Elsewhere, we noted another by-product, the navel. Again, we can be confident of this conclusion because, not only do we lack any evidence for functional design, but we also straightforwardly understand how the belly button is incidentally produced along with a trait with functional design, the umbilical cord. Identification of adaptations does not compete with identification of by-products (cf. Gould & Lewontin, 1979); it assists identification of by-products.

When we are less certain of the nature of functional design, we are often also less certain of whether a complex trait is a by-product. We illustrate this point with the human female orgasm. Symons (1979) concluded that women's orgasm (and the anatomy and physiology that underlie it, including the clitoris) is a by-product of direct selection for the male orgasm and penis. As Gould (1987) later emphasized, we know that the clitoris and the penis develop from the same fetal tissue, the genital tubercle. (This is not to say that all features of the penis and the clitoris are homologues; e.g., the clitoris has no urethra; O'Connell, Sanjeevan, & Hutson, 2005.) Naturally, we can be fairly certain of the function of the penis—the penis itself is for depositing sperm in the vagina (even if the evolutionary agents shaping some components of it, e.g., the coronal ridge, remain a matter of uncertainty; see Shackelford & Pound, 2006). Contrary to Gould's (1987) argument, however, the fact that the clitoris derives from the same tissue as the penis is completely irrelevant to establishing it as a by-product, and for reasons we have already explained: where it developed addresses a question of *origin*, not a question of persistence. Symons argued that women's orgasm is a by-product based on lack of evidence for functional design. We could be even more certain that female orgasm is a by-product if we could establish that all aspects of it quite naturally fall out as incidental effects of penises and male orgasm, in the same way we can understand how a belly button results from an umbilical cord. In fact, however, some researchers have argued that women's orgasm has functional implications that are not readily understood as incidental effects (e.g., it retains sperm in the reproductive tract and hence can bias paternity of offspring; Baker & Bellis, 1995). Others (e.g., Puts, 2006a) catalogue qualities that, although perhaps not strongly suggestive of any particular function, do raise the question of how they are merely incidental effects of male orgasm (e.g., women experience intense pleasurable feelings, apparently not diminished from men's; women can have multiple orgasms; physiological responses such as oxytocin release, not obviously derived from the physiology underlying male orgasm, are triggered by female orgasm; women apparently experience copulatory orgasm at rates that vary with male partner features).

The Case of the Female Orgasm—what is it for, if anything?—was recently thrust into the public spotlight, as Elisabeth Lloyd (2005) published a book by that title. After weighing the evidence, she pronounced in Symons's and Gould's favor: that female orgasm is a by-product of male ejaculation. Unfortunately, as Puts (2006a) details, Lloyd did not systematically apply adaptationist reasoning to evaluate hypotheses that female orgasm is *an evolved adaptation*; instead, she demanded evidence that orgasm is *currently adaptive*. (As we sadly observed already, conflation of these distinct concepts is common indeed.)

In our view, the case of women's orgasm remains in jury, Lloyd's judgment notwithstanding; female orgasm today is the lateral line of fish of another era. Future research will decide whether women's orgasm, similar to the lateral line of fishes, indeed does have functional design or, similar to the belly button, lacks it. We treat women's orgasm in more detail later in the book.

Adaptationism and Phylogenetics Are Not Alternatives

Adaptationism and phylogenetics are sometimes treated as alternatives in the biological literature (for examples, see the review and critique by Reeve & Sherman, 2001). As we have already stressed, however, these two methods for studying ultimate causation offer complementary, not competing, explanations.

Their complementary nature can be illustrated by a contrast between the approaches of two prominent researchers of women's sexuality, Sarah Hrdy and Donald Symons. Symons (1979, 1982) investigated women's sexuality through adaptationist studies—by identifying woman's sexual adaptations and characterizing their functional designs. He furthermore argued that if one wishes to understand the evolutionary history of women's adaptations, one must study women; study of nonhuman primates does not directly inform our understanding of women's adaptations.

Hrdy (1979, 1981), by contrast, studied the sexuality of nonhuman primates and, on that basis, hypothesized about the phylogenetic roots of women's sexuality. Based on evidence that female competitive behavior and multiple mating occur commonly in taxa of extant Old World primates, she hypothesized that women descended from early primate female ancestors who were competitive and who mated often with multiple males (see also Hrdy & Whitten, 1987). That all female primates (and other female mammals) were descendants of females who were competitive for ecological resources is now generally accepted (e.g., Campbell, 2002; Jones, 2003). Multiple mating by females, too, is a widespread homologous trait among primates and, according to the phylogenetic principle of parsimony, likely characterized the ancestral species from which all Old World primates evolved.

Is one approach correct and the other wrong? No. Whereas each contributes in a unique way to an understanding of the evolutionary history of woman, each also informs the other.

What Nonhuman Female Primates Tell Us About Woman's Sexuality

We must be clear, however, about just how each method enlightens our understanding of the evolutionary history of human females. Symons (1982) was right when he emphasized that demonstration of the selection that shaped women's sexuality requires adaptationism applied to *women's* sexuality. Studying female chimps' or yellow baboons' sexuality can identify the selection that produced sexual traits in these primates but cannot identify the selection that made the sexual adaptations specific to women. Women-specific adaptations were designed functionally to solve sexual problems in the hominin lineage.

Naturally, however, not all adaptations (or, for that matter, by-products) of human females arose in the hominin line. Phylogenetic studies yield the necessary evidence for common ancestry when a human female trait is homologous with a trait in other lineages. The comparative study of the sexuality of chimps, baboons, and other nonhuman primates, along with the study of women, can causally identify the original state of sexual traits in shared ancestral species. Sexual behavior does not fossilize, nor do its neural and hormonal bases. But neither do mammary glands. Just as, based on comparative data and the principle of parsimony, phylogenetic research can pinpoint the origin of mammary glands, phylogenetic research can yield strong inferences about the mating behavior of extinct ancestral species (see, e.g., Hall, 2003; Sillén-Tullberg & Møller, 1993).

Let us be clear, however, about what the phylogenetic research on nonhuman primates actually demonstrates. It can show that a certain trait *originated* in nonhuman primates and the time of the origin(s) in the primate tree. By itself, however, it cannot demonstrate that the trait persisted in humans. In principle, it is possible that some traits found universally in all species of nonhuman primates are absent in humans (i.e., did not persist in the hominin line). Demonstration that an adaptation or by-product exists in humans, whether it is uniquely human or not, requires study of humans. Phylogenetic research of nonhuman primates similarly cannot establish that a homologous trait actually found in humans and shown to be an adaptation in nonhuman primates is also an adaptation in humans, much less that it serves the same function in humans. In principle, it is possible that the trait was maintained in the hominin lineage by indirect selection. In addition, it is possible that the trait was modified in important ways in its descent through the hominin line owing to novel direct selection pressures in that line. And once again, demonstration that a trait functions in a particular manner in humans requires adaptationist study of humans.

These reminders of what phylogenetic research can and cannot demonstrate caution against certain interpretations not uncommonly made or implied by primatologists: that study of nonhuman primates offers the deepest possible insight into the events of the evolutionary history of humans, as these animals represent, in a way unadulterated by a deceptive cloak of modern human culture, our ancestral state. One primatologist, Joan Silk (2001, p. 448), offered just this observation about a volume edited by another primatologist, Frans de Waal (2001), but then went on to explain the error of its ways:

> This work is predicated on the assumption that all the important features of human adaptation . . . can be traced back and are rooted in the behavior of our primate relatives. This idea must in some sense be correct. We did not evolve without history. But at the same time, there is little serious consideration of the possibility that some features of modern human life represent derived characters that arose after humans diverged from chimpanzees. . . . The question is not whether studies of other primates

can help us understand our origins; the question is how much they tell us about ourselves.

In this last line, Silk nails precisely the distinction we find useful: Phylogenetic studies can reveal origins; studies of humans are needed to establish persistence in humans. That is, illumination of the nature of selection that favored the individuals who became the ancestors of modern humans demands study of the evolved products of that selection in *humans*.

In this same review, Silk (2001, p. 443) wrote:

> We [primatologists] need to explain to undergraduates why the study of other primates is relevant to introductory courses in human evolution.... Moreover, every grant proposal that a primatologist writes begins (or ends) with an obligatory paragraph about what the study of other primates can tell us about the behavior of humans.... These justifications never come easily to me.

Phylogenetic analysis *is* important, however, and as important as adaptationism, in providing knowledge of the evolutionary history of *Homo sapiens*. Nonhuman primates often tell us everything about the phylogenetic origin of woman's sexuality, even if very little to nothing about the hominin-lineage-specific selection that molded woman's sexuality. Some features of woman are much older than the first primate, and in such cases, phylogenetic comparisons appropriately extend beyond primates. Women's mammary glands are among these features. Estrus, we will show, is probably another. In our view, demonstrations that human mammary glands and women's estrus are indeed phylogenetically very old do shed light on ourselves and our history.

Evolutionary Psychologists and Behavioral Ecologists Often Ignore Origins

Whereas some primatologists misconstrue the implications of their studies for an understanding of humans, many evolutionary psychologists and human behavioral ecologists are overly dismissive of (or, even if not overtly dismissive, uninterested in) phylogenetic studies and studies of nonhuman primates. The study of human mental and behavioral features by behavioral ecologists and evolutionary psychologists has focused almost exclusively on one component of evolutionary causation—maintenance, especially through selection in the last few to several million years; it generally ignores the features' origins. Although this focus has delivered an impressive array of discoveries about human functional design and hence human evolutionary history, it yields only partial knowledge of that history. A comprehensive study of human evolutionary history demands full recognition that both adaptationism and phylogenetics are two distinct and equally important tools for understanding that history. No scientifically legitimate reason argues for one method over the other when the goal is to illuminate the ultimate causes of the features of humans.

Some Comparative Research Can Offer Insight Into and Understanding of Persistence

Thus far, we have discussed phylogenetic research as though it addresses only questions of origins; identification of the causes of maintenance, we have implied, requires adaptationist studies of individual species. We do not mean to imply that causal maintenance processes cannot be identified by comparative data. Although comparative studies are required for phylogenetic reconstructions, they can and often are used in adaptationist research as well. In this adaptationist approach, a researcher regresses variation in a feature (e.g., testis size) among members of a clade (e.g., Old World primates) on a hypothetical selection context (e.g., degree of female promiscuity as a proxy of past intensity of sperm competition). One finding is that human males possess testis size characteristic of mild sperm competition, which suggests that human evolutionary history included this kind of sexual selection (Smith, 1984a,b; see also Gomendio, Martin-Coello, Crespo, Magana, & Roldán, 1998). In such an analysis, nonhuman primates do not identify the history of selection that produced humans; rather, human design (here, testis size) suggests that history in light of other closely related species showing variable testis size, which is predictable from female mating patterns. Recent applications of this comparative method include research regressing inferred traits on inferred ecological contexts across ancestral species inferred from a phylogenetic analysis (i.e., within a constructed phylogenetic tree; e.g., Sillén-Tullberg & Møller, 1993). Adaptationist analysis of the functional design of a single species, however, remains, in general, the most common approach, and it often yields convincing evidence of the function of adaptation and thus the past direct selection that crafted the adaptation.

Darwin's Method of Historical Science Has Triumphed

Phylogenetics and adaptationism elucidate the ultimate causes of the traits of living organisms. As Charles Darwin first argued, when combined and properly applied, they can lend full understanding to these causes. Contrary to arguments by some biologists (e.g., Coyne, 2000) and many creationists, though we cannot directly observe deep-time history, we can know it. As Ghiselin (1969) pronounced, the general method of historical science Darwin invented has "triumphed." Its logic is straightforward and powerful. Scientific hypotheses conjecture possible causation, to be tested by empirical evaluation of predictions or consequences. Evolutionary historical scientific hypotheses conjecture possible causation, such as common ancestry or selection, that acted in the deep-time past. Actual deep-time historical causes imply consequences, the predictions offered by evolutionary historical hypotheses to be evaluated through empirical research. Darwin's method can penetrate vast stretches of deep-time history to identify causation; it is respectfully applied not only in biology but also in all sciences charged with understanding the distant past, including geology and astronomy. In this book, we apply the method to shed light on the evolutionary history of women's sexuality.

Brief Words on Methodologies for Data Collection

Informative research in behavioral biology can take many forms. Researchers can do naturalistic observation. They can introduce experimental manipulations. Those manipulations can take place in a natural field setting. Alternatively, they can take place within a controlled laboratory. For any particular research question, a variety of methods can be useful. Indeed, because of the limitations imposed by any one method, multiple studies using different methods are often necessary to satisfactorily answer an interesting question about the nature of functional design. For instance, whereas laboratory environments allow precise control of extraneous variables, they may lack the ecological validity of field studies of the same organism in its natural habitat.

In the study of people, one potentially controversial methodology is the use of self-report by participants. (Probably, given their training, biologists are, on the whole, more skeptical of this methodology than are psychologists, many of whom use it regularly.) Setchell and Kappeler (2003), for instance, criticized a body of work on human sexual selection on the grounds that it used self-report. Self-reports, they noted, may be biased (sometimes because participants intentionally lie). They may also be flawed due to faulty recall of the events reported.

Later in the book, we describe and refer to a large number of studies on men and women. Many, though not all, use some form of self-report. Not all of these reports are alike, however. Many of the studies we describe merely used the reports to assess people's immediate reactions to stimuli presented: for example, their ratings of the pleasantness of the scent of a T-shirt, of the attractiveness of a person's face (as presented in a photograph), or of the attractiveness of a person's voice. These "self-reports" do not rely on deep memory of events. They merely rely on people's ability to discriminate internal states contingent on presentation of the stimulus and sensitivity of a measure (e.g., a rating scale) to capture meaningful discriminations. The assumption that people can discriminate internal states relating to cognitive or affective reactions is similarly made by anyone who asks another person whether they like the temperature of their bath water.

People may sometimes lie about their reactions. Researchers (and reviewers for journals) must and often do ask themselves whether participants would have had motives to lie in a particular research setting and how any resulting pattern of lying might account for findings. A host who asks her guest whether he likes his meal may have reason to suspect a motive for lying and distrust an affirmative response. In many instances, however, researchers have little reason to believe that participants have a motive to deliberately lie about their reactions to the scent of T-shirts, the attractiveness of faces, and so on. And if they thought that participants did have a motive to lie (e.g., simply to spoil the researchers' results), it is not clear how that motive would account for the results achieved. The criticism that participants could *possibly* lie in a study (with no specification of the motive and its impact on results) does not explain away a study's results. (Naturally, participants might also sloppily fill out questionnaires without reading them closely or even at all. Random responses to questionnaires, however, rarely generate meaningful and interesting patterns of results. Hence, the mere possibility that participants randomly respond to questionnaires does not explain away results.)

Other forms of self-report do demand that an individual can accurately recall and report past events. In some studies we describe, for instance, participants were asked how many times they had intercourse in the preceding 2 days, or how many times their partners called them to check up on them. These methodologies typically use self-report as a substitute for observation of events that human behavioral researchers simply cannot feasibly make themselves. Fairly close correspondence between the reports received and occurrence of events reported is often assumed. The accuracy of the reports may depend on the nature of the events reported and the time depth of recall. Self-reported frequency of copulation in the preceding 2 days is likely to be much more accurate than self-reported frequency of nose twitches in the preceding 2 days or self-reported frequency of copulation in the preceding 3 years. In fact, self-reported frequency of copulation in the preceding 2 days may covary across participants nearly perfectly with actual frequency. (We do not mean to imply, however, that every report of copulation frequency is accurate; see, e.g., Morris, 1993.)

Self-reports of past behavior too may be inaccurate because of deliberate lying. Once again, however, the mere possibility of deliberate lying is not a reason to reject results of a study outright; researchers and critical scholars must ask what motives for lying (or even self-deceit) might exist and how they would explain actual findings. In chapter 9, for instance, we discuss a study that found that women reported greater frequency of attraction to men other than their primary partners in the preceding 2 days when asked on days that they were fertile in the cycle than when asked on days during the luteal phase. As predicted based on theory, they particularly showed this pattern if they shared particular alleles (at major histocompatibilty complex loci) with their primary partners. Participants could have lied, and the true frequencies, if we had been able to observe them, may have shown no such pattern whatsoever. But what motives for lying would have biased the results in *precisely* the way they came out?

Many standard methodologies used in psychological research rely on self-report (e.g., an oral statement during a psychological interview, a numerical rating on a personality questionnaire, a key punch on a computer-administered perception task). The vast majority of studies on humans published in psychological journals used some form of self-report or rating. Naturally, self-report studies are not flawless. If possible, it is best if alternative methodologies are also used to test a hypothesis. We discuss a number of studies that used no form of self-report. Just as an experimental study is not intrinsically better than an observational study—they each have their own strengths for answering particular kinds of questions—studies that do not use self-reports are not intrinsically better than those that do. Multiple studies using different methods can best shed light on the nature of functional design of human sexuality. Again, that is no different for any area of behavioral biology, no matter which species is under consideration.

For detailed discussion of the scientific utility and validity of self-report data in human research, see Daly and Wilson (1999) and Woodward and Richards (2005).

Summary

Phylogenetics and adaptationism are not alternative methods for understanding evolutionary history. They deal with distinct and complementary ultimate causes of traits of organisms. Phylogenetics identifies trait origins in both time and node on the Tree of Life, as well as in the form of the phenotypic precursor of the trait. All traits first occur in evolutionary history as novelties arising from development. In the case of trait origin, development is an ultimate cause. The origin of a novel phylogenetic trait may or may not involve a novel genetic underpinning. When an adaptive novelty versus its available phenotypic alternatives reflects heritable variation, selection may lead to the novelty's persistence. Adaptationism identifies causes that evolutionarily maintained traits after their origin. Depending on the trait, maintenance may involve drift, direct selection, or indirect selection. Mammary glands arose as a developmental deviation of sweat glands and were maintained by direct selection for the function of nourishing offspring. Lineage-specific mammary glands reflect origin events after the first mammary novelty plus lineage-specific selection. The concept of phylogenetic inertia and related notions are erroneous. Phylogenetic reconstruction is based on homology, not convergence, and the principle of parsimony.

Evolved adaptations exist because of their direct selection in the past. They may be currently adaptive, maladaptive, or neutral in their fitness effects. Incidental effects of adaptations are very common traits of organisms. Adaptationism identifies these traits and distinguishes them from adaptations by their absence of functional design. Adaptations exhibit functional design. They are "fitted" (Latin, *aptus*) "to" (*ad*) a specific problem affecting the reproductive success of individuals in past generations. Psychological adaptations and by-products are identified by the same methods of adaptationism as other adaptations and by-products.

Many researchers interested in human behavior have ignored phylogenetic origin, and primatologists have sometimes conflated the two complementary methods for elucidating ultimate causation. This confusion has led to the widespread view that human behavioral adaptations can be identified by studying nonhuman primates. Such studies, in concert with the study of humans, identify the phylogenetic origin of human adaptations and by-products, but cannot in themselves identify evolved adaptations of humans; studies of humans are necessary to identify the features that were directly selected in the hominin lineage. Darwin's method of historical science is the procedure used to identify deep-time historical causation in all sciences that deal with this time frame.

3   Extended Female Sexuality

Definition and Evolutionary Hypotheses

The definition of extended female sexuality is straightforward: Extended female sexuality is female receptivity to sex or proceptivity for sex (through which females seek to copulate with males) during periods other than when females are fertile—sex when they cannot conceive (see also Rodríguez-Gironés & Enquist, 2001). By definition, then, extended female sexuality involves sex with no direct reproductive benefits via conception. By another name, extended sexuality is nonreproductive sexuality. We prefer the term *extended sexuality*, partly so as not to imply that it lacks reproductive benefits. (That is, we wish to avoid the implication that sex that is not about "procreation" is merely about "recreation.") Sex has costs, and these costs must be offset by benefits for sex to evolve. We hence adopt the working assumption that extended female sexuality itself does indeed result in net reproductive benefits for females who engage in it or is an incidental effect of other features that enhance reproductive success. (As well, some forms of sexuality that are nonconceptive are not extended sexuality; see our discussion of polyandrous sex later.) This chapter focuses on the question of what selection pressures directly or indirectly lead to extended female sexuality.

By no means do females of all species exhibit extended sexuality. In many species, females copulate only during fertile phases of their reproductive cycles. In others, however, extended female sexuality is extreme. Carrion beetle females copulate across the reproductive cycle, frequently with a pair-bonded partner and at times with multiple other males (Müller & Eggert, 1989). And female bittacid scorpionflies mate with multiple males every day throughout their entire adult lives (Thornhill, 1976). Copulation outside the fertile phase of the ovarian cycle occurs in many species of birds (Birkhead & Møller, 1992, 1993a) and in some species of nonprimate mammals (Jeppsson, 1986; Kleiman, 1977; Morris & van Aarde, 1985).

No mammalian female known to biology, however, matches the amplified form of women's extended sexuality. Women can possibly conceive 5 or 6 days of their cycles. The chances are pronounced just 2 or 3 days (e.g., Wilcox, Weinberg, & Baird, 1995). Yet women engage in and seek copulation throughout their cycles. In aggregate data, one finds little discernable change in mating frequency across the cycle aside from a drop at menstruation (e.g., Baker & Bellis, 1995; Brewis & Meyer, 2005). Even for women who never use hormonal contraceptives, the proportion of lifetime copulations having more than a negligible chance of resulting in conception is small—even discounting sex during pregnancy and lactation, typically less than one in six. The mean likelihood of conception resulting from a single episode of unprotected, consensual human copulation is estimated to be a mere 2–4% (Gottschall & Gottschall, 2003).

Moreover, nonconceptive cycling frequently occurs in women. In a large random sample of Swedish women, conceptions occurred after a median of 2–3 months of unprotected sex, potentially reflecting a nonconceptive cycle prevalence of over 50%. (Some nonfertility could have been due to lost conceptuses.) Twenty-five percent of women did not conceive until 6 months of cycling (Axmon, Rylander, Albin, & Hagmar, 2006). Women copulate with their primary partners during nonconceptive cycles at rates as high as or higher than the rate at which couples copulate during conceptive cycles (Burleson, Gregory, & Trevathan, 1995).

Obviously, then, any appreciation of the evolutionary history of women's sexuality must elucidate their extended sexuality. When did it originate? What direct or indirect selection pressures have maintained it, and how? Though not matching the same degree of extended sexuality of women, females of a large number of nonhuman primate species surely exhibit it (e.g., Birkhead & Møller, 1998; Dixson, 1998; Hrdy, 1981; Hrdy & Whitten, 1987; Kleiman, 1977; Reichard & Boesch, 2003; Small, 1993; Soltis, 2002). The distribution in nonhuman primates bears importantly on the phylogenetic origin of women's extended sexuality. This chapter takes up the question of why extended female sexuality is ever found. The next chapter homes in on the extended sexuality of women in particular.

Extended Sexuality and Polyandry

Hrdy (1981) highlighted extreme polyandry—female mating with multiple males—in nonhuman primates. In some of these species, females mate with all or nearly all males within their troop during a single period of sexual swelling lasting 10 days or less and multiple times with some males. Individual chimpanzee females, for instance, have been observed to copulate 40 times in a single morning, with a large number of different males (see Konner, 1982). Extended female sexuality and polyandry are clearly distinct concepts. For one, females may, in principle, mate with multiple males within a fertile period and hence not exhibit extended sexuality at all. For another, females may engage in sex outside of a fertile phase and yet mate with only a single male. (Of course, the concepts are also hardly mutually exclusive; females can mate with multiple males *and* exhibit extended sexuality, and, as we shall see, many nonhuman female primates clearly often do.) Although they are distinct, extended female sexuality and polyandry do share one key component. Again, extended female sexuality is, by definition, nonconceptive sex. Because females need not engage in multiple mating to reproduce, polyandry is similarly thought to entail nonconceptive sex. (If a female engages in a single mating with each of 25 males who compete for a single conceptus, then only 4% of her copulations—1 of 25—are conceptive.) More accurately, the function of polyandry—the reproductive benefit that selected it—is often thought not to have to do with its effects on probability of conception or fecundity per se (though, in principle, it could be, as the number of male mates could positively relate to the probability of finding compatible or diverse genes [Zeh & Zeh, 2001] or intrinsically good genes [Simmons, 2005]). Rather, the function of female multiple mating is often thought to be about increasing offspring well-being in ways other than those provided by genetic benefits to them (e.g., as Hrdy [1981] herself first emphasized, through paternity confusion). In this sense, female mating with multiple males often should be understood as a form of evolved "nonconceptive" sex, in the same sense that extended female sexuality must be. Though we focus on extended female sexuality in this chapter, we also discuss forms of polyandry in this light.

High Rates of Female Copulation and Their Evolution

Why do females engage in extended sexuality? This question is a special case of a broader question that biologists have long pondered: the question of why females in many species have evolved to copulate more than once, up to many times, with the same male or with multiple males within a single reproductive episode (see Baker & Bellis, 1995; Hrdy, 1979, 1981; Jennions & Petrie, 2000; Parker, 1979a, 1979b; Stacey, 1982; Thornhill & Alcock, 1983). Once (or at most a few times) should be optimal if the function of mating, from the female perspective, is merely to obtain enough sperm to fertilize all available eggs. Failure to obtain sufficient sperm through a single copulation is not a problem females likely confront commonly and evolve costly adaptations to solve. Sexual selection on males, after all, should strongly favor their ability to deliver adequate numbers of sperm per ejaculate to fertilize available eggs. We can think about this idea within a life-history resource-allocation framework. Males in most species must expend considerable costs (e.g., resources on displays, male-male competition, and the like) that could be expended on other traits (e.g., those enhancing survival) simply to entice a female to copulate with them. Sperm production itself, however, is in all likelihood relatively cheap. Even under conditions of extreme malnutrition, men produce sperm at rates comparable to healthy men (Ellison, 2001). (By contrast, women's fertility is highly sensitive to their current energy budget; Ellison, 2001.) If sperm production is cheap, it is hard to imagine that males would often not benefit from expending (at some small cost to, say, viability, or even access to multiple mates), on top of those expended to achieve copulation, whatever costs are required to produce numbers of sperm sufficient to achieve fertility at close to the asymptotic level reached as a function of sperm quantity (i.e., to near "max-out" on conception rate, as affected by sperm quantity). An apt analogy to the male who fails to deliver adequate sperm numbers is the competitive runner who expends tremendous effort over many months of training for a race but who then, on race day, fails to take the time to tie his shoes, thereby seriously compromising his chances of winning.

Some females (e.g., birds) store sperm. In that instance, selection should operate on males to produce sufficient numbers of sperm that can survive storage (Birkhead & Møller, 1998). In light of strong selection on males to deliver adequate sperm numbers, a female should not be selected to expend considerable costs to bolster numbers of sperm to which she exposes her reproductive tract through an increased rate of mating. (Török, Michl, Garamszegi, & Barna, 2003, showed that female collared flycatchers must be inseminated at least twice to obtain maximum fertility in a single clutch as a result of sperm loss from the sperm storage organ. As they discuss, however, the sperm loss may reflect female adaptation to obtain genetic benefits for offspring, especially given that females can readily obtain additional inseminations from the pair-bond partner or extra-pair partners.)

Biologists have advanced several alternative theories for why females copulate multiple times, even within a single reproductive episode.

First, some theorists propose that sperm competition explains frequent female copulation with multiple males. Specifically, the fitness of females may be enhanced (through the reproductive success of sons) by "running sperm races"—by placing sperm of different males in competition to produce sons that succeed in sperm competition (Baker & Bellis, 1995; Birkhead & Møller, 1992, 1998; Parker, 1970, 1979b; Simmons, 2005).

Second, by mating with multiple males, females may choose sires whose genes increase offspring diversity or fitness (e.g., Andersson, 1994; Jennions & Petrie, 2000; Trivers, 1972; Zeh & Zeh, 2001). In one scenario, through multiple mating a female may be more likely to encounter a male with "compatible genes"—one that, in concert with her own, enhances fertility or offspring fitness (Zeh & Zeh, 2001).

Third, repeated copulation may be a means through which females can incrementally access nongenetic material assistance delivered by males (Alexander & Borgia, 1979; Gowaty, 1996; Hill, 1982; Hrdy & Whitten, 1987; Soltis, 2002; Stacey, 1982; Thornhill, 1976, 1979; Thornhill & Alcock, 1983; Wolf, 1975; for humans, Alexander & Noonan, 1979; Benshoof & Thornhill, 1979; Hill, 1982; Hrdy, 1979, 1981; Strassmann, 1981; Symons, 1979; Turke, 1984). The claim here is not that frequent female sexual activity sets the stage for selection on males to deliver material benefits to females. Rather, the claim is that delivery of material goods and services to females benefits males in some species (through its effects on their access to mates or their offspring's quality) and means of doing so evolve as male adaptations. That is, in the context of *preexisting* male delivery of material benefits, particularly when males do so to gain sexual access to females, female copulation may be selected to obtain these benefits rather than sperm per se. Hence in these species females may evolve to copulate at rates higher than optimal for sperm acquisition (and rate of conception) per se. In principle, the male material benefit garnered by females through copulation could be delivered in any one (or combination) of a wide variety of different specific currencies: food, shelter, status gain, self-protection, protection of offspring. Naturally, for extended sexuality to have effects on male delivery of benefits, males must not be able to perfectly assess when females are fertile. As we argue in chapter 5, females rarely signal their fertility status and, though males can detect fertility status at better than chance rates, they can rarely do so perfectly on that basis. According to this argument, however, extended sexuality does not function to conceal or suppress cues of fertility; rather, it functions to obtain delivery of material benefits, a precondition of which is imperfect detection of female fertility status by conspecifics (see chapters 11 and 12).

Petrie (1992) offered a specific form of this hypothesis. She proposed that females in pair-bonding species may mate multiply (including when infertile) with a pair-bond male to render any attempt he might make to establish a pair bond with another female, simultaneously or sequentially, nonadaptive.

Fourth, frequent female sexual activity may not evolve because of any benefits to females whatsoever. Rather, males may be selected to lead females to engage in copulations at rates higher than that which optimizes female fitness (e.g., Parker, 1984). In this instance, males and females are presumed to have conflicting reproductive interests, with selection for male mating traits that promote male interests leading to compromises in female interests (and what Thornhill and Alcock, 1983, referred to as "convenience polyandry").

Of course, these hypotheses are not mutually exclusive. Each could explain frequent female mating in some species. And yet other hypotheses have been offered. For instance, Lombardo, Thorpe, and Power (1999) proposed that females frequently copulate because they thereby receive sexually transmitted beneficial microbes. To our knowledge, no such microbes are known in humans, and hence we do not entertain this proposal further, but it could apply to some species. For further discussion of these and other hypotheses for high rates of female copulation, see Thornhill and Alcock (1983) and Jennions and Petrie (2000).

What Explains Extended Female Sexuality?

We now turn to the special case of extended female sexuality per se. Two of the theories for frequent mating in females cannot explain the evolution of extended sexuality. Females reap benefits through running sperm races or seeking genetic benefits only when they are fertile. Yet extended female sexuality is, by definition, sex when females cannot conceive an offspring (for additional discussion, see Thornhill, 2006). In some species, females store sperm. In such species, mating during a presumed "infertile" phase is in fact not nonconceptive, for sperm inseminated during the infertile phase is available for conception when females become fertile (see Birkhead & Møller, 1993a,b, for treatments of sperm storage in relation to sexual selection). Women do not store sperm, however; typically, sperm remains viable in women's reproductive tracts for 2 days or less, and rarely up to 6 days (Gomendio, Harcourt, & Roldán,

1998). Indeed, sperm storage apparently is nonexistent in primates and very rare in mammals, restricted to the relatively few species in which mating and ovulation are dissociated seasonally (e.g., certain bats; Birkhead & Møller, 1993b; Gomendio et al., 1998).

From the preceding list, one primary hypothesis that explains female extended sexuality by selection for its benefits in females remains: the male-assistance hypothesis. We next turn to discussing its merits, after which we consider alternatives, including the hypothesis that extended female sexuality is selected because of its benefits to males, against the interest of females.

The Male-Assistance Hypothesis: Modeling and Predictions

The male-assistance hypothesis has been evaluated extensively. On conceptual grounds, one can ask whether the idea is theoretically rigorous. As first conjectured by a number of researchers, the theory was stated as a verbal model. But can one construct a more rigorous quantitative form of the theory, one that forces assumptions to be explicitly stated and evolutionary consequences precisely derived? The answer is yes. Using game theoretical modeling, Rodríguez-Gironés and Enquist (2001) explored whether extended female sexuality as a female strategy and male assistance as a male strategy could stably coevolve. They found that, indeed, extended female sexuality could evolve as a female adaptation for securing male material assistance. In their model, all males were assumed to be of equal genetic quality, and no sperm competition occurred. Hence, neither condition is necessary to explain maintenance of female extended sexuality. Male assistance itself is sufficient to explain extended female sexuality as an evolutionarily stable strategy when that strategy competes with modeled alternatives. Wakano and Ihara (2005), in game-theoretical and two-locus diploid models involving some different parameters than those modeled by Rodríguez-Gironés and Enquist (2001), similarly demonstrated evolutionary stability of co-occurring female multiple mating and male delivery of material assistance to females. (See also related modeling by Stacey, 1982.)

This theory of extended sexuality can also be empirically evaluated. It offers three main predictions, the first two of which are straightforward. First, comparative studies should show a particular pattern of distribution of female extended sexuality: In species in which it is found, males should deliver material benefits to females. Second, studies should show that females reproductively benefit from extended sexuality. A third, and perhaps less obvious, prediction also follows: In species in which females engage in extended sexuality, males they favor when fertile (during the period of "nonextended" sexuality) should not be precisely the same males they selectively mate with during the period of extended sexuality. According to our theory, females get different forms of benefits through mating when fertile than they receive during extended sexuality. When fertile, females conceive, and hence females should evolve to be sensitive to phenotypic traits that correspond to the quality of the packets of DNA they receive from males (in addition to the quality and quantity of nongenetic material benefits males provide).

When infertile, however, females need not be concerned with the quality of genetic benefits for offspring they obtain—for at that time they obtain none—but, so the theory states, only with getting material benefits from males. In some species, as we will see, the males from whom a female can derive material benefits through extended sexuality are precisely those males from whom she does not want genes for offspring. (In other species, we note, during extended sexuality females may choose long-term mates with whom they also mate during fertile periods. In those cases, females should attend to the quality of genetic benefits they could receive from a male even when choosing a mate during extended sexuality. But her extended sexuality per se should still function to obtain material benefits. See chapter 4.)

The Distribution of Female Extended Sexuality Across Species

The comparative evidence for a positive association between extended female sexuality and male assistance is very substantial. In a comprehensive examination of the hypothesis, Rodríguez-Gironés and Enquist (2001) observed this association across a variety of taxa. Stacey's (1982) summary of studies of communally breeding birds and mammals also supports the hypothesis that high rates of female promiscuity occur in species with male assistance to females. In analyses specific to insects, Arnqvist and Nilsson (2000) add comparative data supporting the hypothesis that female extended sexuality functions to obtain male material aid. And in a review of relevant studies of nonhuman primates, Soltis (2002), too, concludes that, in species with multimale and multifemale social organization, the main benefit to females of extensive sexuality may be reduced offspring maltreatment, including infanticide, by a female's mates (see also van Schaik, Prodham, & Van Noordwijk, 2004)—an important form of material benefit from males—and that, in species with paternal care, females may benefit primarily from increased paternal care from multiple mates. Newer studies support similar conclusions. For instance, a detailed study of the pair-bonding European blackbird found that, outside the fertile phase of their reproductive cycles, females solicit matings from and mate with their pair-bond mate. This extended sexuality of females extends and increases mate guarding by the pair-bond partner, which benefits females by protecting them from sexually coercive males, as well as, perhaps, receiving other material benefits from partners (Wysocki & Halupka, 2004). The distribution of extended sexuality across distantly related taxa points to its convergent, independent evolution in distantly related taxa.

We consider extended sexuality in nonhuman primate females in some detail. Extended sexuality in primates is frequent and empirically well established. One of the oldest forms of the male-assistance hypothesis—Hrdy's paternity-confusion hypothesis—was offered to explain extended sexuality in female primates. This hypothesis has been closely scrutinized and heavily debated. Extended sexuality coexists in many nonhuman primate females with sexual swellings that typically reach peak tumescence near peak fertility (Dixson, 1998). The presence of this peak "fertility signal" (a concept with which we take issue), along with extended sexuality, has confused some researchers into thinking that females cannot have it both ways—advertise fertility and yet obtain material benefits from males when not fertile. In discussing these views, we further specify what the male-assistance hypothesis does and does not require. Finally, nonhuman female primates are close relatives of women. In the next chapter, we argue that, despite homology, the form of extended sexuality in women is very different from the form found in most nonhuman primates, and to make that argument we must first establish what forms are found in nonhuman primates.

Hrdy's Hypothesis and Extended Sexuality

Many nonhuman female primates have tremendous potential to gain male-provided nongenetic benefits in exchange for mating. Many live in multimale, multifemale groups. Females hence potentially interact with multiple males frequently, and, therefore, males have ample opportunity to affect, through their actions, female fitness outcomes. Even when males do not actively behave in ways that benefit females (e.g., provide direct paternal care, provisioning, physical protection, or assistance through alliance formation ["friendships"; Smuts, 1985]), males can surely actively impose costs on females through maltreatment of offspring or direct aggression toward females. (For a discussion of the multitude of benefits male nonhuman primates may provide, see Hawkes, 2004; Hrdy, 1981; Silk, Alberts, & Altmann, 2003; Smuts, 1985.)

In a very influential set of writings, Hrdy (1979, 1981) emphasized the ways that female sexual behavior can adaptively regulate the forms of male behavior that harm them or offspring. In multimale-multifemale groups, each male regularly interacts with infants and juveniles that he did not father. Potentially, his fitness could be enhanced if he were to kill offspring not his own. Those offspring would then not be present to compete with his own, and females who lactate those offspring could conceive and invest in new offspring (which the focal male might sire) sooner. Indeed, Hrdy observed infanticide by males in langurs; it subsequently has been observed in a variety of other species. Of course, such actions, though serving male interests, do not serve mothers' reproductive interests. By mating with each male (or, at least, a large number of males), Hrdy proposed, females could prevent maltreatment of offspring (or even promote prosocial behavior toward offspring by males) because each male would assume some probability of paternity. As already noted, after reviewing the evidence, Soltis (2002) concluded that, indeed, in species with multimale and multifemale social organization, the main benefit to females of extensive sexuality (multiple mating) appears to be reduced offspring maltreatment, including infanticide, by a female's mates. (See also Borries, Koenig, & Winkler, 2001; Hrdy, 1981, 2000; Pazol, 2003; van Schaik, 2000; van Schaik et al., 2004.)

A couple of points deserve emphasis here. The first concerns a label: Hrdy's hypothesis argues that the benefits of multiple mating are achieved through "paternity confusion." In fact, however, "paternity confusion" by itself offers no benefits. Any reproductive benefit a female receives from a male must fall into one of two broad categories: The benefit is a genetic benefit that increases her fitness by bolstering offspring fitness, or the benefit is a nongenetic material benefit that directly influences her reproductive success. "Paternity confusion" is merely a way station on the road toward true benefits a female derives from multiple mating, according to Hrdy: benefits that directly enhance her reproductive success and/or the well-being of her offspring. These benefits, not paternity confusion, ultimately give rise to the selection that accounts for female multiple mating. Though getting males not to kill one's offspring may not seem like obtaining male "assistance," by the logic of evolutionary economics it does, of course, qualify as a nongenetic material benefit that increases female reproductive success and is "delivered" by males. In this sense, Hrdy's theory is a prime exemplar of the broader category of the "male-assistance" hypothesis.

Second, Hrdy attempted to explain female polyandry, not extended female sexuality. As we discussed earlier, these two forms of sexuality, though clearly distinct, have a common component. Both are often nonconceptive. Hence, both are often thought to be maintained by nongenetic material benefits. In fact, benefits via paternity confusion often maintain both polyandry and extended sexuality in primate females. The vast majority of mammalian females mate only around the time of peak fertility in their cycles (e.g., Crews & Moore, 1986; Wolff & Macdonald, 2004). By contrast, many nonhuman primate females copulate often when infertile, even when frequency of copulation peaks around fertile periods; that is, they exhibit extended sexuality (e.g., Andelman, 1987; Borries et al., 2001; Deschner, Heistermann, Hodges, & Boesch, 2004; DeVleeschouwer, Heistermann, VanElsacker, & Verheven, 2000; Digby, 1999; Dixson, 1998; Hrdy, 1981; Hrdy & Whitten, 1987; Michael, Bonsall, & Zumpe, 1976; Pazol, 2003; Stockley, 2002). In the Old World primates in particular, in which female sexual swellings occur in multiple taxa, peak fertility and mating are often uncoupled (Anderson & Bielert, 1994; Dixson, 1998; Engelhardt, Hodges, Niemitz, & Heistermann, 2005; Hrdy & Whitten, 1987; Nunn, 1999; van Schaik, Hodges, & Nunn, 2000; Wrangham, 1993).

Female common chimps illustrate this pattern. During about 27–40% of the female's 36-day estrous cycle, females have a pronounced sexual swelling. Of this 10- to 14-day period, however, females can conceive offspring through copulation on only 3–4 days (e.g., Deschner et al., 2004). Yet females not only engage in sex but also actively solicit sex from males during days on which they have little to no chance of conceiving offspring. Indeed, though males solicit sex frequently throughout the period of swelling, female chimps actually initiate more copulation and are less resistant to male sexual solicitations on days of low fertility—during extended sexuality (Stumpf & Boesch, 2005).

Female pygmy chimpanzees (bonobos) exhibit even more extreme extended sexuality. They experience some degree of sexual swelling throughout their entire menstrual cycle and, though copulations are concentrated during peak fertility, they occur throughout the cycle (Reichert, Heistermann, Hodges, Boesch, & Hohmann, 2002; Savage-Rumbaugh & Wilkerson, 1978; Takahata, Ihobe, & Idani, 1996; Wrangham, 1993). Adult pygmy chimp males give food to females more frequently than do common chimp males, including just prior to copulation (Kano, 1980; Savage-Rumbaugh & Wilkerson, 1978; note, however, that common chimp females do appear to use sex to obtain meat;

Stanford, Wallis, Pongo, & Goodall, 1994; Wrangham, 1993). The differences between chimpanzee species are hence consistent with the male-assistance hypothesis.

In some species, the degree of female extended sexuality is conditional and in ways consistent with the male-assistance hypothesis. In a species of langur monkey and the blue monkey, female sexual receptivity and mating have been observed to extend into infertile phases more deeply in multimale groups than in one-male groups. In multimale groups, females appear to mate with all available resident males (Borries et al., 2001; Pazol, 2003). Naturally, multimale groups pose problems of leading nonfathers to refrain from maltreatment of offspring that single-male groups do not. The study of blue monkeys provides particularly compelling evidence for conditional shifts, for it demonstrated the effect using precisely the same females, observed in the two different contexts. In addition, this study documented changes in concentrations in ovarian hormones (estrogen and progesterone) across contexts, which accompanied the changes in extended sexuality. Pazol (2003) proposed that females possess adaptation to alter endocrine hormones to conditionally extend their sexuality.

Some nonhuman primates exhibit unequivocal extended sexuality—when pregnant. At times, female common chimps exhibit sexual swellings when pregnant, leading males to solicit sex from them. Consistent with the paternity-confusion hypothesis, Wallis and Goodall (1993) propose that this tactic is conditionally and strategically used when females encounter new males, who would otherwise be hostile toward them or their offspring after their birth. Indeed, pregnant female hamadryas baboons become sexual after new males take over a troop (Zinner & Deschner, 2000). Female langurs (Hrdy, 1981; Heistermann et al., 2001), golden lion tamarins (Kleiman & Mack, 1977), and sooty mangabey (Gordon, Gust, Busse, & Wilson, 1991) similarly copulate when pregnant. Pazol (2003) reviewed an earlier idea that extended female sexuality during pregnancy is merely an incidental effect of high levels of ovarian hormones at this time, with no direct reproductive benefit. We take up this and related by-product hypotheses in more detail later. For now, we simply note that at least some of these extensions appear strategic, as they selectively occur in the presence of nonresident males presenting a threat to females' future offspring (Pazol, 2003).

Are Males Truly Confused About Paternity?

Hrdy's hypothesis purports that males are confused about paternity. In fact, however, recent data indicate that nonhuman male primates *can* discriminate their offspring from others at better than chance levels (based on the timing of mating or offspring features). Furthermore, primate males are discriminative parental nepotists; they preferentially care for juveniles to whom they are related genetically (for a review, see Buchan, Alberts, Silk, & Altmann, 2003). These findings have led some researchers to question Hrdy's theory. As Sherman and Neff (2003) stated, "male baboons are fairly certain of their paternity (p. 136). This casts doubt on the conventional wisdom that female primates copulate with several males to . . . reduce infanticide and increase the pool of males that provide care for young." Silk (2001) observed that the findings raise a "question about the effectiveness of their [females'] strategy" (p. 446).

Although we agree with Silk that findings raise questions about the "effectiveness" of a female strategy to confuse paternity, they do not, in our minds, seriously threaten Hrdy's theory. Male and female interests do not coincide. Indeed, it is precisely because male interests do not serve those of females that females purportedly mate with multiple males (to confuse paternity). When females succeed in leading males to aid offspring not males' own, however, females have succeeded in getting males to act against their own interests. Males do not benefit from misdirecting parental care or other aid toward nongenetic offspring. Naturally, then, we should expect selection to operate on males to be sensitive to whatever cues of paternity are available to them and, all else equal, for females to conceal cues. When optima of social interactants differ, whether they be males and females or other categories of social players, antagonistic coevolution between players' strategies, often conceptualized as an "arms race," ensues. Player A evolves a strategy against B. B evolves a counterstrategy—to which A evolves a counter-counterstrategy. And so on. Solutions in arms races are often temporary and incomplete. And at any point in time, neither party may be extremely well adapted to the other. Strategies that evolve in coevolutionary arms races, then, may not be deemed particularly "effective." (Our immune systems evolve to defend against pathogens, but we still get sick.) Nonetheless, selection is responsible for the evolution of these less-than-perfectly-effective strategies. Because female adaptation to secure male aid through extended sexuality and male adaptation to detect and favor own genetic offspring antagonistically coevolve, neither party will be perfectly effective, which is the basis for current selection continuing to operate on both sexes and favor new strategies.

Yet another point is raised here. If females are selected to *conceal* cues of paternity, why do they advertise fertility with sexual swellings at all? Why do sexual swellings exist? A strategy in which females, on the one hand, desire to confuse paternity but then, on the other, *advertise fertility* (with a costly signal at that) appears incoherent. (Indeed, in their game-theoretic model discussed earlier, Rodríguez-Gironés and Enquist, 2001, explicitly assume that males lack any cue of fertility other than female copulability and hence explicitly state that the model does not apply to species with female adaptation that functions to signal cycle-related fertility—sex swellings or scents that they assume are directly selected to advertise peak fertility in the cycle.) We take up this question in chapter 5. In brief, we argue there that, in fact, sexual swellings rarely if ever evolve because of benefits females derive from *advertising cycle fertility*. From the female standpoint, they are not fertility signals. They serve other functions in females. Sexual swellings may often provide males with *some* information about female fertility status (i.e., there is some covariation between swelling size and cycle fertility), but that information is *generally no more valid than information males have access to in any case, absent sexual swellings*, and hence adds little if anything to what males can otherwise infer about female fertility. (As noted previously, when females confuse paternity, their sex swellings do not perfectly covary with their fertility status.) Sexual swellings hence do not subvert a female strategy to obtain male assistance through multiple mating and extended sexuality. Again, we discuss these issues in greater detail in chapter 5.

The Effect of Extended Female Sexuality on Female Reproductive Success

The distribution of extended female sexuality across a wide variety of taxa, including primates, supports a first prediction of the male-assistance hypothesis. We now turn to a second prediction: According to the hypothesis, females should benefit from male assistance garnered through extended sexuality, and thereby enjoy enhanced reproductive success.

To our knowledge, this question has been quantitatively, broadly addressed in just one taxon, insects, and only indirectly. Through meta-analyses on the published literature on insects, Arnqvist and Nilsson (2000) examined the relationship between female mate number and lifetime reproductive success. They found a significant predicted average positive effect in species in which males deliver material benefits to females at mating. Although multiple mating in these species may be associated with extended sexuality, we must note that these data do not speak directly to the effects of extended sexuality per se. We note, too, that Arnqvist and Nilsson (2000) found that the result just mentioned also applied to insect species without any apparent male material assistance (though this pattern may be due to a design flaw in these studies; see Harano, Yasui, & Miyatake, 2006). Though no similar quantitative meta-analyses have been performed on studies of birds or other taxa, studies on individual species are consistent with the prediction. We mentioned earlier the finding that, in European blackbirds, female mating with pair-bond males outside of the fertile phase of the reproductive cycle leads these males to protect mates from sexual coercion by other males, as well as other potential material benefits. In this instance, extended female sexuality itself, not just multiple mating, enhances male delivery of material benefits. Other examples may illustrate similar effects. For instance, in dunnocks, multiple males may assist in feeding offspring. Female reproductive success is enhanced by mating with multiple males because males feed young only or primarily if they mate with their mothers (Davies, 1992). Other studies of birds have not examined reproductive success per se but establish that females do obtain material benefits through multiple mating. Similar to the dunnock, female acorn woodpeckers and noisy miners often mate with multiple males, and males who mate with a female are more likely to feed their young (see Stacey, 1982, for a review). In red-winged blackbirds, extra-pair copulation (EPC) partners of females assist those females in nest defense against predators and allow females to feed on their territories (Gray, 1997). If these EPCs involve female sexual motivation when females are infertile in their reproductive cycles, then females obtain male material assistance through extended sexuality. Another possible example is the Adélie penguin, females of which obtain nest material from males in exchange for EPCs (Hunter & Davis, 1998). And great grey shrike females appear to obtain significant amounts of food from EPC partners (Tryjanowski & Hromada, 2005). Again, with the exception of the European blackbird, these studies directly investigated multiple mating rather than extended sexuality itself.

In all, available evidence is consistent with the prediction that extended sexual activity of females, where it occurs, secures male material benefits and often enhances female reproductive success through acquisition of these benefits. At the same time, we acknowledge the paucity of data that directly test the prediction that infertile-phase females benefit from sexual extension through greater access to male-provided material benefits. We suspect that the paucity is due largely to a general lack of explicit recognition among biologists that female sexuality has two distinct functional forms: one functioning when females are fertile in their cycles to obtain good genes and the other functioning outside the fertile phase to obtain male-delivered material benefits.

Variation in Female Preferences Across the Cycle

A third prediction of the male-assistance hypothesis concerns changes in female preference across the cycle. The hypothesis predicts that females with extended sexuality possess *conditional* preferences across the fertile and infertile periods of their reproductive seasons or bouts. In nonhuman primates with extended sexuality but without pair bonding, during the fertile phase females should prefer mates with superior genetic quality; by contrast, they should prefer mates who will deliver material assistance when infertile in their cycles. If, as conjectured, females garner male material benefits through paternity confusion, it follows that males from whom females can obtain material benefits when infertile will *not* typically be those promising superior genes for offspring, for the latter will typically have sexual access when females are fertile.

In pair-bonding species exhibiting extended sexuality, the typical expectation is that females will prefer primary mates (whether possessing good genes or not) when infertile, for those are males from whom pair-bonded females can expect a flow of material benefits. In some cases—those species in which females engage in EPC with males possessing superior genes—pair-bonded females should be expected to particularly prefer males with superior genes when fertile. If the female's primary partner has superior genes, she should prefer him. (Seeking EPC is costly and, when the primary partner has good genes, yields few to no incremental benefits.) If the female's primary partner lacks good genes, she may prefer extra-pair males with superior genes (depending on the costs of possibly losing her primary partner or his assistance).

Evidence speaks to conditional female mate preferences across infertile and fertile phases in a number of nonhuman species with extended female sexuality.

Evidence From Jungle Fowl

The red jungle fowl, wild ancestor of domestic chickens, lives in multifemale, multimale groups. Male-female pairs do not form, and males provide no direct parental care. Hens mate frequently, often multiple times per day, apparently to secure male-provided nuptial food gifts and protection from sexually coercive males (Pizzari, 2003; Thornhill, 1988). Females mate with socially dominant roosters more often than with subordinate roosters in general. This favoritism, however, is greatest when hens are near peak fertility in their reproductive cycles—when they lay eggs (Thornhill, 1988). Hens, furthermore, prefer roosters with large combs (see meta-analysis by Parker & Ligon, 2003), and comb size positively predicts dominance, health, and heritable offspring survival (Ligon et al., 1991; Parker, 2003; Zuk et al., 1990; Zuk, Thornhill, Ligon, & Johnson, 1990).

Evidence hence suggests the pattern predicted by the male-assistance hypothesis of extended female sexuality: Hens prefer males with superior genes when fertile, but mate less selectively with males during infertile phases of their reproductive cycle, in all likelihood to garner male-provided, nongenetic material benefits (food and protection).

Evidence From Common Chimpanzees

Despite being relatively distant phylogenetic relatives, female jungle fowl and common chimps exhibit similar patterns of sexual behavior across their reproductive cycles. When female common chimps are highly fertile, they mate more frequently with socially dominant males than when they are not highly fertile. The reverse pattern typifies female mating with relatively subordinate males (Matsumoto-Oda, 1999). Chimp females appear to often control mating, and hence these patterns are very unlikely to be due to male sexual coercion alone. Indeed, in an important recent study, Stumpf and Boesch (2005) specifically examined patterns of female initiation of sex and found similar, but clearer and more compelling, patterns. When infertile, females actually initiate copulations with more males (i.e., seek to mate more promiscuously) and exhibit less mate choice (i.e., mate less discriminately by rejecting fewer males) than when fertile in the estrous cycle. When fertile, female discrimination and preference was clear: Females initiated sex with males who quickly ascended in social dominance. Stumpf and Boesch (2005) propose, consistent with the male-assistance hypothesis, that female sexuality outside peak estrus functions to accrue material benefits such as protection of offspring (via paternity confusion), access to food, social support, and grooming. These findings point to a shift in preference sets across the cycle of chimps very similar to that of jungle fowl: a preference for material benefits when infertile and a preference for male markers of genetic quality (e.g., dominance-related traits) when probability of conception peaks. An implication for future research is that the same kind of female preference shift should be found in the many other nonhuman primates with female sexuality outside estrus.

Notably, Martin Muller and colleagues (in press) offer an alternative interpretation of Stumpf and Boesch's (2005) findings. Male common chimps regularly coerce matings in some settings and may thereby reduce female choice during estrus. Possibly, changes in female mating patterns are driven by male motivation and ability to coerce matings rather than choice. Though this observation is interesting we note that the presence of male coercion need not imply a lack of female choice in the same mating system. Male sexual coercion and female mate choice evolve in a sexually antagonistic manner and hence may coexist in a system. Stumpf and Boesch (2005) studied chimps under low male density, which may be the ideal setting for female choice to be expressed due to reduced male coercion.

Evidence From Pair-Bonded Species

In chapter 10, we discuss dual mating strategies in pair-bonding birds, specifically in the context of EPC. We briefly illustrate dual mating preferences with a few examples here. First, consider the barn swallow. When females are most fertile in their reproductive cycle (just preceding and during egg laying), they particularly prefer long-tailed males, which have relatively few parasites and strong resistance to parasites. Such males deliver relatively little parental assistance, however, compared with other males that build large nests and provide more parental care. Females prefer highly parental males, but not selectively when most fertile (Møller, 2001; Soler, Cuervo, Møller, & de Lope, 1998). Shifts in preference are reflected in females' EPC behavior (Møller, 2001). During the fertile phase, females solicit more copulations from extra-pair males likely carrying superior genes for offspring survival (males with longer tails, fewer parasites, and enhanced immunity) than at infertile cycle phases (during extended sexuality). Females, however, maintain the same rate of copulation with the in-pair mate during fertile and infertile cycle phases.

Switches in preference of the pair-bonding collared flycatcher are even more clearly documented. Females prefer as extra-pair mating partners highly ornamented males (those who display a large forehead patch) but prefer less ornamented males as primary partners. Highly ornamented males have better genes—they produce offspring with higher viability—than do less ornamented males (Sheldon, Merila, Qvarnström, Gustafson, & Ellegren, 1997). Female flycatchers tend to engage in EPC selectively within the middle part of their fertile periods, despite regular copulations during and outside the fertile period with primary partners (Michl, Torok, Griffith, & Sheldon, 2002). In the bearded tit, a species in which extra-pair mating is common, females also interact more with extra-pair males when highly fertile in their reproductive cycles (Hoi, 1997).

Thus, in a variety of pair-bonding bird species, during the fertile phase of their cycles females prefer sires that are highly ornamented and hence likely to carry genes that increase offspring survival (for a fuller discussion, see chapter 10) but engage in extended sexuality to obtain material benefits.

Females of pair-bonding nonhuman primates (gibbons and callitrichids) also engage in EPC, including during highly fertile periods of their cycles (see review in French & Schaffner, 2000). Primatologists have not yet addressed whether females in these species prefer males inclined toward material assistance during periods of low fertility within their cycles but prefer males with putative markers of good genes during periods of peak fertility.

In chapters 9 and 10, we review abundant evidence that women exhibit dual sexuality. Their preferences when fertile differ predictably from their preferences when infertile.

The Male-Assistance Hypothesis Versus Male-Driven Female Extended Sexuality

The primary alternative explanation for extended sexuality, again, is that it reflects adaptation in males to seek matings in the face of uncertainty about when females are fertile, with an incidental by-product being female copulation during infertile periods. In this view, females do not benefit and have never benefited reproductively from copulation during infertile windows. Rather, they are coerced by or simply acquiesce to ardent males. In Thornhill and Alcock's (1983) terminology, females in this view engage in *convenience polyandry*.

Again, this alternative presumes that males are not fully certain of when females are fertile. In chapter 5, we argue that, in fact, females only rarely evolve signals designed to tell males when they are fertile. In most species, we propose, the most reliable cues to fertility status available to males may well be incidental by-products of physiological changes associated with fertility (e.g., changes in female scent as a function of changes in concentrations of estrogen), not female signals (e.g., sex swellings or estrous behaviors). Given this scenario, it is not surprising that males' detection of females' fertility status is imperfect, with the result that they are not able to fully discriminate matings that could result in conception from those that cannot.

We have already, however, catalogued abundant findings left unexplained by male-driven extended sexuality: the distribution of species with extended sexuality; the fact that, in some Old World primates, females actively solicit copulation when infertile; conditional shifts in female extended sexuality consistent with the male-assistance hypothesis; suggestive evidence from insects that, in species in which males deliver material benefits to obtain matings, female multiple mating is associated with reproductive success; changes in female sexual preferences across the cycle in a variety of species. These findings convincingly argue against the adequacy of male-driven processes to explain extended female sexuality, at least as a widespread phenomenon.

At the same time, there is little denying that, very often, males and females have conflicting interests over the rate of mating (Arnqvist & Nilsson, 2000; Arnqvist & Rowe, 2005; Clutton-Brock & Parker, 1995; Hammerstein & Parker, 1987; Thornhill & Alcock, 1983). In species with extended female sexuality, females should be expected to strive to control selectively mating frequency, mate identity, and mate number in a way that optimizes their net benefits. For females to effectively garner male-delivered material benefits through a strategy of extended sexuality, however, males must lack perfect knowledge and hence can be expected to evolve to value copulations with females, even those who are infertile. Males adaptively accept a high rate of false positives (matings with females that do not lead to conception). With males interested in copulation, females must contend with male solicitations for copulation during extended sexuality in males' presumed interest, not females' (just as they must contend with those solicitations when they are fertile). Coevolutionary races between female strategies to optimize their own mating and male strategies to manipulate females into mating to serve male interests ensue. Consistent with this scenario, Arnqvist and Nilsson (2000) found that female reproductive success was increased by multiple mates but maximized at a smaller number of mates than was male reproductive success. The actual rate of mating can generally be expected to be some compromise between male and female optima, with each sex imperfectly adapted to the other. We hence surely do not maintain that *no* copulations during extended female sexuality occur in males' but not females' interest. We do contend that male interests generally do not account for the existence of periods of extended female sexuality per se.

Do Women Have Extended Sexuality Because Men Are Motivated to Train Their Immune Systems?

In chapter 4, we discuss extended sexuality in women per se. In this chapter, however, we deal with a couple of hypotheses specific to women's extended sexuality related to the idea that men's interests drive it.

It was long thought that women's immune responses must be dampened to tolerate sperm and a conceptus, which contain foreign antigens the immune system may attack. In fact, however, the situation is more complicated than this. Recognition of sperm antigens by the female immune system, based on prior exposure to them, appear to increase probability of conception and successful, adaptive implantation (Robertson, Bromfield, & Tremellen, 2003; Robertson & Sharkey, 2001; but see Hall, Noble, Lindow, & Masson, 2001). Indeed, in couples who have exclusively used condoms prior to trying to conceive, women are more likely to experience preeclampsia during pregnancy (see Robertson et al., 2003), presumably because women have not had prior exposure to their partners' sperm. Men, then, may have evolved to prefer to copulate with women, even when they are infertile, because such copulations facilitate successful reproduction resulting from subsequent copulations. In short, the argument is that men are motivated to adaptively train women's immune systems. In this scenario, women, too, could benefit from copulation outside the fertile phase and hence may have been selected to extend their sexuality to infertile phases (see Robillard, Chaline, Chaout, & Hulsey, 2003).

The training hypothesis encounters a major empirical challenge: It predicts that male and female extended sexual motivation will occur widely, even universally, among mammals, as male antigens are a potential problem for female reproductive success in mammals in general (or at least all placental mammals, in which fetal and maternal tissues interact). Yet extended female sexuality is found rarely in mammals, including placental species. Females of the vast majority exhibit sexual behavior only when fertile and in estrus (e.g., Nelson, 2000; Wolff & Macdonald, 2004).

Indeed, the training hypothesis suggests extended female sexuality should be found too in nonmammalian vertebrates with placenta-like analogs combining tissues of maternal and fetal origin (e.g., live-bearing placental sharks, certain live-bearing snakes). Evidence of female or male sexual motivation outside the fertile phase of female reproductive cycles in these vertebrates is lacking.

Finally, the training hypothesis provides little reason to expect extended female sexuality in birds, as fetal and maternal tissue do not interact in birds. Yet extended female sexuality appears to be quite common in pair-bonding birds.

Another training hypothesis is possible (Corey Fincher, personal communication, October 2004). Copulation with a female outside of the fertile phase may advantage a male's sperm in sperm competition at the fertile phase. That is, nonconceptive matings between a woman and her primary partner may train her immune system to prefer the sperm of her pair-bond mate over that of other men. Because this hypothesis views nonconceptive sex as a means by which males can enhance their chances at paternity, it views nonconceptive sex to be an outcome of sexual selection on men. Of course, for women to benefit from being receptive to sex outside of the fertile period, they must benefit, too. And here, this hypothesis simply becomes another version of the male-assistance hypothesis, not an alternative to it. In the male-assistance hypothesis, as we have presented it, males are sexually selected to engage in nonconceptive sex because they cannot detect with certainty that it is nonconceptive sex. Females benefit through acquisition of benefits that males deliver to attain sexual access. In the view of this training hypothesis, males once again are sexually selected to seek nonconceptive sex (in this instance, to enhance chances at paternity). As Clutton-Brock (1991) and Geary (2000) note, males are particularly likely to engage in tactics that enhance paternity when they deliver material benefits to specific females. Females, then, once again benefit by being receptive to nonconceptive sex because they can thereby garner male-delivered nongenetic material benefits.

Spuhler's Hypothesis

The male-driven hypothesis proposes that female extended sexuality is an incidental effect of male adaptations affecting sexual ardor. Some researchers have proposed that female extended sexuality is an incidental effect of *female* adaptations—that is, adaptations selected not for beneficial effects of extended sexuality but for benefits of correlated traits, which carried extended sexuality along with them. Spuhler's (1979) hypothesis directly pertains to women's extended sexuality. He proposed that women have been selected to excrete high levels of adrenal hormones, which function to increase endurance for walking or running. High adrenal hormone levels carried along with them extended female sexuality.

As we discussed in chapter 2, by-product hypotheses are particularly compelling when the means by which incidental effects (e.g., belly buttons) result from adaptations (e.g., umbilical cords) is straightforwardly appreciated. In this instance, however, no link between hormones and endurance walking as adaptation and extended sexuality as natural incidental effect has been established. Indeed, given the costs of extended sexuality to women (e.g., loss of opportunity to do other fitness-related activity, exposure to contagion), with no offsetting benefits (according to this hypothesis), there is reason to think that selection would favor dissociation of extended sexuality from whatever mechanisms affect endurance walking. As well, this by-product hypothesis cannot account for extended female sexuality in invertebrates that lack adrenal systems (but that are characterized by male assistance; e.g., Arnqvist & Nilsson, 2000).

Summary

In many species, female sexual receptivity and proceptivity occur outside of the fertile phase of the reproductive cycle. We refer to this nonconceptive sexuality of females as "extended female sexuality." This concept is distinct from polyandry, though extended sexuality can co-occur with polyandry. Rarely will females be sperm-limited. Extended sexuality, we argue, typically functions to obtain nongenetic material benefits from males. As predicted, across species, including nonhuman primates, it is associated with male delivery of such benefits. Other predictions of the male-assistance hypothesis for extended sexuality also receive support. In a number of nonhuman species, for instance, females when fertile prefer males who are likely to actually possess high genetic quality but when infertile prefer males who can deliver material benefits. Because females only rarely possess signals of cycle fertility (a major theme of chapter 5), male detection of female fertility status is imperfect, which leads males to be motivated to have sex with females during extended sexuality. Other hypotheses for female extended sexuality are inconsistent with evidence currently available and may account for only a small portion of cases in which it occurs. The general implication for research is that in taxa in which female extended sexuality is common (e.g., pair-bonding birds and nonhuman primates), investigations will benefit from the realization that females possess two functionally distinct sexualities: one manifested at the fertile phase that includes adaptation for obtaining good genes for offspring, the other—extended female sexuality—manifested at infertile cycle phases that includes adaptation to obtain nongenetic material benefits, typically from males.

4   The Evolution of Human Mating Systems and Parental Care

Extended Sexuality in Women: What Is It For?

In chapter 3, we, following others, argued that extended female sexuality typically functions to obtain male-delivered material assistance. As we also noted, no mammalian female known to biology matches the amplified form of extended sexuality exhibited by women. Though women can possibly conceive on 5 or 6 days of their cycles in which ovulation occurs, with pronounced chances occurring just 2–3 days, women engage in and seek copulation throughout these cycles. Indeed, in aggregate data, human mating frequency varies very little across the cycle, aside from a drop at menstruation (see chapter 10). Furthermore, women of reproductive age often have nonovulatory cycles and mate frequently within these cycles. And human females may be sexually active during years of adolescence before establishing reliable ovulatory cycles. Indeed, human female adolescents appear to be more sexually motivated than adolescents in other primates in which adolescent females exhibit sexuality (see chapter 6). Finally, women are proceptive and receptive when pregnant. Naturally, if the male-assistance hypothesis of extended female sexuality applies specifically to women, women benefit through male-delivered material assistance by mating during infertile times of their lives. But precisely how?

In this chapter, we explore answers to this question. The precise nongenetic material benefits that females obtain from males through extended sexuality vary across species. As we saw in the last chapter, many Old World female primates obtain the benefit of reducing male aggression toward offspring by mating promiscuously with multiple males, particularly during periods of extended sexuality. In pair-bonding birds, by contrast, the benefits females obtain through extended sexuality are typically very different: In many species, females enhance the flow of investments their primary pair-bond mates provide to them and their offspring by mating exclusively (or almost exclusively) with their pair-bond mates during periods of extended sexuality. In yet other species, females may directly exchange sex for food. The same hypothesis—the male-assistance hypothesis—applies to all of these instances of extended female sexuality. But the contexts in which females obtain male assistance vary greatly as a function of the mating systems of species and the social and economic relations between males and females these systems entail.

Key questions that women's extended sexuality raises, then, are the following: What is the nature of the mating and parenting system (or systems) in which human mating adaptations have evolved? And, in light of that mating system, what male-delivered material benefits have shaped the nature and form of women's extended sexuality?

Patterns of Marriage: A Cross-Cultural Perspective

Marriage is nearly universal across human societies. (Purportedly, the Na, an ethnic minority living in the Himalayan foothills in China, lack any such institution. Rather, brothers and sisters live together for life. Siblings help women care for offspring. Fathers do not. See Hua, 2001.) Traditionally, many anthropologists and other scientists have inferred human mating or parenting arrangements from marital arrangements (see, for example, Low, 1990, and references therein). These inferences are potentially risky, as institutional arrangements need not directly map onto patterns of mating or parental care (see Leach, 1988; also Hawkes, 2004; Symons, 1979). Moreover, institutional arrangements may lend little insight into key issues pertaining to mating, such as whether men have been selected to invest in offspring due to benefits deriving from paternal care (e.g., Hawkes, 2004). Nonetheless, they are a convenient point from which exploration into human mating patterns can embark.

The Standard Cross-Cultural Sample (SCCS) is a collection of 186 modern and historical human societies selected by Murdock and White (1969) for their distinctiveness. Purportedly, they are weakly redundant representations of human culture, as they do not closely derive from common cultures or possess similarities due to horizontal cultural diffusion. Accordingly, they are thought to be appropriate for study of cross-cultural universals and associations between cultural and ecological variables. Within the SCCS, more than 80% of societies permit polygyny. Less than 20% are completely monogamous, and 1% are characterized by a nonzero level of polyandry (multiple husbands taken by one female simultaneously). In about 60% of the SCCS societies that permit polygyny, however, more than 80% of wives are in monogamous unions (Murdock & White, 1969).

As agriculture, herding, and other relatively recent means of production may alter mating arrangements, Frank Marlowe (2003b) sought to examine the mating arrangements of the 36 foraging societies within the SCCS. (Data on levels of polygyny were available on 30 of them. Foragers, for this study, were defined as groups who attain less than 10% of their diet from cultivated foods or domesticated animals.) All women were monogamously married in just three societies. The median percentage of polygynously married women, however, was just 10%; women's rate of polygyny was 12% or less in almost two-thirds of foraging societies. In about one-quarter of them, by contrast, 35% or more of the women were married polygynously.

Are humans, then, a socially polygynous or a monogamous species? Clearly, both kinds of marital arrangements exist. Polygyny has commonly been permitted in human societies. By far, however, most marital arrangements across human societies, as represented by the SCCS, have been monogamous unions. Patterns of marital arrangements suggest, then, that humans facultatively mate: In most instances, males and females form monogamous marital bonds. But in many societies, minorities of women enter into polygynous relationships, wherein a woman shares one husband with multiple wives. In very rare instances, a woman may have multiple husbands simultaneously.

The Human Adaptive Complex, Biparental Care, and Pair Bonding

The Role of Male Hunting

A traditional view in anthropology is that human pair bonding derives from the importance of male provisioning for offspring—a form of biparental care (e.g., Lancaster & Lancaster, 1983; Lovejoy, 1981; Westermarck, 1929). In most primate species (including our closest relatives), individuals of both sexes are largely responsible for their own subsistence after at most a few years of care following birth. Though males provide a variety of material services to females, including sometimes sharing food in exchange for sex (e.g., Dunbar, 1987), in these species mothers harvest the overwhelming majority of calories consumed by offspring during pregnancy and lactation. In contrast, in the majority of human foraging populations studied to date, the average adult male generates more calories than he consumes. These food resources yield benefits for reproductive women and juveniles by making calories and macronutrients such as protein available to them to consume. Marlowe (2001) estimated that, on average, men produce 64% of the calories in the 95 foraging societies on which sufficient information is available. In Hillard Kaplan and colleagues' (Kaplan, Hill, Lancaster, & Hurtado, 2000) analysis of studies that carefully measured produced foods in nine hunter-gatherer societies, men generated on average about 66% of all calories consumed.

The primary activity through which men generate surplus calories that subsidize women and children's diets is hunting (which, for purposes here, is broadly defined to include any activity aimed to harvest animal meat, including fishing). Though women forage and extract roots (and, in a meaningful minority of societies, produce more calories than men produce), only rarely do they hunt to a substantial degree (for an exception, see Hart, Pilling, & Goodale, 1987, on the Tiwi of Australia). Human foragers, according to this view, are adapted to a diet consisting of high-quality, calorically rich foods. Compared to close phylogenetic relatives, humans consume large amounts of high-quality but difficult-to-extract resources such as animal protein, roots, and nuts. Whereas chimpanzees obtain about 95% of their calories from collected foods requiring no extraction (e.g., fruits, leaves), for instance, only about 8% of calories consumed by modern hunter-gatherers are from foods requiring no extraction. Vertebrate meat in particular accounts for, on average, 30–80% of human hunter-gatherer caloric intake but just 2% of chimpanzee diets (Kaplan et al., 2000). Male subsidization of female and juvenile diets is largely achieved through men's hunting of animal meats (see figure 4.1).

Building on earlier views (e.g., Lancaster & Lancaster, 1983; Westermarck, 1929), Kaplan et al. (2000) explain male hunting as the outcome of selection for male parental effort. According to these authors, male hunting functions to harvest nutrients not only for self but also to foster the viability and health of reproductive partners and offspring. Kaplan et al. (2000), however, expand on earlier views and situate human biparental care within a larger set of human coevolved features, which they refer to as the human adaptive complex (see also Kaplan, Lancaster, & Robson, 2003).

First, humans are characterized by a very long period of juvenile dependence. Whereas chimpanzees are responsible for their own foraging by age 5, human children in foraging societies are dependent on subsidies to their diet until they are almost 20. Children's need for subsidies (caloric consumption minus self-production) peaks when they are about 11 (Kaplan et al., 2000).

Second, people also live relatively long lives. If a human forager lives into adulthood, he or she has a very good chance of living into his or her 60s (see, e.g., Hill & Hurtado, 1996, the best documented demography on a foraging people, the Ache of Paraguay). Longevity in humans is fostered by substantial investment in somatic repair and immune systems, as well as risk-averse strategies to avoid predation (see Robson & Kaplan, 2003).

Third, human productivity continues well into adulthood. Indeed, Kaplan et al. (2000) estimate that male Ache foragers reach peak productivity (net rate of energy capture) when they are in their 40s. And, in a study of hunting within the Tsimane Indians of Bolivia, Gurven, Kaplan, and Gutierrez (2006) found that direct encounter with important prey items and successful capture of prey involve skills not fully developed for 10–20 years following onset of adulthood, despite strength peaking in the early 20s. Fourth, Kaplan et al. (2000) argue that the expansion of the brain during human evolution, as well as an extended, subsidized period of learning, have made possible the high rates of productivity characteristic of humans. Uniquely human capacities to engage in innovative thinking and flexible problem solving, in their view, make it possible for humans to subsist in virtually every terrestrial environment found on earth. Humans also have evolved capacities to engage in highly cooperative forms of social alliances that support high rates of productivity. (Full consideration of the debates about what forms of selection led to brain expansion during the course of human evolution is beyond the scope of this book. For a recent overview of ideas and controversies, see contributions in Gangestad & Simpson, 2007.)

These features purportedly constitute an "adaptive complex" in that each contributes to the effectiveness (or possible adaptiveness) of others. If, for instance, humans

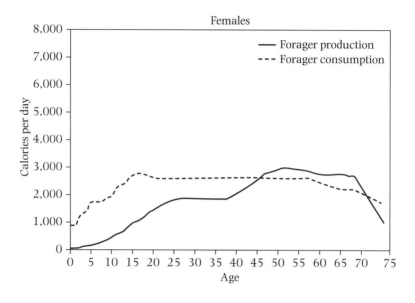

Females

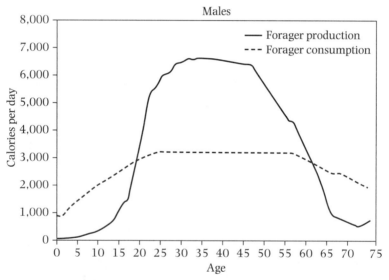

Males

Figure 4.1 The energy production and consumption of females (top) and males (bottom) in relation to age averaged across three hunter-gatherer societies (Ache, Hiwi, and Hadza). Females do not show a net positive production until they reach 45 years of age, which is sustained until about age 70 years. Males achieve net productivity in the age range of about 17–60 years. Based on figure 3 of Kaplan et al. (2000); reprinted with permission of Wiley-Liss, Inc., a subsidiary of John Wiley & Sons, Inc.

did not live long lives and produce surplus calories well into adulthood, they simply could not afford to have their long, sustained period of juvenile dependence. The math is simple: On average, individuals must produce as much during their life spans as they consume. To overcome the very substantial net caloric deficit they build up during their first 20 years, then, humans must be adapted to live long, productive lives. (Put otherwise, no species could afford to simultaneously possess the extended period of childhood of humans and the adult life history profile of chimpanzees, particularly male chimps, whose production never meaningfully exceeds their consumption throughout their entire lives.) At the same time, the extended period of juvenile dependence may very well permit the acquisition of skills, knowledge, and ability to learn ("embodied capital"; e.g., Robson & Kaplan, 2003) that fosters high rates of productivity well into adulthood.

These features render a human life course very different from the life histories of our closest living relatives, the two chimp species, as well as, presumably, our common ancestors with those relatives, who lived 5–8 million years ago. Biparental care purportedly is yet another key feature in the coevolved human adaptive complex.

The Evolution of Biparental Care: Complementarity

Biparental care is relatively uncommon in the biological world. In theory, selection could favor any mixture of parental care by mothers and fathers. Empirically, however, the typical evolved solution to parenting that characterizes species is one in which members of one sex—usually females—are fully responsible for parental effort. The other sex—typically males—incur by far the greatest costs of mating effort, costs of seeking and competing for mates. This parenting solution characterizes most mammals. A recent conceptual analysis suggests that differences in the strength of sexual selection across the sexes themselves lead this solution to be, by far, the modal one (Kokko & Jennions, 2002). When females particularly prefer a small subset of males (often for their intrinsic good genes; see chapter 7), the cost of caring for young is great for those males; they give up precious mating opportunities if they exert effort to invest in offspring. The large cost of parental effort that the most desired males pay leads them not to care for offspring in most cases, which means that females alone are typically responsible for parental care. (For further discussion of these issues, see chapter 5.)

Exceptions to the typical solution do exist, of course. Biparental care characterizes the majority of bird species, some rodents, and some primates, among other species. Recent modeling suggests that complementarity of each sex's parental investment may be an important factor leading biparental care to be favored over the typical division of reproductive efforts (Kokko & Johnstone, 2002). Complementarity of the sexes' parental efforts exists when the total beneficial effect of the sexes' efforts exceeds what would be the sum of the individual beneficial effects of males and females were they investing in offspring separately. (Put otherwise, complementarity implies nonadditive, multiplicative effects of each parent's investments.) It may be critical to the evolution of biparental care because, with complementarity, a father's investment not only has its own fitness benefits but also ratchets up the fitness benefits of the mother's investment. Within the aerial niche occupied by most bird species, complementarity may partly exist because, while one parent flies away to gather food for offspring, the other guards the nest. If one parent alone were to leave the chicks in order to forage, the chicks could be easy prey for predators.

What gave rise to biparental care in humans? The traditional argument is that humans, too, evolved to occupy a niche in which complementarity of parental efforts exists. Dependent juveniles demand caretaking. They simultaneously require substantial nutritional subsidies. Women cannot effectively engage in many forms of foraging while simultaneously caring for and protecting offspring (particularly infants). Male foragers, then, perform most of the hunting, which functions to subsidize the diets of dependent children and caretaking mothers. Women with small children forage in ways that are compatible with child care. Though women and their kin (e.g., postreproductive women) may subsidize offspring to some extent, the bulk of the subsidies to children in modal foraging societies are calories that men harvest (e.g., Marlowe, 2001, 2003a). Male subsidization, then, directly increases female reproduction by increasing offspring quality (lowering the childhood mortality rate) and/or increasing the rate of reproduction (decreasing the interbirth interval) that females can possibly achieve. (For a comparative analysis of biparental care as a function of trade-offs between maternal care and foraging, see Ember & Ember, 1979.)

Obviously, women could not have evolved to become dependent on subsidies achieved through men's hunting (and children could not have become dependent on these subsidies) without men first providing some measure of subsidy. The argument, then, is that male efforts that led to subsidy and the remaining elements of the human adaptive complex (e.g., the long period of juvenile dependence) coevolved over time, in increments, as did human entry into and deepening commitment to an ecological niche requiring capture of high-quality food items. On average, men and women who entered into codependent relationships in which men subsidized the diets of their partners and children, according to this perspective, outreproduced those who did not.

An analysis of close to 100 foraging societies by Marlowe (2001) is consistent with the view that women can and do turn the surplus of calories generated by men into production of offspring and thereby reproductively benefit from this surplus generated through male hunting. Though, on average, men generate about two-thirds of the calories in foraging societies, the degree of male contribution to the diet varies considerable across foraging societies (from about 40% to close to 100%). If women and offspring directly benefit from male subsidies, women's fecundity should be relatively great in societies in which male contribution to subsistence is relatively large. Marlowe (2001) found precisely this association. The effect of men's contribution of subsistence on female fecundity appears to be at least partly mediated through the interbirth interval: In societies in which men contribute more calories to the diet, the delay between the birth of one offspring and the same woman's next offspring appears to be shorter. (For an overview of the energetics of human pregnancy and lactation, see Dufour & Sauther, 2002; see also Ellison, 2001.)

According to this male-hunting-as-parental-effort theory, the nuclear family is a key economic unit in the evolution of human mating relations. For subsidies generated by male hunting to function as parental effort, nutrients that men generate must flow from them to mates (and then to offspring) or directly to offspring. Male-female pair bonding importantly cements the psychological foundations of these resource flows. As we discuss later, doubts that male-generated subsidies flow within the nuclear family in this manner has led some scholars to question the plausibility of the male-hunting-as-parental-effort theory.

When biparental care and mutual mate choice exists, the sexes may choose each other on similar grounds and, as a result, sexual selection may favor the same traits in both sexes. Alternatively, the sexes may choose on different bases, in which case mutual mate choice and sexual selection on each sex may lead to sex-specific exaggerated traits. Within primates, there are well-established associations between sexual dimorphism of body and canine size and mating system (e.g., Plavkan & van Schaik, 1999). In species that pair bond, the sexes are more similar in body size and possess canines of more similar size, relative to species whose mating is characterized by promiscuity or high levels of polygyny. Again, however, mutual mate choice need not imply that the sexes are sexually selected to be similar in all ways.

Though many studies have investigated mate preferences of college students and community samples (e.g., Buss, 1989a,b), few have investigated mate preferences in foraging societies. A rare exception is a study by Marlowe (2005) on mate preferences in the Hadza of Tanzania, a foraging society in which men generate about 40% of the calories and 12% of women are mated polygynously. Consistent with the view that men and women engage in a division of labor within a family, Hadza women highly value men's foraging abilities and intelligence, whereas Hadza men value age-related fertility in a mate more than women do. Marlowe did not detect sex differences in preferences for a partner's personal character and physical attractiveness.

Why Polygyny?

A standard explanation for polygyny in human societies, where it occurs, is Orians's (1969) polygyny threshold model. In this model, males differ in their resource holdings. Polygyny is favored if a female would do better for herself by becoming the second mate of the male with the greatest resource holdings than by becoming the first mate of the best unmated male available to her. This model has been successfully applied to an understanding of polygyny within some human societies (e.g., the Kipsigis of Kenya; Borgerhoff Mulder, 1990).

A more complex model is one that derives from the idea that mates provide more than resources. They provide genes as well. If males vary considerably in their genetic quality and, hence, in the extent to which they can provide good genes to a female's offspring, a female could be better off by becoming the second mate of a male of high genetic quality (despite having to share male-provided resources with his first mate) than by becoming the first mate of the best available unmated male (in terms of a composite of material resources and genes; Weatherhead & Robertson, 1979).

In chapter 7, we treat the sources and nature of variation in genetic quality in more detail. There, we discuss the idea that the presence of parasitic organisms can lead to greater expression of genes that influence fitness and thereby to greater genetic variance in fitness (i.e., greater phenotypic variance in condition as a function of genetic variation). In turn, greater genetic variation in fitness among males should increase the chances that some portion of females will find it worthwhile to enter polygynous unions. Low (1990) coded the prevalence of seven different parasites in regions occupied by the societies of the SCCS. She predicted and found that, indeed, in cultures in which people are exposed to relatively great levels of parasite stress, polygyny is relatively common. Marlowe (2003b) found the same association in foraging societies.

In societies in which women are responsible for a relatively large share of their own subsistence and their offspring's diets, one might similarly expect women to be more likely to enter into polygynous unions. In such instances, the marginal gains a monogamously mated woman enjoys by receiving all of a mate's provisioning to offspring, as opposed to receiving half of his total provisioning to two wives, are less than if men bring in a larger percentage of calories for offspring. Consistent with this prediction, Marlowe (2003b) found that, across forager societies, women's contribution to subsistence is positively correlated with the level of polygyny.

As Marlowe (2003b) also noted, parasite prevalence and men's contribution to subsistence account for less than 30% of the variance in levels of polygyny in foraging societies. A variety of other factors may be important (e.g., threats of sexual coercion, which demand effective male protection). As well, the interests of parents and grandparents may be in conflict with those of their daughters and granddaughters (discussed later); in some cultures, polygynous marriages may be arranged in the interests of these family members, not in those of the daughters who marry (see Marlowe, 2003b).

Have Men Been Selected Directly to Invest in Offspring?

Critique of the Male-Hunting-as-Parental-Effort Theory

The male-hunting-as-parental-effort theory has been seriously challenged over the past two decades. The fundamental difficulty that this theory faces, according to critics, is that nuclear families are, in fact, not the potent economic units in foraging societies that this theory implies (Hawkes, 1991, 2004; Hawkes, O' Connell, & Blurton Jones, 1991, 2001). In the Hadza of Tanzania and the Ache of Paraguay, for instance, hunters have little control over the distribution of meat generated through their efforts, as meat is shared widely across community members. This pattern particularly characterizes the sharing of meat from large game. A Hadza hunter's own family receives no more meat from a large game animal he kills than the meat they receive from the same-sized animal killed by a neighbor. Yet men allocate a substantial portion of their time to hunting large game. Large-game hunting cannot be explained as parental effort if, in fact, a man's hunting does not differentially advantage his own offspring over the offspring of other men. (Indeed, as Hawkes et al., 2001, also claim,

Hadza men actually generate fewer calories per unit time hunting large game than smaller game, an added reason that it does not effectively or efficiently generate calories for a nuclear family.)

According to Kristen Hawkes (2004), men's hunting functions as mating effort—effort to gain access to mates—rather than as parental effort. Men garner prestige through successful hunting exploits, particularly big-game hunting. Ultimately, prestige translates into mating opportunities (including mating with other men's wives; see also Kaplan & Hill, 1985). As Marlowe (2003a) notes, compared to poor hunters, good hunters among the Hadza are more likely to obtain second or new wives once their first wife has reached the age of menopause. Put otherwise, hunting is a form of "showing off" (e.g., Hawkes, 1991).

Of course, the diets of women and their offspring are subsidized by male hunting, as the male-hunting-as-parental-effort emphasizes. But these subsidies, in the male-hunting-as-mating-effort view, are not generated directly by women's own mates or by children's own fathers. Rather, they are generated by the efforts of the community of men to gain mates. In economists' terms, the substantial surplus calories generated by male hunting that benefit women and offspring are "positive externalities" of men's showing off—windfalls they enjoy, not benefits men's efforts were designed to achieve. In adaptationist terms, the surplus calories that men's hunting generates for their community are fortuitous by-products. Men do not gain fitness directly as a result of generating food surpluses that enhance the fitness of mates or offspring. They gain fitness from hunting, especially large-game hunting, because success at it leads to mating opportunities.

Hawkes et al. (2001) do argue that the diets of women and their children are subsidized through the efforts of family members, but husbands do not play the primary role in this regard. Rather, maternal kin—most important, mothers of mothers (i.e., children's grandmothers)—work to directly subsidize the diets of women of reproductive age and their offspring. Hawkes et al. (2001) explain the long life span after the end of women's reproductive years as an extended period of productivity affecting women's own fitness by enhancing the fitness of offspring and grand-offspring (see also Hawkes, 2003).

According to this framework, then, the nuclear family is *not* an economic unit that is key to understanding the evolution of human male-female relations. Men and women do have offspring together and form marriages. But the extended maternal family (a grandmother, her offspring, and her grand-offspring) is of greater economic importance, according to this view, than the nuclear family. Because a man invests time and effort to hunt to gain access to mates in general and not one mating partner in which he invests exclusively, sexual monogamy is not highly important to either men's or women's fitness outcomes from this perspective. According to the male-hunting-as-parental-effort theory, by contrast, the fitness of both sexes can be harmed through a spouse's infidelity. A cuckolded man invests substantially in another man's offspring. And a woman whose husband has offspring with another woman may suffer because a portion of her husband's parental effort is diverted away from her own children and toward other children.

Hadza women do prefer to marry good hunters. (Indeed, Hadza women mention this attribute more often than any other when asked what characteristics they desire in a husband; Marlowe, 2005.) And good hunters have more surviving offspring than poor hunters do. But, Hawkes and others have argued, women do not prefer good hunters for their foraging returns. Nor do good hunters have more surviving offspring because they bring more food to their families. Rather, good hunters possess social prestige, which may benefit their wives indirectly. As well, effective hunting may be an honest signal of a man's genetic quality, which may benefit women's offspring (see Hawkes & Bliege Bird, 2002; Smith, 2002; see also chapter 7). Finally, good hunters obtain better wives, who work harder and more proficiently than the wives of other men. In one analysis, offspring nutritional status covaried positively with women's foraging returns but not with men's (see Hawkes, 2004).

A Blended View

The male-hunting-as-parental-effort and the male-hunting-as-mating-effort theories can be presented in extreme forms (and, indeed, they sometimes are—if not by their proponents, then by their critics). But a blended position or mixed model is also possible. Men's hunting may function as parental effort *as well as* showing off; historically, men may have benefited reproductively from hunting in currencies of enhanced viability of offspring *as well as* mating opportunities. Accordingly, men's hunting may arise from psychological adaptations with two different functions (indeed, at least partly served by distinct adaptations)—parental effort and mating effort.

In such a mixed-model view, different hunting endeavors may differentially benefit men through parental investment and mating effort. Hawkes et al. (2001) emphasize that men's large-game hunting is not an effective or efficient means of provisioning offspring. But large-game hunting (in at least some foraging societies) may benefit men substantially in the form of mating effort. By contrast, men in foraging societies have much more control of the distribution of captured small game and may preferentially direct it toward primary partners and offspring, such that hunting of small game functions as an effective means of paternal investment. Hawkes et al. (2001) observe that Hadza men spend much more time hunting large game than returns to their families warrant, which implies that male Hadza foragers do not allocate their time to foraging in ways that maximize gains through parental investment So long as opportunities for mating with women other than current primary partners are available to men (in the form of extra-pair mating or additional wives) and can be achieved through foraging, however, men should not be expected to allocate foraging time in ways that maximize the fitness benefits of parental effort *alone* (see, for instance, Wedell, Kvarnemo, Lessells, & Tregenza, 2006). Of course, that need not imply that male foraging often *does* function as parental effort.

Indeed, Marlowe (2003a) offers data on Hadza foraging rates and activities strongly supporting a blended view. Overall, married Hadza women produce as many calories as do married Hadza men. Women with small offspring, however, do not. Compared with all other married women, women whose youngest children are 3 years of age or younger harvest about one-third fewer calories. And women with infants 1 year of age or younger harvest only about half as much. Women's child care, it seems, does interfere with effective foraging. When women have young children, however, their husbands make up for the shortfall. Hence, whereas in couples without children 8 years of age or younger, wives produce more calories than husbands do (approximately 3,300 vs. 2,900), in couples with infants less than 1 year of age men produce almost 70% of the calories (approximately 1,700 by wives vs. 3,800 by husbands). Hadza men, then, appear to facultatively adjust their work efforts (and perhaps the prey items they target) in response to the direct food production of wives, as it varies with the presence or absence of young children. The view that men's work functions solely as mating effort has a difficult time explaining this pattern (see Marlowe, 2003a, for a discussion of possible alternative explanations). As Marlowe (2003a) also emphasizes, men's production in this study was not achieved exclusively through hunting. Collection of honey accounts for about 30% of the calories that Hadza men produce. Men typically have substantial control over the distribution of honey they collect and, according to informal reports, they try to direct what they do not eat themselves to their families. Men who are good hunters also tend to harvest more honey and other foods than other men do (possibly due to good hunters' overall vigor). Hence women married to good hunters truly do benefit directly from their husbands' foraging skills.

Finally, separation of men into fathers and stepfathers provides additional evidence that men's production functions partly as parental effort. Approximately 30% of Hadza children have stepfathers. In contrast to genetic fathers, stepfathers do not enhance food production in response to the presence of young stepchildren in the household (see also Marlowe, 1999).

The Hadza are at the low end of the cross-cultural distribution of men's contributions to subsistence (with men producing only about 40% of the calories compared with, on average, about 64% of men across foraging societies; Marlowe, 2001). In foraging societies in which wives produce lesser amounts of food (e.g., in colder climates), men might be expected to engage in even more parental efforts through food production than Hadza men perform. At the same time, there is little reason to doubt that even men with children do allocate time to activities that function purely or largely as mating efforts, including forms of hunting.

Consider as well recent work by Robert and Marsha Quinlan. Across societies of the standard cross-cultural sample, pair-bond stability (a low divorce rate) positively predicts older ages of infants at weaning (Quinlan & Quinlan, 2008). And in a rural village on the Caribbean island of Dominica, women with coresident mates weaned children at later ages than did women without coresident mates (Quinlan, Quinlan, & Flinn, 2003). As lactation interferes with women's ability to produce food, male subsidy purportedly permits women to invest in young offspring through nursing. (At the same time, we note, male contributions to subsistence are actually predictive of shorter interbirth intervals in foraging societies; Marlowe, 2001. Taken together, these results imply that male subsidies increase the total amount of time a woman allocates to reproductive effort through both gestation and lactation.)

Trade-Offs in Allocation of Effort to Parenting and Mating

In species in which males and females form pair bonds and share parenting duties but in which both sexes also engage in extra-pair copulation (EPC), males face a trade-off between parental effort and mating effort. Parental effort increases male fitness by increasing offspring quality or increasing the rate of offspring production by a mate. Mating effort can increase male fitness should he succeed in obtaining extra-pair matings. How much males allocate to parental effort and mating effort, respectively, should hence depend partly on whether males can successfully obtain extra-pair matings through mating effort. Through comparative analysis of pair-bonding birds, Møller and Thornhill (1998a) demonstrated how this trade-off may work. The prevalence of extra-pair paternity (percentage of offspring sired by males other than females' social partners) varies considerably across species (or across populations even within single species). In species in which the extra-pair paternity rate is relatively high, male attractiveness (which affects males' ability to obtain EPCs) negatively covaries with male parental feeding. In these species, males can potentially obtain EPCs, and those males whose efforts to obtain extra-pair matings are purportedly most profitable (attractive males) do less parental care, arguably to allocate greater effort toward pursuit of extra-pair matings. By contrast, in species in which the extra-pair paternity rate is low, attractive males do no less (and, in some instances, do more) feeding of young than unattractive males. In those populations, neither attractive nor unattractive males have much chance to obtain extra-pair matings. Hence, both attractive and unattractive males allocate much time to parental care. (See also Møller & Jennions, 2001. The negative relationship between male quality and male assistance to females is also observed in some insects; for a review, see Bussière, 2002.)

Several lines of evidence suggest that men face this same trade-off. Among the Hadza, fathers with relatively many mating opportunities engage in less parental care than do fathers with few mating opportunities (Marlowe, 1999). And among the Tsimané Indians of Bolivia, men without dependent children have more EPCs than when they do have dependent children (Winking, Kaplan, Gurven, & Rucas, 2007). These studies yield evidence consistent with men facultatively adjusting the amount of effort they put toward mating and parenting, respectively, as a function of the (historical) payoffs to each, where payoffs vary with men's personal characteristics (their attractiveness) or life circumstances (having a dependent child). (See also chapter 7. For related discussion of and additional evidence for trade-offs men face, see Gangestad & Thornhill, 1997a; Gangestad & Simpson, 2000.)

Do Men Possess Design That Functions to Allocate Effort to Parenting?

We argued in chapter 2 that compelling evidence that particular selection pressures have effectively shaped an organism's phenotype historically is to be found in the nature of the organism those selection pressures shaped. That is, effective selection on an organism leaves its signature in the design of the organism. The most compelling evidence that men were historically shaped to allocate effort to parenting, then, should be found in evidence that men possess design that functions to allocate effort to parenting. Studies showing that men respond to circumstances that, in theory, affect the payoffs to parenting and mating effort do, in fact, suggest that men possess design to engage in parental effort. Related work may shed light on the nature and design of physiological mechanisms that underpin adaptive modulation of effort: investigations of factors that affect men's testosterone (T) levels.

Across a wide range of taxa, T appears to facilitate male mating effort by channeling energetic resources to features particularly useful in male-male competition (e.g., muscles) and, due to necessary trade-offs, away from other targets of allocation (e.g., immune function; see Ellison, 2001, 2003; see also chapter 7). In some species in which males invest in offspring (e.g., marmosets, some birds), male T levels drop after the birth/hatching of the mates' offspring (e.g., Nunes, Fite, & French, 2000; Nunes, Fite, Patera, & French, 2001). Men's T levels too appear to drop when they become mated or have offspring (e.g., Berg & Wynne-Edwards, 2001; Booth & Dabbs, 1993; Burnham et al., 2003; Gray et al., 2004; Gray, Kahlenberg, Barrett, Lipson, & Ellison, 2002; Gray, Yang, & Pope, 2006; Mazur & Michalek, 1998; Storey, Walsh, Quinton, & Wynne-Edwards, 2000). In birds, marmosets, and men, drops in T may facilitate paternal investment. In fact, men who have lower T levels respond more prosocially to infant cries than do men with higher levels of T (Fleming, Corter, Stallings, & Steiner, 2002).

One set of studies further illustrates the facultative nature of men's allocation of effort to mating. As mentioned, in Western samples, men who are mated in serious dating or marital relationships typically have lower T than single men do. In two studies, McIntyre et al. (2006) found this same difference in men's T as a function of mating status. The effect of mating status, however, was moderated by men's interest in pursuing extra-pair relationships with women other than primary partners. Men who claimed to have little interest in and history of extra-pair relationships revealed the typical drop in T when mated, as compared with being single. Men who claimed interest in and had a history of extra-pair relationships, by contrast, showed no difference: Such men had T levels just as high when they were in relationships as when they were single. (Perhaps of related interest, across species of birds, male T covaries with the total extra-pair paternity rate, but not well with overall levels of polygyny; see Garanszegi, Eens, Hurtrez-Boussès, & Møller, 2005.)

Though these data offer strong hints that men have adaptation shaped for the function of parental investment (trading off against mating effort), more research is needed. One study found that, when polygynously mated, Kenyan Swahili men's T levels remain high (perhaps because maintaining multiple mates requires sustained mating effort; Gray, 2003). And alternatives must be ruled out. For instance, perhaps men simply have lower opportunity to engage in male-male competition when they have offspring as a result of uniquely modern social practices, leading to lower T levels. Or females may manipulate men's T levels in their own interests and against male interests, such that changes do not reflect male adaptation (e.g., Gray et al., 2002).

Additional Features That May Speak to the Nature and Function of Human Pair Bonding and Paternal Care

Mutual Mate Choice   Studies of mate preferences strongly point to mutual mate choice in modern societies. In Buss's (1989b) classic study of mate preferences in 37 cultures, both men and women, on average, rated "kindness and understanding" as their number one preference. And a study by Buston and Emlen (2003) found that people tended to prefer valued traits in others that they perceived themselves to possess. The sexes differ with respect to particular mate preferences as well. For instance, men particularly prefer physical attractiveness in women, and this sex difference appears to be cross-culturally widespread (Buss, 1989b; but see Marlowe, 2005). The fact that both sexes possess strong preferences for long-term romantic partners, however—even if they differ in some respects—is additional evidence that humans exhibit mutual mate choice.

It is not merely that men and women prefer many of the same traits in mates. Men and women are similarly choosy when it comes to long-term partner choice. Kenrick, Sadalla, Groth, and Trost (1990) describe this phenomenon in terms of their "qualified parental investment model." (See also Trivers, 1972, who described humans' sexual selection system similarly.) In studies on college students, men and women claim to be nearly equally choosy when it comes to evaluating people as long-term mates (in that they identify, on average, a near equal minimum acceptable level of desired traits in a long-term mate). By contrast, when they evaluate people as potential sex partners—mateships lacking either party's commitment to an enduring relationship—men are considerably less picky than women (Kenrick et al., 1990; Kenrick, Groth, Trost, & Sadalla, 1993). This pattern is consistent with men and women having evolved to engage in mutual mate choice in contexts in which partners cooperate to provide biparental care (while also being open to opportunistic mating outside of stable pair-bonds).

Female Mate Preferences in Long-Term and Short-Term Mating Contexts   Evidence on what characteristics people prefer to see in mates and the mates of their children is consistent with an ancestry of paternal care. As we discuss at greater length in chapter 7, men's degree of facial masculinity is purported to signal (or, ancestrally, to have signaled) men's vigor and genetic quality. People tend to think that men with masculinized faces, however, are less likely to be good investors in romantic partners and offspring (see review in Penton-Voak & Perrett, 2001). Kruger (2006) presented women with two male faces—one masculinized through computer technology, the other feminized—and asked them to pick which they would prefer as an affair partner and as a marriage partner. On average, women preferred the man with the masculinized face for an affair partner (chosen 66% of the time) and the one with the feminized face for a marriage partner (chosen 63% of the time). This pattern is consistent with the idea that, because men invest in partners and offspring, women are willing to trade off some degree of genetic quality for greater willingness to care for offspring when choosing a long-term partner (see also Buss & Schmitt, 1993; Penton-Voak et al., 2003).

Preferences Parents Have About Their Daughters' Mates   Kruger (2006) also asked men and women which of the two men they would prefer to have as a son-in-law. As expected, both sexes preferred the feminized man as a son-in-law. Interestingly, these preferences were even stronger than women's own preferences for a feminized man as a marriage partner (75% vs. 63%). This difference suggests a conflict of interest between daughters and parents over daughters' ideal mates. Such a conflict would arise if fathers invest in offspring, but to variable degrees, and if grandparents invest an amount that depends on the level of paternal investment. That is, grandparents invest more in grand-children if paternal care is minimal or absent than if paternal care is substantial (Buunk, Park, & Dubbs, 2008). In such a case, a daughter should not be willing to trade off as much genetic quality for paternal investment qualities in a mate as her parents should want to see her trade off. (This conflict of interest is, naturally, a particular instance of parent-offspring conflict; see Trivers, 1974). The fact that parents and offspring have conflicts of interest over the qualities of an optimal mate purportedly contributes to the common practice of parentally arranged marriages (though, we recognize, is not the sole reason for arranged marriages; see Buunk et al., 2008).

Romantic Love   Both sexes have the capacity for romantic love, a capacity that, to our knowledge, can be found across cultures (e.g., Jankowiak & Fischer, 1992; see also Fisher, Aron, & Brown, 2005). The function of romantic love is not clear. One possibility is that love functions as a signal of intent to another person of commitment to a long-term interest in a relationship with the person (see Gangestad & Thornhill, 2007; for related and other views, see also H. Fisher, 2004; Frank, 1988).

Proprietariness and Threats of Infidelity   As we noted earlier, if men invest directly in offspring, infidelity by both men and women threatens the other partner. In spe-cies in which males do not invest in offspring or provide substantial material benefits to females, multiple mating by males need not impose costs on females. After examin-ing the cross-cultural record, Jankowiak, Nell, and Buckmaster (2002) concluded that, across cultures, sexual propriety within marriages and love relationships is a presumed right of both sexes.

Sensitivity of Investment Decisions to Levels of Paternal Uncertainty   That men's calibration of paternity certainty affects their investment in the offspring available to them for investment is strong evidence for psychological adaptation that functions as paternal care. In a large Western sample, Anderson, Kaplan, and Lancaster (2007) found that men directed more assistance toward offspring that they report are likely to be their own genetic offspring than toward offspring they suspect may be the product of their mate's infidelity. Indeed, as we discuss in chapter 12, men's perception of resem-blance to offspring positively affects their willingness to assist them. Finally, abundant evidence indicates that men invest less strongly in their stepchildren than in their puta-tive genetic children (see partial review in Anderson et al., 2007; additional evidence that men's paternity confidence positively affects their investment is also reviewed by Anderson et al., 2007).

Cross-cultural patterning of "avuncular nepotism" also speaks to paternal invest-ment patterns sensitive to paternity certainty (Alexander, 1979). In a minority of human cultures, men pass heritable wealth to their sisters' sons rather than to their own putative children (their wives' offspring). In some societies, men may invest time and effort into helping train or assist in other ways their sisters' sons as well. In cultures with matrilineal inheritance patterns, the extra-pair paternity rate (typically estimated from ethnographic investigation of patterns of sexual behavior) tends to be relatively high (e.g., Flinn, 1981; Gaulin & Schlegel, 1980; Morgan, 1877). In turn, high extra-pair paternity (and hence matrilineal inheritance) tends to co-occur with particular means of subsistence. Cultures in which members depend on coastal fishing (e.g., cultures located in the insular Pacific) or horticulture are overrepresented among matrilineal societies. Coastal fishing may render it more difficult for men to guard mates. Women may be more efficient producers through horticulture and, hence, women in horticul-tural societies may be less dependent on men as providers. Matriliny is rare among pas-toralists and agropastoralists.

As a number of scholars have observed, the extra-pair paternity rate would have to be exceptionally high (73%) for men to maximize inclusive fitness by investing in sisters' offspring rather than their own offspring (e.g., Greene, 1979; Hartung, 1985; Holden, Sear, & Mace, 2003). (As the extra-pair paternity rate increases, a man's mean relat-edness to offspring obviously diminishes. But so too does his mean relatedness to his sister's offspring.) Decisions about inheritance patterns, however, are not made solely by individuals whose resources are passed down. Men's parents and grandparents may also exert influence over these decisions (Hartung, 1985). And, for it to pay parents and grandparents to prefer to invest in daughter lineages over son lineages, the extra-pair paternity rate need not be exceptionally high, particularly if the marginal benefit of wealth to sons is relatively modest (Holden et al., 2003). Whereas parents and grand-parents can exert influence over decisions concerning transfer of heritable wealth, they may have a difficult time exerting influence over decisions about how a son allocates time. Perhaps for that reason, men's decisions to allocate time to caring for their own off-spring do not appear to meaningfully covary with estimated extra-pair paternity rates (Gaulin & Schlegel, 1980). (See also Chrastil, Getz, Euler, & Starks, 2006, for a review of literature on effects of paternal uncertainty on grandparental investment patterns.)

In sum, examination of the psychological makeup of modern humans reveals sugges-tive evidence that humans have been selected to pair bond, express mutual mate choice, and biparentally care for offspring. In these respects, we appear to have diverged sub-stantially from our closest phylogenetic relatives (see also Geary, 2000).

The Male-Assistance Hypothesis of Extended Sexuality in Women

Male Assistance That Led to Selection for Women's Extended Sexuality

We now turn to the male-assistance hypothesis of extended sexuality in women. The form of this hypothesis that applies to humans is, purportedly, at least very similar to the form that applies to other pair-bonding species. Men deliver material benefits and services (e.g., food, protection, shelter) to primary partners (and vice versa, though benefits women deliver to their partners are not key to understanding extended sexuality). Women's copulability outside of the fertile phase of the cycle coevolved with male delivery of benefits to facilitate their flow, yielding a variety of forms of extended sexuality: sexual motivation during infertile phases of ovulatory cycles, during anovulatory cycles, during adolescence, and during pregnancy and lactation. That is, following the logic of Rodríguez-Gironés and Enquist's (2001) model, female ancestors of modern women who possessed extended sexuality outreproduced other females specifically because they outperformed them in the realm of garnering male-provided material benefits. Men who paired with females with whom they could copulate regularly delivered greater flow of benefits (partly to achieve continued sexual access) than did men paired with females with whom they could copulate only during the fertile phase. In addition, men paired with women with extended sexuality may have been less likely to seek extra-pair partners or to pair-bond with women other than current mates (Petrie, 1992).

The argument, of course, is not that men benefit from copulation per se, simply for the sake of the "pleasure" of copulation. No doubt, men typically find sex pleasurable, but the pleasure that men (or women, for that matter) find in sex was selected, in theory, because sex has, on average, reproductive benefits. Men should not have evolved to find a class of sex pleasurable if that class of sex had no benefits or has net negative reproductive consequences. (Hence, men, as well as women, typically find the thought of incestuous sex disgusting; see Lieberman, Tooby, & Cosmides, 2003, 2007.) In Rodríguez-Gironés and Enquist's (2001) model, males must possess imperfect knowledge of their mates' fertility status, as it changes across the cycle. They need not be completely ignorant of females' cycle-related fertility (and, as we discuss in chapter 11, men are not completely ignorant of the times that women are fertile in their cycles). Males simply need be unable to completely rule out possibility of conception. When males cannot completely rule out that a female has some risk of conception, they generally will be sexually selected to be motivated to copulate with a female (under appropriate circumstances). Men's interest in copulating across the cycle, even in the absence of female interest, coupled with the female's copulability across the entire cycle and at other infertile times, satisfies the assumption in the game-theoretic model that males do not have unambiguous direct cues of peak fertility. Presumably, it is the lack of unambiguous fertility cues that has selected for men's sexual interests in women throughout their cycles and in adolescent women. (Later, we treat in greater detail how female adolescent sexuality functions to gain male material benefits in the context of the human mating system of pair bonding, including long-term pair bonding; see chapter 6.)

That women's extended sexuality is extreme, in comparative perspective, is consistent with the view that pair bonding and male delivery of associated material services have been highly important to women's fitness in human evolution. Table 4.1 lists a number of these purported benefits. As we and others have emphasized, they include male paternal care (e.g., Alexander, 1979, 1990; Alexander & Noonan, 1979; Geary, 2000; Geary & Flinn, 2001).

Table 4.1 The Contexts Proposed for the Direct Selection of Female Extended Sexuality in Human Evolutionary History

I	Female–female competition for paternal care of offspring (protection, teaching, etc.) fueled by extremely altricial offspring, requiring intensive and extended parental care.
II	Female–female competition for male-provided calories for offspring and the female herself, given the relatively low caloric provision by females because they are pregnant or lactating or otherwise engaged in child-care activities that restrict their caloric access.
III	Female–female competition for males capable of and willing to protect the females and the females' daughters from capture by raiding males and sexual coercion.
IV	Offsetting child maltreatment by adult males.

Another service that male partners purportedly provided to women ancestrally is protection of mates and their female relatives from capture during raids and warfare and from sexual coercion by other men in the same group. (See Smuts, 1992; Smuts & Smuts, 1993; Thornhill & Palmer, 2000, for treatment of the importance cross-culturally to women of protection from sexually coercive males. See also Mesnick, 1997, and Wilson & Mesnick, 1997, on the bodyguard hypothesis for the evolution of human pair bonding.)

The benefits that women ancestrally garnered from men and that led to their extended sexuality need not have been delivered solely by primary partners. Male mating effort, leading them to deliver resources to women in hopes of gaining sexual access, may also have selected for women to possess extended sexuality (see, e.g., Hill, 1982; Symons, 1979). Though we do not dismiss the potential importance of these benefits, we suspect that women's reliance on a continued flow of material benefits delivered by primary partners typically meant that it was not worth the risks of losing those benefits by being unfaithful for exchange of a single meal. Hence it would not commonly benefit women to be unfaithful to an investing primary partner (at least one she wished to retain) during extended sexuality (particular if she had small children; see Marlowe, 2003a), unless the material benefits gained through infidelity were considerable. In some bird species, females rarely or never engage in extra-pair copulation during extended sexuality. (Some of these same female birds, by contrast, do so during the fertile phase of their cycles, but not to garner material benefits; rather, they often appear to do so to obtain genetic benefits for offspring. See chapters 9–12.)

Do Women Possess Extended Sexuality to Confuse Paternity Through Multiple Mating?

As we discussed in chapter 3, Hrdy's paternity confusion hypothesis is one important variant of the male-assistance hypothesis for female extended sexuality. Reduction of maltreatment of offspring by resident males through promiscuous mating has probably been a very important benefit leading to the evolution of female extended sexuality in a number of nonhuman primate species (Dixson, 1998; Hawkes, 2004; Heistermann et al., 2001; Hrdy, 1981, 1997, 2000; Hrdy & Whitten, 1987; Palombit, 1999; Pazol, 2003; van Schaik, 2000; van Schaik & Dunbar, 1990; van Schaik et al., 2004; Wrangham, 1993).

Might human females have been shaped to confuse paternity through mating with multiple males during a period of extended sexuality as well? Possibly—though there is reason for doubt. In pair-bonding species in which females' fitness is importantly affected by a flow of material benefits delivered by primary partners, female tactics to increase paternity confidence in these partners may be more successful than tactics to confuse paternity. Hence, once again, in a variety of bird species, females rarely or never engage in extra-pair copulation during extended sexuality, though they copulate with a social partner frequently. Similarly, women's extended sexuality may be directed primarily, even if not exclusively, toward primary partners (see chapter 10).

Nonetheless, perhaps a strategy of dispersing paternity confidence widely by mating with multiple men during extended sexuality has been adaptive in women conditionally, in restricted sets of circumstances (see Hrdy, 2000; see also Beckerman et al., 1998).

Women's Extended Sexuality Versus Concealed Ovulation

A number of writers have conflated extended sexuality and concealed ovulation (e.g., Strassmann, 1981; Symons, 1979; Turke, 1984). Specifically, some scholars have claimed that extended sexuality has functioned to garner material benefits by concealing ovulation (e.g., Alexander, 1979). The argument is that extended sexuality renders women's sexual interests continuous—effectively unchanging—across the cycle. If males discern females' fertile phase largely through females' sexual proceptivity and receptivity, continuous sexuality leaves males (as well as females themselves) ignorant of when cycling females are fertile. Lack of knowledge of female fertility status, in turn, may alter males' optimal strategy of allocating time and effort in a way that fosters paternal care. If males know which females are fertile at any point in time, they may be best off pursuing mating opportunities with those females. By contrast, if males do not know whether their mate or any other female is fertile at any point in time, many males may be better off investing in one female (with whom he mates and guards) and her offspring (see chapter 11).

Lack of perfect knowledge of female cycle-related fertility status is a necessary condition for the coevolution of female extended sexuality and male-delivered benefits during extended sexuality (Rodríguez-Gironés & Enquist, 2001). For female extended sexuality to function to obtain nongenetic material benefits or services from males, males cannot be able to rule out that a female in extended sexuality is fertile; hence males must lack perfect discrimination of female fertile phases from nonfertile phases for extended sexuality to evolve. Nonetheless, it is a mistake to conflate extended sexuality and concealed fertility within the cycle. They are two distinguishable and, at least potentially, functionally distinct phenotypic traits (or suites of traits). As we discuss in detail in chapter 5, sexual selection strongly favors male capabilities to discern and act on valid cues of female fertility status. Female receptivity may or may not be a valid cue. Typically, males attend to by-products of alterations in female chemistry across the cycle indicative of fertility status, though these cues may not be completely unambiguous cues of cycle-related fertility. *Concealment* of fertility status implies adaptation to suppress the cues to which males have evolved to attend and by which they discern female fertility status. Extended sexuality can evolve with *or without* concealed fertility. If the cues of fertility status that males attend to are sufficiently ambiguous, such that it pays them to copulate with (and deliver material benefits to) nonfertile females (whose fertility status, of course, males cannot perfectly gauge), female extended sexuality can evolve (in response to fitness gained through male delivery of material benefits) without any female adaptation for concealed ovulation whatsoever. For an expanded discussion of this theme, see chapter 11.

Women's extended sexuality should also not be conflated with continuous sexuality. As we noted in chapter 3, females of many species are sexually receptive outside their fertile period and hence possess extended sexuality. As we have emphasized, from a comparative perspective women's extended sexuality is extreme; women are sexually receptive throughout their cycles and during a variety of other nonfertile times. By no means do these claims imply, however, that females of species who exhibit extended sexuality, including women, express sexual preferences and motives that are *unchanging* (or continuous) throughout their sexually receptive periods. Indeed, we argue that, across all species with extended sexuality, instances in which female sexual interests do not change across their reproductive cycles are relative rarities. Instead, females of these species almost always express *dual sexuality*—sexual preferences and motives during extended sexuality that differ from those exhibited during fertile cycle phases (see chapter 8).

Summary

Women's sexuality is extended to a degree not matched by other female mammals. This amplification, similar to female extended sexuality in other species with this adaptation, was favored by direct selection because those females with extended sexuality received material benefits and services from males by exchanging infertile sex for them. Women's extended sexuality evolved in the context of the human mating system, which reveals important adaptations that served to affect amplification of extended sexuality. These adaptations account for, at least partly, facultative marriage and associated pair-bond arrangements, most commonly social monogamy with small to moderate levels of polygyny and rarely polyandry, and biparental investment in offspring. Comprehensive studies of foraging societies reveal that men subsidize the reproduction of their partners and their offspring, at least during critical times. Subsidies of calories and other nutrients are primarily achieved through men's hunting, which increases female reproductive success by reducing interbirth interval and/or offspring mortality.

Two primary explanations for the evolution of men's hunting activity have been offered and defended: that it functions as parental effort and that it functions as mating effort. Both receive some empirical support. They are not exclusive alternatives, and a blended view is most likely correct. Men hunt both as parental effort and mating effort; possibly, some adaptations for hunting serve one function and not the other.

Paternal care coevolved with the evolution of highly extended juvenile life, demanding intensive parental investment from conception through adolescence, and high levels of productivity that subsidize offspring well into the life span. Humans are very different from other mammals in the degree of these coevolved life history characteristics.

Evidence suggests that pair-bonded men possess psychological adaptation that conditionally alters the amount of effort allocated toward mating and parenting based on the ancestrally adaptive value of each. This evidence, as well as other forms of evidence suggestive of psychological design, is consistent with the reasonable view that men in pair bonds express design for both parental and mating effort.

Although paternal care was an important material benefit selecting for extended female sexuality in human evolutionary history, other material benefits from pair-bonded males in exchange for mating, such as protection from sexual coercion, were likely important as well. Ancestral women may have gained material benefits that affected selection for extended sexuality through extra-pair mating, but large benefits would have been required to offset possible lost investment from main partners, particularly for women with dependent children. Hence, though in many species of nonhuman primates mating with multiple males to reduce their maltreatment of offspring has importantly selected for extended female sexuality, women's extended sexuality probably does not significantly reflect this function.

Women's extended sexuality is distinct from concealed ovulation. Extended sexuality can evolve with or without concealed ovulation. Furthermore, women's sexuality is not continuous in the sense of unchanging sexual motivation across the menstrual cycle. Women possess dual sexuality across their ovulatory cycles: estrous sexuality when fertile and extended sexuality when not.

5   Female Ornaments and Signaling

Distribution of Female Ornaments Across Species and Patterns

Sexual selection researchers refer to a variety of morphological features that capture the visual attention of conspecifics (notably, those of the other sex) but do not function as weapons or shields but as *ornaments* (Bradbury & Vehrencamp, 1998; the Latin root of the word means "to adorn or decorate"). Examples include patches of skin that are brightly colored, gatherings of colored feathers, feather plumes, and elongated or elaborated fins and tails. The same term, however, aptly applies to olfactory and acoustical traits that attract attention from conspecifics, notably in sexual contexts. In most animal species, males are the more ornamented sex (Andersson, 1994). Nonetheless, females in sex-role-reversed species, some socially monogamous species, some polygynous species with male territoriality or other mating systems, and some Old World primates possess ornaments. In Old World primates, female ornaments are referred to as sexual swellings—enlarged, often differentially colored anogenital areas. Old World primate females with multimale, multifemale social arrangements (e.g., most macaques, mandrills, baboons, and chimpanzees) typically possess sexual swellings, but so, to lesser degrees, do a monogamous gibbon and a harem polygynous langur (Dixson, 1998; cf. Pagel & Meade, 2006). Female sexual swellings occur in other mammalian taxa and at least one species of bird (see review in Dixson, 1998). Relatedly, some female primates and other species display a sexual skin, a differentially colored anogenital region without accompanying edema (swelling).

In primates, the color and size of ornaments typically peaks when females are near peak conception probability during estrus, though even pregnant females or otherwise infertile females (adolescent and anovulatory cycling females) may display ornaments (see Anderson & Bielert, 1994, and Dixson, 1998, for reviews; Deschner et al., 2004; Mohle, Heistermann, Dittami, Reinberg, & Hodges, 2005; Gesquiere, Wango, Alberts, & Altmann, 2007). Female sexual ornaments in primates and many other species generally are not displayed outside of a breeding context. (Throughout the remainder of the book, when we write of "conception probability," we refer to probability of conception contingent on sexual intercourse—in humans, unprotected sexual intercourse.)

A wide range of findings on female ornaments exemplifies their important role in mating behavior. In the sex-role-reversed pipefish (in which females compete for male parental care), females develop a colorful venter during their brief breeding season. Males prefer more colorful females as mates, as color positively predicts female condition (Berglund, Rosenqvist, & Bernet, 1997; Berglund & Rosenqvist, 2001). Females of the lizard *Sceloporus virgatus*, a polygynous species, develop an orange throat patch during the breeding season. Males prefer more colorful females, which are in better phenotypic condition, have fewer parasites, and produce eggs of higher quantity and quality (Weiss, 2002, 2006). In the domestic chicken, another polygynous species, roosters prefer highly ornamented females, those with large combs. Though hens possess smaller combs than roosters, comb size in hens positively predicts condition, reproductive value, egg size, and yolk size (Pizzari, Cornwallis, Levlie, Jakobsson, & Birkhead, 2003). In the bluethroat (Amundsen, Forsgren, & Hansen, 1997) and barn owl, two socially monogamous bird species, large female ornaments are associated with superior female condition. In the owl, offspring of mothers with greater ornamentation inherit better immunocompetence, parasite resistance, and developmental stability than do offspring of less ornamented mothers (Roulin, 2004; Roulin, Jungi, Pfister, & Dijkstra, 2000). Emery and Whitten (2003) reported that swelling size of female common chimpanzees is positively correlated with ovarian function, measured as levels of ovarian steroids (most notably, estrogen). Finally, Domb and Pagel (2001) found that the size of sex swellings in female olive baboons positively predicts female condition, early age of maturity, number of offspring, and number of surviving offspring. Males, they also reported, prefer females with large swellings: Males particularly compete for, sometimes with costly aggression, and groom females with large ornaments. (See, however, other views and responses in Domb & Pagel, 2002; Nunn, van Schaik, & Zinner, 2001; Zinner, Alberts, Nunn, & Altmann, 2002; Zinner, Nunn, van Schaik, & Kappeler, 2004.) Male mate choice in savanna baboons (Hausfater, 1975; Packer, 1979; Smuts, 1985) favors sexual swellings of females that are maximally turgid and may be based, in part, on individual variation in swelling size (see Bercovitch, 2001). Moreover, laboratory experiments demonstrate that male baboons are more sexually motivated (as measured by masturbation frequency) by females with artificially exaggerated swellings than by females with normal swellings (Bielert & Anderson, 1985; Girolami & Bielert, 1987; also see Waitt, Gerald, Little, & Kraiselburd, 2006). (Other evidence of male mate choice in Old World nonhuman primates is summarized by Shahnoor & Jones, 2003; see also Alberts, Buchan, & Altmann, 2006; Gesquiere et al., 2007. Male mate choice in primates and other animals has not received as much attention from researchers as female choice. Yet it is known that mate choice by males can have dramatic positive effects on male reproductive success; see, e.g., Gowaty, Drickamer, & Schmid-Holmes, 2003, on house mice.)

These examples illustrate four generalizations about female ornaments. First, they are sexually selected as signals—they function to communicate information. Second, they are sexually selected through female-female competition for mates, male choice, or both. Third, they are rarely seen in species in which males transfer no material benefits to females; compared with other females, highly ornamented females often obtain more male material benefits. Fourth, the degree of ornamentation across females of a species often positively predicts condition or health in addition to number and survival of offspring. (For further evidence in a wide variety of species, see Amundsen, 2000; Amundsen & Forsgren, 2001; Amundsen & Pärn, 2006; Cuervo & Møller, 2000; Cuervo, Møller, & de Lope, 2003; Daunt, Monaghan, Wanless, & Harris, 2003; Griggio, Matessi, & Pilastro, 2003; Griggio, Valera, Cassas, & Pilastro, 2005; Hill, 2002; Jawor, Gray, Beall, & Breitwisch, 2004; Krebs, Hunte, & Green, 2004; LeBas, Hockman, & Ritchie, 2003; Siefferman & Hill, 2005; Swenson, 1997; Weiss, 2002, 2006.)

These generalizations are by no means hard and fast rules; some exceptions are notable. Ornaments of female red-winged blackbirds and barn swallows, for instance, may not have evolved through sexual selection (Cuervo, de Lope, & Møller, 1996; Muma & Weatherhead, 1991). In captive mandrills with food provisioning, evidence that female sexual swellings predict condition or are sexually selected is mixed (Setchell & Wickings, 2004, reported largely negative findings; Setchell et al., 2006, reported largely null findings in low-power analyses, but largely in a predicted direction; Dixson & Anderson, 2004, reported positive findings).

More generally, after a partial review of the literature on ornamentation across species, Cotton, Fowler, and Pomiankowski (2004) question whether special associations between degrees of ornamentation and individual health or condition have often been demonstrated. They point to the need for better experimental designs than have often been used to test these relationships. Though many studies reveal that ornaments hypothesized to function in sexual competition for mates are condition-dependent, few show that they are associated with condition more than nonornamental traits are (Cotton et al., 2004).

Though cognizant of the limitations imposed by the current literature and mindful of the need for more and better research, we defend the view that many, if not most, female ornaments have evolved as signals of quality or current reproductive condition. Though the view that ornaments function as honest signals of quality is widely represented and argued within behavioral ecology and the study of animals generally (see review by Searcy & Nowicki, 2005), the view is in fact not the traditional explanation of sexual swellings in Old World primates. Instead, female sexual swellings have traditionally been viewed as signals of fertility within the reproductive cycle. Much of the current chapter describes this view and argues that it is largely if not fully mistaken.

Female Signaling of Fertility Within the Reproductive Cycle

Darwin's theory of female ornaments was that they are nonfunctional by-products of ornaments favored directly by sexual selection acting only on males (Darwin, 1871; see Andersson, 1994). That is, just as men purportedly have nipples because these features have been selected in women for specific functions and are nonfunctionally expressed in men, ornaments that have been selected in males may be nonfunctionally expressed in females. (We note here that we do not completely discount the possibility that nipples in men have been maintained because of their tactile sensitivity. Nonetheless, the idea that one sex may possess some features that are functional only in the other sex appears to be noncontroversial; see, e.g., Rice & Chippendale, 2001.) This theory may apply to instances in which female ornaments are muted forms of male ornaments (again, as male nipples are muted forms of female nipples). The model, however, cannot explain female ornaments in general (Cuervo & Møller, 2000; but see Cuervo et al., 1996; Muma & Weatherhead, 1991). Certainly, it cannot explain cases in which only females of a species exhibit an ornament—for instance, the anogenital sexual swellings in various Old World primates.

The conventional explanation of female primate sexual swellings, as well as the behaviors and odors produced by females in other taxa at fertile times of their reproductive cycle to which males sexually respond, is that these traits function to signal high fertility in the reproductive cycle (e.g., Aldridge & Duvall, 2002; Burt, 1992; Deschner et al., 2004; DeVleeschouwer et al., 2000; Hamilton, 1984; Hrdy, 1981; Jacobson, 1972; Liley & Stacey, 1983; Mason, 1992; Nelson, 1995; Scott & Vermierssen, 1994; Szalay & Costello, 1991). That is, females are presumed to be functionally designed to produce sex attractants, lay scent trails, or otherwise be conspicuous to assist males in finding them; otherwise, males might not be interested in mating, and females would not be inseminated. More generally, estrus in mammals and reptiles has often been similarly interpreted (on mammals, see Nelson, 2000, or general mammalogy textbooks; on reptiles, see Aldridge & Duvall, 2002, and the earlier references to this view cited therein). In chapter 8, we discuss and critique this view of estrus.

Women's sexual ornaments (e.g., their enlarged breasts), of course, are not specific to the fertile phase of their cycles. The conventional interpretation of fertility signals nonetheless pertains to traditional explanations of women's ornaments, as well, for the retention of women's ornamentation across the cycle and their entire reproductive life spans has been interpreted by some scholars to mean that women have been selected to *deceptively* signal permanent cycle fertility (e.g., Szalay & Costello, 1991). Though, in some modified (but highly qualified) form, this idea may have some merit (see chapter 6), we argue that it is founded on a fundamentally incorrect view of female signals in general. Specifically, we argue that female sexual signals rarely evolve because they let males know when females are fertile. Put otherwise, female sexual signals rarely *function* to signal fertility in the reproductive cycle.

The appeal of the idea that females evolved to signal cycle-related fertility with sexual swellings is understandable (even if, as we argue, largely misguided). As we noted, in many species with such swellings, females do tend to exhibit them when fertile in their cycles. Though there is not, by any means, perfect correspondence between sexual swellings and cycle-based fertility status (as many primatologists have emphasized; e.g., Deschner, Heistermann, Hodges, & Boesch, 2003; Deschner et al., 2004; Engelhardt et al., 2005; Mohle et al., 2005; Reichert et al., 2002), positive covariation exists. Males who selectively seek copulations with females with sexual swellings therefore direct their efforts toward females capable of conceiving their offspring at better than chance levels. That is, males could have adaptive reason to pay attention to a signal correlated with fertility. Furthermore, the explanation proceeds, because females who are fertile need sperm to conceive, females could benefit by letting males know they are fertile. Hence females have been selected to produce a signal that males have been selected to attend to, as a means of gaining male attention (and sperm) just when they can use it.

Though appealing on the surface, on deeper reflection this view is almost certainly wrong, at least in this simple form.

Parental Effort, Mating Effort, Females, and Males

Biologists widely recognize that, because of their different roles in reproduction, males and females typically have behavioral and physiological adaptations surrounding reproduction with different functions. Typically, females, as the sex that usually has a greater obligate investment in offspring (and, in mammalian species, always does), have adaptations that function to adaptively allocate their parental investment through control of that investment. By contrast, males typically have adaptations for fertilizing ova and thereby accessing parental investment held by the opposite sex (Alexander & Borgia, 1979; Bateman, 1948; Charnov, 1979, 1982; Eberhard, 1996; Emlen & Oring, 1977; Kodric-Brown & Brown, 1987; Queller, 1997; Parker, Baker, & Smith, 1972; Thornhill & Alcock, 1983; Trivers, 1972; Williams, 1966). This pattern holds true even in species in which males do invest substantially (though discriminately) in offspring, including in so-called sex-role-reversed species such as pipefish (Berglund, Widemo, & Rosenqvist, 2005) and tettigoniid katydids (Gwynne, 2001). Despite its nearly universal acceptance, this insight has implications that are often unappreciated or overlooked when it comes to interpreting female sexual behavior and ornamentation. The traditional interpretation that female primate sexual swellings and other female traits signal fertility is a prime case in point.

Parental effort is any effort allocated by an individual that increases the quality of offspring produced (Alexander & Borgia, 1979; Hirschfield & Tinkle, 1975; Low, 1978, 2001; Trivers, 1972). This effort obviously includes investment of time, energy, risk, and resources to support the growth and survival of offspring after they are born. But parental effort also includes investment of resources prior to birth, beginning with investment of resources into gametes. By definition, females produce larger gametes than males, and hence offer a larger initial investment in offspring. Though the initial asymmetry in female and male investment may disappear or reverse (see Gwynne, 2001; Trivers,

1972; Williams, 1966), it typically becomes exaggerated with investment beyond gamete production. In all eutherian mammalian species, internal fertilization and gestation require obligate parental effort on the part of females that far exceeds that of males. Indeed, in about 95% of mammalian species, males engage in very little or no parental effort (Clutton-Brock & Parker, 1992).

That is not to say that reproduction is not costly for males, as well, of course. In most species, however, males' reproductive effort consists largely of mating effort, which is the male effort to find and attract mates and inseminate them, thereby increasing number of fertilizations. In most mammals, males expend far more mating effort than do females.

Two reasons that the sex difference in initial investment in parental effort becomes exaggerated are key to understanding most females and most males. First, as the sex that makes the greater parental investment, females have reason to be very discriminating about how they use that investment. They should be selected to make decisions that lead that investment to go far. If females mate with males who provide a poor complement of DNA, they have squandered that investment. Hence, females should be selective about whom they will mate with. When female mating is selective, the number of males that are eligible to mate (based on female preferences) is limited. Eligible males, then, can expect a relatively high future reproductive rate, which leads them to engage in mating effort—efforts to find and attract mates—rather than parental effort. Males who might benefit by parenting (because of a low expected future reproductive rate derived from mating effort) do not get the chance because females do not select them (Kokko & Jennions, 2002).

In some circumstances, biparental care has evolved. As we argued in chapter 4, humans appear to be one such species. In these species, females partly select males for their willingness to invest in parenting. Most or all mated males may engage in parental effort, and there is a smaller sex difference in allocation of effort toward mating and parenting. Nonetheless, some sex difference in parental effort typically remains. Males still have a smaller *obligate* investment in parenting with the production of each offspring. If they can attract females to mate with them without providing parental investment, they may benefit from engaging in mating effort in a way that females do not (e.g., Symons, 1979).

A second reason that the sex difference in initial investment typically becomes exaggerated is that the sexes often differ in parental certainty. The difference in parental certainty itself arises because of the asymmetry in initial investment. As females typically benefit from control over fertilization and additional investment into the offspring to an extent greater than males do, in many species fertilization takes place within a female's reproductive tract (where she can, for instance, better exert control over who sires her offspring). In such instances, males are less certain about who their offspring are than are females. This asymmetry leads males to further devalue parental effort (which could be squandered on offspring not their own; Alexander & Borgia, 1979; Kokko & Jennions, 2002; Queller, 1997). They may engage in efforts to increase paternity certainty. This reason that parental investment becomes increasingly sexually asymmetric applies even when males do engage in substantial amounts of parental effort.

In sum, selection on females generally leads to adaptations designed to foster the *efficient expenditure of parental effort*. Females are hence designed to assess ecological circumstances that affect efficient expenditure of their parental investment. They are, furthermore, designed to differentially allocate parental investment depending on how circumstances affect its efficient use. These decisions should be sensitive to variations in season, resource availability, and mortality risks (e.g., diseases) that affect optimal allocation of parental effort. Females should also be designed to assess the quality of sires for their parental investments, for sire quality affects the profits females derive from their investments. When males do provide parental investment, females should be designed to assess and promote paternal investment into their offspring, which once again enhance the profits garnered from females' own investments.

Selection on males, by contrast, typically leads to adaptations designed to *secure the parental investment held by females*. In many species, male reproductive success is maximized solely through ability to gain female parental investment; males succeed directly as a function of their ability to get females to invest in offspring they sire. Males in some species can also enhance fitness through their own parental efforts, but they get the opportunity to do so only if they are able to satisfy criteria by which females offering parental investment select sires; before they can engage in paternal investment, they must become fathers. (Obviously, it is also true that females cannot mother before they become mothers, but females typically need not satisfy stringent criteria of male choosiness simply to become mothers.) Hence males compete with other males to capture the parental investment offered by females. Naturally, selection on males should lead to phenotypic design for obtaining and assessing information about levels of parental investment females offer (whether it be condition-dependent, age-dependent, or cycle-dependent), including female attributes reflecting female phenotypic quality, and to allocating their mating efforts when those efforts are most likely to pay off. If they engage in parental effort at all, they should seek out and be sensitive to information pertaining to whether they have, indeed, successfully conceived an offspring (i.e., assess paternity certainty; see chapter 12).

Why Females Should Rarely Be Selected to Signal Fertility in Their Reproductive Cycles

Females Should Not Bear the Costs of Fertility Signals

As just noted, males should be designed to assess females' ability and readiness to expend parental investment. Indeed, knowing when a female is about to expend parental investment is among the most valuable pieces of information a male can possibly possess. Sexual selection on males hence strongly favors abilities to perceive this information and act on it to secure female parental investment. These selection pressures are not recent; they have operated since the first females and males appeared in the history of life, that is, since the phylogenetic origin of anisogamy, and continuously thereafter (on the origin of anisogamy, see Parker et al., 1972; Kodric-Brown & Brown, 1987).

To appreciate, then, why females are unlikely to be selected to signal fertility in their reproductive cycles (at least with any signal that is costly to them), consider two females, one who signals and the other who does not. According to conventional thinking, the one who signals benefits because she lets males know when they should compete for her parental investment. Implicitly, this argument presumes that males do not know when the nonsignaling female is fertile (and furthermore that their not knowing hurts her reproductive success). A lone male who pursued copulations with the female who did not signal, however, could potentially outcompete males who ignored her, and, for that reason alone, great advantages to signaling fertility are not obvious. More fundamentally, however, if there are any incidental cues associated with the nonsignaling female's fertility status (such as by-products of female fertility status), a male who perceived and acted on those cues would clearly be strongly advantaged over other males. Hence strong sexual selection on males should lead to adaptations to detect any cues of fertility status that exist (Thornhill, 1979; Thornhill & Alcock, 1983; West-Eberhard, 1984; Williams, 1992). We consider the evidence that such cues typically do exist and, therefore, that males typically know when females are fertile. (Instances in which selection favors female *suppression* of incidental cues—including, we argue, in humans—are exceptions. See chapter 11. Obviously these cases are not ones in which females purportedly *signal* fertility.)

Though females might be able to provide even better information to males through signaling, the marginal benefits that females can accrue through signaling may be slight (as males are generally already tuned in to incidental fertility cues) and very unlikely to offset the energetic costs of large signals. And that surely applies to most sexual swellings in primates, which are not particularly accurate cues of female ovulation (not obviously better than cues males have available anyway; e.g., Deschner et al., 2004; Engelhardt et al., 2005; Mohle et al., 2005; Reichert et al., 2002), yet costly. (We do not push the extreme argument that female swellings could not provide *some* incremental information about fertility status in the cycle to males and that females could not derive *some* small benefit from doing so. We do argue that the benefits of advertising fertility in the cycle per se are very unlikely to pay for the costs of sexual swellings. On the whole, other benefits most likely render those costs worthwhile.)

Based on precisely this line of reasoning, Pagel (1994) criticized the traditional view of sexual swellings of female primates as cycle-related fertility signals. In a quantitative game-theoretic model, he found that the sole evolutionarily stable strategy for males is to find and compete for fertile females—females that are at the point in their reproductive cycle at which conception can occur. Accordingly, females are not under selection to evolve costly traits that signal cycle-related fertility, such as sexual swellings.

Males Do Possess Adaptations to Detect Female Fertile States

If females emit incidental cues of fertility status, and if selection shapes male adaptations to detect and act on them, comparative data should yield much evidence for such adaptations (e.g., Thornhill, 1979). Females, by contrast, should rarely possess adaptations to detect reproductive readiness in males. In fact, these expectations are confirmed. Across

a wide variety of species, males possess behavioral and related morphological adaptations to detect female fertile state and pursue females that are in that state. Females
typically lack comparable adaptations.

Many of these adaptations are specialized sensors for olfactory detection, often at a
great distance, of females that emit by-products of reproductive maturation. Male silk-
worm moths, for instance, have specialized antennae that can detect minute concentrations of a chemical that females release as a by-product during egg production (Jacobson,
1972). Bulls use chemicals in the urine of estrous cows, apparently breakdown products
of the biochemical progression to peak fertility across the estrous cycle, to detect fertile
females (Kumar, Archunan, Jeyaraman, & Narasimhan, 2000). Estrous female vervet
monkeys excrete in urine prenanediol-3 alpha-glucuronide, a chemical related to ovarian function, which appears to be sexually attractive to conspecific males (see Szalay &
Costello, 1991). In the long-tailed macaque, males' sexual interest in females correlates
strongly and positively with female estrogen levels and peaks at the most fertile phase
of females' menstrual cycles. Indeed, males prefer fertile-phase females having cycles
that lead to conception over females in the comparable phase of nonconception cycles.
Apparently, males perceive scent cues of estrogen levels (or their by-products) across
the menstrual cycle (Engelhardt, Pfiefer, Heistermann, & Niemitz, 2004). In the chacma
baboon, though males require several months of residency with a female to develop an
ability to discern effectively her conceptive from her nonconceptive cycles, they ultimately do so, probably at least partly based on estrogen-related scent cues (Weingrill,
Lycett, Barrett, Hill, & Henzi, 2003). Male chimps, too, can detect conceptive from nonconceptive cycles, presumably based on differences in estrogen levels (Deschner et al.,
2004). So can male mandrills (Setchell, Charpentier, & Wickings, 2005), baboons
(Alberts et al., 2006; Gesquiere et al., 2007), and stump-tailed macaques (Cerda-Molina,
Hernandez-Lopez, Chavira, et al., 2006; Cerda-Molina, Hernandez-Lopez, Rojas-Maya,
Murcia-Mejia, & Mondragon-Ceballos, 2006).

Functionally equivalent male adaptations that detect and track olfactory cues of
female fertile states have been reported in algae, annelid worms, various arthropod
taxa, salamanders, fish, lizards, snakes and multiple taxa of mammals (Aldridge &
Duvall, 2002; Ferris et al., 2004; Liley & Stacey, 1983; Mason, 1992; Pickard, Holt,
Green, Cano, & Abaigar, 2003; Preston, Stevenson, & Wilson, 2003; Rajanarayanan
& Archunan, 2004; Roberts & Uetz, 2005; Scott & Vermierssen, 1994; Shine, Phillips,
Waye, LeMaster, & Mason, 2003; Thornhill, 1979; Thornhill & Alcock, 1983; West-
Eberhard, 1984; Williams, 1992). Behavioral adaptations to seek out this information
are also widespread. Males of the vast majority of species of mammals, including most,
if not all, Old World primates (excluding men in typical circumstances), exert considerable effort to touch, investigate, monitor, and smell the genital region of females
(DeVleeschouwer et al., 2000; Dixson, 1998; Hoogland, 1995; Michael et al., 1976;
Michael & Zumpe, 1982; Nelson, 1995; Takahashi, 1990; Weingrill et al., 2003). Once
again, functionally similar adaptations to attend olfactorily, visually, or tactually to
females' genitalia or gonopores are found in males of nonmammalian vertebrates and
even arthropods. (See Thornhill & Alcock, 1983, for a partial review of these male
adaptations in insects.)

Signal Versus By-Product Cue

By-product cues must be distinguished from signals. Organisms respond to a great variety of stimuli or cues in their environments, only some of which are signals. A signal is a *communicative adaptation*—that is, a trait that has been selected directly because of its communication effect (Burghardt, 1970; Otte, 1974; also see Bradbury & Vehrencamp, 1998; Liley & Stacey, 1983). Put otherwise, a signal evolved because of information value it provides to other organisms (if sexual signals, members of the other sex), which those members act on in a way that benefits reproductively individuals who exhibit the signal. Communication—transmission of information to other organisms, which benefits the transmitter—is the *function* of a signal. Incidental or by-product cues can be emitted with no direct selection on emission of the cue for communicative value. These cues, then, are not signals. As Williams (1992) argued, chemical and other phenotypic stimuli emitted by fertile-phase females are rarely signals; instead, they are incidental by-products of changes to internal states of reproductive readiness. That is, these features are outcomes of physiological processes and structures that were shaped by selection for functions *other than* their communicative effects (e.g., in several cases discussed previously, structures designed to regulate estrogen levels as a means of affecting reproductive outcomes). Put otherwise, these processes and structures would have been favored by selection even if the incidental by-products detectable by males were completely absent. As we noted in chapter 2, to say that a feature is a by-product is to say that it had no role in bringing about the evolutionary persistence of the trait that gives rise to it as a by-product.

This argument does not imply that females do not benefit from by-products emitted incidentally at egg maturation or that correspond to high fertility in the reproductive cycle. They may, in fact, benefit through the attraction of suitable mates. But not every benefit of a trait is an evolutionary function, one that played a role in effectively shaping the trait (Symons, 1979, 1987; Thornhill, 1990, 1997; Williams, 1966, 1992; see also chapter 2, this volume). The benefit of attracting males typically played no role in bringing about the processes that led to by-products that attract males. In an evolutionary sequence, females emitted these by-products even before males used them as cues. Males then evolved specialized physiological, morphological, neural, and behavioral adaptations for finding and effectively inseminating fertile females on the basis of these by-products. The new benefits that accrued to females as a consequence of male adaptation typically had no effect on changes in frequencies of alleles underlying the development and operation of mechanisms regulating how those by-products were emitted and hence played little or no role in the evolution of these mechanisms. Females, in some sense, obtained these benefits for free, with no additional adaptation required. Hence females typically possess no specialized adaptation that functions to let males know they are reproductively ready.

Obviously, the claim that these stimuli are mere incidental effects is not to imply that they do not play important roles in male-female interaction and mating. They often play absolutely crucial roles. These roles, however, emerge through adaptation in males, not females (see also Williams, 1992). The fact that female emissions do play terribly important roles in mating unfortunately mislead many researchers into thinking that they *must* be signals. But they of course need not be and typically are not.

In the literature on nonhuman animal sexual behavior, writers frequently use the terms *pheromone*, *signal*, and *advertisement* to refer to stimuli emitted by sexually receptive females (for recent examples on primates, for instance, see Cerda-Molina, Hernandez-Lopez, Chavira, et al., 2006; Cerda-Molina, Hernandez-Lopez, Rojas-Maya, et al., 2006). All are inaccurate and misleading, as all imply that these stimuli function to attract males by communicating a message. In fact, these stimuli typically have no function in communication. Indeed, they have no function whatsoever.

Though writers typically imply that the communicative functional message of female stimuli that males track alerts males to reproductive readiness, other communicative functions are also claimed, most notably functions of communicating sex and species identity appropriate for reproduction (e.g., Dixson, 1998; Halpern, 1992; Mason, 1992). Though males may sometimes identify sex and species on the basis of scents and other fertility-related cues, these discriminations are by-products of selection for male abilities to detect females that are fertile and of high reproductive value. For instance, male garter snakes can sometimes discriminate the sex and species of other garter snakes based on olfactory cues (though evidence is mixed; see Mason, 1992). Males are highly attracted to scent of a female garter snake in her fertile phase, probably a by-product of the formation of mature eggs and associated estrogen (Halpern, 1992). They furthermore possess the ability to discriminate female body size and condition and prefer the scent of large fertile females in good condition (Shine et al., 2003). As males do not emit these same odors and females of different species have slightly different scents, males can discriminate sex and species based on the same cues. But these effects emerged as by-products of sexual selection on males to find and be attracted to fertile females (of their own species) in good condition. (For additional discussion of the reasons that sex and species identity are unlikely to be widespread functions of pair formation and courtship signaling systems, see Thornhill & Alcock, 1983; West-Eberhard, 1984.)

Our argument against female signaling of reproductive readiness has been largely conceptual. Females have no reason to expend large amounts of effort on signaling their readiness. Selection on males ensures that they will find fertile females based on by-products (or, at most, minimal efforts to disperse scents). Empirical evidence for female adaptation to signal, however, is also lacking. As we emphasized in chapter 2, an adaptation is a trait showing a specific functional organization/design because of a history of direct selection for that design. A directly selected trait—an adaptation—is one that solves a problem limiting individuals' reproductive success and that, because of the solution it provides, is favored directly by selection. Demonstrating that a trait is an adaptation is an onerous task. It requires evidence for functional design in the trait to solve the problem. There is no compelling evidence, however, for functional design in typical vertebrate females to emit a vaginal scent at the fertile phase of the reproductive cycle to solve a problem that limits their reproductive success. Again, we have argued that females do not face a problem of getting access to limited sperm. Sperm for females is rarely limited (though selection, of course, does favor optimal allocation of sperm by males; see Wedell, Gage, & Parker, 2002). Furthermore, however, there is, for most species studied, no evidence of functional design to emit scents. When scents are signals, individuals typically have specialized machinery to manufacture and/ or emit them, the chemicals produced typically involve complex molecules, and they are produced in relatively large quantities (see Bradbury & Vehrencamp, 1998). But, with few if any exceptions, females possess no specialized machinery to manufacture their vaginal scents. Their scents result from simple chemical-breakdown products of physiological changes associated with the onset of female reproductive status. They are emitted in small quantities. And females possess no specialized features to store, emit, or disperse scents.

Showy Flowers Are Not Female Fertility Signals

Female signaling of fertility should be rare not only in animals but also in plants. Indeed, a classic case illustrates this point. The showy flowers of the angiosperms (flowering plants) do appear to be specialized structures that function to attract pollinators. A traditional interpretation of hermaphroditic flowers such as the rose or daylily was that they are functional analogs of female animal adaptations for signaling reproductive-cycle fertility, female adaptations designed to ensure fertilization of their ova. The interpretation shares those analogs' theoretical problems. As Charnov (1979) first emphasized (see also Bateman, 1948), female reproductive success in plants is very unlikely to be limited by available pollen—by fertilization rate—because sexual selection on males ensures high rates of fertilization of female gametes. Rarely will females in flowering plants experience a net benefit from developing costly structures that function to attract pollinators. Showy flowers, then, are typically sexually selected *male* adaptations that increase male "mating success" by attracting pollinators (Andersson, 1994; Bell, 1985; Charnov, 1982; Willson & Burley, 1983).

Knight et al. (2005) thoroughly reviewed effects of pollen limitation on female reproduction. Many cases have been documented. Knight et al. (2005) suggest, however, that these instances may typically arise from evolutionarily novel circumstances. For instance, introduction of a plant into a new ecology (a common practice) may also introduce novel distributions of and densities of plants and pollinators, which then give rise to pollen limitation on female reproduction. Pollen limitation in novel circumstances, however, by no means implies that the same plants evolved under circumstances of pollen limitation.

Indeed, a classic purported case of sperm limitation has recently been shown to be an outcome of evolutionary novel conditions. Reproduction of the roundworm, *Caenorhabditis elegans*, a commonly studied species, is sperm-limited in laboratory conditions. Females commonly lay many unfertilized eggs; males do not deliver adequate sperm numbers to fertilize all eggs. As the title of their article announces, Goranson, Ebersole, and Brault (2005) have resolved the "adaptive conundrum" this phenomenon poses. In favorable lab conditions, females achieve unusually large body size, which leads them to lay very large numbers of eggs. In harsher natural conditions, females grow to smaller size, they lay fewer eggs, and available males fertilize them. Females' reproductive success is simply not sperm-limited in natural conditions.

Similarly, evolutionary novelty may explain examples of sperm limitation in broadcast spawning invertebrates and algae (Yund, 2000). In these instances, mobile male gametes seek out female gametes, typically using chemical cues. Industrial pollutants may kill or disrupt adaptive chemotaxic orientation of male gametes (e.g., see Meric et al., 2005).

Polyspermy, Not Sperm Limitation, Is a Problem Females Typically Must Solve

Male gametes are ardent and, similar to males themselves seeking females, sperm often use chemical cues to find eggs for fertilization in the female reproductive tract. Polyspermy (entry of multiple sperm into an egg at conception) is widespread in organisms and occurs in humans. It is maladaptive for both sexes, as it leads to death of the zygote or maladaptive development. Polyspermy can be understood as a maladaptive by-product of male adaptation to conceive before sperm (1) die, (2) are outcompeted by sperm of another male, or (3) are rejected by the ovum or female reproductive tract. Selection on females of various taxa has crafted counteradaptations that function to prevent polyspermy (e.g., defenses against multiple sperm entering the egg, reduction of sperm numbers in the reproductive tract; for evidence of adaptation of eggs against polyspermy in humans, see Patrat et al., 2006). These counteradaptations may well have maladaptive by-products themselves: infertility. As Arnqvist and Rowe (2005) note, the line between a female's effective prevention of polyspermy and her infertility due to overprotection of eggs against polyspermy can be a thin one. High rates of female infertility in species (not atypically 10–15%; Arnqvist & Rowe, 2005) is not due to insufficient quantities of sperm; it is more likely due to female adaptation in response to the overabundance of sperm, which functions to prevent polyspermy. (See also Gomendio et al., 2006.)

Misinterpretations of Female Fertility Signals in Animals

If, in fact, females rarely signal cycle-related fertility, the many instances in which researchers have explained phenomena as outcomes of female adaptation to signal cycle-related fertility to males typically reflect misinterpretation. Two kinds of misinterpretation are common. The first kind is one that sees female adaptation to signal when no such adaptation exists. Again, males are adapted to respond sexually to incidental-effect cues of female fertility. The second kind of misinterpretation is one that correctly recognizes female adaptation to signal but misinterprets the function of the signal, viewing it as a signal of fertility in the reproductive cycle when, in fact, it is a signal designed to honestly communicate information about quality.

Female Scent as a By-Product   As already noted, researchers commonly interpret female by-products to which males are attracted as signals, when in fact they have no function. Aldridge and Duvall (2002), for instance, interpret the function of the scent of fertile-phase female rattlesnakes derived from egg maturation as "advertising female receptivity" and "assisting males in locating" females. Elsewhere in the same article, however, these authors document that males are elegantly designed to find estrous females (also see Duvall & Schuett, 1997).

In many species of insects and the Asian elephant, females at peak fertility produce and release a particular molecule to which males are attracted. The chemical has been hypothesized not only to signal peak female fertility but also to function to sexually motivate males and synchronize their mating behavior with that of females (Rasmussen, 1999). The widespread occurrence of the chemical across taxonomically distant females has been interpreted as evolutionary convergence. That is, the same selection pressure purportedly acted independently to create the same female adaptation involving this chemical in distantly related species (Rasmussen, 1999; Rasmussen, Lee, Roelofs, Zang, & Daves, 1996). Moreover, the compound's volatility has been argued to be a key reason that it serves the function of communicating peak fertility well and, as a result, has been selected to do so in different taxa (Rasmussen et al., 1996).

A simple by-product hypothesis accounts for the widespread distribution of this chemical just as effectively. This chemical may simply be a common by-product of physiological achievement of peak female fertility, which males have evolved to detect and be attracted to. As predicted by this hypothesis, the chemical has a simple structure, and females possess no specialized apparatus to emit it. Indeed, in the Asian elephant it appears to be "manufactured" in the liver (Rasmussen, 2001). The liver, of course, is where many by-products of metabolic processes are broken down, released into the bloodstream, and excreted in urine. Males do, in fact, detect this chemical in urine; it cannot be detected in the female vaginal area. The liver need not possess a special adaptation to create this chemical so that males can detect it in urine. It merely needs to do its jobs, which involve preparing waste products for elimination. Rasmussen et al. (1996) are probably right when they suggest that the chemical's volatility is a reason that males use it as a cue to discriminate the fertility status of females, but for the wrong reason. Instead of females having been shaped to produce a "signal" that, due to its volatility, tracks female fertility, males evolved to detect and be sexually motivated by a chemical that, due to its volatility, affords effective tracking of female fertile states.

In the same species of elephant, bulls emit a diverse array of specialized chemicals that regulate sexual competition among males and are associated with males' relative status in the hierarchy of local males. The traits responsible for production and dispersal of these chemicals do reveal evidence of functional design for communication as honest signals of male quality (Rasmussen, 1999; Rasmussen, Krishnamurthy, & Sukamar, 2005; Schulte & Rasmussen, 1999). The chemically simple sex attractant of female elephants, however, does not provide evidence of any function.

Recently, Rasmussen and colleagues have identified a purported "pheromone" in the urine of female African elephants, one that is shared with a species of beetle (Goodwin et al., 2006). It too permits males to detect a female's preovulatory phase (Bagley, Goodwin, Rasmussen, & Schulte, 2006). Once again, this chemical is probably simply a by-product of a metabolic process. It is not a signal or a pheromone, and its production has no function.

Female Ornaments as Signals of Quality   In some species, females do appear to produce chemicals that function as signals. In these instances, however, these chemical signals are unlikely to have evolved because they signal peak female fertility. Rather, their function is probably to signal individual female quality. In tamarin monkeys, both sexes parentally invest. Because males invest in offspring, females compete for male parental care. Males have accordingly evolved to prefer to mate with and invest in the offspring of high-quality females. Put otherwise, in this mating system, mutual mate choice has evolved due to selection on the preferences of both sexes (e.g., Kokko & Johnstone, 2002). Heymann (2003) argues that sexual selection on females in tamarin and related New World monkeys has led to specialized scent glands and scent marking behavior that function to honestly signal quality. (At the same time, full evaluation of these claims demands additional research. See Dixson, 1998, for a review of literature on scent glands and scent marking in female primates.)

Female isabella moths produce a chemical sex attractant. In contrast to most other female moths, however, females in this species produce copious quantities of the chemical in unusually elaborate glands, which are then released in visibly aerosol form. In contrast to most male moths, which respond to a few molecules of sex attractant, male isabella moths sexually respond only to large amounts of the scent. Furthermore, females in this species actively court males more than males court females, atypical of moths. Indeed, the courtship pheromonal system used by males, elaborately designed to communicate with potential mates in most moths, is vestigial in the isabella moth.

The mating system of this moth appears to involve sex-role reversal with mutual sexual selection, but greater sexual selection on females than on males (that is, more selective mate choice by males than by females; see Gwynne, 2001, for a discussion of sex-role reversals in insects). As should be expected if females are sexually selected to attract males, males invest substantially in offspring. Specifically, they offer nuptial gifts of pyrrolizidine alkaloids, toxins that function to provide aposematic predator defense. As males may collect and provide differing amounts of alkaloid, females may compete for males who provide much. Indeed, in a closely related species, females prefer males with relatively large loads of the alkaloid. Males able to deliver large loads of alkaloids can therefore afford to be more choosy than typical male moths, which, we suggest, has led to adaptations underlying male mate choice and female competitive signaling of quality, including through costly production of an aerosol sex attractant (see Krasnoff & Roelofs, 1988, 1990; Krasnoff & Yager, 1988.) Consistent with this interpretation of sex-role reversal and female competitive signaling of quality, Lim and Greenfield (2006) recently reported the remarkable discovery that displaying females in this moth form leks.

Honest-signal female ornamentation in the form of pheromonal communication probably occurs in other moth taxa. Honest signaling should be seriously entertained whenever females possess complex structures to emit or disperse scents, as seen, for instance, in certain female arctiid and noctuid moths. We note, however, that mere presence of female glands that produce or store sex attractants is not evidence sufficient to conclude that females possess ornaments. Glands and their secretions function in numerous contexts. Glands that function to eliminate waste, for example, may excrete products that attract males if those products are produced primarily or only by fertile females. In such cases, no adaptation for female signaling exists. Indeed, females of the leafminer moth, a parthenogenetic species (and hence totally lacking males), engage in scent-dispersing behavior, possibly due to excretion (Mozuraitis, Buda, Liblikas, Unelius, & Borg-Karlson, 2002). Obviously, the scent does not attract males, as none exist. In a phylogenetically closely related two-sex species, however, it does. Despite the evolution of parthenogenesis in the leafminer moth, scent-dispersing behavior was retained, probably because it functions in a context other than mate attraction.

Circumstances That May Select for Female Signals of Cycle-Related Fertility

We have emphasized that female signaling of fertility within the reproductive cycle should be rare in animals, at least in any conspicuous form. Under very limited circumstances, however, it may evolve. Specifically, if males are rarely encountered and highly dispersed, females who signal fertility and thereby reduce time to search for a mate could be favored by selection (e.g., when populations are small, individuals are widely dispersed, and the adult sex ratio is female-biased due to high male mortality). *Linyphia litigosa* spiders may be an example. Females in this species release a chemical that may function as a pheromone that signals fertility in the reproductive cycle to males. Interestingly, they have been observed to release pheromones only under unusual circumstances—when they are unmated, full of eggs, and isolated experimentally from males for several days, situations that may simulate the natural atypical circumstances of small, marginal populations late in the year after most males are dead (Paul Watson, personal communication, October 15, 2004; 1986). The possibility that the putative sex pheromone of female *Linyphia* is a by-product scent produced as a result of physiological changes associated with the absence of mating and/or with advanced egg age inside the female has not yet been ruled out. The alternative hypothesis, that *Linyphia* females engage in competitive signaling of individual quality, seems unlikely, as males have little opportunity to compare females and thus choose mates under the circumstances in which the scent is released (specifically, very low population density).

Females Resist Unsuitable Mates But Do Not Signal Nonreceptivity

We have emphasized that, despite rare exceptions, females do not commonly possess costly adaptation that functions to signal their fertility in the reproductive cycle. Similarly, we anticipate that females will lack costly adaptation that functions to signal nonreceptivity. A signal of nonreceptivity in the cycle is equivalent to a signal of receptivity (in that, in the former case, males know that a female can conceive when not signaling). This is not to say that females lack adaptation to resist males when infertile in their cycles, as well as when fertile. They do, as documented across animal taxa (Arnqvist & Rowe, 2005). For example, nonreceptive female tenebrionid beetles spray a noxious, disabling spray on persistent suitors. Sprayed males fall into a "coma" for several hours, which permits females to escape. The spray and spraying behavior are adaptations, though primarily used against predators. Female houseflies have a spine on their middle legs that they jab into the wings of unwanted mates. Overly persistent males may pay for their rudeness by having their wings shredded. This female adaptation may function to serve mate choice and/or to reduce costs of insemination (see Thornhill & Alcock, 1983, for these and other examples in insects). Female alpacas spit, kick, and bite males that try mating outside peak estrus. Both sexes engage in the same agonistic behaviors toward other llamas and even toward humans who bother them (Vaughn, Macmillan, Anderson, & D'Occhio, 2003). Female domestic dogs and cats are notorious for their outright hostility toward ardent males that attempt intromission outside peak estrus. (See Arnqvist & Rowe, 2005, for a review of resistance and rejection by females in the context of sexually antagonistic interactions.)

These adaptations function not as signals of nonreceptivity but instead to reduce harassment by ardent males and other unwanted social interaction, to lower risk of predation and costs of insemination when infertile (see Arqvist & Rowe, 2005; also see chapter 7, this volume), and, when fertile, to avoid insemination by unsuitable males. Resistance or rejection is not a "signal" to males of nonreceptivity or infertility, as some biologists have conjectured (see Ringo, 1996; Schuett & Duvall, 1996). That is to say, its costs are not (largely) paid for by *information* transmitted to males. Instead, its primary benefit is the benefit of rejection of males itself.

Fertile Females Typically Do Not Have Adaptation to Incite Male Competition

The Hypothesis That Females Signal Fertility to Incite Male Competition

Females, we have claimed, rarely if ever have adaptation that functions to signal fertility in the reproductive cycle to thereby ensure insemination and conception. Sexual selection on males ensures insemination and conception of fertile-phase females. An alternate but related form of the hypothesis is that females signal fertility not to obtain *any* insemination but, rather, to ensure that males with the best genes inseminate them. The basic idea is that fertility signals function to incite male-male competition. Therefore, males with the best genes compete to inseminate a signaling female. Though a female could surely be inseminated without signaling, this argument acknowledges, she improves the quality of her offspring by inciting competition. This idea was, in fact, first introduced as an explanation of female sexual swellings in primates (Clutton-Brock & Harvey, 1976).

This hypothesis encounters precisely the same problem that any fertility signal hypothesis has: It assumes that males do not know which females are fertile without an overt, costly signal. Just as sexual selection on males ensures that males will detect extant by-product cues of female fertility status as it changes across the cycle, so it ensures competition between males to inseminate females when they are fertile. Males do not typically require a special signal of fertility to be interested in competing for copulation with fertile females.

Indeed, according to this hypothesis, dominant males should not dominate matings with fertile females in primate species lacking female sexual swellings, as they lack a key sign of fertility to motivate them. In fact, precisely the opposite is found: Dominant males overrepresentatively mate with females at peak fertility even in nonhuman primates that lack sexual swellings and skins (Baker, Dietz, & Kleiman, 1993; see a review of other studies in Dixson, 1998). This same pattern exists in nonprimate mammals in general, despite the absence of sex skins and swellings in the great majority of these species (e.g., Ginsberg & Huck, 1989; Preston et al., 2003).

As Pagel (1994) keenly pointed out, this idea also faces a problem explaining the *costliness* of a purported fertility signal. Suppose males *did* respond to and compete for females displaying a costly fertility signal. Females who possessed less exaggerated forms of the signal would presumably lose out. Imagine now, however, a dominant male who does not compete for females with costly signals but rather copulates with females that exhibit weaker cues. If, in fact, the cue tells nothing more than fertility status (i.e., it does not relate to female quality or reproductive condition), this male wins out, for he has paid minimal costs for competition relative to other successful males. And these females win out, for they obtained a dominant sire but paid fewer costs for their signals. Mere incitation of competition cannot sustain costly fertility signals. Put otherwise, costly signaling of fertility status to incite male interest is not an evolutionarily stable strategy. Fertility cues should hence typically be cheap—and, indeed, as we have just seen, in many species females pay no special costs for them whatsoever; they are mere by-products of adaptations with different functions.

Other purported fertility cues in nonprimate species have also been mistakenly interpreted as fertility signals designed to incite competition. Females of some avian species (such as alpine accentors) sing only when fertile, thereby attracting males (see Montgomerie & Thornhill, 1989). Similarly, female bearded tits exhibit "chase-flights" during the fertile phase to attract males. Though often interpreted as fertility signals (e.g., Cox & Le Boeuf, 1977; Hoi, 1997; Langmore, Davies, Hatchwell, & Hartley, 1996), these signals, we propose, may be ornamental signals of individual female quality. Indeed, chase-flight behavior by bearded tit females is energetically demanding. As expected if it is a quality signal, females in good condition engage in it more than do females in poor condition (see "Female Ornaments as Honest Signals of Quality," below).

Consider another female behavior claimed to function to incite male-male competition: homosexual mounting behavior of estrous female cattle and estrous females in certain other mammals (Adkins-Regan, 2005; Parker & Pearson, 1976), including nonhuman primates (O'Neill, Fedigan, & Zeigler, 2004). Parker and Pearson (1976) argued that, by mounting other females, females let males know they are in estrous and thereby garner bulls' attention. Bulls are indeed attracted to female homosexual mounting of females (Geary & Reeves, 1992). Once again, however, this behavior is better inter-preted as an ornament signaling quality by displaying the ability of a female to domi-nate another female. In fact, only some estrous females exhibit the behavior (VanVliet & VanEerdenburg, 1996).

Female Copulation Protests Do Not Function to Incite Competition Per Se

One of the traits first offered as a female adaptation for inciting male-male competition was described by Cox and Le Boeuf (1977). Female northern elephant seals protest the vast majority of copulations males attempt. They are particularly likely to protest copu-lation attempts by subordinate males. When a female protests, males dominant to the offending male often intervene (at times, competing with each other to do so). As Cox and Le Boeuf note, the effect of the protest is indeed to perturb male behavior. Accordingly, they suggest that the function of a protest is that "it literally wakes up sleeping males and prompts them to live up to their social obligations" (1977, p. 328). As a result of interventions initiated by protests, females end up with sires that offer better genes for offspring.

As Cox and Le Boeuf also note, however, if females did not protest, "males would still compete and interfere with each other's copulations" (1977, p. 328). And, indeed, it makes little sense to think that males are, in a sense, asleep on the job; surely, males who were not so lax would be selected over "sleeping" males. It makes more sense to think that, in fact, males are highly vigilant to ongoings but that, in large aggregations in which a few males can garner most matings, they must simultaneously monitor many events. The fact that they *do* respond to protested copulations with intervention sug-gests that males do indeed monitor their surroundings for key fitness-relevant events. A protested copulation brings attention to an event that calls for an adaptive response (intervention)—but it does not *initiate* a motive to compete with other males in an otherwise stupefied male.

Female protests in this species are not signals of fertile states. Indeed, females protest virtually every copulation attempt prior to their fertile phase. As they near ovulation, they selectively do not protest mounts by dominant males. Their protests have benefits of reducing the direct material costs of copulation (during which females can be injured; see Cunningham, 2003) and increasing genetic quality of offspring (partly through the mere effect of rejecting a low-quality male, partly through intervention by self-interested dominant males). It does not do so, however, by waking up sleeping males.

Primate Copulation Calls

In some species of Old World monkeys and apes, females vocalize during or soon after mating (Dixson, 1998; Maestripieri & Roney, 2005; Pradhan, Englehardt, van Schaik, & Maestripieri, 2006). The prevalence of female calls varies greatly across species and studies, ranging from 6 to 99% of all copulations. Some researchers have claimed that female calls function to incite male-male competition (see Maestripieri & Roney, 2005, for a discussion). Maestripieri and Roney (2005) specifically proposed that these calls function to promote postcopulatory mate guarding by preferred males and thereby reduce the likelihood of mating with nonpreferred males when females are fertile in their cycles. Indeed, they claimed that copulation calls themselves are honest signals of fertility. Consistent with this hypothesis, the occurrence and intensity of female calls are associated with high levels of sexual swelling (i.e., at peak estrus and fertility), matings with dominant rather than young or subordinate males, and the timing of male ejaculation. Males, furthermore, are most likely to respond to calls when females are in estrus. In Barbary macaques, males are more attracted to recorded calls of females at peak fertility than calls emitted outside peak fertility in the cycle (Semple & McComb, 2000).

Pradhan and colleagues (2006) offer related but distinct interpretations. Copulation calls function by alerting other members of a primate group—notably males—that a female has copulated. (Naturally, many will have attended to the act, but some may have been attending to other females, feeding, or engaged in other activities.) Copulation calls do typically lead other males to approach a calling female and, should she not be attended by her consort, reduce the amount of time passed before she mates again. A copulation call, hence, can, in effect, increase the benefit of a dominant male to attend to a female he has just inseminated, which can increase his confidence of paternity in the offspring and provide her subsequent material benefits (e.g., his protection of the offspring), as well as reduce her cost of resisting inferior males. Naturally, this strategy will work best if the male suspects that the calling female is fertile. But by no means need that imply that the copulation call itself is a fertility signal. Males can detect fertility cues. If a copulation call occurs in the context of the calling female emitting such cues, the payoffs to a male guarding her postcopulation are greater than if she emits no fertility cues. Hence, the strategy of calling is more likely to pay off when a female is fertile. Co-occurrence of calling with peak fertility is not an indication that calling signals fertility. Indeed, as Pradhan et al. (2006) stress, females may call when infertile or pregnant, as well.

Pradhan et al. (2006) speculate that, in species in which females call after nearly every copulation (not just those with dominant males), females may benefit by increasing sperm competition. In such cases, dominant males do not tend to guard after a female calls and she mates again more quickly. The argument is not that females are sperm-limited (or even limited by good DNA) but, rather, that females may benefit by running sperm races or selecting compatible genes in utero (e.g., Zeh & Zeh, 2001; see also chapter 7, this volume). Just as female elephant seals call other males' attention to fitness-relevant events by protesting copulations, these female primates may call males' attention to fitness-relevant events to which they may adaptively respond. No assumption that females "wake up" sleeping males, however, need be made. Whether adaptive in utero selection for sperm occurs in primates is unknown, and, hence, this hypothesis requires additional evaluation. In any case, however, in these species copulation calls occur throughout females' receptive periods (both fertile and infertile phases) and hence clearly do not function as fertility signals.

Clarifications

As we discuss later, female ornaments possibly do influence male interest in competing for females and thereby affect the quality of genes of her offspring's sire. They do so, however, not because they signal fertility. Rather, under special circumstances, they could influence male choices about how much and what kind of mating effort to exert to mate with their female bearers because they signal female quality.

We do not rule out the possibility that *low-cost* signals that garner attention from males when females are fertile could potentially evolve, particularly when males are widely distributed spatially. Some female mammals vocalize during estrus. In one well-studied example, female African elephants emit the "estrous call" at peak fertility and before mating. Dominant, but not subordinate, males are attracted by the call (Poole, 1999). These calls may be low cost to females; by no means can one safely generalize this example to instances of *costly* signals such as primate sexual swellings. The alternative hypothesis that these calls communicate information about female quality should also be evaluated.

Finally, to say that females rarely signal cycle-related fertility to incite competition between males to garner sperm or good genes is not to say that females lack adaptations that function to efficiently bring about reliable contact with and conception by males of superior genetic quality while reducing contact and mating with inferior males. Such adaptations are straightforwardly expected by basic understandings of female mate choice. Female resistance against certain ardent males, as discussed earlier, is one kind of such adaptation. In addition, fertile-phase females may possess adaptations that affect their habitat preferences and movement patterns to increase contact with males of high genetic quality and simultaneously reduce or eliminate contact with low-quality males. Female guppies, for example, select microhabitats that ensure reduced contact with low-quality males and enhanced contact with high-genetic-quality males when in their fertile phase (Kodric-Brown & Nicoletto, 2005). The microhabitats preferred by fertile females are more costly for males, especially low-quality ones, than for females to inhabit. But these microhabitat preferences do not function to incite male-male competition. Male guppies compete for fertile-phase females in any natural habitat and are highly sexually motivated by a by-product scent emitted by these females (see chapter 8 for additional discussion). Rather, these habitat preferences affect females' relative rates of encounter with males of high and low quality.

Female Ornaments as Honest Signals of Quality

We began this chapter by noting that, across a wide variety of species, many traits deemed attractive and the perceptual adaptations that deem them so appear to have evolved as components of honest signaling of quality. Quality can refer to either phenotypic traits or genotypic features. In the context of mate choice, phenotypic quality refers to features that enhance the value of an individual as a mate (e.g., ability or willingness to provide current investment, as affected by current condition or indicators of future condition; condition-based fertility; reproductive value; ability to provide genetic benefits to offspring). Genetic quality specifically refers to ability to deliver genetic benefits to offspring. The differential cost to bearers of honest signals and/or the differential utility of signals across the life span renders the degree of a signal elaboration an honest or reliable and stable display of quality (see review by Searcy & Nowicki, 2005; also Getty, 2002, 2006; Grafen, 1990; Kodric-Brown & Brown, 1984; Zahavi & Zahavi, 1997). Sexual signals often are costly in terms of energy, predation risk, or the fact that they involve or invite social competition from same-sex others. Though much more attention has been directed to understand male signals of quality (for the reason that they are much more common than female signals), signals of quality may evolve in females as well. If how much females pay for or get out of signals of a certain size (in currencies of fitness gains) depends on their condition, how "big" (large, intense, or otherwise costly) a signal it is worth females to pay for will vary systematically with their condition; for those in better condition, it will pay to build a bigger signal (e.g., Getty, 2002, 2006; Grafen, 1990). Of course, costly signaling cannot evolve without perceivers' responses to pay for signals' costs. And, naturally, perceivers' responses will not be directly selected unless they promote perceivers' reproductive success. As males evolve to be attracted to females with bigger signals (because they covary with female condition), the signaling system becomes stably reliable. At that point, females possess adaptations to signal as a function of condition. And males have adaptations guiding their mate choice based on female signals' degree of elaboration.

How reliable signaling systems become stabilized once signals discriminate individuals' quality is not difficult to appreciate: Once that condition is met, it pays mate choosers to choose on the basis of the signal, which reinforces the signaling sex to possess a big signal and stabilizing the signal as a reliable indicator of quality. Harder to explain is how the signal *becomes* predictive of quality. It cannot simply be that mate choosers prefer the big signal; if they did, it would *be* an indicator of quality. Mate choosers, after all, cannot evolve to prefer the big signal as a quality indicator until it has *already* become predictive of quality. Obviously, mate choosers cannot know ahead of time that the signal will become a signal of quality *if only* they would prefer it sufficiently (and, even more important, will not be benefited for doing so until it *is* a signal correlated with quality). For a trait to evolve as a signal of quality, it must somehow be correlated with quality before it actually qualifies as a signal of quality (a trait that evolved because mate choosers preferred it as a quality signal). Sexual selection theorists propose two main scenarios by which traits become associated with quality and thereby become signals of quality.

The first is the route of preferred signal through sensory bias. In this scenario, a trait first becomes preferred not because of any fitness advantage to the mate chooser for preferring it. Instead, the preference for the trait is merely a by-product of a sensory adaptation of the mate chooser that has a function unrelated to mate choice. For instance, perhaps mate choosers are drawn to red objects because ripe fruits are red (i.e., "attend to red" reflects a food preference adaptation). Potential mates that exhibit some redness, though not edible, get attended to, which gives those potential mates some advantage on the mating market. As the preferred trait becomes exaggerated through mating advantages of those who have it (e.g., as signals become increasingly red or large), however, individuals in best condition are able to effectively display it in its most preferred form. At that point, the trait has become an indicator of quality, and preferences for the trait no longer need to be maintained as by-products of nonmating adaptations (see Kokko, Brooks, Jennions, & Morley, 2003; for a recent discussion of sensory bias and other sexual selection models, see Fuller, Houle, & Travis, 2005).

The second route is that the preferred trait is associated with quality prior to its evolving as a signal. Individuals in better condition generally have greater amounts of energy to allocate to traits important to survival and reproduction. Hence, a wide variety of traits may actually discriminate individuals of different quality. In many species, however, as the amount of available energy increases, a larger proportion is allocated to traits that foster immediate reproduction. Individuals have but one life to give. When circumstances threaten survival, it often makes adaptive sense for individuals to protect that one life by engaging in mortality reduction efforts (e.g., sequestering energy reserves to maintain survival). When individuals' condition is more favorable, they may reproductively benefit from putting a greater amount of energy into reproductive traits. That certainly happens in human females, whose estrogen levels and fertility levels are very sensitive to their energy stores and energy balance (e.g., Ellison, 2001). Only when their caloric intake consistently exceeds demands of energy for maintenance of survival functions do estrogen levels increase sufficiently to render successful conception and implantation likely. Certain traits—often reproductive ones—thus may tend to *particularly* differentiate individuals varying in condition, simply due to the way that individuals have been shaped to optimally allocate energy to their different traits. At this point, the traits that covary with condition do not do so because they have evolved through sexual selection as signals. Because they *do* covary with condition, however, selection can lead to adaptations in mate choosers to attend to and prefer mates who exhibit these traits. When such adaptations evolve, the quality-discriminating traits are sexually selected for their signal value, as well as selected for functions they had prior to being valued by mate choosers. Their added benefit as signals should lead their bearers to allocate more effort into developing them than they would otherwise. These traits thereby become exaggerated as displays of condition through sexual selection (see also Rowe & Houle, 1996).

To say that a signaling system is reliable or honest is not to imply a complete absence of deceptive signaling as a component of the system. Signaling systems that have been investigated in detail reveal largely honest signaling with an admixture of deceit (see review in Searcy & Nowicki, 2005). Signaling systems require honesty to be stable because only honest signals lead selection to maintain perceivers' responses, which in turn maintain benefits to signalers. In light of the complexities of some signaling systems, *some* signalers can gain advantages by signaling dishonestly (i.e., the magnitude of their signals does not correspond with their quality). As well, some signals may be errors: An individual may produce too small or large a signal for its current quality, but not for an earlier ontogenetic state when the signaling trait was developed. Furthermore, signals may be honest in some contexts but not others (e.g., a signal may be an indicator of quality in younger but not older individuals; Kokko, 1997). But even in the presence of deceptive signaling, perceivers' responses to signals must promote their own fitness, *on average*, for the signaling system to persist (e.g., Kokko, 1997). Put otherwise, dishonest signaling in animal communication can only persist in the context of a largely reliable signaling system (Searcy & Nowicki, 2005).

In chapter 7, we discuss in greater depth the literature examining whether male sexually selected traits are indicators of condition. At that time, we also delve into issues pertaining to whether these indicators of condition benefit choosers through good genes or direct benefits, as well as variations on the evolution of honest signals of good genes (e.g., the Fisherian process).

Sexual Swellings as Signals of Quality

Sexual Swellings as Ornaments

If female sexual swellings are not signals of cycle-based fertility or instruments designed to incite competition between males by advertising fertility, how did they evolve? Pagel (1994; also Bercovitch, 2001) hypothesized that the swellings are adaptations that function to display female phenotypic and genetic quality. Males that preferred the exaggerated forms of display experienced a net reproductive benefit, and hence the male preference for exaggerated displays evolved. Domb and Pagel (2001) subsequently found that degree of ornamentation in olive baboons covaries positively with lifetime female reproductive success. Furthermore, males compete for and groom highly ornamented females more than they do their less ornamented counterparts.

We suspect that Pagel's ideas, with some modification, are correct about many sexual swellings in females. We recognize, however, that these ideas have been controversial (e.g., Nunn, 1999; Nunn et al., 2001; Zinner et al., 2002; Zinner et al., 2004). We suggest that some of the questions about Pagel's hypothesis are due to misunderstandings of its implications. Other criticisms can be deflected by modification of the hypothesis. At the same time, we fully acknowledge that much additional research testing these ideas, as well as a variety of alternative hypotheses for sexual swellings, is needed (Deschner et al., 2004; Dixson, 1998; Dixson & Anderson, 2004; Emery & Whitten, 2003; Hrdy, 2000; Nunn, 1999; Setchell et al., 2006; Setchell & Wickings, 2004; Stallmann & Froehlich, 2000; Zinner et al., 2004).

Effective Selection for Female Ornamentation

Just as female signaling of fertility status should be observed only under restrictive conditions, so should female ornamentation advertising quality—albeit conditions not uncommonly found. To be favored by selection, female ornamentation must accrue benefits to offset its costs. Males deliver those benefits, and they do so in currencies of nongenetic material benefits and genetic benefits (better genes) for offspring. Through signaling, females may be able to obtain greater benefits from males than they would otherwise. They can do so, however, only if males are constrained to have to make choices about which females to deliver benefits to—that is, when they cannot (or it does not pay for them to) provide equal benefits to all females. Under those conditions, high-quality females may be able to entice males to provide benefits to them and their offspring rather than to female competitors by signaling their quality. The greater the extent to which males must decide to whom to direct benefits, the greater the potential for female ornaments to evolve. And the greater the extent to which females differ in their phenotypic and genetic quality (e.g., as a result of disease prevalence or differential accrual of resources), the greater the potential for female ornaments to evolve. Put otherwise, selection can favor female advertising of quality only when males pay substantial opportunity costs if they direct mating effort or parental effort toward low-quality females and their offspring that they could more profitably direct toward trying to mate with or parent the offspring of high-quality females (either currently or in the future). These conditions, of course, are also the conditions under which male mate choice evolves (e.g., Bercovitch, 2001; Bonduriansky, 2001; Pagel, 1994; Parker, 1983; Shine et al., 2003).

The circumstances in which these conditions are most clearly met are ones in which males deliver important fitness-affecting, nongenetic material benefits to females, in which these benefits are not shared across females (e.g., females cannot collaterally benefit from male efforts to assist other resident females), and in which females exhibit substantial variation in phenotypic and genetic quality. In those circumstances, males may be limited in the amount of valuable resources or services they can accrue or deliver to females. And, even if they can deliver services (e.g., physical protection for offspring) to all, it may pay males to differentially allocate those services across females as a function of females' quality (or quality of their offspring). Sex-role-reversed species, in which males provide greater parental investment than do females, are obvious examples. Still, sex-role-reversed species are uncommon (e.g., Andersson, 1994; Gwynne, 2001; Thornhill & Alcock, 1983). In most species in which males provide material benefits to females, female ornamentation nonetheless remains absent, perhaps because thresholds on the conditions that favor female ornamentation are not often reached.

Other circumstances in which female ornamentation may evolve are (1) when dense overlap in female fertile periods causes males to have to choose between multiple fertile females to pursue or (2) when even dominant males risk injury in male-male competition for females, particularly when females vary substantially in quality. In such circumstances, high-quality males may be selected to forgo mating with lower quality females because, under (1), they are forced to make choices through time constraints or because, under (2), the costs of competition (and loss of future reproduction) are not offset by the benefits of additional matings with lower quality females currently. In either set of conditions, female signals of quality could be correlated with the fitness of offspring and thereby evolve. (Under (2), females who signal quality mate with dominant males— but dominant males would presumably also win competitions and mate with them even if females displayed no quality-discriminating signal. Their advantage stems from the fact that females who cannot signal quality mate with less dominant males than if there were no signal. Hence, the offspring of quality signalers do have better genes than the offspring of female competitors, which leads genes for the signal to spread.)

Pagel (1994) proposed a form of the latter hypothesis to explain female sexual swellings. Specifically, he proposed that, because male mating effort in many Old World primates is costly (largely due to costs associated with aggressive combat), males can evolve to choose how to allocate their mating effort based on a signal of female quality. Females with preferred signals hence obtain genetic advantages for offspring. In fact, though this scenario appears to be plausible, whether female signals of quality actually have evolved in absence of male delivery of nongenetic material benefits conditional on female signal size is an unanswered question. (Though the male baboons in Domb and Pagel's [2001] study apparently provide little to no paternal care, females may nonetheless derive material benefits from mating with dominant males, such as reduced levels of aggression toward offspring.) As we show, the nonmutually exclusive hypotheses that, on the one hand, female signals of quality derive benefits through male delivery of material benefits or, on the other hand, female signals of quality derive benefits through genetic benefits for offspring by regulating male mating effort offer some different predictions.

The Phylogenetic Distribution of Female Sexual Swellings

Female ornamentation can evolve in a variety of different mating systems, as illustrated in Old World primates. Female sexual swellings are most commonly observed in species living in multimale, multifemale groups (Clutton-Brock & Harvey, 1976; Dixson, 1998; Hrdy, 1997; Nunn, 1999). Indeed, the phylogenetic record shows that they tend to evolve in species that have these social structures and have multiple origins in the phylogenetic tree (e.g., chimpanzee swellings originated in their lineage after it diverged from ancestors shared with humans and gorillas and independently of its evolution in baboons and macaques; e.g., Dixson, 1998; Pagel & Meade, 2006). In some of these species, males care for their own genetic offspring more than for unrelated juveniles. In addition, males protect estrous mates from other males, including those that may sexually coerce estrous females (e.g., Maestripieri & Roney, 2005). Males delivering material benefits to females should be selected to prefer high-quality over low-quality mates, due to cost to the male of providing the benefits (e.g., a male's lost time foraging associated with his mate guarding, as in baboons; Alberts, Altmann, & Wilson, 1996).

Yet female sexual swellings are found in other primate mating systems, as well (although these swellings are generally of smaller size and thus, presumably, are less costly). In a langur monkey, which lives in spatially separate, polygynous one-male groups, swellings may have evolved through intense competition for disproportionately large shares of the harem master's time-limited material services and delivery of benefits (Tenaza, 1989). Females of a socially monogamous gibbon also display them (Dahl & Nadler, 1992). These females may compete for pair-bond partners who are good providers or material benefits from extra-pair males. Females of New World primate species in which males invest considerably in offspring (e.g., marmosets and tamarin monkeys) lack sexual swellings, but may have evolved alternative ornaments for signaling quality, notably through scent displays (see Heymann, 2003).

In some species of Old World primates, female swellings are particularly pronounced and, as a result, costly (see brief review in Emery & Whitten, 2003). Nunn (1999) found evidence that exaggerated female sexual swellings in primates are associated with three conditions. First, primate species with large sexual swellings live in groups with more males than do those without large sexual swellings. We suggest that this pattern arises because females in groups with many males face relatively great risks of aggression toward offspring and, hence, gain considerable material benefits by mating with males who can defend (and will not aggress against) their offspring. Second, female primates with large sexual swellings have sex for longer periods of time. In our terms, they have longer periods of extended female sexuality. As we argued in chapter 3, female extended sexuality, too, is favored when females can garner male material benefits through sex. In many Old World primates with sexual swellings, benefits of extended sexuality are achieved through reduction of aggression toward offspring via paternity confusion. Covariation between costly female signaling of quality and duration of extended sexuality is hence to be expected. We emphasize, however, that these adaptations are independent and require different explanations (e.g., costly signals of quality could evolve in absence of greatly extended sexuality). Third, exaggerated female sexual swellings tend to be found in species that lack seasonal breeding. By chance alone, females of seasonal breeders will tend to be fertile at the same time that other females are. Nunn (1999) argued that Pagel's (1994) hypothesis should expect that seasonal breeding should favor costly signaling of quality to attract male attention and mobilize male mating effort; in seasonal breeders, females often come into estrus simultaneously and hence compete for dominant males' attention when fertile. Thus, he argued, the relative absence of large swellings in these species is inconsistent with the idea that swellings are honest signals of quality. As we noted earlier, however, one must separate the honest signaling hypothesis that argues that benefits are achieved at least partly through delivery of male material benefits over a temporally extended period (including periods of time in which females do not display swellings) from Pagel's specific hypothesis, which proposes that benefits are achieved solely by mobilizing male sexual interest when females are in estrus.

In fact, we suggest that honest signaling theory can account for the exaggerated ornamentation in nonseasonal breeding primates. Seasonal versus nonseasonal breeding is distinguishable partly on the basis of what most affects optimal timing of female reproduction: immediate availability of resources (for females or offspring), as it varies seasonally, or female condition and resource accrual (which may peak during any season). In species that breed seasonally, variation in female condition and resource accumulation matter less than variation in resources across seasons (a reason they breed seasonally). In species that breed nonseasonally, female condition and resource accumulation (including accumulation of social resources) matter more. In long-lived primates, females acquire, over the long term, nutrients, energy stores, social status, protective alliances with males, and other social resources, all of which may affect outcomes of female reproduction. Again, some of these factors probably matter more in nonseasonal breeders than in seasonal breeders, on average. And some of these resources (e.g., alliances) are delivered by males. Females of nonseasonal breeding primates, then, may be under stronger selection to signal their quality and condition to males, so as to accumulate male-delivered benefits.

Consider, for instance, female baboons (*Papio*). Females reproduce throughout the year, though successful reproduction may depend on ecological factors such as rainfall, ambient temperature, and group size (Beehner, Onderdonk, Alberts, & Altmann, 2006). Female baboons have moderate to large swellings. Swelling size may reveal to males when an individual female is in a condition favorable to successful reproduction (independent of season). Furthermore, through honest signaling of condition, females can gain some of the male-delivered benefits that further promote successful reproduction (e.g., through alliances with males). During their periods of sexual swelling, female baboons form alliances with males (Dixson, 1998). And, indeed, females often exhibit swellings well in advance of becoming consistently ovulatory in their cycles, purportedly to impress males and gain social alliances. (Female baboons may swell up to 18 times prior to conceiving; Altmann, Altmann, Hausfater, & McCuskey, 1977; see also chapter 6, this volume.) More generally, when nonseasonal reproduction reflects the fact that females of the species rely on accumulated social and other resources for successful reproduction, females may exhibit large sexual swellings, which function to honestly signal condition and thereby, when exaggerated, lead to acquisition of resources.

At the same time, nonseasonal reproduction is clearly not necessary for large female sexual swellings to be directly selected. Female Barbary macaques, for instance, are seasonal breeders and yet possess large sexual swellings (Mohle et al., 2005).

Correspondence Between Size of Female Ornaments and Fertility Within the Cycle

As we have acknowledged, female sexual swellings in Old World primates are often largest when their probability of conception is near peak within the estrous cycle (for review, see Dixson, 1998; also Zinner et al., 2004; Deschner et al., 2004). We have emphasized, too, that females also often exhibit swellings when infertile (e.g., Anderson & Bielert, 1994; Mohle et al., 2005). Again, we have argued that the link between swelling and estrus does not imply that females have benefited through signaling fertility per se. The link must nonetheless be explained. According to our view, it arises because at this time females optimally advertise quality and males optimally attend to signals of quality.

Consider signaling of quality from the female's perspective. Females may benefit most from signaling when fertile for two reasons. First, at this time they may most benefit from male delivery of material benefits (e.g., protection from sexually coercive males). Second, at this time males attend to them and are most interested in mating with them, based on preexisting by-product cues of cycle-related fertility. Hence female efforts to signal quality are most likely to be recognized at this time. Relative to the traditional view, this idea reverses a key sequence of events in the evolution of signaling. In the traditional view, female ornaments *lead to* male attention to females at peak fertility. By contrast, the idea we propose is that male attention to females when near peak fertility *leads* females to display ornaments at that time (though, we emphasize, they obviously do signal at other times as well).

From the male's perspective, too, female signaling of quality optimally occurs near peak fertility. At that time, males are most focused on fertile females. Competing demands on male attention are minimized when quality signaling occurs simultaneously with female fertility (for compatible arguments, see Pagel, 1994; see also Bercovitch, 2001; Domb & Pagel, 2001).

Finally, another potentially relevant fact is that female sexual swellings are estrogen dependent (Dixson, 1998). As we mentioned earlier in this chapter (and discuss in further detail in chapter 6), women's estrogen level is highly sensitive to and related to energy storage and balance and, likely, to overall condition (particularly as it relates to readiness to reproduce). These associations may be true of Old World primates in general (Ellison, 2001). Hence the honest information provided by elaboration of female sexual swellings may be anchored fundamentally in the temporal correspondence between the fertile period of the cycle and females' production of high estrogen levels. The fertile phase, then, may be the time when females can most validly display their quality to males. This is not to say that males depend heavily on swellings themselves to assess cycle fertility (or that female swellings were directly selected because they provide such information). Males primarily use by-products to assess cycle-related fertility status, just as males of mammalian species in which females lack swelling do. Rather, the argument is that males assess female condition and readiness to reproduce and that females can often best provide information about those qualities (through swelling displays) when they near maximum fertility.

The Graded Nature of Sexual Swellings and Their Role in Evoking Male Sexual Interest

Despite the common positive covariation between size of sexual swellings and cycle-relevant fertility status among female primates, individual females exhibit even stronger associations. Sexual swellings of individual female common chimps change across approximately 10 days of swelling (in a 36-day reproductive cycle), for instance, and are maximally enlarged during the 4 days of the cycle at which conception probability peaks. Dominant males respond by increased mating rate and mate guarding of highly fertile females (Deschner et al., 2004). One explanation of this pattern is the graded-signal hypothesis (Nunn, 1999). Females in species with swellings, Nunn suggested, face the dual problems of biasing paternity toward dominant males on the one hand and confusing paternity on the other. They solve these problems, he argued, by signaling maximal fertility when at peak fertility, thereby drawing attention and mate guarding from the most dominant males, and mating with submissive males when less fertile but still sexually swollen (albeit to a less exaggerated degree).

According to Nunn's hypothesis, dominant males should use swelling size itself to decide whether to mate with females. In our way of thinking, fertility status per se, as reflected through by-product cues of fertility status (e.g., scents) and not swelling size, should largely affect male decisions about how to direct sexual effort. In fact, Deschner et al. (2004) found evidence for the latter, but not the former, in chimpanzees. Though, on average, females tend to show maximum swelling around ovulation, for individual females swelling may peak several days before or after peak fertility. Attention to by-product cues of fertility status, not swelling size, explains why and how dominant males focus sexual attention on and often monopolize females at peak conception risk. The same applies to long-tailed macaques, another species in which maximum sexual swelling does not reliably correspond to peak fertility in the cycle (though estrogen does): Males focus their mating effort on females in peak fertility per se (Engelhardt et al., 2004; Englehardt et al., 2005)

These findings regarding chimps and macaques are also inconsistent with one version of the swelling-as-quality-signal hypothesis—namely, Pagel's hypothesis that swellings function to mobilize male mating interest. The findings are fully consistent, however, with the hypothesis that these signals of quality at least partly benefit females through male delivery of material benefits. Males' attention to female fertility in the estrous cycle reflects male mating effort to inseminate females. It need not reflect any effort to offer particular females material benefits. In many circumstances, males should be sexually interested in fertile females who are not particularly of high quality. When multiple swollen females are at peak fertility, males may have to decide which females to pursue on the basis of swelling size per se. As well, even dominant males may also adaptively decide not to compete for copulations with low-quality females if competition risks injury.

Put otherwise, according to the hypothesis that female signals of quality function to obtain male delivery of material benefits, female swellings do not necessarily motivate male interest in copulation per se (cf. Deschner et al., 2004). They motivate male material benefit delivery to females. In chapter 6, we discuss exaggeration of sexual swellings in Old World primate adolescents, whereby subfecund females exhibit sexual swellings to a degree greater than older fecund females. These swellings do not stimulate male sexual motivation. Arguably, they instead function to attract delivery of material benefits from males.

This is not to say that males should not pay attention to female swelling size when deciding how to allocate mating effort (due to its costs), as well as delivery of material benefits. As just noted, they should. Sexual interest, however, need not be evoked by swellings per se.

Tests of the Hypothesis That Swellings Are Quality or Condition Indicators

As noted previously, Domb and Pagel (2001) found strong support for the hypothesis that swelling size signals female reproductive quality in a wild population of olive baboons. Swelling size positively predicted earlier maturation, lifetime number of offspring produced, and number of surviving offspring. (See Zinner et al., 2002, for discussion of study limitations.)

Emery and Whitten (2003) reported a positive association between female ovarian function (notably, estrogen levels) and swelling size measured in the same reproductive cycles of chimpanzee females. In humans, estrogen levels vary on a graded continuum with fertility of cycles; cycles in which estrogen is highest tend to be most fertile (e.g., Ellison, 2001; Lipson & Ellison, 1996; see also chapter 6, this volume). The same may be true of other primates. If so, swelling size may be an indicator of current reproductive condition. Although most of the comparisons that Emery and Whitten (2003) had available were *between* individual females, they also found evidence suggesting that, across different cycles of the *same* female, swelling size predicts ovarian function. Consistent with this claim, Deschner et al. (2004) found that, during cycles close in temporal proximity to an actual conceptive cycle of the same female, female chimpanzees had larger swellings than they developed during cycles several months prior to an actual conceptive cycle (cf. Setchell & Wickings, 2004, who did not find a difference in swelling size across conceptive and nonconceptive cycles in mandrills). Naturally, a valid indicator of quality or condition may not only discriminate between females but also may discriminate reproductive condition as it varies across cycles within individual females. Hence, these findings are consistent with a version of the quality-signal hypothesis.

That need not mean, however, that swelling size cannot also reveal long-term individual differences in ability to reproduce (Emery & Whitten, 2003). Based on observations of a small number of females of a semi-free-ranging group of mandrills, Dixson and Anderson (2004) argued that, due to physiological challenges, low-ranking females experienced difficulty maintaining estrogen levels sufficient to trigger ovulation, as revealed by swellings. A larger (but still small) study of the same group yielded mixed findings (Setchell & Wickings, 2004). Though results hinted that males compete more for females with large swellings, swelling size did not reliably relate to quality variables, such as the mean number of cycles to conception. A subsequent study that examined disease loads and health indicators also yielded ambiguous results (Setchell et al., 2006). Of 20 correlations between swelling size and parasite loads, 18 were in the predicted negative direction. As sample size was only 10, however, no individual correlation was statistically reliable.

Critics have claimed that Pagel's hypothesis is challenged by a number of additional findings: Males are sexually attracted to fertile females, not females with the largest swellings; infertile females (e.g., pregnant females) sometimes display swelling, and in some species adolescent females have the maximal swelling size; exaggeration of female swellings does not coevolve with increased male canine size (see Nunn et al., 2001; Zinner et al., 2002.) Our proposal—which, unlike Pagel's, does not argue that swellings function to incite competition between males—is not challenged by these findings. Moreover, because swelling size may vary across females' life history in patterns different from their age-specific fecundities (e.g., young females may have maximal swelling size prior to having reached peak fertility), a lack of association between this purported quality indicator and female reproductive success in short-term studies (e.g., Setchell & Wickings, 2004) is not inconsistent with our proposal. Neither is a lack of association between swelling size and immediate sexual interest (see chapter 6 on the function of adolescent swellings). Domb and Pagel's (2001) work remains the only study that has examined associations between female swelling size and lifetime reproductive success in any species, and it found strong associations.

Clearly, additional studies are needed.

Summary

In the world of animals, female sexual ornaments are widespread. Based on theoretical arguments, we suggest that they are typically honest signals of female personal quality and condition, not of fertility within females' reproductive cycles—and this interpretation applies specifically to female sexual skins and swellings of Old World primates. Males are expected to be under strong sexual selection to find fertile females, and, hence, costly signals of fertility within the cycle should be rare. Generally, cues that females emit when fertile in their cycles have been interpreted as signals designed to assure insemination. One version of this notion claims that females signal to incite male–male competition, which assures insemination by males with good genes. Females typically do not have signaling adaptations that function to ensure insemination by males in general, though they do have other adaptations that function to improve female choice. Female primate copulation calls may well be adaptations, but are not fertility signals or adaptations to incite competition per se.

Conditions that favor the evolution of female ornaments include circumstances in which females vary in quality and females can obtain nongenetic material benefits from males who discriminatively deliver those benefits. Variation in these conditions may explain the distribution of sexual swellings and other ornaments (such as scent markings) among nonhuman female primates. Female sexual swellings may tend to be large when females are at peak fertility in their cycles because females may optimally signal quality and males may optimally assess females' quality at this time. When the size of a female's sexual swelling does not exactly peak when fertility in her cycle is maximal but the presence of by-products of female fertility (such as estrogen-related scents) does, male sexual interest is best predicted by actual fertility status, not swelling size. Furthermore, in some nonhuman primates, adolescent females display maximal swellings but do not attract male sexual interest (see also chapter 6). These patterns are consistent with the view that swelling size is not a primary determinant of male mating interest but rather functions as a signal of individual quality to gain material benefits from males.

6   The Evolution of Women's Permanent Ornaments

Particular female features arouse the sexual attraction of men. Wherever around the globe researchers have looked, men are attracted to hyperfeminine facial features and dimensions. Men are sexually attracted to young women's breasts. In most geographic locations, they prefer women who possess a small waist relative to hips. Smooth feminine skin in general can attract men, as can feminine voice qualities. (See reviews in Feinberg, Jones, B. C., DeBruine et al., 2005; Fink, Grammer, & Thornhill, 2001; Grammer, Fink, Juette, Ronzal, & Thornhill, 2002; Schaefer et al., 2006; Scheyd, Garver-Apgar, & Gangestad, 2007; Singh, 2002b; Symons, 1995; Thornhill & Gangestad, 1999a; Thornhill & Grammer, 1999.)

A primary question arises about these features: Are they ornaments? Or are they by-products that are sexually attractive to men? As we describe in this chapter, a variety of women's sexually attractive features are facilitated by the actions of estrogen. Women's estrogen levels in turn support their reproductive capacities. According to the hypothesis that women's sexually attractive qualities are ornaments, sexual selection has favored these morphological features directly and exaggerated them because they signal to men that their bearers possess valued mate qualities, with selection directly operating on men to attend to these signals and reproductively benefit women who possess them. According to the hypothesis that they are mere by-products, men have been directly selected to track cues of women's reproductive capacities, but selection has not operated on women to display these cues; the cues men find attractive are merely reliable side effects of women's estrogen levels.

If women's sexually attractive features are ornaments, the question of what they function to signal arises. A variety of hypotheses have been offered. Cant (1981) and Gallup (1982) argued that women's ornaments reflect current age-related fertility levels. Symons (1979) and Buss (1989b) proposed that they largely reflect youthfulness and, therefore, future reproductive value. Marlowe (1998) argued that breasts in particular signal residual reproductive value (future reproductive capacity) more generally, reflective of both age and individual quality. These theories explain women's ornaments as honest signals of valued qualities. A variety of alternative theories explain them as deceptive signals, qualities that deceive men by falsely signaling to men valued states. Specifically, breasts and other features have been claimed to deceptively signal youth (Jones, 1996; Low, Alexander, & Noonan, 1987) and pregnant state (Miller, 1995; Smith, 1984a). Because these traits are permanent and change little across the ovulatory cycle, they have also been argued to be deceptive signals of permanent cycle-related fertility status or, relatedly, to conceal ovulation (e.g., Szalay & Costello, 1991).

We argue that many, if not most, of women's attractive features are indeed ornaments. Though some of these features were first selected to support reproduction, they were subsequently exaggerated and shaped by sexual selection for their signal value. Primarily, they function to honestly signal residual reproductive value. The telling evidence is revealed through their functional design.

The Reproductive Endocrinology of Women's Fertility

The Role of Estrogen as a Modulator of Women's Reproductive Effort

Human mothers transfer large amounts of energetic resources and micronutrients to fetuses during gestation and, in traditional settings, women typically support human infants through breast feeding for a year to several years as well (e.g., Dufour & Sauther, 2002; Ellison, 2001). Because of the energetic demands to produce an offspring, women's reproductive capacities have been shaped to be highly sensitive to features of their energetic intake. Women's bodies "decide" whether to allocate effort (e.g., energetic resources) toward supporting current reproduction (rendering women fertile and reproductively capable at present) or to delay reproduction to the future, a time when they might be able to produce a higher quality offspring at lower cost to their own viability and fecundity. The physiological mechanisms that underlie these decisions presumably have been shaped by selection to allocate effort optimally in the face of energetic constraints and mortality risks (specifically, to maximize adaptively total expenditure of reproductive effort across the life span, at least within environments in which these mechanisms evolved; Charnov & Berrigan, 1993; Hill & Hurtado, 1996).

As thoroughly documented by Peter Ellison (2001; also Jasiénska, 2003), these physiological mechanisms that modulate allocation of effort centrally involve endocrine hormones, notably estrogen. Endocrine hormones are chemical messengers in distributed communication systems. Given the distribution of hormone receptors in various somatic tissues, the production and release of endocrine hormones can simultaneously upregulate and downregulate multiple functions. These systems therefore can modulate the allocation of effort across a variety of domains simultaneously based on immediate circumstances (e.g., under threats of specific pathogens, they can increase effort toward specific forms of immune function at a cost to other functions; see, e.g., McDade, 2003). In theory, selection has shaped the systems that regulate the release of these hormones, as well as the distribution of receptor sites, in ways that modulate effort adaptively depending on cues in the external and internal environment of the individual.

Reproductive hormones of the sexes appear to primarily modulate allocation of energy and other somatic resources (including neural ones) toward or away from specific forms of reproductive effort (Ellison, 2001). Though both testosterone and estrogen play roles in both men and women, testosterone is the most important modulator of reproductive effort in men, whereas estrogen is most important in women. In chapter 7, we discuss the role of testosterone in men. Here, we discuss estrogen in women. Following the vertebrate literature on reproductive physiology and behavior, we use the term *estrogen* as a shorthand for a number of hormones that are chemically similar and that affect female reproduction, with estradiol being a prominent player (e.g., Adkins-Regan, 2005; Dixson, 1998; Lombardi, 1998; Nelson, 2000; Whittier & Tokarz, 1992).

Within women's reproductive cycles, estrogen plays a fundamental role in determining fertility levels (Ferin, Jewelewicz, & Warren, 1993). Secretion of estradiol by the growing egg-containing follicle during the follicular phase of women's cycles triggers the luteinizing hormone surge, which, in turn, triggers ovulation (Ferin et al., 1993). Estrogen level is a valid index of female fertility and reproductive health, not only because of estrogen's role in this ovulatory process but also because estrogen levels covary with follicle size, egg quality, and, after ovulation occurs, the endometrium's (uterine lining's) capacity for successful implantation and gestation of the zygote if fertilization occurs (Lipson & Ellison, 1996; Tchernof, Poehlman, & Despes, 2000; Kirchengast & Huber, 2001a; Ellison, 2001; Jasiénska et al., 2004). Estrogen from the corpus luteum ("yellow body"; the ovarian follicle emptied by ovulation), in concert with progesterone from the same source, determines the suitability of the endometrium for acceptance and nourishment of an embryo. Hence, just as level of estrogen during the menstrual cycle is predictive of successful pregnancy, so too is level of progesterone; moreover, as progesterone level depends on estrogen level, the two levels positively covary (e.g., Jasiénska et al., 2004)

Women's cycle-based fertility is highly sensitive to features of energy intake. Fertility is not an all-or-none characteristic of women's somatic states; it varies along a continuum (Ellison, 2001, 2003; Jasiénska, 2003). Women's level of fertility is compromised under three conditions: (1) when women's energy balance is unfavorable to reproduction—when women consume fewer calories than is required to satisfy energetic demands of survival functions; (2) when women's energy status is unfavorable to reproduction—when they have relatively few stores of fat; (3) when women's energy flux is unfavorable to reproduction—when the total energy utilization (production and consumption, independent of balance) is either very high (with many calories taken in and burned because energy demands are great) or very low (with few calories burned because consumption is very low). These changes in fertility levels are implemented primarily through changes in estrogen levels. Under conditions of unfavorable energy balance, status, and flux, women withdraw reproductive effort by decreasing production and release of estrogen, which leads to decreased probability of ovulation, egg conception, implantation, and successful pregnancy.

The Role of Fat Deposition in Women's Reproduction

To support the energetic demands of gestation and lactation, women store fat. Mammalian females in general (including primates) store more fat than do males. In humans, however, this sex difference is greater than in other mammals (review in McFarlane, 1997; also Pond, 1978, 1981). On average, muscle constitutes 43% of men's body weight, compared with just 36% of women's. By contrast, fat stores account, on average, for 14% of men's weight, but make up over a quarter of an average 16- to 18-year-old woman's weight (Forbes, 1987).

That women store fat to support reproduction is demonstrated by the fact that they possess special design for allocating fat to reproduction. Women store two kinds of fat, android fat and gynoid fat. The two fats differ in their constitution. Gynoid fat is relatively rich in long-chain polyunsaturated fatty acids, which may be particularly important to the development of healthy brains in fetuses and newborns (and may account for enhanced cognitive abilities of breast-fed babies; Agostoni et al., 1999; see review in Lassek & Gaulin, 2006). Whereas android fat is readily mobilized by deficits in energy balance, gynoid fat in women is mobilized to nourish offspring during pregnancy and lactation; it is used to support survival functions more reluctantly than is android fat (Lancaster, 1986; Weisfeld, 1999). For these reasons, gynoid fat is referred to as "reproductive fat."

Gynoid and android fats are stored in different depots. Android fat is accumulated in the trunk, abdomen, and internal fat depots. Gynoid fat is stored in the breasts, hips, buttocks, and thighs. The latter, then, are specialized fat depots dedicated to support reproduction. Human females accumulate large quantities of gynoid fat during adolescence. Given adequate nutrition, women's total body fat stores increase by over 200% within just a few years following puberty. Most notably, young women accumulate stores of gynoid fat, which, together with alterations in pelvic shape, account for the "shapely" adult feminine body form that adolescent girls attain (Kirchengast, Gruber, Sator, Knogler, & Huber, 1997; also Forbes, 1987; Lancaster, 1986; Pond, 1978).

The precise distribution of women's reproductive fat stores may be adaptive. Storage of fat on the hips, buttocks, and thighs leaves women's center of gravity low, which may be particularly important to stability of women carrying infants. Storing fat in core areas rather than peripheral regions minimizes costs that "dead weight" imposes on locomotion (e.g., energy required to overcome inertia to initiate limb movement). Fat in the breasts (augmented during pregnancy) may be stored where it can be efficiently mobilized during lactation (Pawlowski & Grabarczyk, 2003; Pond, 1981).

Just as estrogen is important to modulating fertility, estrogen critically affects storage of fat reserves to be allocated to future reproduction. Whereas testosterone facilitates deposition of android fat to the trunk and abdomen, estrogen facilitates deposition in hip, buttocks, thighs, and breasts (Forbes, 1987; Ibáñez, Ong, de Zegher, et al., 2003;

Kirchengast et al., 1997; Pedersen, Kristensen, Hermann, Katzenellenbogen, & Richelsen, 2004). Indeed, estrogen supplementation, administered orally or through transdermal patches, accelerates fat deposition in these structures in girls with delayed puberty (review in Pozo & Argente, 2003) and male-to-female transsexuals (Elbers, Asscheman, Seidell, & Gooren, 1999; Gooren, 2005; Kanhai, Hage, van Diest, Bloemena, & Mulder, 2000). As a result, measured ratios of estrogen to testosterone in women predict their ratios of gynoid fat to android fat (Singh, 1993, 1995, 2002a,b; Kirchengast et al., 1997). As should also be expected, women with higher ratios of gynoid to android fat are more fertile than their counterparts with lower ratios (Kirchengast et al., 1997; Singh, 1993, 1995, 2002). In fact, high gynoid-to-android fat ratios also predict reduced cardiac disease, less cortisol reactivity (possibly indicative of increased ability to cope with stress), and overall better health (Epel et al., 2000; Kirchengast & Huber, 2001a, 2001b, 2004; Nieschlag, Kramer, & Nieschlag, 2003). Androgenization in women, as partly reflected by relatively low gynoid-to-android fat ratios, appears to lower women's fitness overall, at least in modern settings (Nieschlag et al., 2003).

Do Gynoid Fat Depots Function as Signals?

As we have described them, gynoid fat depots function in reproduction. They store energy critical to women's parental effort. And their distribution makes adaptive sense, given trade-offs between women's storage of reproductive fat and their costs to efficient locomotion. We suspect that the initial function of these fat stores was to support reproduction (Lancaster, 1986).

One can ask why women accumulate these stores of fat more than other female primates. Again, the benefits of storing fat for reproduction trade off against costs they impose on locomotor efficiency. As we described in chapter 4, in human foragers, men typically generate a surplus of calories through hunting (that is, more calories than they consume), which subsidizes reproductive female and offspring energetic demands. Male subsidization of female caloric demands possibly reduced the premium on female locomotor efficiency, leading females to trade off additional amounts of efficiency for fat storage. As well, fetal and neonate brain development demands more fat in humans than in close phylogenetic relatives, and these demands may have coevolved with women's gynoid fat depots (Lassek & Gaulin, 2006).

Storage of fat for reproduction, then, was selected in women. Just as it does not pay women to be fertile under many conditions, however, it does not pay women to allocate energy to future reproduction through storage of gynoid fat under all conditions. In particular, women's allocation of energy to gynoid fat depots should be sensitive to their energy balance and energy flux. When women do not reliably consume calories in excess of those required to sustain survival functions, they should not divert energy to storage of energy reserved for future reproduction. When their energy consumption is very high or very low due to high demands or low intake, they similarly should not store energy for future reproduction. Consistent with these expectations, women who engage in "weight cycling" ("yo-yo" dieting) appear to accumulate more android fat than gynoid fat during upward swings of their weight cycling (Wallner et al., 2004).

More generally, women's optimal storage of gynoid fat may be affected by a variety of environmental and heritable factors that influence its benefits or costs: their nutritional status, their current health, their health history, their ability to efficiently utilize energetic resources, or the degree to which fat storage compromises their locomotor efficiency.

Men are attracted to women who display particular forms of gynoid fat depots, as we detail later. A by-product hypothesis explaining men's attraction to these features readily follows from what we have stated thus far. Ancestral women who possessed gynoid fat deposits "decided" to allocate energy to reproduction. They were, on average, more fertile than women who lacked gynoid fat. Moreover, because, on average, these women were in better condition than were women who lacked gynoid fat, they also possessed greater residual reproductive effort to expend during their lives. Men who favored mating with women who possessed gynoid fat therefore outreproduced men who did not, and male attraction to female gynoid fat was selected.

This scenario, we suspect, did occur. At the same time, we think it fails to fully explain men's attraction to women's estrogen-facilitated features. A pure by-product hypothesis is inadequate for both conceptual and empirical reasons. We consider conceptual reasons first. (We discuss empirical reasons later in the chapter.) In chapter 5, we noted that honest signals of quality must overcome a threshold of being attractive, at which time they can then become sexually selected as signals. As we described, individuals who are valued as mates may differ from those less valued as mates in a variety of ways. For one, they are often in better condition and hence have more energy to allocate in general. For another, each individual's optimal allocation of energy, as shaped by selection, may lead individuals in better condition to allocate more energy to reproductive traits than individuals in poorer condition. Mate choosers of the other sex, then, may be selected to favor as mates those individuals with particularly well-developed reproductive features. Because these features attain added value due to their sexually selected benefits, their bearers should then be selected to allocate additional resources into them and their display value. Hence, these features become exaggerated through sexual selection.

In chapter 5, we described this process as a means by which features *can* become selected as signals. We make a bolder statement here: If a feature becomes attractive as a by-product because it relates to condition, as in this scenario, benefits accrued through sexual selection *will* lead the trait to become exaggerated, rendering it a signal (if not always, nearly always). The feature was valuable prior to picking up benefits accrued through sexual selection (via preferences of the other sex). Sexual selection adds to the marginal benefits of further investing in the feature. Hence it will pay individuals to allocate greater costs (energetic expenditures) to the feature. Because those individuals who can best afford to do so are those in best condition, the trait remains a good indicator of condition—but now functions (at least partly) as a signal of condition (see, e.g., Kokko et al., 2003).

In the by-product hypothesis, men are attracted to gynoid fat not because gynoid fat functions to *signal* reproductive capacity, but, rather, because it happens to be a correlate of women's reproductive capacity. Once gynoid fat stores were attractive to men, women should have been selected to allocate additional energetic resources to them.

As well, selection should have shaped means by which these traits are displayed to men in attractive forms. These features should have evolved to function, partly, as honest signals of valued reproductive capacity. More generally and for similar reasons, men will have found features of women that reflect production of estrogen attractive. These features, furthermore, should have been shaped to function as signals.

The Signaling of Reproductive Value

Women's Features Are Maximally Developed During Adolescence and in Young, Nulliparous Women

As noted earlier, an individual's residual reproductive value (residual RV) is the individual's future reproductive potential or total expected reproductive success from the present time forward (Fisher, 1930; Williams, 1966). For any given woman, RV is age-dependent. It increases throughout childhood as she successfully passes through a period during which death but not reproduction is possible, reaches a maximum at the beginning of the reproductive period (in traditional populations, the late teens; e.g., Hill & Hurtado, 1996), and steadily declines thereafter, reaching zero at the onset of menopause. RV varies across women of the same age, however. Those women who, based on phenotypic or genotypic features, can expect to live longer or experience higher age-specific fecundity have greater RV than those who possess features associated with higher mortality rates or lower fecundity. RV is distinct from age-specific fertility (capacity to successfully conceive), of course. Women's maximum fertility typically occurs during the early 20s.

A number of authors have argued that men are sexually attracted to features associated with RV (e.g., Barber, 1995; Buss, 1989b; Marlowe, 1998; Singh, 1993; Symons, 1979; Thornhill & Gangestad, 1993; Thornhill & Thornhill, 1983; Weisfeld, 1999). In particular, some authors have argued that men are attracted to features associated with women's age of maximum RV, late adolescence (e.g., Marshall & Tanner, 1974). And, indeed, many studies show that sexually attractive features are maximally developed in women at these ages. Women's breasts, for instance, develop at puberty, reaching adult size by late adolescence. Men are particularly attracted to breasts that are firm, upright, and characterized by relatively reduced nipple pigmentation. These features peak during adolescence and in young, nulliparous women (see reviews by Grammer et al., 2002; Grammer, Fink, Møller, & Thornhill, 2003; Schaefer et al., 2006). Women's waist-to-hip ratio (WHR) is a phenotypic indicator of the ratio of gynoid fat distributed throughout the hips and buttocks to android fat around the abdomen. In many modern and traditional populations (though exceptions exist; see later in the chapter), men find women's bodies with relatively low WHRs (around .7) particularly attractive (e.g., Singh, 1993, 1995, 2002a,b; Singh & Randall, 2007; Singh, Renn, & Singh, 2007). WHRs reach minimum values during adolescence and, on average, rise as a function of women's age and parity (e.g., Lassek & Gaulin, 2006).

Accounts that argue that men evolved to value these features as signs of female youthfulness (and age-based RV) are incomplete, however, for at least two reasons. First, they take for granted female adolescent features. Any complete account must explain *why* these features emerge during this period. Second, these accounts do not explain how and why men's preferences discriminate between women of the same age. The view that sexually attractive features evolved to function as signals of RV addresses both issues.

Adolescent Gynoid Fat Depots Were Exaggerated by Sexual Selection and Hence Are Ornaments

Marlowe (1998) proposed that women's breasts evolved as signals of personal quality, as it pertains to women's residual RV. That is, adolescent women's breasts develop prior to their reproductive capacities not merely to prepare for reproduction itself. Rather, the full extent and timing of adolescent gynoid fat accumulation is partly the outcomes of competition between women to display to men their quality at the time at which they begin reproduction. (Indeed, they develop even before reproduction, for reasons we discuss later.) We extend this view to a fuller set of women's estrogen-facilitated features. In general, adolescent displays of gynoid fat and other estrogen-facilitated features (e.g., features of skin, face, and voice) function as ornaments. Men evolved to find them attractive not merely because they indicate youthfulness. Instead, women's sexually attractive displays (or their exaggeration during adolescence) and men's attraction to them coevolved because these features were ancestrally associated with quality and residual reproductive value.

This theory explains why, in traditional populations, women's breasts and other gynoid fat depots develop well before their reaching full reproductive capacity (see Lancaster, 1986; Weisfeld, 1999; and references therein). Downy labial hair and breast buds are the first visible signs of puberty in girls. Menarche, the first menses, follows. Anovulatory cycles occur for up to 2 years after menarche, but low fertility cycles and irregular ovulation are the norm for several years, particularly in traditional populations. Even in the United States, full ovulatory function is not reached until ages 18–20. Yet gynoid-fat depots are fully developed by 15 or 16 years of age (see Ferin et al., 1993; Forbes, 1987; Frisch, 1990; Lancaster, 1986; Marshall & Tanner, 1974; Reynolds, 1951; Weisfeld, 1999).

By no means is this pattern an inevitable consequence of design for maturation. The age distribution of ornamentation in vertebrates (including sexual swellings in primates—though see exceptions discussed later; Dixson, 1998) is typically reversed, with young adults possessing ornaments less developed or colorful compared to older adults. (For additional examples of ornamental feathers or combs in male and female birds, see Andersson, 1994; Weisfeld, 1999.)

Men do not ignore adolescent women with estrogen-faciliated traits. In fact, men are unusual and extreme among male primates in the extent to which they are sexually attracted to very young semiadult and adult females; other male primates typically prefer females comparable in age to middle-age women (see Anderson & Bielert, 1994; Hrdy, 1997; Dixson, 1998). Male common chimpanzees, in fact, possess a strong sexual preference for old females, not young or middle-age females (Muller, Thompson, & Wrangham, 2006). Men are also different from women, who are not typically attracted to pubescent and adolescent males (Jones, 1996; Quinsey & Lalumiere, 1995; Quinsey, Rice, Harris, & Reid, 1993; Symons, 1979). Although men are most attracted to women in their late teens to late 20s, on at least some measures (including self-report and plethysmography), heterosexual men are almost as sexually responsive to nude pubertal and early adolescent females (for reviews, see Quinsey & Lalumière, 1995; Quinsey et al., 1993).

The alternative hypothesis that sexually attractive adolescent features evolved merely to support future reproduction does not appear adequate. Even if human fetuses do require substantial energetic support, relative to close primate relatives, it is by no means clear why human females should gain advantages in currencies of future reproduction by depositing adult levels of gynoid fat well before their reaching reproductive age, absent any effects of sexual selection. The benefits that boosted the development of sexually attractive adolescent features to levels observed in traditional and modern populations today, we propose, included sexually selected benefits.

This theory also predicts that women of the same age will differ in attractiveness and that these differences relate to residual RV. Research on breast symmetry shows that, as expected, symmetric breasts are more sexually attractive than asymmetric ones, even when breast asymmetry is small in magnitude (Singh, 1995). Furthermore, in three different samples, women with relatively symmetric breasts were found to have had more children across the lifetime, with age controlled (Manning, Scutt, Whitehouse, & Leinster, 1997; Møller, Soler, & Thornhill, 1995; see chapter 7 for a fuller discussion of symmetry and the underlying trait of developmental stability). Later in this chapter, we review additional evidence that individual differences in estrogen-facilitated attractiveness of young women relate to women's residual reproductive value.

Function of Signaling Reproductive Value

The Context of Long-Term Pair-Bonding

Women's ornaments appear early in adolescence and are retained throughout women's reproductive lives. Accordingly, we refer to woman's ornamentation as *permanent-residual-reproductive-value ornamentation.*

Women's ornaments were probably selected in the context of long-term mating and pair-bonding (on long-term human pair bonding, see Alexander, 1979, 1990; Buss and Schmitt, 1993; Geary & Flinn, 2001; Sillén-Tullberg & Møller, 1993; Symons, 1979; see also chapter 4, this volume). In long-term pair bonding, men would have benefited by mating with females who are not only fertile but who, as repeated reproduction occurs over the long-term, were capable of repeated reproduction and production of high-quality offspring. Selection hence operated on females to signal residual reproductive value and on males to assess it and provide greater amounts of nongenetic material benefits and services to females with markers of superior future reproductive value. Given substantial but discriminative male provisioning and care (chapter 4), women who signaled high residual reproductive value could garner large direct benefits for production of offspring.

In contexts in which men do not anticipate long-term investment in a sexual partner and her offspring, in fact, it makes sense that they will be most sexually attractive to women who are at maximal current fertility, not maximum residual RV. We expect, however, that women's investment in ornamentation profited ancestrally through garnering material benefits delivered by males (see chapter 5) and, hence, ornamentation itself (even if not sexual attractiveness per se) will be maximal close to the time of peak residual RV.

Ornamentation as a Form of Female–Female Competition

Ornamentation must be understood, then, as a means by which females compete with other females for forms of male-delivered material benefits. As one writer, Nancy Etcoff (1999), put it in shorthand, "beauty contests" among females occurred repeatedly over generations of human evolutionary history. "Winners" honestly signaled their superior quality and condition, as it affects their reproductive value, a major feature of a mate that has repeatedly affected male reproductive success. They benefited in currencies of male-provided material benefits such as protection, shelter, food, and status and, in addition, could have gained access to quality genetic sires, who generally may be very capable of providing material benefits but often are unwilling to do so except when paired with a high-quality mate (see chapter 7). (We emphasize that, as also described in chapter 7, not only women but also men have been subjected to "beauty contests" over the course of hominin evolution, in the sense that women too have historically evaluated men's suitability as mates partly based on physical features.)

As other scholars have reviewed and concluded, vast evidence supports the claim that attractive women do secure more male-provided material benefits and services because they are attractive to men: attractive female adolescents are far more likely to end up in marriages with resource-rich men of high status than are unattractive female adolescents (reviewed in Buss, 1994); physically attractive women gain substantial social benefits (e.g., job advancement, salary) in the primarily male-controlled Western workplace (see review by Jackson, 1992); women's facial attractiveness positively covaries with the number of long-term, but not short-term, sexual partners (whereas men exhibit the reverse pattern; Rhodes, Simmons, & Peters, 2005); the nature of men's attraction to ornamented women by itself strongly implies that men, in general, prefer to benefit them over less ornamented women. Though most research has examined these phenomena in modern Western societies, research examining traditional and foraging groups demonstrates similar patterns, even if based on different forms of material benefits (e.g., Borgerhoff Mulder, 1988; Jones, 1996).

Naturally, the claim that women's ornamentation has functioned in female-female competition over a long history within hominins should not be confused with the claim that women compete with women *only* through ornamentation. Nor should it be confused with the claim that men value women *only* for their attractiveness. Men care about many attributes in a mate, as do women (e.g., kindness, intelligence, cooperativeness, status, skill and interest in raising children, and [for men] paternity reliability; see review in Buss, 2003b, and chapter 4, this volume). Women compete with other women to impress men in these respects as well. And, naturally, female-female competition functions not only in realms of mating and pair bonding but also in other contexts to secure resources, social allies, and control over outcomes for women and their offspring. Though highly important to understanding women's adaptations in general, these latter forms of female-female competition and the adaptations that function within these contexts largely fall outside the scope of this book (for reviews, see Campbell, 2002; Low, 2001).

Signaling of Reproductive Value in Other Species

When reproductive unions between males and females last for a season or less, there is typically no selection on males to assess future reproductive value of mates and no selection on females to signal it. In most of these instances, female ornaments, if they exist at all, are temporary and coincide with seasonal reproduction. (We discuss exceptions of adolescent exaggeration of swellings in nonhuman primates later.) They advertise potential reproduction for the current season. Some authors have argued that women's ornaments signal current reproductive potential (Cant, 1981; Gallup, 1982). A prime reason that these accounts are insufficient is that they should not expect women's ornaments to remain permanent. Very few species pair bond for the long term. Perennial female ornaments, then, are rare.

Nonetheless, some nonhuman instances of perennial ornamentation do appear to exist. Consider the colorful female throat ornament in the lizard *Sceloporus virgatus*. Its size and hue correlate positively with its owner's health and general condition. Males accordingly prefer to mate with more ornamented females. A female develops her ornament when she becomes sexually receptive and maintains it fully after her receptive phase throughout egg maturation and until oviposition in her mate's territory (Weiss, 2002, 2006). Retention of ornamentation may be an honest signal of residual reproductive value, possibly important for attracting and maintaining the interest of a male possessing a territory with high-quality food and oviposition sites. If female ornamentation were manipulated artificially (see, e.g., Baird, 2004), females with reduced ornamentation may be peripheralized by male territory holders, whereas females with enhanced ornamentation may retain or improve their location.

Long-term pair-bonding bird species are candidates for female perennial ornaments. In some of these species, males and females pair bond for many years, even their entire reproductive life spans. This pattern is especially common in tropical passerines. Females of these species also exhibit, relative to temperate passerines, more vocalizations during pair bonding (Freed, 1987). Female vocalizations in these species may be the analog of women's perennial ornamentation.

Adolescent Exaggeration of Signals in Nonhuman Primates

Humans are not the only primate with exaggerated ornamentation in adolescent or semiadult females. Indeed, adolescent ornamentation is fairly common among primates.

Anderson and Bielert (1994) reported its presence in about one-quarter of Old World monkeys and apes (catarrhines; see table 6.1). It takes on two forms. In one form, adolescent females possess sexual swellings or skin, but older females lack it. In the other form, both adolescents and adults are ornamented, but adolescents more so. Women are of the second type (Anderson & Bielert, 1994). Anderson and Bielert (1994) propose that, in species in which adolescent signaling occurs, it functions to signal reproductive value to males to obtain fitness-enhancing male material assistance. In our view, signaling in these species largely reflects competition between same-age females, with individual differences in signal quality (affected by estrogen levels; Dixson, 1998) reflecting personal quality and hence reproductive value.

Males in virtually all the catarrhines have a sexual preference for postadolescent females (Anderson & Bielert, 1994; Dixson, 1998; Hrdy, 1997; Muller et al., 2006). As we discussed earlier, despite the significant degree of sexual attractiveness of pubertal and adolescent females to men, men appear to be maximally attracted to females of ages 18 years through their 20s (Jones, 1996; Quinsey & Lalumiere, 1995; Quinsey et al., 1993). Such women, if they have not reproduced, will retain most or all of their full ornamentation acquired a few years earlier (Forbes, 1987; Lassek & Gaulin, 2006).

Some researchers have hypothesized that female adolescent ornamentation functions to obtain sexual attention from males who would otherwise sexually focus on

Table 6.1 Primate Species With Adolescent Exaggeration

Species	Preference for Post-Adolescent Females[a]	Adolescent Reproductive Success[b]	Female Transfer[c]
Macaca Sylvanus	+	L	O
M. mulatta	+	L	O
Papio anubis	+	L	O
Macaca assamensis			
M. arctoides	+		C
M. fascicularis	+	L	O
M. fuscata	+	L	O
Erythrocebus patas	O	E	N
Theropithecus gelada	+	L	C
Gorilla gorilla	+	L	C
Homo sapiens	+	L	C

Based on table 1 of Anderson and Bielert (1994).
[a]+, male sexual preference for older females over adolescent females; 0, no preference between adolescent and older females.
[b]Adolescents' reproductive success is lower (L) or equal (E) to that of older females.
[c]C: common occurrence; O: occasional; N: never observed.

older females (see references in Anderson & Bielert, 1994). This hypothesis argues that adolescent females deceive males into acting toward them as though they are fertile, thereby extracting material benefits from them against males' reproductive interests. Anderson and Bielert (1994) note two critical shortcomings of this hypothesis. First, adolescent females actually engage in relatively little sexual activity and receive little sexual attention from males (especially older males). Second and relatedly, males sexually prefer postadolescent females who have relatively small ornaments, or none at all, over more ornamented adolescent females.

It appears, then, that ornamented adolescent females seek material benefits that males, especially older ones, can provide, but not primarily mediated through sexual attention. Instead, males may value information about females' reproductive value as females enter reproductive age and, in hopes of future reproduction with them, deliver material assistance to ones most promising. Benefits males provide include access to resources, protection, and enhanced status and social alliances. Though many adolescent females are subfertile, some may have offspring benefited through male protection (see Anderson & Bielert, 1994). Older males in particular deliver benefits, both because they are more capable of doing so (as younger males are subordinate) and because they themselves benefit more from doing so (as younger males typically disperse from their natal group at the end of adolescence).

Adolescent catarrhine females often transfer between groups or otherwise commonly find themselves in a nonnatal group in which they will reproduce (see Anderson & Bielert, 1994). Females who transfer face problems. They must learn the whereabouts of resources in their new territory, as well as form effective alliances with residents of both sexes, typically without close kin available to assist them. These circumstances in particular may have favored female competitive signaling of quality. Older males may favor females who signal high quality, enabling them to transfer without injury, ostracism, or death, and enhancing female reproductive success in the new group. Females may also have been selected to signal quality to adult females that are potential alliance partners. As expected, then, Anderson and Bielert (1994) found a positive association across catarrhine species between the presence of female transfer between groups and adolescent exaggeration in female ornamentation: In species in which females transfer between groups, adolescents tend to possess exaggerated ornaments.

High levels of ornamentation in adolescents with little or no fertility bolster the case that these ornaments signal residual RV. We suggest that even in species in which adolescents have smaller ornaments than older females, adolescent ornamentation still functions to honestly signal future reproductive value (on the common chimp, see Wolfe & Schulman, 1984).

Adolescent Exaggeration of Ornaments in Humans

If women's ornaments signal personal quality and RV because humans engage in long-term pair bonding, as we have argued, it makes sense that their ornaments would be fully developed when RV is maximal—at the beginning of the reproductive period. As we have noted, however, women's ornaments are fully developed prior to the beginning of reproduction. We suggest that adolescent human females possess exaggerated ornaments for much the same reason that adolescent female nonhuman primates do.

Adolescent females in foraging or traditional societies sometimes transfer into new ecological and social group settings through arranged marriage or capture (e.g., Chagnon, 1992; Geary, 1998; Hill & Hurtado, 1996). Well-developed ornaments in adolescents connote health, reproductive fat stores, and possibly general phenotypic and genetic quality, thereby promising reproductive value. Though men find adolescent females sexually attractive, they typically prefer as sexual mates women who are several years older (and fertile). Men's interest in adolescent females, whether sexual or otherwise affiliative, may have been selected as investments in future paternity. Adolescent ornamentation may also function to signal personal quality to kin, who may invest differentially as a function of quality.

Individual Variation in Peak Ornamentation

Menarche occurs when women are within a few years of maximal ornamentation— typically, around age 14 in traditional societies (see Weisfeld, 1997), and 1 to 2 years earlier in modern societies (see Herman-Giddens et al., 1997). At this time, adolescent girls develop sexual interest as well (Weisfeld, 1999). Within modern societies and across traditional societies, the age of menarche and development of female ornamentation varies considerably (Herman-Giddens et al., 1997; Kanazawa, 2001; Marshall & Tanner, 1974; Reynolds, 1950). According to life-history theory, as adult age-specific mortality rates increase, individuals typically maximize lifetime reproductive effort by beginning reproduction earlier. Variation in age at menarche may reflect adaptive, conditional shifts depending on cues that ancestrally would have predicted adult survival. Females experiencing conditions indicative of reduced resource availability and likelihood of survival, such as unpredictable resource levels and father absence (Ellis, McFadyen-Ketchum, Dodge, Pettit, & Bates, 1999; Ellis et al., 2003; Ellis & Garber, 2000; Kanazawa, 2001), do tend to achieve relatively early menarche. So too do girls whose mothers repartner with unrelated males (stepfathers or boyfriends). Possibly, early menarche and development of ornamentation may have allowed ancestral women without strong kin support or living with unrelated older men to garner male-provided benefits, as well as nepotistic benefits. Just as with adolescent exaggerations in nonhuman primates, these benefits need not be generated by sexual interest of older men but rather, in ancestral populations at least, increased probability of their future reproduction.

Age of Peak Ornamentation and Female Sexual Competition

If human adolescent ornamentation reflects female-female competition for male-delivered and nepotistic benefits, as we have argued, it makes sense that adolescent females should also behaviorally compete for male attention at this time. They do (Weisfeld, 1999;

Weisfeld & Woodward, 2004). In one study on the Tsimané, South American foragers and horticulturalists with natural fertility unaffected by hormonal or other modern contraception, Rucas (2004; Rucas et al., 2006) found that female–female arguments over mates are most frequent well before females reach maximum age-related fertility (but close to maximum reproductive value), around 16 years of age. Female–female competitions over resources other than mates (e.g., food and friends), by contrast, increase as women age. The Tsimané engage in long-term, socially monogamous pair bonding and experience low divorce rates. Adolescent females who successfully compete for males (e.g., with more attractive estrogen-based ornamentation) may garner substantial benefits through high-quality providers or mates who can enhance their social status.

Extended Sexuality and Adolescent Exaggeration

In traditional and modern societies, young women experience a period of low fertility in which they are nonetheless sexually active. Sexual activity during this period, we suggest, serves the same function that extended sexuality outside of the fertile phase in fully reproductive, normally cycling women does: It functions to garner nongenetic material benefits and services, mostly from men (see chapter 3). Adolescent ornamentation enhances women's ability to do so.

We noted earlier that adolescent females with exaggerated ornaments in nonhuman primates exhibit and elicit from males relatively little sexual interest (Anderson & Bielert, 1994). Human adolescent females appear to differ in these respects. This difference is likely another manifestation of the profound implications of long-term pair bonding in humans. Human female lifetime reproductive success has historically been influenced by ability to attract male attention during adolescence. Men's sexual interest in adolescent females reflects the fact that, typically, their reproductive success achieved through pair bonds was not maximized by attending solely to cues of current fertility but also to cues of reproductive value.

In many human societies, adolescent female sexuality may involve "sexual experimentation"—a period of relatively relaxed standards of discrimination of mates, followed by increased selectivity (Weisfeld & Woodward, 2004). At the same time, adolescent females are clearly capable of establishing a strong bond with a single male (Weisfeld & Woodward, 2004). Limited selectivity with little risk of conception may allow adolescent females to obtain material benefits from multiple mates, whereas capacity for long-term bonding may promote fidelity to an outstanding mate, if found.

Women's Bodies Exhibit Redundant Estrogen-Facilitated Ornaments

We have emphasized women's ornamentation in the form of gynoid fat depots in the breasts, hips, buttocks, and thighs. These ornaments no doubt importantly affect women's attractiveness. In fact, however, other features similarly function as women's ornaments. To date, available evidence points to their being facilitated by estrogen.

A number of excellent reviews have recently appeared: Feinberg, Jones, et al., (2005); ink et al. (2001); Grammer et al. (2002); Grammer, Fink, et al. (2003b); Rhodes (2006); Schaefer et al. (2006b); Singh (2002); Symons (1995); Thornhill & Gangestad (1999a); Thornhill & Grammer (1999). For a full review of the literature, we refer readers to these sources. Next, we sketch important findings and interpretations of this literature.

Breasts

Cant (1981) and Gallup (1982) proposed that breasts are indicators of fat reserves to be used in gestation and lactation, relevant to both current fertility and future reproductive value. Fat per se, however, is not sexually attractive in breasts or anywhere else in the female body. Rather, gynoid fat specifically stored in secondary sexual traits is attractive (Kaplan, 1997; Lancaster, 1986; Pond, 1978; Singh, 1993). Women may have evolved ways to display this fat just under the skin so that it has distinctive visual and tactile qualities. Android fat deposited in the abdomen but also in the breasts, buttocks, or thighs of women is not considered attractive (e.g., Singh, 1993).

Large breasts on women per se are not attractive. Men prefer firm and upright breasts over sagging breasts and adult breast form over juvenile breast form (Grammer, Fink, Møller, & Manning, 2005). Adult breasts contain varying proportions of android and gynoid fat deposits, however; those containing large amounts of android fat are less attractive. Perhaps because android fat deposits may account for much variation in breast size, breast size does not strongly predict milk quantity and quality (see references and discussion of the correlation in Low et al., 1987; Marlowe, 1998; Møller et al., 1995). In theory, gynoid fat depots in breasts should predict positively all aspects of female lifetime reproductive capacity (conception probability, probability of successful pregnancy, offspring quality, lactation quality, and so on).

Jasiénska et al. (2004) provide evidence relevant to this point. They measured progesterone and estrogen in saliva samples each day of one entire menstrual cycle in a sample of 119 Polish women between 24 and 37 years of age, all of whom were healthy, ovulating normally, and of normal weight. They also measured women's breast size and WHR. Because all women were of normal weight, android fat levels should not have been less variable in this sample than in unselected samples. Both large breast size and low WHR—in this normal-weight sample, reflective of gynoid fat deposits—independently predicted higher estrogen and progesterone levels. But these variables also interacted to predict hormone levels: Breast size was particularly strongly associated with estrogen levels among women with low WHR and therefore lacking android fat. Indeed, women with large breasts and low WHRs had 26% higher mean estrogen across the cycle and 37% higher mean mid-cycle estrogen than women from any of the other groups (see figure 6.1), reflective of perhaps a threefold difference in conception probability (Jasiénska et al., 2004).

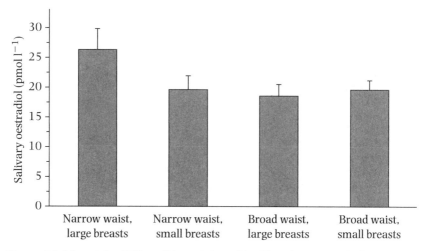

Figure 6.1   Mean (plus 95% confidence intervals) mid-cycle estrogen in four groups of women with different combinations of waist-to-hip ratio and breast size. Women with both low waist-to-hip ratio and large breasts have greater salivary levels of estrogen at mid-cycle than women in the other three groups. Based on figure 1 of Jasieńska et al. (2004). Reprinted with permission of The Royal Society.

Waist-to-Hip Ratio

As expected, women's WHR predicts hormone levels. WHR is correlated negatively with serum estrogen (Singh, 2002b) and positively with serum testosterone (Ibáñez, Ong, de Zegher, et al., 2003). As well, WHR negatively predicts women's fertility (Singh, 1993, 2002a,b), including neonate birth weight, which in turn positively predicts infant survival (Pawlowski & Dunbar, 2005). As discussed earlier, men typically find women's bodies with lower ratios (ratios around .7) more attractive than bodies with higher ratios (Jasiénska et al., 2004; Marlowe et al., 2005; Singh, 1993, 1995, 2002a,b; Streeter & McBurney, 2003; Sugiyama, 2004). WHR appears to be associated with women's RV: at its minimum in young, nulliparous women and increasing with age and with parity (see Lassek & Gaulin, 2006; Singh, 1993). WHR is, furthermore, a good indicator of women's ratio of android to gynoid fat (e.g., Singh, 1993; Ibáñez, Ong, de Zegher, et al., 2003; Pederson et al., 2004). Lassek and Gaulin (2008) recently reported that, with other major predictors statistically controlled, women with low WHRs and their offspring tend to have relatively high scores on tests of cognitive performance, which the authors attributed to the availability of omega-3 fatty acids for brain development during pregnancy and lactation (though, we note, WHR may be a marker of quality and hence may predict other indicators of quality, too, including—through heritable effects—quality indicators in offspring).

In some foraging or traditional societies (e.g., the Hadza of Tanzania; Marlowe & Wetsman, 2001—but see Marlowe, Apicella, & Reed, 2005; the Matsigenka of Peru;

Yu & Shepard, 1998), men may prefer women with moderate rather than low WHR. In nutritionally stressed populations, men may prefer stored fat in general, without regard to it being gynoid or android fat (Marlowe & Wetsman, 2001). In fact, Pawlowski and Dunbar (2005) found that body mass index (BMI; effectively a measure of weight for height) is a better predictor of neonate birth weight than is WHR in the relatively thin subset of a large sample of Polish women. Alternatively, less feminine women may be preferred where sexual selection strongly favors testosteronized sons. Through gestational maternal effects, testosteronized mothers may produce such sons (Manning, Trivers, Singh, & Thornhill, 1999) and more sons (e.g., Manning et al., 1999, and references to other studies therein; see also chapter 7, this volume, on intralocus sexually antagonistic conflict).

Women's Facial Ornamentation

One of the most robust findings on physical attractiveness is that men are particularly attracted to women's faces with hyperfeminine proportions. Men's and women's faces differ in a number of ways. Most notably, the lower portions of women's faces are smaller, relatively to total face length. Women's jaws and chins in particular are less strongly developed. Because men develop stronger brow ridges as well, women's eyes tend to appear larger and more "open." Men's cheekbones and mid-facial region grow to be more protruberant, such that, in profile, women's faces are flatter and their cheekbones more gracile. These sex differences appear to reflect sexual selection acting on facial features in sex-specific ways (Thornhill & Gangestad, 1993; Weston, Friday, & Liò, 2007). Wherever it has been examined, men prefer females faces that, proportionately, are more feminine than average: the United States (e.g., Johnston & Franklin, 1993), the United Kingdom and Japan (Perrett, May, & Yoshikawa, 1994), Russia (Jones & Hill, 1993), and two foraging groups, the Ache of Paraguay and the Hiwi of Venezuela (Jones & Hill, 1993). (For reviews, see Jones, 1996; Rhodes, 2006; Scheyd et al., 2007; Symons, 1995; Thornhill & Gangestad, 1999a.)

Feminine faces are estrogenized. During puberty, estrogen caps facial bone growth and thus leaves the lower face relatively small, with a delicate lower jaw, reduced eyebrow bone ridges, flat face with shallow-set eyes, enlarged lips, and often specialized, apparently gynoid fat deposits over the cheek bones (see figure 6.2). Indeed, Law Smith et al. (2006) found, in a sample of young, normally ovulating women, that ratings of their facial attractiveness, femininity, and health (whether assessed by men or women) each correlated positively with current estrogen levels, as assayed using urinary metabolites of estrogen. Interestingly, the same associations were not observed when women wore makeup, perhaps because women partly use makeup to feminize features that are relatively unfeminine.

Facial femininity in women may also vary with health and resistance to infection. In one age-controlled study, women's facial femininity (measured from facial photographs) predicted negatively self-reported incidence of respiratory (though not gastrointestinal) infection over the preceding 3 years (Thornhill & Gangestad, 2006). More generally, however, evidence pertaining to whether women's facial femininity or attractiveness is

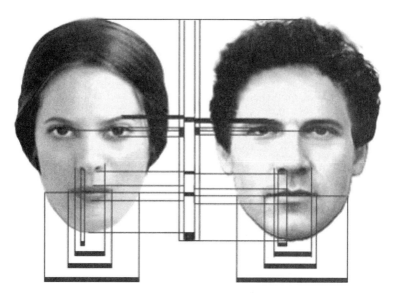

Figure 6.2 Sexual dimorphism in the human face. The faces are averages made by combining numerous same-sex faces. Note in particular the sex difference in fullness of the lips, width of the mid-face (wider in men), and the size of the lower face, especially lower jaw (larger in men). Note, too, that the reduced eyebrow ridge growth and limited mid-face projection in the woman's face makes the female face appear flat and the eyes enlarged. Pictures by and courtesy of Victor Johnston.

associated with current or developmental health is mixed, with some, but not all, studies showing positive associations (for reviews, see Gangestad & Scheyd, 2005; Grammer et al., 2005; Langlois et al., 2000; Weeden & Sabini, 2005). No study has examined facial femininity in relation to health in traditional societies exposed to ecological circumstances reasonably similar to ones encountered by ancestral humans, in which childhood mortality rates may typically have been 30–50% (e.g., Hill & Hurtado, 1996) and energy budgets were constrained.

Women's facial features take on their characteristic adult form during adolescence—well before peak age-dependent fertility and perhaps even before the age of maximum RV in traditional populations (again, late teens). As women age and their RV (as well as age-specific fertility; e.g., Dunson, Colombo, & Baird, 2002) declines, women's facial features change; the lower face becomes longer, the eyebrow ridges more prominent, and the lips smaller (Enlow, 1990; Johnston & Franklin, 1993; Jones, 1996; Symons, 1995). Similar changes appear to occur with increasing parity (Symons, 1995; Thornhill & Gangestad, 1999a). Women's facial ornamentation hence appears to signal RV, based on age, parity, and personal quality.

Skin

Skin morphology is sexually dimorphic (Fink, Grammer, & Matts, 2006). The sexes differ in melanin pigmentation and hemoglobin-based color, with men having darker and ruddier faces than women (Frost, 1988, 1994). Women have less hairy, smoother, and more finely textured skin containing more subcutaneous fat (Montagna, 1985a, 1985b; Pond, 1978). Structurally, women's smooth, fine-textured skin reflects more shallow and narrow crevices than are seen in men's skin. All feminine features of skin appear to be facilitated by estrogen, to be maximally developed in young women, and to decline in attractiveness with age (Fink et al., 2001; Manning, Bundred, Newton & Flanagan, 2003; Montagna, 1985a, 1985b; also Frost, 1994; Jones, 1996). (We anticipate, however, that female skin is most attractive during female adolescence.) Like women's facial ornamentation, the estrogen-facilitated features of young women's skin may well be honest signals of residual reproductive value.

Recent studies experimentally manipulated the uniformity of skin color and skin smoothness in digitized photographs of young women's faces. As expected, these manipulations affect rated facial attractiveness (Fink et al., 2001; Fink et al., 2006), independent of the effects of female facial shape (and hence estrogen-facilitated structural features), facial symmetry, and age.

In some nonhuman Old World primates, size of sexual swellings covaries with female quality and attractiveness to males (Domb & Pagel, 2001; Dixson & Anderson, 2004). To our knowledge, however, no one has examined the sexual attractiveness of other features of sex skin and their possible relationships to female quality among conspecific females in these species. Based on findings on women, one might well expect that relatively subtle but discernable estrogen-facilitated features of female sex skins in nonhuman primates and other taxa signal female quality.

Multiple authors have noted parallels between sex skins of female, nonhuman, Old World primates and women's skin (e.g., Montagna, 1985a, 1985b; Szalay & Costello, 1991). Both are estrogen facilitated. There may well be homologies with respect to smoothness and related fine structure, though additional research is needed to ascertain degrees of histological and developmental similarity and to assess whether similarities are due to homology or convergence. One possible scenario suggested by phylogenetic comparisons is that females of the species ancestral to both the *Pan* lineage (the two species of chimps) and the hominin lineage perhaps had slight sex skin development in the anogenital region (and surely not exaggerated sexual swellings; Sillén-Tullberg & Møller, 1993 [see figure 6.3]; see also Pagel & Meade, 2006). Perhaps sexual selection led this skin region to expand to cover women's bodies generally. In any case, women's skin does appear to function in sexual signaling and thus is appropriately referred to as sex skin. (The point here, of course, is not that skin has no other functions. Obviously, skin has many functions. Some features of women's skin nonetheless may have been selected or accentuated by selection for their signal value, the critical feature of any sexually selected signal.)

This view contrasts with a common interpretation in the literature: that women lack sexual swellings and sex skin (see, for instance, Dixson, 1998). More generally, however, it appears that women do possess estrogenized sexual "swellings" (albeit ones that have

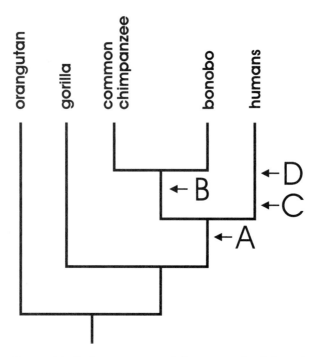

Figure 6.3 Phylogenetic tree of the Hominidae (adapted from the "Tree of Life" web project; www.tolweb.org) depicting the evolutionary events indicated by recent phylogenetic analyses of Catarrhini (see Sillén-Tullberg & Møller, 1993). Females of the most recent common ancestor (at A) of *Pan* (common chimp and bonobo) and *Homo* possessed slight sexual swelling or mere pinkness of the perineal area. Prominent sexual swellings evolved anew in the species that gave rise to *Pan* (B). Female genital ornamentation like that at A was lost in the most recent lineage leading to *Homo* (C), and in that same lineage, permanent estrogen-facilitated sexual ornamentation of woman arose (D).

direct reproductive functions as well: gynoid fat deposits) and features of skin that have been under sexual selection. The female sexual swellings of nonhuman primates are estrogen-facilitated water depots, not fat depots (Dixson, 1998). Another contrasting feature is duration: Women's ornamentation is perennial, not temporary as in females of other Old World primates.

Voice

Women's voices may also contain qualities that have been selected as permanent residual reproductive-value ornaments—honest signals of residual reproductive value.

Certain voice qualities (e.g., pitch) are sexually dimorphic (e.g., women speak at higher pitches) and appear to be influenced by sex-specific ratios of sex hormones (androgen and estrogen; Abitbol, Abitbol, & Abitbol, 1999). As expected, given the age-related decline in estrogen levels, women's voices become less attractive with age, on average (Collins & Missing, 2003). Evidence furthermore hints that, in women, an estrogenized voice is associated with longevity. In a study of about 300 female opera singers, those with more estrogenized voices (sopranos) lived longer than those with more androgenized voices (Nieschlag et al., 2003). And, in a study with age controlled, the voices of relatively symmetric women were more attractive than those of asymmetric women (Hughes, Harrison, & Gallup, 2002; on symmetry and developmental stability, see chapter 7, this volume). No research, to our knowledge, has examined the relationship between the sexual attractiveness of women's voices and age across a wide range of ages, including adolescence—the age at which vocal attractiveness should peak if it reflects RV. More generally, additional work is needed to assess whether women's voice qualities have been sexually selected as signals or are incidental by-products of estrogenization and developmental stability (also see later discussion).

Other Ornaments

Other estrogen-mediated secondary sexual characteristics of women are candidate honest signals of reproductive value. Women's locomotion is characterized by greater fluidity than is men's (Grammer, Keki, Striebel, Atzmüller, & Fink, 2003, and references therein). More generally, studies that capture and characterize movement patterns document a variety of differences in men's and women's gait. For instance, women tend to carry their elbows inward (and, indeed, women's forearms angle outward from their upper arms more than do men's; Grammer, Keki, et al., 2003). Women walk in a relative straight line, one foot directly in front of the other, more than do men (Grammer, Keki, et al., 2003). These features probably reflect effects of estrogen and, if so, highly estrogenized gait may signal good reproductive condition (Grammer, Fink, et al., 2003). A recent experimental study supports this hypothesis, as it found that degree of feminine gait (as well as separately manipulated WHR in ambulatory animated women) positively affects women's attractiveness (Johnson & Tassinary, 2007).

As we discuss later, the proportional lengths of men's and women's digits of the hands differ due to sex-specific hormonal effects, with men's index fingers typically being shorter than their ring fingers, women's less so (or with the index finger longer; Manning, 2002). One study found that people rate hands with sex-typical digit ratios more attractive than hands with digit ratios more typical of the other sex (Saino, Romano & Innocenti, 2006). Another candidate is relatively small foot size (Fessler et al., 2005). Women have smaller feet than men, even when women's smaller body size is accounted for (Barber, 1995). Estrogen may cap growth of foot bones in the same way it caps facial growth. Fessler et al. (2005) found that men from multiple societies rate depictions of small female feet more attractive than large feet. Further research is needed to assess whether these features are attractive because they are incidental by-products of estrogen or because they have been sexually selected to signal quality.

Two additional features deserve mention. First, averageness (or nondistinctiveness) of facial features predicts women's attractiveness (see review by Rhodes, 2006), possibly because it is associated with developmental stability, lack of mutations, or heterozygosity at loci at which there is heterozygote advantage (Thornhill & Gangestad, 1993). Zebrowitz & Rhodes (2004) found that facial averageness predicts pubertal intelligence and adolescent health in women, but only in the lower half of the averageness distribution. They therefore proposed that averageness discriminates average from "bad" genes (or poor condition) but does not connote "good" genes. In fact, however, this effect may not be robust, as the regression slopes for high and low averageness groups did not significantly differ. Moreover, a plausible alternative is that prediction at the high end of averageness is compromised because there do exist some nonaverage features indicative of good condition (e.g., feminine features in women). There is no reason to think, however, that facial averageness is highly associated with estrogenization.

Second, women's BMI predicts women's attractiveness, such that moderate (or low moderate) values are preferred (e.g., Tovee, Maisey, Emery, & Cornelissen, 1999). BMI in women does not appear to be highly associated with estrogen levels. Women with high BMI values (hence overweight or obese) vary with respect to their ratios of gynoid-to-android fat (Ragoobirsingh, Morrison, Johnson, & Lewis-Fuller, 2004). As we mentioned earlier, women's BMI is positively predictive of the neonate birth weight of their offspring at the low end of the range of BMI, a possible reason why men in certain traditional societies (who live in conditions of energetic constraint) prefer heavier women. In the West, high BMI is associated with low socioeconomic status (e.g., Laaksonen, Sarlio-Lahteenkorva, & Lahelma, 2004), which could contribute to its lack of attractiveness. (We recognize, however, that the direction of causality could run the other direction, with lack of attractiveness contributing to low socioeconomic status.)

Multiple Signals Covary Positively

We and others have argued that women's phenotypes reflect multiple, redundant signals of quality and RV (Thornhill & Gangestad, 1993; Thornhill & Grammer, 1999). Signaling across multiple channels is observed in many species. One explanation—the redundant-signal hypothesis—states that each signal provides imperfect information about quality, and hence signals are redundant (though, of course, not perfectly so) indicators of this same quality (Candolin, 2003; Hebets & Papaj, 2005; Johnstone, 1995, 1996; Møller & Pomiankowski, 1993). Aggregation of information yielded by multiple redundant imperfect signals provides an index of quality much better than each individually (in the same way that, say, a 10-item test provides information about a person's ability better than that provided by a 1-item test). In addition, redundant information provided in multiple modalities may facilitate processing of each (in the way that hand gestures may facilitate comprehension of verbal instructions to someone wanting directions to a location). An alternative explanation—the multiple-message hypothesis—for multiple signals is that different signals carry information about different desired qualities (Møller & Pomiankowski, 1993). For instance, separate signals may evolve to reveal information about body size, parasite resistance, and foraging ability. (Indeed, some signals could reveal useful information about features other than quality—e.g., the signaler's species, sex, location; Hebets & Papaj, 2005.)

Each theory probably explains some instances of multiple signals in some species. The theories offer different predictions with regard to covariation between signals. Whereas the redundant-signal hypothesis predicts that signals ought to covary with one another (such that the presence of one attractive feature is associated with the presence of other attractive features, though not perfectly), the multiple-message hypothesis predicts that signals should be relatively uncorrelated. In the case of women's ornaments, the redundant-signal hypothesis has been supported.

Thornhill and Grammer (1999) asked men to independently rate the attractiveness of 96 nude young women based on pictures revealing only certain features. For instance, one set of pictures revealed only women's faces. Others showed women's fronts (with their faces covered). Yet others showed women's backs. Men could not match the photo of one feature of a woman with the photo of another feature of the same woman. Ratings of attractiveness of any one feature correlated positively (though weakly to moderately) with ratings of attractiveness of any other feature. Hence, on average, women who have attractive faces have attractive breasts, thighs, hips, buttocks, and backs. (See also Penton-Voak et al., 2003, who found that women's facial attractiveness negatively predicted their waist-to-hip-ratio.) In theory (but also supported by available evidence), all features are partially redundant signals of estrogen, aging, and personal quality. In studies using these same photos, women's facial and body symmetry positively correlated with one another and with attractive manifestations of facial and body estrogen-facilitated traits (Grammer et al., 2002; Schaefer et al., 2006).

Attractive vocal qualities may similarly relate to women's facial and bodily attractiveness. Collins and Missing (2003) and Feinberg, Jones, DeBruine, et al. (2005) found that, with age effects statistically controlled, women's voice attractiveness positively predicted their facial attractiveness, independently rated (and, in Feinberg et al.'s study, facial femininity measured from photographs as well). Moreover, voice attractiveness among young women negatively predicts their waist-to-hip ratio (Hughes, Dispenza, & Gallup, 2004).

Multiple Signals Should Reveal Quality

If, indeed, women's multiple ornaments redundantly reflect personal quality and RV, all should not only covary with each other; each should covary with indicators of quality and RV.

Correlations With Estrogen   Estrogen reflects women's readiness to conceive and ability to carry through a pregnancy to a successful outcome, as reflected by her condition and past ability to prepare for the energetic costs of gestation and lactation. As noted in our brief review, multiple signals reflect current estrogen levels, even when age is controlled: qualities of breasts (Jasiénska et al., 2004), waist-to-hip ratio (Jasiékska et al., 2004; Singh, 1993, 2002a, b); and facial femininity (Law Smith et al., 2006). We anticipate that women's skin and voice qualities similarly covary with current estrogen levels, even when age is controlled. Estrogen, as we discussed earlier, affects women's gynoid-to-android fat ratio; high ratios tend to be attractive. Degree of feminine digit ratio positively predicts current estradiol level in young women (McIntyre, Chapman, Lipson, & Ellison, 2007). (On a possibly related association between feminine digit ratios and menstrual cycle regularity, see Scarbrough & Johnston, 2005.)

Correlations With Health Status   Studies have examined associations of attractive features with actual and perceived health. Generally, positive associations (age controlled) between ratings of women's facial attractiveness and ratings of perceived health have been reported (see Jones et al., 2001; Rhodes, 2006). The correlations between attractive features with estrogen levels (reviewed earlier) reflect associations with an important component of women's health: their fertility and reproductive health. Studies of other components of actual health of women have yielded mixed results: positive associations with attractiveness in some studies but not others (see reviews in Hönekopp, Bartholomé, & Jansen, 2004; Langlois et al., 2000; Rhodes, 2006; Rhodes, Chan, Zebrowitz, & Simmons, 2003; Scheyd et al., 2007; Thornhill & Gangestad, 2006; Thornhill & Møller, 1997). One additional study found a positive relationship between facial attractiveness and physical fitness (as measured by a standard test that assesses a combination of endurance, strength of different muscle groups, dexterity, and other variables; Hönekopp et al., 2004). Henderson and Anglin (2003) found that the facial attractiveness of 17-year-old females positively predicted their longevity.

As we noted earlier, women's gynoid-to-android fat ratios predict cardiac health and cortisol responses. Feminine ornamentation may positively relate to general health and lifespan (see also Nieschlag et al., 2003). Future research may investigate whether these associations are specific to Western societies in which cardiac and metabolic diseases importantly affect health and longevity or, in fact, are also found in traditional and foraging societies in which energetic constraint and infectious disease importantly affect health.

As mentioned earlier, women's developmental stability correlates positively with ratings of the attractiveness of their bodies, as reflected in estrogen-facilitated traits. In chapter 7, we review evidence on associations between female ornamentation and developmental stability, a component of developmental health reflected in bilateral symmetry. As will be seen, though mixed findings exist, a number of important positive links have also been documented.

Not All Female Manifestations of Estrogen Are Signals   The fact that multiple signals reflect estrogen levels should not be confused with the mistaken claim that all manifestations of estrogen are signals. Indeed, not all manifestations of estrogen that men detect and act on are signals. In chapter 11, we propose that women's estrogen (or related physiology) generates by-products, which relate to their fertility, as it changes across the cycle. Some of these by-products (e.g., scents) may sexually stimulate and attract men. Others arise incidentally from women's perception of increased desirability or attractiveness at ovulation, when estrogen levels are near their peak (e.g., women wear sexier clothing at this time). As we argued in chapter 5, these by-products do not function to signal cycle-related high fertility. Indeed, they have no function at all. They are simply by-products of adaptations with function unrelated to them (in this context, largely adaptations that facilitate women's fertility and successful conception; Ellison, 2001).

Ornaments Are Not Deceptive Signals

Some authors have argued that women's ornaments are not honest signals but rather are deceptive. Low et al. (1987) argued that breast, buttock, and thigh fat deceptively signal female quality (see also critiques by Anderson, 1988; Caro & Sellen, 1990). Jones (1996) argued that women's feminine facial features deceptively signal youth. Miller (1995) suggested that women's breasts deceptively signal pregnancy. If men had a preexisting bias to provision pregnant women with developed breasts, they may have been deceived into provisioning nonpregnant large-breasted females. Smith (1984b) similarly conjectured that breasts deceptively signal a state of pregnancy, but purportedly for the function of relaxing male partners' vigilance and mate guarding, permitting women greater freedom to copulate outside the pair bond.

These hypotheses cannot explain a variety of facts. First, ornaments are maximally manifested before the onset of full adult life and decline in attractiveness with age and parity, as expected if they function to honestly signal reproductive value. Second, ornaments reflect estrogen levels and thereby fertility levels. Third, evidence suggests that individual variation in ornament quality or attractiveness covaries with condition, general health, and possibly developmental stability, and thereby with phenotypic and possibly genetic quality variation; this variation clearly covaries with reproductive health. Fourth, there exists an overall positive manifold of correlations between the quality of different ornaments (e.g., breasts, hips and buttocks, facial features, voice). These hypotheses that women's ornaments are deceptive also encounter serious theoretical problems, as selection should disfavor males preferentially attracted to dishonest signals and favor males with ability to discern true quality (e.g., Kokko et al., 2003; Searcy & Nowicki, 2005).

A particularly common view in the literature is that women's bodily ornaments deceptively signal peak cycle-related fertility throughout the cycle (but manifested permanently; Szalay & Costello, 1991)—that is, that these ornaments signal to men that their bearers are continuously fertile throughout the cycle, when in fact they are not. This view stems from the more general view that sexual swellings in nonhuman primates signal peak cycle-related fertility. As we argued in chapter 5, this general view is probably mistaken. Females ancestral to humans did not possess features whose function was to signal to males their fertile state in the cycle, and males ancestral to humans did not infer females' cycle-related fertile states on the basis of signals. Rather, they did so on the basis of by-products. The view that women could have deceived men into treating them as though they were fertile throughout the cycle by retaining ornaments permanently is therefore also erroneous. As well, this view suffers from the theoretical problems and inability to explain key empirical findings about women's ornaments that challenge other theories that women's ornaments are deceptive.

As we discussed in chapter 5, deceptive signaling can occur, but deception cannot persist over evolutionary time unless it occurs within a fundamentally honest signaling system (Searcy & Nowicki, 2005). Hence, for instance, women may undergo cosmetic surgery and deceptively alter their ornamental features. But that deception occurs in the context of a fundamentally honest signaling system. If, due to deception, the system loses its fundamental honesty, then over many generations selection should result in degradation and loss of the system.

Outstanding Conceptual and Empirical Issues

Evidence compiled to date strongly suggests that many features of women that men reliably find attractive are honest signals of reproductive value, as reflected by age, parity, and personal quality. Furthermore, evidence strongly supports the hypothesis that the function of these signals—the benefit that led selection to favor these signals ancestrally—was to obtain male-delivered material benefits, largely provided through long-term pair bonding. At the same time, a number of conceptual and empirical issues demand further attention.

What Are the Costs of Estrogen?

Women's ornaments, evidence indicates, reflect an ontogenetic history of estrogen effects, as well as current estrogen levels. If women's ornaments are honest signals of quality, so too must estrogen honestly reflect quality (Thornhill & Gangestad, 1993, 1999a; Thornhill & Møller, 1997; Thornhill & Grammer, 1999; Manning, 2002). What are the costs to estrogen that keep it honest? That is, what prevents women of lower quality from achieving the same estrogen levels as women of higher quality?

The idea that we presented in the first section of this chapter is that life history trade-offs play key roles in keeping estrogen levels honest indicators of quality and current condition. Estrogen modulates women's allocation of effort to reproduction (both current and future) and away from somatic maintenance. If estrogen increases conversion of energy embodied in glucose into stored gynoid fat, that energy is not available for allocation to daily metabolic demands, immune function, or storage into android fat (readily available for maintenance later). Optimal allocations of energy into reproduction vary across women. In theory, women who are unhealthy, who have a history of fluctuating (and sometimes negative) energy balance, and whose metabolic demands are higher because they less efficiently use energy cannot afford to convert as much energy into reproduction as women who are healthy, maintain a consistently positive energy balance, and efficiently utilize energy. Female estrogen levels are honest signals of quality and condition because optimal levels are condition-dependent. It does not pay women of lower quality to cheat by producing more; they maximize their own fitness by maintaining levels lower than those that maximize the fitness of women of higher quality.

Other ideas about estrogen's costs have been proposed. Estrogen may metabolize into toxic by-products that trigger cancers in women and hence negatively affect some physiological systems (Service, 1998). Estrogen may also specifically target and diminish bearers' immunocompetence, above and beyond the effects of energetic trade-offs (review by Roberts, Walker, & Alexander, 2001; also Schuster & Schaub, 2001). Estrogenization, then, may honestly signal ability to bear these costs. In theory, it would seem that females could evolve means to avert the toxic effects of estrogen or specific targeting of immune function and, hence, it has been argued that these kinds of costs should not be evolutionarily stable (unless there are unknown constraints to the evolution of means to avert these costs, such as constraints in genetic variation; Kokko et al., 2003). Future research may explore more precisely the costs that keep estrogen an honest indicator of quality and condition.

What Keeps Ornaments Honest Indicators of Quality and Condition?

Ornaments presumably reflect estrogen levels. If estrogen levels honestly reflect quality and condition, it may seem that ornaments must do so as well. In fact, however, one must entertain the possibility that ornaments could develop by a means other than through influences of estrogen. If so, women could evolve attractive ornamentation in absence of estrogen levels. What maintains the honesty of phenotypic ornaments themselves?

In the instances of gynoid fat storage, one basis for honest signaling seems clear: Storage of calories into fat allocated for reproduction is itself costly. There is simply no way to fake storage of fat without allocating calories to it. Some features of gynoid fat storage may nonetheless have evolved as means of ensuring or bolstering honesty of the signal. Gynoid fat tends to be stored just below the surface of the skin. In vertebrates, including primates, fat in both sexes is typically stored internally rather than concentrated in specialized organs under the skin. In humans, men store much fat internally as well (Pond, 1978, 1981). By depositing fat on the surface where it can be seen (even if not perceived as "fat" but rather by the smooth, silky appearance gynoid fat generally affords), women better reveal that they do, indeed, have fat stored for reproduction. Fat stored on the surface may also reveal inabilities to effectively and efficiently deposit it. Cellulite is gynoid fat, but it creates the appearance of rippled, not smooth, skin where it appears (particularly in the thighs and buttocks). It is considered unattractive. Gynoid fat that is deposited in and below the dermis rather than just below the epidermis creates the appearance of cellulite (for a review of cellulite research, see Rossi & Vergnanini, 2000). Possibly, women who can consistently deposit fat just below the epidermis possess quality (e.g., in ways similar to how developmental stability and symmetry can reflect quality; see chapter 7). Additional research can empirically address this speculation.

Other hypotheses for why women deposit surficial fat must be entertained. Pawlowski (1999b) proposed that women's storage of fat in permanent breasts originated in the hominin lineage as a side effect of hominin-specific adaptation for subcutaneous fat deposition that functioned in retention of body heat against low nocturnal temperatures in the cold nights of *Homo*'s African environment of evolutionary adaptedness. The same argument could be applied to all gynoid fat storage (see Pawlowski, 1999b). This view is not inconsistent with the view we propose. As we discussed in chapter 2, evolutionary origin must be explained in terms other than selection, and initial functions can differ from subsequent functions. Indeed, Pawlowski (1999b) argued that surficial gynoid fat depots may have been selected as signals through sexual selection subsequently. Possibly, they were selected as signals partly because surface fat reveals condition particularly well. At the same time, Pawlowski's hypothesis does not adequately explain why gynoid fat deposition is facilitated by estrogen. His hypothesis takes this fact as a given.

What costs keep facial feminization and voice qualities honest are less obvious. Specifically, it is not clear what prevents the evolution of a means by which women with relatively low RV could dishonestly signal high RV by developing or maintaining feminine qualities or voice pitch through a means other than costly estrogenization. One possibility is that, for some reason, means of doing so given current developmental systems are simply unavailable (that is, there is no possible variant in current developmental systems underlying the growth of the larynx, mouth, or other facial features affecting voice qualities that gives rise to femininity *except* variations in estrogen). If so, the reason for that developmental constraint remains unknown.

Another possibility is related to the expectation that there is always selection for individuals to deceptively signal, where they possibly can, and for perceivers to assess true quality and hence disregard deceptive signals. It may be that women's facial femininity and voice are, at this point in time, sufficiently dependent on estrogen that they can serve as reasonably honest signals of quality. Over time, deceptive signaling of facial and vocal estrogenization (through means other than estrogen) may evolve, in which case this system will not persist. Future theoretical and empirical work may address this issue.

In addition, future work may more fully address whether women's ornaments are particularly sensitive to variation in their quality or condition. As Cotton et al. (2004) noted, though much research shows that nonhuman sexual ornamentation covaries with individual condition, little research shows that ornament quality is affected more dramatically by condition than is the quality of other features. Experiments on red jungle fowl are exceptions. Disease negatively influences the development of the rooster's comb more than that of the rooster's nonornamental, ordinary traits, strongly suggesting that the comb is an adaptation that functions to signal individual quality, specifically disease resistance and health in general (Zuk, Thornhill, Ligon, & Johnson, 1990). Do women's ornaments decline in attractiveness under increasing food reduction or deprivation with slope greater than that seen with functional declines in other systems (vascular, digestive, immune systems, etc.)? Anecdotally, cases of anorexia are consistent with this being so. But solid empirical work addressing this question is needed.

What Factors Influence Variation in Women's Quality and Condition?

If, indeed, women's ornamentation is affected by quality and condition, what factors, precisely, influence and account for variation in "quality" and "condition"? Nutritional status surely appears to be one. As we noted at the outset of this chapter, women who experience negative energy balance, low energy load, and high energy flux maintain lower levels of estrogen and hence fertility. These factors should hence also affect women's ornamentation.

Disease is almost certainly another. Infectious disease demands allocation of effort and energetic resources to immune function, taking away energy available for reproduction. Whether a history of infection may lead individuals to adaptively allocate more energy to immune function (in anticipation that, if infection has occurred in the past, pathogens are more likely to be encountered in the near future) and, in women, away from allocation to storage of reproductive fat is unknown (see McDade, 2003).

Some variation in ornamentation and fertility almost surely reflects genetic variation. In chapter 7, we discuss at greater length reasons why genetic variation in fitness may be maintained. Here, we simply note that mutation-selection balance and coevolutionary races are two major processes that may maintain heritable variation in fitness. Possibly, much variation in female ornamentation is due to genetic variation maintained by these processes. Sexually antagonistic selection, of both interlocus and intralocus varieties, may have particularly important implications for understanding female ornaments. We discuss these issues as they pertain to women's attractiveness further in chapter 7.

Ornaments as Indicators of Quality and Indicators of Age-Based Reproductive Value

Ancestral women who displayed ornaments, we presume, were favored by sexual selection because they gained male-delivered material benefits through signaling their quality and hence their reproductive value. In the scenario we have sketched, men were selected to benefit women who displayed their quality because these women, on average, had a longer and more successful reproductive career ahead of them. In this view, then, women competed against other women of similar age for material benefits.

As we have emphasized, however, women's ornaments also vary as a function of age and parity. Hence, any particular individual woman will typically display more attractive ornamentation at certain times during her life course (e.g., when entering the reproductive phase) than at other times (e.g., when 20 years older or after giving birth to five offspring). Ornaments, we have noted, display RV as a function not only of personal quality but also of age and number of existing offspring. This fact is surely convenient for men. A man choosing a mate for her RV need not necessarily be concerned about the precise reasons for a mate's RV (though, we note, heritable reasons for high RV may translate into advantages for offspring). Why, however, should women have evolved to signal quality with ornaments that are also sensitive to age- and parity-based RV? That is, why would selection have favored women who signaled with such ornaments? Why would selection not have favored women who displayed their quality with ornaments not terribly sensitive to aging and childbirth? (Indeed, it would seem that a woman who could display her qualities in this manner would be advantaged over one whose ornaments "decayed" over time; the former, but not the latter, may be valued by men even when well past her prime.)

The reason, we suspect, has much to do with the fact that ornaments sensitive to aging and parity *do* signal information that is useful to men. High-quality women could indeed win out by displaying a trait that accurately reflected their quality but did not decay over time—if men would evolve to pay attention to it. The problem, however, is that men are unlikely to evolve to pay attention to this ornament for a simple reason: They *cannot* trust it to honestly advertise RV. Men *can* trust ornaments that do reveal personal quality but also decay over time in ways that accurately reflect RV. Honest signals tend to win out over dishonest ones. It is probably not merely happenstance, then, that the ornaments that function to display women's personal quality also reflect age-based and parity-based RV. (See Marlowe, 1998, for a similar argument pertaining specifically to women's breasts.)

The question of what precise mechanisms lead women's ornaments to be affected by age and parity is not fully answered. Physiological trade-offs between attractiveness and other features, which change as a function of age and parity, combined with or arising from senescence, are probably involved.

Why Is There Greater Emphasis on Female Attractiveness in *Homo sapiens*?

In most species, only males have sexual ornaments (scents, behavior, or morphology) and only females choose based on evolved responses to these signals (their "attractiveness"; Andersson, 1994). Humans exhibit a strikingly different pattern: Female facial and bodily attractiveness generally appear to be more important than male physical attractiveness (or at least as important; Marlowe, 2005), particularly for long-term mate choice. *Homo sapiens* is what Gwynne (1991; see also Bonduriansky, 2001) refers to as a partially sex-role-reversed species. Both sexes exercise mate choice, as well as compete intrasexually for the opposite sex. But why do males typically value physical appearance in mates more than females do? This question has been posed in the literature numerous times without a fully satisfactory answer (Gottschall et al., 2005). Symons (1979) and Buss (1994, 2003b), for instance, posed and answered it by arguing that men value youth and women value status and resource holding. The perspective we offer in this book provides a more complete answer.

The first reason concerns the long-term nature of human pair bonds. Female appearance should be important to males in any female-ornamented species if ornaments function to compete for nongenetic material benefits provided by males. In contrast to most female-ornamented species, however, women's ornaments are enduring and affect their attractiveness to men throughout their reproductive life course, as well as throughout their ovulatory cycles.

A second reason has less to do with female attractiveness and more to do with men's attractiveness. Women do care about men's appearance. As we discuss in chapter 9, however, evidence reveals that women particularly value men's physical attractiveness during the fertile phase of their ovulatory cycles, and particularly when they evaluate men's "sexiness" as opposed to their value as long-term mates. They appear to care about physical attractiveness of mates less than men do when evaluating mates as long-term partners (as well as, perhaps, even when evaluating men's sexiness when infertile in their cycles). A reason that women weight men's attractiveness (or, at least, signals of quality) is, we suspect, the trade-offs women face when choosing long-term partners: all else being equal, men who possess phenotypic indicators of good condition and, we suggest, genetic quality are less willing to invest in relationships and offspring than men who lack these indicators, partly because these men *are* valued by women as sex partners (see chapters 7 and 9). Ironically, then, a reason for the sex difference in the extent to which physical attractiveness is valued in *long-term mates* partly derives from the fact that women highly value indicators of quality—including aspects of physical appearance—in sexual partners. Once again, it is also because of the important role of long-term pair bonding in human reproduction. Because women do find some male physical features particularly sexually attractive during the fertile phase of their cycles, however, the importance of men's physical attractiveness to women's mate choice is probably underestimated within the general literature.

Women's Ornaments Are Not for Extended Sexuality or Concealed Estrus

Both women's extended sexuality and their permanent ornaments, we have argued, were selected because they enhanced acquisition of male-delivered material benefits and services. Despite possessing similar functions, they are separate adaptations. Permanent ornaments are not adaptations of extended sexuality. Extended sexuality is the psychological and behavioral readiness to conditionally mate when, during the cycle or life course (e.g., when pregnant or during adolescence), fertility is at or near zero. Permanent ornaments are simply ornaments permanently displayed. The distinctiveness of these traits can be appreciated in comparative data. Extended female sexuality typically occurs in complete absence of female sexual ornaments. Vervet monkeys and some langurs possess no female sexual ornaments, for instance, but do exhibit extended female sexuality (see Andelman, 1987; Hrdy, 1981). And extended female sexuality is far more prevalent in pair-bonding bird species than is female ornamentation. Perennial female ornaments are rarely found and have not evolved to facilitate extended sexuality. They function in women to display RV, important because of humans' long-term pair bonding.

Women's perennial ornamentation similarly did not evolve because it concealed cycle-related high fertility. It is true that women are not fertile most of the time they are ornamented, but ornaments were not selected merely because they disguise cycle-related fertility. Instead, these ornaments' functional design strongly suggests them to be signals of phenotypic and genetic quality pertaining to RV. In chapter 11, we explicitly discuss adaptations designed to conceal cycle fertility, at which time we clarify and expand the arguments supporting our claims here.

Summary

Most of women's facial and bodily features that men find attractive are not merely by-products of estrogen that men have been sexually selected to attend to, given estrogen's association with fertility and reproductive health. Rather, they are condition-dependent ornaments that function to honestly signal residual reproductive value. The ratio of gynoid fat relative to android fat is an honest correlate of women's ability to expend future reproductive effort. Ancestral men accordingly preferred women possessing depots of gynoid fat, and these preferences exerted sexual selection on females to display exaggerated fat depots (e.g., in the breasts, buttocks, hips, and thighs) and, in turn, sexual selection on males to increasingly attend to and be sexually motivated by these depots. Put otherwise, these ornaments reflect the evolution of signaling systems. Women's gynoid-fat ornaments reach maximum size in adolescence, prior to their becoming highly fertile, and are permanently retained throughout their reproductive lives. This model of female ornaments—as signals of residual reproductive value—can be applied to other estrogen-facilitated traits in women. Adolescent exaggeration and permanence of women's ornaments were selected in the context of long-term pair bonding involving discriminative provisioning (partly based on quality of ornaments) of nongenetic material benefits to females by males. Possibly, kin, too, provide more benefits to attractive women than to less attractive women.

In the various nonhuman Old World primates in which adolescents display ornaments (sometimes in exaggerated form), female ornaments may be adaptations that function to obtain material benefits from males and, possibly, kin. In humans, adolescent females appear to have greater sexual interests than do adolescents in other Old World primates, which may be due to greater importance of male-delivered material benefits to female human ancestors, relative to other Old World primates, particularly early in the reproductive life course.

Women's attractive features tend to covary positively with each other, and evidence suggests relationships between a number of them and fertility, health, developmental stability, and longevity. Women's ornaments hence appear to be redundant signals of overall fitness, not signals with distinct messages. Empirical data and theory cast serious doubt on a variety of proposals that women's ornaments are primarily dishonest signals of fitness, fertility, or pregnancy.

The chief means by which women's estrogenization acts as an honest signal is likely the trade-off between reproductive effort and somatic effort. If women optimally allocate effort, those in better condition allocate more effort to reproduction than do women in poorer condition. The maintenance of the very substantial variation in the attractiveness of young women's ornaments probably reflects, in part, outcomes of mutation-selection balance and coevolutionary races, including sexually antagonistic selection (see chapter 7).

7   Good Genes and Mate Choice

An Overview of Good Genes and Mate Choice

In several chapters to this point, we have referred to "good genes" and "genetic quality." These topics play more important roles in chapters to come. This chapter explains the nature of good genes, considers the evolutionary processes that give rise to and maintain variation in genetic quality in natural populations, and discusses literature on potential markers of quality. In chapter 6, we argued that women's ornaments have evolved, through sexual selection, as signals of superior quality and condition, which, in part, may reflect genetic quality. We elaborate that argument here. We also discuss potential signals of genetic quality that men possess by virtue of evolution via sexual selection.

The processes that maintain genetic variation in fitness compose a broad topic within evolutionary biology. For present purposes, we are interested in specific implications of this genetic variation: implications for mate choice. In sexually reproducing species, individuals receive two kinds of resources from their mates that affect their own fecundity or that of offspring, and thereby their own fitness: DNA (which combines with mate choosers' own in offspring) and nongenetic material benefits (delivered either as a result of adaptation in mates to offer benefits or as by-products of mates' adaptations designed for functions other than to deliver benefits). Selection favors choice of mates that, all else equal, provide DNA that promotes offspring (and hence mate choosers' own) fitness. In his classic paper on sexual selection, Trivers (1972) introduced the term "good genes" to refer to choice of individuals who possess alleles that could benefit choosers' offspring.

Sexual selection theorists and researchers now distinguish three different types of genetic benefits relevant to mate choice (Jennions & Petrie, 2000):

First, mates can offer *intrinsic good genes*. Individuals who offer intrinsic good genes possess alleles that are associated, on average, with relatively high fitness (over several

144

generations or more). On a continuum of genetic variation in fitness, these individuals are at the high end of the distribution. Put otherwise, they possess alleles that are currently favored by directional selection. They hence can pass on alleles that provide fitness advantages to any mate chooser's offspring (and subsequent descendants). Meaningful variation in the extent to which mates offer intrinsic good genes implies meaningful *additive genetic variation in fitness* (i.e., some alleles are in fact favored by directional selection, and others disfavored by it). In the next section, we discuss how much genetic variation in fitness persists in natural populations and the processes that maintain it.

Second, mates can offer *compatible* (or complementary) *genes*. These individuals possess specific alleles that work well with the mate chooser's own alleles (either at the same locus or at different loci) to promote fitness in offspring but do not work well with all mate choosers' alleles (e.g., Zeh & Zeh, 2001). Variation in the extent to which mates offer compatible genes implies meaningful *nonadditive genetic variation in fitness*. That is, it implies that, at some influential genetic loci, heterozygotes possess mean fitness not equal to the mean fitness that homozygotes possess (e.g., heterozygotes possess higher fitness than homozygotes) or that there exist epistatic effects on fitness—nonadditive interactive effects on fitness involving allelic variation at different loci. In the third section of this chapter, we discuss candidate exemplars of these kinds of effects.

Third, mates can offer *diverse genes*. As a number of scholars have noted (e.g., Ellison, 1994), lineage extinctions are evolutionarily important events. A lineage extinction occurs when all of a focal individual's descendants (whether at the first generation or the *n*th generation) fail to reproduce. If the environments to which descendents are exposed are unpredictably variable in nature, individuals may enhance fitness by diversifying offspring, including through genetic diversification. (More to the relevant point, an allele that predisposed an individual to genetically diversify offspring could be favored by selection over multiple generations of descendents.) Diversification is a form of bet hedging. Given environmental uncertainty, one's descendents could get "lucky" (possess alleles favored in future environments) or "unlucky" (possess alleles disfavored by future environments). Genetic diversification is purportedly a means of reducing variance of the reproductive outcomes of descendents in the face of such uncertainty, which can be favored through avoidance of lineage extinction, even if the mean of those outcomes is not altered or slightly lower (e.g., Gillespie, 1977). In the fourth section of this chapter, we consider these and other arguments.

Maintenance of Genetic Variation in Fitness

How Much Genetic Variation in Fitness Persists in Natural Populations?

The Additive Genetic Coefficient of Variation   Intrinsic good genes, once again, are alleles favored by directional selection (or, in recent evolutionary history, have been favored). Collectively, their aggregate effects on fitness contribute to and account for additive genetic variation in fitness. One fundamental question concerning intrinsic good genes and their importance in mate choice is how much additive genetic variation in fitness exists in natural populations.

For any quantitative trait measured on a ratio scale, additive genetic variance can be quantified with the additive genetic coefficient of variation, or CV_A. (A ratio scale is a measure that has a meaningful zero point—a zero that implies no quantity of the trait— and units of measurement that linearly relate to quantities of the trait. Height measured in centimeters is a ratio-level scale.) The CV_A is the square root of the additive genetic variance (in a sense, the "additive genetic standard deviation") of the trait divided by the trait mean. Typically, this value is multiplied times 100, which converts a proportionate measure into a percentage measure. (The additive genetic variance is the variance in the trait associated with simple additive effects of alleles, aggregated across all loci at which allelic variation affects the trait. It does not reflect all genotypic effects on traits, for it does not reflect dominance [e.g., heterosis] and most epistatic effects. Because dominance and most epistatic effects are not heritable in a strict sense [effects in the parental genotype do not predict effects in the offspring genotype], they are not relevant to measuring genetic variation pertinent to mate choice for intrinsic good genes. See Fisher, 1930, and Falconer & Mackay, 1996.)

A simple example can illustrate the calculation and meaning of the CV_A. Suppose men's height averaged 70 inches in a population, with a standard deviation of 3 inches. Suppose further that the heritability of men's height in that population is 0.85. The trait variance, then, is 9, and the additive genetic variance is 7.65 ($9 \times .85$). The CV_A is the square root of 7.65 divided by 70, times 100: approximately 4. This means that the standard deviation in height due only to differences associated with allelic effects on height is 4% the mean of height. (Though we have chosen hypothetical values, men's height does appear to have a CV_A close to 4, at least in Western populations; see, e.g., Miller & Penke, 2007.)

The implications of the size of the CV_A for individual differences can be appreciated further in terms of a ratio of the values on the trait possessed by individuals near the top of the distribution (say, 2 standard deviations above the mean) and the values on the trait possessed by individuals near the bottom of the distribution (say, 2 standard deviations below the mean). In this instance, it would be approximately 1.17 (most readily calculated as $[100 + 2(4)]/[100 - 2(4)]$). That is, individuals who, by virtue of the independent effects of their alleles, are near the top of the distribution in terms of being predisposed to being tall are 17% taller than individuals who are near the bottom of the distribution in terms of being predisposed by their alleles to being tall. One can also say that those near the top of the distribution in being predisposed to being tall are 8% taller than the average person ($[100 + 2(4)]/100 = 1.08$).

What Is the Additive Genetic Variance of Fitness?   The size of the CV_A of fitness has major implications for the importance of selecting a mate for intrinsic good genes. If the CV_A of fitness is 4, similar to that of height, then a mate at the top 2% of the distribution in genetic fitness, paired with a randomly chosen individual, would produce offspring that are, on average, 4% more fit than offspring produced by two randomly chosen individuals or two individuals of average genetic fitness. (Here, "genetic fitness" is short-hand for variation in predisposition to high fitness by virtue of the additive effects of individuals' alleles. Because individuals contribute only half their alleles to an offspring, the mean fitness advantage of one's offspring due to additive genetic effects, when mates are individuals randomly chosen from the population, is only half that possessed by the individual him- or herself; see Falconer & Mackay, 1996.) That advantage might be enough to drive selection for adaptations for choice of mates who possess good genes—but perhaps not, if the costs of that mate choice (e.g., in currencies of waiting time or costs of resisting other suitors) are considerable.

Until the early 1990s, one major reason to question whether mate choice for intrinsic good genes could be profitable was serious doubt that the amount of genetic variance in fitness in natural populations could render choice for good genes profitable. Fisher's (1930) fundamental theorem of natural selection states that directional selection on a trait reduces its additive genetic variation. Fitness, by definition under directional selection, should thus have most of its additive genetic variation exhausted by selection, "which creates a serious difficulty for the good genes hypothesis" (Charlesworth, 1987, p. 22; for similar expressions of skepticism, see Maynard Smith, 1978; Partridge, 1983; Taylor & Williams, 1982). Naturally, if there is no additive genetic fitness variation in a pool of potential mates, there is no reason to choose one mate over any other for additive genetic benefits. And even if some small amount of genetic variation in fitness persists, the costs of good-genes mate choice may not exceed the benefits of obtaining a mate with relatively good genes.

Fitness itself is a trait difficult to measure. One way to estimate the genetic variance in fitness is to estimate the genetic variance of fitness-relevant traits (fitness compo-nents) purportedly under strong directional selection. Two major fitness components are fecundity and longevity. Fifteen years ago, David Houle (1992) first brought atten-tion to what was at the time a very surprising fact: Fitness components appear to actu-ally have CV_As that are substantially *larger*, not smaller, than those of many ordinary traits (e.g., morphological traits such as height). To draw this conclusion, Houle (1992) examined scores of studies in the literature, many on *Drosophila*. Whereas ordinary morphological traits or traits under stabilizing selection have CV_As less than 5, Houle found that fitness traits typically had CV_As of 10+. Sgro and Hoffman (1998) reported similar findings on *Drosophila*: Whereas fecundity had a mean CV_A of 20, wing length had a $CV_A < 2$ (see also Hughes, 1995). Using direct and indirect means of estimation based on observations of a number of different species, Burt (1995) estimated the CV_A of fitness itself to typically be 10–30. (For other relevant findings, see Gardner, Fowler, Barton, & Partridge, 2005.)

Overall, more recent longitudinal studies on natural populations of relatively long-lived organisms reveal similar patterns. In males, lifetime reproductive success (RS) has an estimated CV_A of 17, 16, and 32 in collared flycatchers (Merilä & Sheldon, 2000), great tits (McCleery et al., 2004), and red deer (Kruuk et al., 2000), respectively. (Merilä and Sheldon, 2000, noted that the value for flycatchers is probably an underestimate, as extra-pair paternity was not assessed in the study. But Brommer, Kirkpatrick, Qvarnström, & Gustafsson, 2007, reported a lower estimate.) Female RS in these same species was estimated to have a CV_A of 29, 6, and 0. Sampling variability renders each individual estimate very imprecise; the variation in the estimates may not be highly meaningful. The means across studies of 22 (for males) and 12 (for females)—considerably more stable—are substantial, however, and in the range anticipated by Houle (1992) and Burt (1995). In red deer, male body size is a strong predictor of reproductive success and also had a substantial CV_A (32; Kruuk et al., 2000). By contrast, ordinary morphological traits not predictive (in a linear fashion) of RS in these species (e.g., tarsus length and wing length in birds, jaw length in red deer) had low CV_A values (always < 4 and, on average, close to 2).

These values mean that there could be very substantial benefits to choice based on intrinsic good genes. If the CV_A of fitness is 20, mating with an individual two standard deviations greater than the mean of genetic fitness translates into 20% greater expected fitness of offspring, relative to mating with an individual at the mean of genetic fitness (all else equal). But, of course, for mate choice for good genes to actually work in this way, mate choosers need to be able to rely on a reasonably good *indicator* or *signal* of intrinsic *genetic* fitness (*not* merely overall fitness, which contains much variation due to nongenetic and nonadditive genetic sources) to identify individuals who possess good genes. Later, we discuss this critical issue in more detail.

Mutations

What evolutionary processes cause and maintain genetic variation in fitness? One major cause is mutation. Deleterious mutations in germ cells caused by DNA copying errors occur at each genetic locus at some small probability. The mean effects on fitness of mutations vary; whereas some deleterious mutations are lethal, most probably have very minor effects (no more than a few percentage points' effect on fitness in the heterozygous state; e.g., Eyre-Walker, Woolfit, & Phelps, 2006). Mutations that cause death prior to reproduction (or that prevent reproduction for any other reason), even in the heterozygous state, are eliminated by selection immediately. For all other deleterious mutations, there is a nonzero probability that they will pass through to the following generation, with the probability proportional to the mutation's mean harmful effect on fitness. Those mutated alleles with weak effects on fitness may persist for many generations, affecting many individuals, before being eliminated by selection. (Indeed, for mutations with very weak effects, there is some small probability that they will become fixed in the population through random drift, particularly when the size of the population is small; see Lynch et al., 1999.) In an idealized population at equilibrium, the rate at which negative effects on fitness due to mutation are eliminated by selection per generation is equal to the rate at which negative effects on fitness are introduced by fresh mutations per generation; the population is said to be in mutation-selection balance (e.g., Fisher, 1930). At equilibrium, there exist a certain number of mutations in the population (aggregated across all loci), which have a distribution of fitness effects. Not all individuals, however, have precisely the same number of mutations or mutations with precisely the same effects on fitness. (Naturally, nearly all mutations are ones several to many generations old, which individuals presently in the population inherited, not fresh ones that originated in the germ cells that produced current individuals; as Keller & Miller, 2006, put it, "Most mutations are a family legacy, not an individual foible"; p. 397.) The variation across individuals in the effects of the mutations they possess at equilibrium produces a characteristic standing genetic variation in fitness due to deleterious mutation (Fisher, 1930; Houle, 1992). Genetic variation in fitness can also be generated by positive selection for rare favorable mutations not yet at fixation, but, because favorable mutations are rare, they probably account for much less variation than do deleterious mutations (see, e.g., Smith & Eyre-Walker, 2002).

How much of the CV_A in fitness in natural populations is due to mutation-selection balance? This has turned out to be a question difficult to answer. Laboratory studies have produced variable answers. Charlesworth (1990) estimated that mutation could account for a CV_A in fitness in *Drosophila melanogaster* of 17. Charlesworth and Hughes's (2000) estimate for the same species was 8. These estimates rely on assumptions. Furthermore, the amount of fitness variation due to deleterious mutation could vary across species, or even across populations of the same species. It is thought that it is probably larger in vertebrates than nonvertebrates. Though the mean mutation rate per locus in humans appears to be only 60% of what it is in *Drosophila* (despite many more germline cell divisions per generation in human males; Drost & Lee, 1995; see also Crow, 1997), the number of amino-acid coding genes in the mammalian genome is 4–5 times larger. Hence the number of new deleterious mutations per genome per generation in humans is likely at least double what it is in *Drosophila* (see Lynch et al., 1999).

The genetic variance in fitness maintained by mutation-selection balance is partly a function of U, the deleterious mutation rate per diploid genome per generation, which is itself a function of the mean rate of mutation per locus and the total number of functional loci in the genome. Eyre-Walker and Keightley (1999) (using indirect methods comparing the genomes of humans with those of close phylogenetic relatives) estimated U to be 1.6 in amino acid coding regions of the human genome (which Eyre-Walker et al., 2006, later revised to be 1.8), a value identical to the minimum value Lynch, Latta, Hicks, and Giorgiana (1998) conjectured based on estimated mutation rates per genome per germline cell division. (Keightley, Lercher, and Eyre-Walker, 2005, note, however, that many mutations of relatively weak effect individually—but with possibly meaningful aggregate effects—have also accumulated in noncoding, regulatory gene regions of the human genome.) Nachman and Crowell (2000) estimated U to be at least 2.0 and possibly greater in humans (see also Kondrashov, 2001).

The genetic variance in fitness maintained by mutation-selection balance also partly depends on the distribution of fitness effects of those mutations and interactions between mutations' effects. Eyre-Walker et al. (2006) recently estimated that most mildly deleterious mutations in humans have historically had small effects on fitness, with a mean effect no more than a few percentage points. Their estimate of 4.3% negative effect, they note, may overestimate the true value by as much as threefold. (A value of about 2% would be close to estimates for other species; Lynch et al., 1999.)

If U is 2 and the mean effect of mutations on fitness has been 1–3% in humans, the population degrades in fitness by 2–6% per generation by fresh deleterious mutation alone. For a population at mutation-selection equilibrium (as humans may have been historically), with removal of mutations per generation having a net effect on fitness equal (but opposite in direction) to that owing to new mutations, selection against mutations would *increase* mean fitness 2–6% per generation (Burt, 1995). According to Fisher's fundamental theorem, the proportional increase in fitness in one generation due to selection (the evolvability or I_A of fitness) is equal to the additive genetic variance standardized by mean fitness squared (see Houle, 1992). The square root of this value times 100 gives the CV_A (Houle, 1992). In this instance, $\sqrt{.02} \times 100 = 15$ and $\sqrt{.06} \times 100 = 24$. That is, empirical estimates of U and the mean effect of mutation, together with idealized assumptions, yield an estimated CV_A of fitness due to mutation alone of approximately 20 ± 5—values similar to what Burt (1995) estimated for the CV_A of fitness (due to all causes) itself.

This estimate does rely on assumptions. More precise estimates await further research (see, for instance, Gardner et al., 2005; Kondrashov, 2001). Even if the true CV_A of fitness due to variation in effects of mutations on fitness alone were only half of this value (10), however, it would be substantial. In all likelihood, a considerable proportion of the genetic variance in fitness in human populations (as well as, most likely, natural populations of other sexual species) has historically been due to variation in mutational effects on fitness across individuals (see, for instance, Lynch et al., 1999).

In light of selection on fitness, how does mutation produce and maintain so much variation in fitness? Houle et al. (1996) argue that the amount of variation generated and maintained in a trait by mutation alone is predicted by its mutational target size, the number of loci at which mutations could affect the trait (see also Houle, 1998). The mutational target size of fitness itself is very large—effectively, the entire functional genome. Though allelic variation due to deleterious mutation at any particular locus has a very small effect on fitness, the cumulative effects, aggregated over many loci at which mutations have effects, add up to a substantial total effect on fitness.

Host-Pathogen Coevolution

As noted earlier, selection, absent any countervailing process, would eliminate all additive genetic variation in fitness (Fisher, 1930). Deleterious mutation (changes in the genome that cause lack of adaptation of individuals to their environments) is one countervailing process that prevents the complete elimination of genetic fitness variation. Rapid and recurring change in the selective environment (which similarly causes individuals to become less well adapted to their environments) is another. If the direction of selection on allelic variation at an individual locus changes at a sufficiently rapid and recurring rate, at least some meaningful proportion of the time no one allele will be fixed (or nearly fixed) in the population at that locus; allelic variation translating into fitness variation will often be found. If many loci are affected by rapid and recurring changes, the sum total effect on genetic fitness variation could be considerable (e.g., Eshel & Hamilton, 1984). In recent investigations examining the relative fitnesses of wild-type chromosomes in *Drosophila*, Gardner et al. (2005) found very substantial fitness differences. Perhaps even more notably, however, they found that, through time, patterns of selection for individual chromosomes systematically varied. These patterns imply meaningful gene × (temporally variable) environment interactions. Fluctuating selection, they suggested, may maintain a meaningful proportion of the genetic variation they observed. Some theorists strongly suspect that processes other than simply mutation-selection balance maintain meaningful variation; contrary to expectation of the mutation-selection model that deleterious alleles will be represented in rare frequencies, many polymorphisms found in nature involve multiple alleles of at least intermediate frequency (though their effects on fitness is typically unknown; see Turelli & Barton, 2004, and references therein).

Hamilton and Zuk (1982; Hamilton, 1980, 1982) famously proposed that antagonistic coevolution of hosts and pathogens entails that both hosts and pathogens are sources of relatively rapid changes in the selective environments of the other. That is, evolution of pathogens, in response to host adaptation to them, entails changing selection pressures on hosts. And evolution of hosts, in response to pathogen adaptation to them, entails changing selection pressures on pathogens. (More generally, see also Van Valen, 1973, on Red Queen processes, which he named after the character in *Alice in Wonderland* who needed to keep running simply to stay in the same place.) Recurrent fluctuating selection on host organisms could maintain genetic variation in fitness. Based on simulations, Eshel and Hamilton (1984) proposed that coevolution with pathogens possessing intermediate intergeneration intervals—such as human macroparasites—can, in theory, best maintain heritable fitness variation in hosts. (Hamilton and Zuk, 1982, also claimed that negative frequency-dependent selection—selection tending to favor rare alleles—operates and plays a critical role in maintaining allelic variation in this particular Red Queen process.)

In theory, any host allele coding for a protein to which a pathogen could adapt could be subject to fluctuating selection (e.g., Tooby, 1982). In practice, selection on host alleles coding for components of immune defense seems the most likely candidate to be "yanked around" by coevolution with pathogens. Major histocompatibility complex (MHC) alleles code for cell-surface markers that components of the immune system use to detect self- and non-self-peptides. At particular points in time, individual alleles may be beneficial in defense against particular pathogens (e.g., Lohm et al., 2002). Temporally varying dynamics of host and pathogen populations could lead to changes in the prevalence of individual MHC alleles, as well as maintenance of MHC genetic variation, particularly in concert with frequency-dependent selection (Hedrick, 2002; though, as we discuss later, heterozygote superiority may well importantly contribute to the maintenance of MHC diversity in humans as well; see Hedrick, 1998; Black & Hedrick, 1997; see also Geise & Hedrick, 2003). Other components of the immune system could similarly be involved in host-pathogen coevolution.

Intragenomic Coevolutionary Processes

Some of the most important coevolutionary processes maintaining genetic variation in fitness may involve antagonistic coevolution at different loci within a single species—what Rice and Holland (1997) refer to as interlocus contest evolution, or the intraspecific Red Queen.

An example of an intraspecific Red Queen process is maternal-fetal coevolution (see Rice & Holland, 1997, for discussion of other instances). Fetal genes maximally benefit from a greater flow of nutrients from the mother than the level maximizing maternal fitness (Haig, 1993; Trivers, 1974). Hence, a newly arising allele that, when expressed in fetuses, increases the flow of nutrients from mothers to the fetus may be selected and spread. And a newly arising allele at a different locus that, when expressed in mothers, decreases flow of nutrients (e.g., because it undermines a new fetal adaptation) may be selected and spread. At any point in time, most loci involved in the maternal-fetal conflict may be monomorphic (with one allele spread to fixation). But a nonnegligible proportion of them could be in a state of transition in which a recently and currently favored allele has not yet fully spread to fixation. If so, some mothers are expected to be better adapted to the conflict than others. And similarly, some fetuses should be better adapted to the conflict than others. In sum, recurrent genetic variation in fitness could be an outcome of maternal-fetal coevolution.

The intraspecific Red Queen process that has received the most attention to date is sexually antagonistic coevolution (e.g., Arnqvist & Rowe, 2005). In a sexually reproducing species, sexual conflicts of interest typically exist. That is, what gives a male advantage in competition with other males may actually decrease the fitness of his female mates relative to other females, and what gives a female an edge in competition with other females may disadvantage her male mates. Hence males may evolve adaptations at the expense of their female mates. In turn, females may evolve counteradaptations at the expense of male mates. Loci at which alleles code for components of male sexually antagonistic adaptations coevolve with loci at which alleles code for components of female sexually antagonistic coevolution (and vice versa). Just as in the scenario involving maternal-fetal coevolution, even if most loci are monomorphic at any point in time, a meaningful proportion may be in a state of transition in which a recently and currently favored allele has not yet spread to fixation. The resulting variation implies within-sex genetic variation in fitness.

In *Drosophila*, as in many species, males and females have conflicting interests over the mating rate. Males can benefit from inducing a female to mate when it is not in the interests of the female to do so (e.g., when she has already mated or mated with a male of higher quality). Female adaptations to resist male attempts may fuel selection for counteradaptation in males, leading to selection for counter-counteradaptation in females, and so on. Rice (1996) demonstrated this antagonistic process in a laboratory experiment on *Drosophila* in which a female line was not permitted to evolve, but a male line could evolve in response to females. After 30 generations, males in this line were better able to induce females to remate and, partly as a result, outreproduced males drawn from the original control stock when both types competed for the same females. The fitness of females, however, was lower when they interacted with these males compared with when they interacted with control males (see also Lew & Rice, 2005). Males had evolved adaptations antagonistic to female interests, to which females were prevented from evolving counteradaptations.

Subsequently, Lew, Morrow, and Rice (2006) and Linder and Rice (2005) estimated the standing genetic variance in female ability to resist male courtship. It was significant and, on average across studies, accounted for a remarkable 40% of the genetic variance in female fitness under normal mating conditions. Other studies provide evidence that male abilities to induce females to mate and effectively supplant stored sperm (offensive tactics), as well as enforcing fidelity of mates and preventing displacement of their own sperm (defensive tactics), are also heritable (e.g., Friberg, Lew, Byrne, & Rice, 2005). Some (perhaps much) of the genetic variation in traits involved in antagonistic interactions between males and females and contributing to fitness is probably due to variants (e.g., mutations) affecting overall vigor (see Kokko et al., 2003). The fact that portions of the genetic variance in males' offensive and defensive tactics were unique, however, suggests that these traits are affected by some genetic loci evolving independently of overall vigor, possibly through sexually antagonistic coevolution (Friberg et al., 2005).

Genetic variation at a locus that results from persistent sexually antagonistic coevolution may affect the fitness of one sex only: that sex whose sexually antagonistic tactics are affected by variation at the locus. Some heritable fitness variation, then, is expected to be sex-specific—transmitted from a parent to same-sex offspring but not to offspring of the other sex.

In chapters 9–12, we offer hypotheses about ways in which men and women have been involved in sexually antagonistic arms races. Possibly, a meaningful amount of genetic variation in male and female fitness (or fitness in ancestral environments) was produced as an outcome of interlocus sexual conflict. How much variation that might be is unknown at this time.

We have described interlocus sexual conflict: coevolution of alleles at *different* loci within an organism's genome fueled by conflict between the sexes. Sexually antagonistic selection can also operate at a *single* locus. In this instance, selection operates on allelic variants in contrary ways when inhabited by the two sexes. That is, whereas one allele is favored over another when carried in males, another allele at the same locus is favored when carried in females. Selection can lead to stable equilibria maintaining variation. Alternatively, one allele may spread due to stronger selection on one sex but impose fitness costs on the other sex carrying it (e.g., Chippendale, Gibson, & Rice, 2001; Rice & Chippendale, 2001). Both sexes may be pulled away from their phenotypic optimum as a result—but, within the sexes, to variable degrees. On traits affected by a single locus at which sexually antagonistic selection operates, "masculine" males may be more fit than "feminine" males, and "feminine" females may be more fit than "masculine" females. For example, variation in prenatal androgen exposure and resultant variation in digit ratios within both sexes (see chapter 6) may be maintained partly by intralocus sexually antagonistic selection (Manning et al. 2000; see also Saino, Leoni, & Romano, 2006). In chapter 6, we suggested that interlocus and intralocus sexual conflicts may importantly cause the maintenance of genetic variation in human ornamentation.

Sexually antagonistic genes cause genetic variation in fitness within the sexes. Because sexually antagonistic alleles are favored in one sex but not the other, however, these genes may not strongly covary with overall heritable fitness: Parents whose sons benefit from the alleles produce daughters that are harmed by them, and vice versa.

The presence of sexually antagonistic genes can favor alleles that affect the sex ratio of offspring, such that masculine parents produce relatively more males, and feminine parents produce relatively more females (e.g., Rice & Chippendale, 2001). Evidence indicates that people relatively masculine in certain ways do have more sons and relatively feminine people have more daughters (see Kanazawa & Vandermassen, 2006). An alternative adaptive explanation for sex ratio of offspring effects is that the sex ratio is biased in favor of males when the father has indicators of good genes. In species in which reproductive skew is greater for males than for females (e.g., humans), males are benefited by alleles promoting general fitness more than are females. Hence, men who, in the Kinsey study on sexual behavior, reported that they had a large number of premarital sex partners (perhaps reflecting their attractiveness to females) tended to have more sons than those reporting that they had few premarital partners (Gangestad & Simpson, 1990). (For evidence of similar adaptive sex-ratio alterations in zebra finches, see Burley, 1986; in collared flycatchers, see Ellegren, Gustafsson, & Sheldon, 1996; in reindeer, see Røed et al., 2007. But a meta-analysis on birds yielded no mean effect; Ewen, Cassey, & Møller, 2004.) Future studies may be able to tease apart the alternative explanations.

Because intralocus sexually antagonistic selection can result in genetic differences affecting fitness in opposite directions for these sexes, these differences are thought to possibly obscure mate choosers' abilities to pick out mates who possess intrinsic good genes for both sexes. (A male who appears to have good genes offers genes benefiting sons but not daughters.) Large amounts of variation in fitness due to sexually antagonistic genes, then, may interfere with selection for mate choice for good genes (Pischedda & Chippendale, 2006).

Sexually antagonistic loci drive the genetic correlation of fitness for the two sexes (the correlation between the genetic differences in each sex affecting reproductive success) to be negative. Several recent studies have estimated these correlations. Chippendale et al. (2001) estimated a genetic correlation of −0.16 for *Drosophila melanogaster*, Foerster et al. (2007) estimated it to be −0.48 for red deer, and Brommer et al. (2007) found a correlation of −0.85 for collared flycatchers. In red deer, male variance in fitness is much greater than female variance in reproductive success, and, hence, it pays females to choose fit males despite the cost of doing so to female offspring. In collared flycatchers, however, Brommer et al.'s (2007) estimates provide little evidence for good-genes sexual selection.

The degree to which the genetic fitnesses of the sexes are correlated—positively or negatively—in traditional humans remains unknown.

Mate Choice and Intrinsic Good Genes

Adaptive mate choice for intrinsic good genes requires additive genetic variance in fitness. Houle's (1992) and Burt's (1995) broad generalization that fitness components and, presumably, fitness itself tend to have large additive genetic coefficients of variation in animal populations is almost certainly true, even if occasional exceptions exist. Less well understood are the relative contributions of the evolutionary processes that maintain heritable fitness variation. Mutation-selection balance probably accounts for a substantial amount of this variation, but how much (30%, 60%, 90%?) is unknown and probably varies across species.

Some comparative analyses indicate a role for mutations driving mate choice for good genes. As it happens, male gametes typically have higher rates of fresh mutations due to the greater number of cell divisions in sperm than in eggs (see Miyata, Hayashida, Kuma, Mitsuyasu, & Yasunaga, 1987), which yields stronger sexual selection for good genes in males than in females. In fact, sexual selection on males through female choice appears to be stronger in species with strong male mutational biases (Bartosch-Härlid, Berlin, Smith, Møller, & Ellegren, 2003). And in pair-bonding birds, the extra-pair paternity rate is predicted by the mutation rate (Møller & Cuervo, 2004). As we discuss in chapter 10, extra-pair copulation may be one means by which females in pair-bonded species exert sire choice favoring good genes. (See Petrie and Roberts, 2007, however, for a model that suggests that strong sexual selection can increase the mutation rate, rather than vice versa.)

Antagonistic coevolutionary processes probably fuel meaningful variation as well, at least in some species. How much they have contributed to fitness variation during hominin evolution is not known. Nor are the relative contributions of interspecific (e.g., host-pathogen) and intraspecific (e.g., mother-fetus, male-female) conflicts of interest that fuel antagonistic coevolution. Other processes we have not discussed (e.g., other forms of antagonistic coevolution, spatially variable selection, epistatic variance that may be released and expressed as additive genetic variance during periods of rapid change [see Pomiankowski & Møller, 1995]) may play some role as well.

The big picture we deal with in this chapter is mate choice for good genes. The most relevant question in that regard is how much genetic variation in fitness exists, not what causes it. And in that regard, the received view today is very different from what it was 20 years ago. Large reservoirs of genetic variance in fitness typically persist in natural populations. That is not to say that mate choice for good genes will evolve in every species—not even every species in which large amounts of genetic variance in fitness exist. Byrne and Rice (2006) found evidence that males of a laboratory population of *Drosophila melanogaster* prefer mating with larger, more fecund females. In the same population, however, genetic benefits of remating with a high-quality male could not pay for the direct costs of remating (Orteiza, Linder, & Rice, 2005). And, hence, an allele that increased female resistance to remating at a cost of missing out on genetic benefits garnered through remating was favored (Stewart, Morrow, & Rice, 2005). Females in this population may still favor high-quality males as first mates. But there is no evidence that female remating tactics are effectively designed to obtain good genes; remating behaviors appear to be driven by male and female antagonistic adaptations (see also Byrne & Rice, 2005). In light of the genetic variation in fitness, however, it should not be surprising that good-genes sexual selection occurs in many species.

The Sexual Selection Continuum

The Viability-Indicator Model Versus the Fisherian Model   The goodness of good genes, once again, refers to their positive influence on offspring reproductive capacity. Intrinsic good genes have this effect independent of the mate chooser's own genes; their effects are additive.

For most of the past several decades, two different good-genes models were thought to be separate. The viability-indicator model (see Andersson, 1994) held that mate choosers prefer mates who possess indicators of viability. These individuals live longer than other potential mates (e.g., they specifically have greater juvenile and/or adult survival, as well as correlates such as better phenotypic condition or resistance to disease). Because viability is partially heritable, these mates' offspring also live longer. All else being equal (e.g., equal age-specific fecundities), long lives promote reproductive success.

The second good-genes model was the Fisherian model. This model proposes that individuals prefer mates who are attractive but live no longer than others (or may die even earlier than others). Because their offspring are also attractive, however, their offspring have greater reproductive success. Particularly when males are the highly ornamented sex and have more variable reproductive success, the effect of a male mate's attractiveness on his son's attractiveness and reproductive success is particularly strong (e.g., Kirkpatrick, 1982; Lande, 1981). Hence this model is sometimes referred to as the "sexy son" hypothesis (Fisher, 1930). The Fisherian model, like the viability-indicator model, is a good-genes model. Because this model does not assume that preferred mates or their offspring have any greater ability to survive than other potential mates and their offspring but, rather, have greater reproductive success merely because they are "attractive," this model has sometimes been posed as one that presumes that preferred mates are not "intrinsically" better than other mates. This has particularly been true when the preferred trait is presumed to be an "arbitrary trait" that is first preferred solely because of a sensory bias (through a process we described in chapter 5). Individuals preferred by a sensory bias are no better than any others in terms of fitness initially and only become "better" because of spread of genes for the arbitrary preference, which occurs due to linkage with genes for the preferred trait (see, e.g., Andersson, 1994).

As we described in chapter 5, a trait initially preferred because of a sensory bias can also ultimately become a viability indicator. As the trait preferred due to sensory bias becomes exaggerated over time, it becomes increasingly dependent on overall condition. Individuals in the best condition can afford to "pull off" growing ornamentation that is most exaggerated. In Rowe and Houle's (1996) terms, as the trait becomes increasingly dependent on condition, it "captures" the genetic variance in condition and, thereby, genetic variance in fitness. What was initially a trait preferred solely because of a sensory bias has, over time, become a viability indicator.

The Viability-Indicator Model and the Fisherian Model Merely Define Two Ends of a Continuum of the Same Sexual Selection Model   Kokko, Brooks, McNamara, & Houston (2002) recently offered an important insight about the viability-indicator and the Fisherian models. They are not two distinct models. Rather, they are variations on the same fundamental model. Both are intrinsic good-genes models, and equally so. They differ only in their assumptions about what an individual who possesses good genes does with the advantage: being particularly good only at attracting mates or being particularly good at surviving, as well.

Selection will shape individuals' allocation of resources (e.g., energy, time) to fitness-promoting activities (e.g., growth, somatic repair, immune function, developing ornaments) in ways that maximize their fitness. Two individuals who have different quality presumably differ in the resources available to them for fitness-promoting activities (or the efficiency with which they can allocate those resources to fitness-promoting activities). And, as we discussed in chapter 5, an individual of good quality may optimally allocate those resources in ways different from the way an individual of poor quality optimally allocates its resources.

The viability-indicator model presumes that individuals allocate resources in a way such that an individual of good quality survives, on average, better than does an individual of poor quality. (For instance, the former may dedicate greater resources to immune function or somatic repair.) Because high-quality individuals are more resourceful, they presumably can simultaneously allocate more resources to developing traits that attract mates (e.g., ornaments), as well.

The Fisherian model presumes that individuals allocate resources in a way such that an individual of good quality survives, on average, *no* better than does an individual of poor quality (or, in fact, may actually survive *less* well). That is, the former may dedicate *fewer* resources to immune function or somatic repair than does the latter. (A high-quality individual may, then, be *more* disease-prone and *less* healthy than a poor-quality individual; see Getty, 2002). Obviously, for high-quality individuals to truly be high quality, in this scenario they must allocate more resources to developing traits useful for attracting mates and, to offset the fact that they survive less well, be much better at attracting mates than low-quality individuals.

What circumstances would lead individuals of high quality to put so much of their resources into competing for mates, such that they actually die earlier, on average, than individuals of poor quality? In theory, these are circumstances in which mate choosers particularly value good genes and pay minimal costs (e.g., in search time) for maintaining very high standards for mate choice (e.g., leks). In short, species for which the Fisherian pattern fits are ones in which sexual selection (typically on one sex's attractiveness only) is extreme (see Kokko et al., 2002, 2003).

The viability-indicator model and the Fisherian model anchor two ends of a continuum. At one end of the continuum, individuals of high quality (and good genes—hence also the offspring of high-quality individuals) are much better at surviving, and fairly minimally better at attracting mates, than low-quality individuals. At the other end, individuals of high quality (and good genes—hence also their offspring) are much better at attracting mates, and even worse at surviving, than low-quality individuals. But these are merely two ends of a continuum; the actual pattern of how good genes translate into fitness (and the fitness of offspring) within a particular species can fall anywhere along the continuum (Kokko et al., 2002).

Where species actually tend to fall on the sexual selection continuum is an empirical question. Jennions, Møller, and Petrie (2001) conducted a meta-analysis of all animal studies, reporting a correlation between a sexually selected male trait and male survival. On average, more highly ornamented males survive better than less ornamented males. The mean across the species studied, then, falls toward the viability-indicator end of the continuum. That result, of course, does not rule out some species clearly falling at the sexy-son end of the continuum. Relatedly, Møller and Alatalo (1999) performed a meta-analysis of the literature of animal studies examining an association between a sexually selected male trait and offspring survival. Consistent with Jennions et al.'s (2001) analysis, they found a mean positive relationship between male ornamentation and offspring survival: In general, more ornamented males sire offspring with greater survival than do less ornamented males (and, in fact, this association is similar in size to that between male ornamentation and degree of male care; Møller & Jennions, 2001). (For other work on ornamentation of sires and health or overall phenotypic condition, see, e.g., Johnson, Thornhill, Ligon, & Zuk, 1993; Zuk, Thornhill, Ligon, & Johnson, 1990; Zuk, Thornhill, Ligon, Johnson, Austad, et al., 1990.)

These meta-analyses examined only male ornaments. In theory, because females are typically under less intense sexual selection than are males, attractive females should be better not only at attracting mates but also at producing high-viability offspring and, perhaps, at surviving (i.e., female ornaments of quality should typically be explained by a model at the viability-indicator end of the sexual selection continuum; see chapters 5 and 6).

Genetic Compatibility

Once again, adaptive mate choice for intrinsic good genes requires additive genetic variance in fitness. By contrast, adaptive mate choice for compatible genes requires nonadditive genetic variance in fitness (see Neff & Pitcher, 2005). Choice for compatible genes involves choosing a mate to create a combination of maternally and paternally derived alleles in offspring that promote fitness independent of the individual fitness effects of the alleles (their additive effects). Nonadditive genetic variance due to fitness effects of favored *combinations* of alleles can come in two forms. Dominance deviations are unique effects associated with combinations of two alleles at single loci (e.g., heterozygote superiority). Epistatic effects are nonadditive effects associated with combinations of two or more alleles at different loci (e.g., Falconer & Mackay, 1996).

The MHC has been investigated in multiple species, including humans, as a possible target of mate choice for compatible genes. The MHC exists in all vertebrates. Loci within the MHC are often polymorphic, sometimes extremely so. In humans, for instance, three loci (the A, B, and DRβ loci) are highly polymorphic, with up to hundreds of different possible alleles at an individual locus and, typically, no allele being more than 20% of the alleles at a locus in a population. MHC alleles are codominantly expressed, and hence heterozygotes can potentially present a wider range of foreign peptides to T-cells and thereby effectively defend against a broader array of pathogen strains. Studies on chinook salmon (Arkush et al., 2002) and mice (McClelland, Penn, & Potts, 2003; Penn, Damjanovich, & Potts, 2002) found evidence for MHC heterozygote superiority (but see Lohm et al., 2002). In humans, MHC heterozygote superiority has been found in resistance to hepatitis B infection (Thurz, Thomas, Greenwood, & Hill, 1997). Furthermore, human couples that possess a common MHC allele produce an underrepresentation of homozygotic offspring, indicating in utero selection against homozygotes (see, e.g., Hedrick & Black, 1997b). (One-quarter of the offspring of couples who share one allele at a MHC locus should be homozygotic at that locus. In fact, less than one-quarter are homozygotes.) Their conceptions more frequently end in spontaneous abortions, as well (Ober, Hyslop, Elias, Weitkamp, & Hauck, 1998). Heterozygote superiority is probably one reason why MHC loci are highly polymorphic (Hedrick, 1998). (For one study showing that pathogen diversity leads to MHC diversity in humans, see Wegner, Reusch, & Kalbe, 2003.)

To choose a mate with whom an individual will produce offspring that are MHC heterozygotes—a mate that possesses compatible MHC alleles—the individual should choose one who shares no (or few) MHC alleles. Mice can detect MHC identities in scents of other mice (Yamazaki, Beauchamp, Curran, Baird, & Boyse, 2000). Based on scent, they prefer mates who possess dissimilar MHC genotypes (Penn & Potts, 1999). Preferences for MHC dissimilarity (or other forms of self-referenced preferences; see, e.g., Milinski et al., 2005) also exist in species of birds (Freeman-Gallant, Meguerdichian, Wheelwright, & Sollecito, 2003), fish (Aeschliman, Haberli, Reusch, Boehm, & Milinski, 2003; Milinski et al., 2005), and lizards (Olsson et al., 2003), though not all species possess them (e.g., Sommer, 2005).

Studies examining MHC preferences in humans have yielded generally supportive but mixed evidence. Preferences for the scent of opposite-sex individuals with dissimilar MHC genotypes have been detected in three of four studies of normally ovulating women (Santos, Schinemann, Gabardo, & Bicalho, 2005; Wedekind & Füri, 1997; Wedekind, Seebeck, Bettens, & Paepke, 1995; cf. Thornhill et al., 2003) and two of three studies of men (Thornhill et al., 2003; Wedekind & Füri, 1997; cf. Santos et al., 2005). (In another study, women preferred the scent of MHC-similar men, but its preference measure may not tap sexual attraction; Jacob, McClintock, Zelano, & Ober, 2002).

A study of Hutterites found married couples to be more MHC-dissimilar than expected by chance (Ober et al., 1997); studies of South American Indian and Japanese couples did not (Hedrick & Black, 1997a; Ihara, Aoki, Tokumaga, Takahashi, & Juji, 2000). A study of romantically involved couples in the United States detected disassortative mating at MHC class-I loci but not at a class-II locus (Garver-Apgar, Gangestad, Thornhill, Miller, & Olp, 2006). (Class I alleles are expressed on nearly all cells and function in resistance to intracellular infection. Class II alleles are expressed only on leukocytes and function in resistance to extracellular infection. Because they are expressed on skin cells, Class I alleles may be more readily detected through scent; see Leinders-Zufall et al., 2004.)

Despite the substantial amount of research done on MHC, it is not the most thoroughly documented case of choice for compatible genes in humans; incest avoidance is. Avoidance of inbreeding is beneficial not because siblings (and other close relatives) possess poor genes. On average, of course, individuals who are opposite-sex siblings to at least one other person have fitness close to the mean in the population. (Because their parents had at least two offspring of different sexes, they may even have slightly higher than mean fitness.) Siblings are poor mate choices because of nonadditive genetic effects expressed in offspring produced by siblings who mate with one another. A sibling may have about the same number of mutations, on average, as a randomly chosen individual but, relative to two randomly chosen individuals, two siblings are much more likely to have mutations at the same locus. In the heterozygous state, mutations at coding sites typically have nonzero deleterious effects on fitness (i.e., mutations are not completely recessive; as noted earlier, the deleterious effect on fitness averages a few percentage points). But two mutations at the same locus typically have a joint effect on fitness much larger than double their effect in a heterozygous state (perhaps, on average, five times the effect; Lynch et al., 1999). Put otherwise, there are large nonadditive components to the effects of mutations. These nonadditive genetic effects drive selection for adaptations that lead individuals to avoid mating with closely related individuals. Adaptations that function to avoid incestuous mating, then, are classic instances of adaptations for mate choice to acquire compatible genes for offspring. (For research on the cues that people use to discriminate siblings from nonsiblings and thereby avoid sex with individuals likely to be siblings, as well as sex differences in use of these cues, see Lieberman et al., 2003, 2007.) Whether humans have adaptation to avoid inbreeding by refraining from mating with individuals more distantly related than first-degree relatives is unknown. Preference for MHC dissimilarity may function as a mechanism of inbreeding avoidance rather than to increase heterozygosity per se (e.g., Penn & Potts, 1999).

Aside from preferences for dissimilar MHC alleles and incest avoidance, we know of no other well-documented adaptations for mate choice for compatible genes in humans. As Zeh and Zeh (1996, 2001) note, both interspecific and intraspecific antagonistic coevolutionary processes create genetic incompatibilities that may be targets of adaptive mate choice (or, as it may be, mate avoidance). For instance, a mother who can effectively suppress a fetus's attempts to extract maternal resources may actually be best off selecting a mate who provides a fetus with capabilities of keeping its own in a tug-of-war with its mother. (See Zeh & Zeh, 1996, 2001, and Jennions & Petrie, 2000, for discussions of other possible nonadditive genetic effects fueling compatible gene choice.)

As Zeh and Zeh (1997) also argue, however, adaptations that function for compatible gene choice may often operate postcopulation in sperm or zygote selection. Individuals typically are not selected to advertise traits related to compatible genes; as Jennions and Petrie (2000) note, males have little reason to advertise features that females might find compatible, as males may be selected to mate as widely as possible. (Whether that should be true in a system involving biparental care and mutual mate choice is unclear.) Indeed, MHC recognition is probably achieved through adaptation that detects incidental by-products of MHC identities (e.g., self-peptides shed by skin cells; e.g., Leinders-Zufall et al., 2004), not a system involving targets' adaptation to signal MHC identities. Many compatible gene effects may pertain to single loci that, unlike MHC identities, do not produce by-products that perceivers can detect. By contrast, postcopulatory mechanisms may be particularly effective at executing female choice for compatible genes; females may evolve ways to assess DNA's "identities" (and hence compatibilities) once sperm is in the female reproductive tract, which is simply not possible before copulation. As well, a conceptus that is not highly fit due to incompatibilities may be selectively aborted (Haig, 1991). The underrepresentation of MHC homozygotes in the offspring of human couples who share MHC alleles (see previous discussion; Hedrick & Black, 1997b) may be the result of postcopulatory adaptation to select for compatible genes. (For reviews of postcopulatory mechanisms of choice for compatible genes, see Jennions & Petrie, 2000; Simmons, 2005.)

Zeh and Zeh (1997, 2001) suggest that females in some species possess adaptation to mate polyandrously, which functions to allow postcopulatory mechanisms to select compatible genes among different males' sperm. In chapter 10, we discuss the possibility that women mate polyandrously to achieve these or other effects garnered by "running" sperm competitions. Though we are skeptical that, in humans, the benefits of polyandrous mating to achieve sperm selection have historically outweighed the costs, the possibility that women possess adaptation to do so under some conditions cannot be ruled out.

One possible additional form of mate choice for compatible genes in humans could be driven by intralocus sexually antagonistic selection. As noted earlier, this form of selection leads each sex to be compromised by what is adaptive and selected for in the other sex. Individuals differ, however, in the extent to which they are moved away from the optimum phenotype for their sex by this form of selection. If male reproductive success is skewed, male "masculinity" may relate to reproductive success in a concave-upward curvilinear fashion. A female who is highly feminized and one who is relatively masculine may produce sons who differentially benefit from receiving genes from a masculine sire. The fitness of the son of the feminized female will be compromised by his mother's feminizing genes. The son of the masculine female will not be. Because masculinity relates to fitness in a rising, curvilinear fashion, the benefit of receiving masculine genes from a father is greater for the son of the masculine female. Hence, perhaps, females masculinized by sexually antagonistic genes adaptively prefer more masculine men, as a function of adaptive choice for compatible genes.

The ratio of the lengths of the 2nd and 4th digits, once again, may be affected by sexually antagonistic genes (Manning et al., 2000). Relatively low digit ratios are masculine, whereas relatively high digit ratios are feminine. Scarbrough and Johnston (2005) asked women to rate the attractiveness of male faces, which were manipulated to vary in their masculinity. In general, women's choice of the man most attractive for a short-term relationship (a sex partner) was more masculine than the man rated most attractive for a long-term relationship. The man most preferred as a long-term partner by women with masculine digit ratios, however, was more masculine than the man most preferred as a long-term partner by women with feminine digit ratios—consistent with the compatible genes hypothesis we outlined. The question of whether these differences in women's preferences truly do reflect adaptation for choice of a mate with compatible genes requires further study.

Diverse Genes

As noted previously, genetic diversification of offspring has traditionally been thought of as a form of bet hedging in the face of future uncertainties: reducing variance in outcomes, even at a cost of mean outcomes (e.g., Gillespie, 1977). This model of the selective advantages of genetic diversification has been questioned (Yasui, 1998). In a nutshell, the argument against this model of selection is that a gene that promotes diversification within families at a cost (e.g., costs of polyandry) will not spread through selection if an alternative allele can achieve the same level of diversification *across* families, but without a cost (for a discussion, see Jennions & Petrie, 2000).

Nonetheless, there are other avenues through which genetic diversification can be advantageous—not only by decreasing variance in fitness outcomes but also by increasing mean fitness outcomes (e.g., Yasui, 1998). In some interaction contexts, full siblings may negatively affect each other's outcomes more than half siblings do. Thus, for instance, two siblings are likely to be exposed to pathogens that each carries. If a pathogen adapts to and thrives within one sibling (e.g., because it has evolved counter-adaptation to the individual's immune defenses), the likelihood that the pathogen will transmit to and thrive within the other sibling is a function of the similarity between the siblings. Full sibs are more similar to one another than are half sibs. Hence, a female may benefit from producing half sibs (with different fathers) rather than full sibs (with the same father). (We note that sibling-sibling conflicts of interest about parental investment, which are greater in half sibs than in full sibs, may more than offset these benefits in some species; see Trivers, 1974.)

A mate chooser can diversify offspring, however, even when producing full sibs. Specifically, a mate who is heterozygous at a particular locus will produce offspring more diverse at that locus than are the offspring produced by a mate who is homozygous. If females particularly benefit from diversifying offspring at loci involved in pathogen recognition and immune defense, it may pay them to prefer a mate who is heterozygous at MHC loci. In a study performed with colleagues, we found precisely this effect. Women strongly preferred the scent of men who were heterozygous at all three loci we assayed (A, B, DRβ) over the scent of men who were homozygous at one or more loci (Thornhill et al., 2003). Possibly, women possess adaptation that functions to diversify their families' MHC alleles by leading them to prefer, as long-term mates (with whom they have, historically, typically had multiple offspring) men heterozygous at MHC. We discuss this finding and interpretation further in chapter 9.

Developmental Instability, Fluctuating Asymmetry, and Genetic Variance

Developmental Instability and Fitness

Evidence that sexual selection has operated on women to shape preferences for mates who possess intrinsic good genes has been indirect. Researchers have asked whether women prefer male traits that in ancestral conditions may have been associated with intrinsic good genes. These traits may not be associated with fitness in many modern environments. The extraordinary health care and lifestyle changes that have occurred in the past century, for instance, may have substantially altered correlations between phenotypic features and fitness.

One candidate trait that may have been related to fitness in ancestral environments is developmental instability. Developmental instability is an individual's proneness to the imprecise expression of developmental design due to genetic and environmental perturbations. These perturbations, importantly, include mutations and pathogens and, hence, major factors that contribute to genetic variation in fitness. Because of its conceptual link to maladaptation (see Leung & Forbes, 1996; Møller, 1997, 1999; Parsons, 1992), developmental instability became a focal point of research on sexual selection in the 1990s (Møller & Swaddle, 1997; Møller & Thornhill, 1998b).

The primary measure of developmental instability used by biologists is fluctuating asymmetry (FA). FA is deviation from perfect symmetry on a bilateral trait that is, on average, symmetrical in the population. For instance, the breadth of people's right and left wrists is, on average, very similar. Some people, however, have slightly larger right wrists. Others have slightly larger left wrists. FA of wrist breadth is calculated as the absolute deviation of the right and left wrist breadths (in some applications, standardized as a proportion of wrist size by dividing by the individuals' own or the population average wrist size). In research conducted at the University of New Mexico, colleagues and we have measured a number of asymmetries in humans—of the ears, elbows, wrists, ankles, feet, and fingers (see, for instance, Furlow, Armijo-Pruett, Gangestad, & Thornhill, 1997; Furlow, Gangestad, & Armijo-Prewitt, 1998; Gangestad, Bennett, & Thornhill, 2001; Gangestad & Thornhill, 1997a, 1997b, 1998, 2003 a,b; Gangestad, Thornhill, & Yeo, 1994;Thoma, Yeo, Gangestad, Lewine, & Davis, 2002; Thoma et al., 2005; Thornhill & Gangestad, 1994, 1999a, 2006; Thornhill, Gangestad, & Comer, 1995; Thornhill et al., 2003). Other researchers have measured similar sets of traits (e.g., Livshits & Kobylianski, 1989; Manning, 1995; Manning, Scutt, & Lewis-Jones, 1998; Manning & Wood, 1998; Rikowski & Grammer, 1999; Waynforth, 1998). The asymmetry that exists in these traits is very small, the mean being 1–2 mm, so small that you cannot detect it reliably through normal social interaction. Hence, the asymmetries we measure cannot serve as cues by which individuals assess others' developmental imprecision (though it is possible that small asymmetries can influence observable performance). The reason we measure them, then, is because they purportedly are markers of underlying developmental imprecision, which may substantially affect the overall phenotypic fitness of individuals. We aggregate across many traits (typically 10) because, as we discuss later, each individual trait's asymmetry very poorly reflects overall developmental instability (Gangestad & Thornhill, 1999; Leung, Forbes, & Houle, 2000).

Many studies have examined associations between FA and fitness components in a wide variety of species. Møller (1999) reviewed studies available at the time and found that, indeed, on average low FA is associated with relatively high fitness (e.g., survival, fecundity). This broad conclusion must be tempered with a caveat: Not all results have supported a link between FA and fitness. Though many null results may very well be due to low power and sampling variability (Type II error; see Gangestad & Thornhill, 2003b), in some species there may be no link. As we noted earlier in this chapter, as well, in some species individuals of high quality may survive no longer than individuals of low quality (e.g., Kokko et al., 2002; Getty, 2002).

One specific component related to fitness that has received much attention in work on FA is disease infection and parasite loads. Møller (2006) recently reviewed this literature. Once again, a robust, mean positive association between asymmetry and disease or parasite load exists (see also Thornhill & Møller, 1997). This literature includes studies on humans. In one study, we found that more symmetrical individuals reported lower frequencies of infections in the past three years (Thornhill & Gangestad, 2006). In a Mayan population, Waynforth (1998) found an association between FA and incidence of major disease. Overall, however, results on humans have been mixed (see Rhodes, 2006). Whether the mixed results are due to modern health care obscuring associations between developmental instability and health is unknown.

A number of studies have specifically examined associations between FA and immune activation (e.g., immunoglobulin levels) during development. Once again, links do, on average, exist: Activation of the immune system during development is associated with higher levels of asymmetry (Møller, 2006). Future research may examine these links in humans.

In humans, associations between body FA and features of the brain have specifically been examined. Human brains are characteristically asymmetrical in particular ways (e.g., the planum temporale is larger in the left hemisphere than in the right hemisphere), partly due to specialization of function in the two hemispheres (e.g., in the case of the planum, specialization of language function in the left hemisphere). One can calculate an individual's atypical brain asymmetry (deviation from mean left vs. right differences). In two studies, Thoma and colleagues (Thoma et al. 2002; Thoma et al., 2005) found that body asymmetry predicted atypical brain asymmetry; individuals with asymmetrical bodies also tended to have brains that were atypically asymmetrical. Atypical anatomical asymmetries may reflect atypical brain organization. Neuropsychological tasks can detect the extent to which some specific functions (e.g., perception of phonemes) are processed primarily in the right or left hemisphere. Individuals with high FA tend to perform simple cognitive tasks in the two hemispheres in an atypical fashion (Yeo, Gangestad, Thoma, Shaw, & Repa, 1997). A number of studies have demonstrated a negative association between FA and intelligence (as assessed by standard psychometric procedures; Bates, 2007; Furlow et al., 1997; Luxen & Buunk, 2006; Prokosch, Yeo, & Miller, 2005; Rahman, Wilson, & Abrahams, 2004 [though not in women]; Thoma et al., 2005; but for a study that failed to find an association, see Johnson, Segal, & Bouchard, 2007). Thoma et al. (2005) found that FA predicts intelligence independently of brain size. Furthermore, FA predicts slow processing in a simple reaction time task (Thoma et al., 2006). More generally, developmental instability appears to be associated with a variety of neurodevelopmental disorders (e.g., schizophrenia, schizotypy, ADHD, dyslexia; see Yeo, Gangestad, Edgar, & Thoma, 1999). Possibly, developmental instability compromises adaptive organization of functions in the brain and/or, at a more molecular level, neural or metabolic function (e.g., Yeo, Hill, Campbell, Vigil, & Brooks, 2000).

Associations between fluctuating asymmetry and functional operation of other organ systems in humans have not been examined. One might expect, however, that, just as developmental instability compromises the integrity of brain development, it also affects the precision with which other important systems are developed. As the human brain is composed of a particularly complex and energy-demanding set of features, with a large proportion of all genes expressed in brain tissue, it may particularly reveal deleterious effects of developmental instability (see, e.g., Gangestad & Yeo, 1997; Yeo et al., 1999).

What Causes Associations Between Fluctuating Asymmetry and Fitness?

The precise developmental processes by which fluctuating asymmetry arises are not well understood (e.g., Van Dongen, 2006). In theory, FA is the outcome of developmental noise—random expression caused by mutations, infection, toxins, and the like—and a developmental system's sensitivity to those perturbations (e.g., lack of buffering or repair mechanisms). Asymmetry, then, can result from high levels of perturbations (e.g., high levels of mutations), high levels of sensitivity to perturbations, or both. Part of the difficulty of interpreting precisely what FA means about an individual is that we do not know the extent to which it reflects the many possible variations that could cause it. And, indeed, the relative impact of causes need not be the same across different species (or, within a species, across different environments).

In general, asymmetry results when populations of cells on one side of the body stop growing while corresponding populations of cells on the other side continue to grow. Babbitt (2006) recently proposed, "variation in fluctuating asymmetry is in large part due to the random exponential growth of cell populations that are terminated randomly around a genetically programmed development time" (p. 258). One major issue concerns whether systematic processes affect termination of development and, if so, what those might be. One possibility is that genetic mutations disrupt adaptively programmed timing of development. Kjaer et al. (2005), for instance, compared individuals from four families with long polyalanine expansions in the HOXD13 gene (known to affect ontogeny of the digits) to individuals from the same families without mutated HOXD13 alleles. Those with long expansions had greater FA of the hands and feet than those without long expansions.

A more pervasive systematic process leading to asymmetry might be oxidative stress. Respiration naturally produces, within cells, unstable reactive oxygen species (ROS) as by-products. ROS can react with proteins or lipids in a cell or, indeed, DNA itself, thereby damaging cell structures. Organisms possess adaptive systems to quickly stabilize ROS (e.g., through production of antioxidants such as superoxide dismutase and catalase). When production of ROS outpaces antioxidant activity, a condition referred to as oxidative stress occurs and cell structures are often damaged. A wide variety of conditions can cause oxidative stress. In fact, one adaptive inflammatory response to pathogens in vertebrates is to produce ROS directed against the pathogens. A cost of this response is potential cell damage. (For instance, destruction of the gastric epithelium as a result of Helicobacter pylori infection in humans—the creation of ulcers—is at least partly due to extracellular ROS released by phagocytes directed against the bacteria; e.g., Ramarao, Gray-Owen, & Meyer, 2000.) Cellular damage may affect regulation of cell replication or lead to cell death. As already noted, infection during development appears to be associated with high FA (Møller, 2006). Oxidative stress may partly or fully mediate this effect. (A possibly interesting side note is that, as some pathogens have adaptations that combat oxidants directed toward them—e.g., Ramarao et al., 2000—host-pathogen coevolution may revolve partly around production of antioxidants. This coevolution may maintain genetic variation in pathogen resistance, as well as developmental instability, in hosts.) More generally, any stressor (e.g., mutation, biotoxin) that causes oxidative stress or interferes with antioxidant adaptations may cause cell damage and replication. Possibly, then, cell damage caused by ROS is a common mediator of the effects of perturbing events and FA (see also von Schantz, Bensch, Grahn, Hasselquist, & Wittzell, 1999).

These specific ideas are speculative; obviously, they demand empirical tests. But a broader point of emphasis here is that, in general, more research is needed before we will understand well the processes that lead to FA. And only after we have a better understanding of these processes will we have a good idea of why (and when) FA taps fitness-related traits (e.g, Van Dongen, 2006).

A pertinent issue in this regard, still unresolved, is the extent to which the asymmetries of different characters are caused by systematic factors shared across characters or ones specific to characters. For instance, are the systematic individual differences that underlie variation in asymmetry of human ear dimensions the same individual differences that underlie variation in asymmetry of human finger lengths? Or are these individual differences largely unshared? The correlation between any two traits' FA tends to be very small—about .05 on average in a large survey of over 1,000 correlations (Gangestad & Thornhill, 1999; see also Gangestad et al., 2001). This finding has led some observers to claim that FA of any two characters largely reflects character-specific causes of developmental imprecision, not organism-wide causes (e.g., Van Valen, 1962).

In fact, however, this conclusion is almost certainly wrong, at least for many species. One must take into account the extent to which systematic individual differences contribute to a single trait's FA. The effects of developmental error on asymmetry have large, unsystematic, random components. Hence, even if the same individual were to grow *the same trait* twice under precisely the same conditions and experience precisely the same amount of developmental error, the individual would not grow precisely the same amount of asymmetry. (Random developmental errors on the right and left side on one occasion might happen to be in opposite directions and accentuate asymmetry, whereas on another occasion they might happen to cancel each other out, yielding little asymmetry.) The amount of variance in a single trait's asymmetry due to systematic individual differences places a limit on the correlation between the asymmetry of two different characters (or, in fact, the correlation that would be observed if individuals could grow the same trait twice). A number of methods can be used to estimate this value, typically referred to as the repeatability of FA (see, for instance, Van Dongen, 2006; Gangestad & Thornhill, 2003b). The median estimated repeatability across a large number of data sets is less than 0.10 (Gangestad & Thornhill, 1999, 2003b). In our most recent data on human traits, we found it to be 0.076 (Gangestad et al., 2001). In 12–14 measurements of skeletal bones of eight species of nonhuman primates, Hallgrímsson (1998) found a remarkably similar average value: 0.072. Though higher values are occasionally observed (e.g., Lens & Van Dongen, 1999), in mammalian skeletal or skull measurements values larger than 0.10 are hardly ever seen (Gangestad & Thornhill, 2003b). (Recently, Graham, Shimizu, Emlen, Freeman, & Merkel, 2003, have questioned the assumptions behind estimations of repeatability and propose alternative assumptions that, if true, imply that true values are even smaller than our estimates. Van Dongen, Talloen, & Lens, 2005, however, found that estimates from real data are very similar based on analyses that assume a range of different models, including alternatives posed by Graham et al., 2003.)

The implication is that, even if two traits' asymmetries shared 100% of their systematic individual differences in propensity to develop asymmetrically, they would correlate, on average, only about 0.07–0.08. The fact that they correlate, on average, around .05 (or, in the human data we collected, about 0.045; Gangestad et al., 2001) means that over 50% of the systematic variance in two traits' asymmetry is shared. That is, most of the systematic individual differences underlying a trait's asymmetry are, in fact, organism-wide variations in developmental instability (see also Lens & Van Dongen, 1999). (If our estimated repeatability of .07 is itself slightly overestimated due to improper assumptions [Graham et al., 2003; Van Dongen et al., 2005], the amount of shared effects could near 100%.)

In fact, this result is not terribly surprising, given that fluctuating asymmetry does relate to fitness, on average. If each individual trait's asymmetry merely reflected components of developmental integration and stability specific to that trait, it would not seem likely that the trait's asymmetry would covary with fitness. (An exception might be made if the asymmetry itself directly affected performance, but that is not the case with the very small asymmetries in finger lengths, ear size, etc., that we measure.) For trait asymmetry (or even a composite of different traits' asymmetries) to plausibly predict fitness, those asymmetries should typically reflect systematic variation across individuals in their propensity to develop precisely in an organism-wide fashion. (Indeed, the correlations between body asymmetry and atypical brain asymmetry and organization require shared variance in developmental processes across traits of the body and brain; cf. Polak & Stillabower, 2004.)

We present one final note on the association between developmental instability and fitness. Though, all else being equal, zero developmental imprecision is optimal, in most instances, processes that maintain developmental precision (e.g., cellular repair mechanisms, antioxidation) do demand energetic resources. Hence allocation to these processes demands trading off allocation to other adaptive processes. In some circumstances, individuals that are most fit in the population may actually be best off allocating fewer resources to maintenance of developmental precision (or compromising it, e.g., through rapid growth) than to the allocation that maximizes the fitness of less fit individuals. (Related to our earlier discussion of the sexual selection continuum, this may be true particularly if sexual selection is very strong, in which case extreme sexual displays by highly fit individuals may compromise developmental precision and other viability-enhancing traits; e.g., Kokko et al., 2002, 2003.) Some of the inconsistency in associations between FA and fitness components reported in the literature probably does reflect true differences across species in the extent to which fit individuals do well by maintaining developmental precision. As we have already noted, in some species there may simply be very little association between developmental instability and fitness.

Mate Preferences and Developmental Instability

Perhaps the most intensively studied correlate of FA to date is mating success. Do symmetric individuals achieve greater mating success than less symmetric individuals? And if so, is their greater success at least partly due to the other sex's preference to mate with relatively symmetric individuals? A decade ago, Møller and Thornhill (1998b) performed an exhaustive meta-analysis of the studies that had been done at that time. Their conclusions were clear: Symmetric individuals do indeed have relatively high mating success. And in many instances, it is because they are attractive to members of the other sex. These conclusions were challenged by Palmer (1999, 2000), who questioned whether the literature might be misleading due to publication bias (in favor of supportive studies). But subsequent analyses that assess bias support the original conclusions (Møller, Thornhill, & Gangestad, 2005; Thornhill, Møller, & Gangestad, 1999). Once again, however, mean positive effects do not imply that positive effects exist in all species or, within species, all environments; in some species no association may exist (at times because there are, in these species, no associations between FA and fitness; see preceding discussion and Møller & Cuervo, 2003).

Over a decade ago, we began asking whether FA predicts number of sex partners in college students, similar to associations examined in nonhuman species. We have now studied over 500 college students of each sex. In sum, we find that, in this population, men's FA does, whereas women's does not, reliably do so (see Gangestad & Thornhill, 1999, for an overview; see also Gangestad & Thornhill, 1997b; Thornhill & Gangestad, 1994). We (with Kevin Bennett; Gangestad et al., 2001) estimated the correlation between men's developmental instability and number of sex partners using latent structural equation modeling, which uses each individual trait's FA as an independent marker of underlying developmental instability. In a sample of over 200 men, the estimated correlation was −0.4 to −0.5 (with body size and age controlled), a sizable effect.

We have also collaborated with two anthropologists, Mark Flinn and Rob Quinlan, in work in a rural village on the Caribbean island of Dominica. There, too, we find that FA predicts number of romantic partners in men (Gangestad, Thornhill, Quinlan, & Flinn, 2007). In that study we measured number of romantic partners by asking other villagers, not target individuals themselves. The correlation between FA and men's number of partners was about −.4.

We suspect that female preferences for symmetry per se have little if anything to do with the causal process that drives these associations. Again, FA is *our* measure of developmental health. It correlates with a variety of physical and behavioral features that may mediate these associations between FA and number of sex partners because women prefer these features, not FA directly. Consider the following examples.

1. In Dominica, FA negatively predicts men's peer status, as assessed through interviews with men. More symmetric men are seen to be better coalition partners than less symmetric men. Female preferences for men with favorable peer status could lead more symmetric men to have more romantic partners (Gangestad, Thornhill, Quinlan, & Flinn, 2007).
2. U.S. college men who have greater body symmetry may have more masculine faces, as assessed by a variety of sexually dimorphic facial dimensions (Gangestad & Thornhill, 2003a; but see Koehler, Simmons, Rhodes, & Peters, 2004). Even the association between facial symmetry and attractiveness may be partly mediated by other facial features (Scheib, Gangestad, & Thornhill, 1999). (See figures 7.1 and 7.2.)
3. Simpson, Gangestad, Christensen, & Leck (1999) had U.S. college men interviewed for a potential lunch date with an attractive female. As part of the interview, each man was asked to tell the woman, as well as a male competitor—someone else she was purportedly interviewing—why he should be chosen for the lunch date over the other. More symmetric men were more likely to engage in direct intrasexual competitive tactics—directly compare themselves with the other and state that they were the

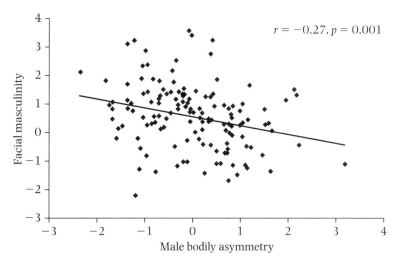

Figure 7.1 Scatterplot of the association between facial masculinity and bodily FA in men, showing a significant, linear association ($r = -0.27$, $p = 0.001$, $N = 141$). Variables on both axes have been standardized (converted to z-scores). The line is the least squares linear regression line. The facial masculinity measure is a composite of sexually dimorphic effects (e.g., testosteronization; see Gangestad & Thornhill, 2003a). From figure 1 of Gangestad and Thornhill (2003a).

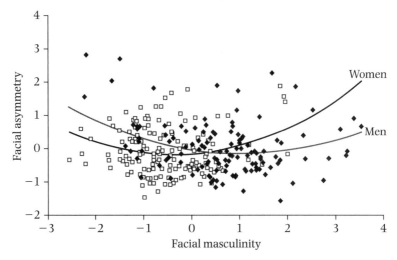

Figure 7.2 Scatterplot of the associations between facial FA and facial masculinity in men (filled diamonds) and women (open squares). For men, there
is a significant linear association as well as a significant curvilinear association ($N = 139$, linear $r = -0.20$, $p < 0.05$; quadratic r [with linear effect
partialled out] $= 0.39$, $p < 0.001$). For women, there is a significant curvilinear association ($N = 151$, linear $r = 0.03$, ns; quadratic r [with linear effect
partialled out] $= 0.23$, $p < 0.01$). The lines are the least-squares regression
lines. Note that, as facial masculinity increases, men tend to have greater
facial symmetry than women; as facial masculinity decreases, women tend
to have greater facial symmetry than men. From figure 2 of Gangestad and
Thornhill (2003a).

better choice—than asymmetric men were. Other research shows that, compared to
asymmetric men, symmetric men are less willing to back down from threats to their
status (Gangestad & Thornhill, 1997a) and, perhaps as a result, more likely to get into
physical fights, particularly those that they escalated into a fight (Furlow et al., 1998;
Manning & Wood, 1998).

4. In college samples, symmetric men appear to be more muscular and vigorous
   (Gangestad & Thornhill, 1997a) and have lower basal metabolic rates (Manning,
   Koukourakis, & Brodie, 1997) than asymmetric men.
5. In a study conducted in Jamaica, Brown et al. (2005) found that men's symmetry predicted the attractiveness of their dance movements. In that study, target individuals
   shown in video clips were reduced to stick figures through computer technology and,
   hence, all information aside from that reflected in dance movement was removed.
6. Symmetric men appear to have more attractive voice qualities than asymmetric men
   (Hughes et al., 2002).
7. As already noted, more symmetric individuals may be more intelligent than less symmetric counterparts.

Again, we suspect that these features (or some set of them) mediate FA's associa-tion with sexual history, partly because of female preferences for them. Intrasexual competitive advantages, independent of female choice, may also play some role. And, as we describe in chapter 8, female preference for the scent of symmetrical men may play a role as well. As we also discuss in chapter 8, women particularly prefer this scent during estrus—when they are fertile in their cycles. Indeed, women particularly prefer a number of these correlates of men's symmetry—intrasexual competitiveness, facial masculinity, muscularity, vocal masculinity—during estrus. Indeed, we argue that women's preferences for symmetric men, which lead symmetric men to have a greater number of sex partners than asymmetrical men, is a noteworthy component of women's estrous sexuality but not a dominant component of women's extended sexuality.

Does Developmental Instability Reflect Additive Genetic Variance in Fitness?

When Møller (1990) and Thornhill (1992a, 1992b) first demonstrated female prefer-ences for symmetrical males in barn swallows and scorpionflies, respectively, their favored interpretation is that male symmetry reflects individual variation in intrinsic good genes (see also Watson & Thornhill, 1994). Symmetrical males purportedly pos-sess, on average, greater heritable fitness than do asymmetrical males. Female pref-erences for symmetrical males, then, function to obtain intrinsic genetic benefits for offspring. The heritable variation in developmental instability could be maintained by mutation-selection and/or coevolutionary processes (e.g., host-parasite coevolution). But, for females to prefer symmetrical males to obtain good genes, this heritable varia-tion should relate substantially to general heritable variation in fitness.

Many studies have examined the heritability of FA, dating to work in the 1960s. Amazingly, however, we still know very little for sure about the heritability of devel-opmental instability (for a recent review, see Leamy & Klingenberg, 2005). We perhaps do have a pretty good sense of the average heritability of FA for single traits. In the sub-sample of studies reviewed by Møller and Thornhill (1997) examining single trait FA, the mean was .041 (see Gangestad & Thornhill, 1999). In a subset of those studies that fully controlled for maternal effects, the mean was .025 (Whitlock & Fowler, 1997). Van Dongen (2000) applied Bayesian hierarchical analysis to 66 heritability estimates from 12 studies of single traits' FA and found a mean value of 0.046. A more recent review reported a mean of approximately 0.03 (Fuller & Houle, 2003). And very recent studies are consistent with these estimates. For instance, Kruuk, Slate, Pemberton, and Clutton-Brock (2003) reported a mean h^2 of .041 across four antler traits' FA in red deer; Stige, Stagsvold, and Vøllestad (2005) found a mean h^2 of 0.065 for two plumage traits in pied flycatchers; Roff and Reale (2004) estimated the FA of four traits in a cricket to have a mean h^2 of 0.031. Though some variation in estimates across characters and species are observed, the estimates are remarkably consistent on the whole. Sampling error may in fact account for most of their variation.

A mean h^2 of 0.03 is in fact very small. (Indeed, in most studies, a value this small is not statistically significant.) Single trait FA, then, has very low heritability. Just as correlations between single trait FA measures must be interpreted in light of the total variance accounted for by meaningful individual differences, however, so too must heritability. As we noted earlier, the median estimated proportion of variance in a single trait's FA (at least in mammals) due to individual differences in developmental instability (the repeatability of FA) is about 0.07. Hence, if 100% of the individual differences in developmental instability were due to additive genetic variance, the heritability of the single trait FA would be just 0.07. To estimate h^2 of underlying differences in developmental instability, one should simply divide the h^2 of single trait FA by its reliable variance. On average, we expect a h^2 of approximately 0.03/0.07 ($\approx$ 0.4). If the mean correlation between two traits' FA (about 0.05) is a better estimate of the proportion of variance in single traits due to systematic differences in FA, then the mean h^2 of organism-wide developmental instability could be closer to 0.030/.05 (= 0.6).

These estimates can be illustrated with data from single studies. Again, Stige et al. (2005) estimated the h^2 of asymmetry of two plumage features in pied flycatchers. The mean h^2 estimate (not statistically significant) was 0.065. Stige et al. (2005) also estimated the repeatability of the traits' FA, which averaged 0.14. The estimated h^2 of developmental instability in this study, then, is 0.065/0.14 = 0.46. The authors nonetheless stated in the title of the article that fluctuating asymmetry in these features is "not heritable" (despite acknowledging that the 95% confidence intervals around their estimates of the h^2 of developmental instability contain a value of 1.00!). (Van Dongen, 2000, estimated lower h^2 for developmental instability, but the repeatabilities in the studies he analyzed were unusually large. We suspect that his results are not representative, though further empirical work must ultimately decide this issue. See Gangestad and Thornhill, 2003b.)

In our view, the available data are consistent with developmental instability typically being moderately heritable, contrary to the conclusions offered by some observers (e.g., Leamy & Klingenberg, 2005; Van Dongen, 2006). Admittedly, few single studies provide clear, strong conclusions. As Fuller and Houle (2003) note, most studies simply have very little power to detect meaningful heritable variation in developmental instability, particularly when they investigate a handful of traits or fewer. Few reports estimate h^2 of underlying developmental instability. Nonetheless, in our view the values typically found for h^2 of FA of individual traits and the reliable variance in FA of individual traits points to a fairly clear expectation: an h^2 of developmental instability of about 0.4.

Naturally, this value may vary across species or, within species, across environments. Only two studies have estimated the h^2 of FA in human skeletal and morphological traits (as opposed to dermatoglyphic FA, which appears to have different causes due to these traits being set by the second trimester of gestation). Livshits and Kobylianski (1989) estimated a composite of eight traits to have an h^2 of 0.31 in an Israeli sample. More recently, Johnson, Gangestad, Segal, and Bouchard (in press) examined heritability of FA based on samples of monozygotic and dizygotic twins reared apart. They estimated a heritability of a 10-trait composite to be about 0.3. As even aggregates of the FA of 8 and 10 traits do not measure organism-wide developmental instability extremely well, the heritability of developmental instability in these populations is almost surely substantially greater than 0.3. (As we observed elsewhere [Gangestad & Thornhill, 1999], Livshits & Kobylianski's estimate assumed that all parent-offspring correlation was due to heritable effects, but Johnson et al.'s did not. See also Sangupta & Karmakar, 2007.)

Leamy and Klingenberg (2005) suggest that FA and developmental instability may have large amounts of nonadditive genetic variance, particularly in the form of epistatic variance. In some systems (e.g., mice) that might be the case, though more research is needed before generalizations can be made. Epistatic variation can reflect a history of selection on additive genetic effects (Hansen, Alvarez-Castro, Carter, Hermisson, & Wagner, 2006). As developmental instability probably has been under directional selection in many systems (Møller, 1999), perhaps it should not be surprising if it does indeed have meaningful amounts of epistatic variance. Nonetheless, for reasons already discussed, we suspect that it also typically has additive genetic variance.

One can also estimate the CV_A of developmental instability. If developmental instability is a fitness trait (and taps heritable variation related to fitness), it should, of course, have a high CV_A. And, indeed, if h^2 of single trait FA is 0.03, the CV_A of developmental instability is indeed high—at least 14 and possibly higher. (These estimates do not vary much with the proportion of variance in single trait FA accounted for by developmental instability; see Gangestad & Thornhill, 2003b, for details). Developmental instability, then, appears to have the signature of a fitness trait.

Our working hypothesis is that developmental instability in human populations has indeed historically been a fitness trait (even if, given modern medicine and reliable birth control, it no longer is). We furthermore work with the hypothesis that it has historically tapped meaningful additive genetic variance in fitness. Though we think that much data are already consistent with this view, we realize that additional work on FA, developmental instability, and fitness is needed before firm conclusions can be drawn.

Signals of Women's Quality Revisited

In chapter 6, we suggested that women's ornaments have been under sexual selection to signal heritable quality in condition, as well as age- and condition-based reproductive value. Ancestrally, females in better condition, for reasons deriving from both their genetic makeup and environmental circumstances, were better able to store gynoid fat, specialized for reproduction, compared with females in worse condition. Differential readiness for reproduction as a function of condition did not require sexual selection. Male preferences were shaped by sexual selection as a result of covariance between female readiness to reproduce and condition. Males who preferred young females who stored gynoid fat and otherwise showed evidence of reproductive readiness (e.g., had features indicative of estrogenization), all else being equal, had greater fitness than those who did not, and potentially for a variety of reasons: Their mates had a longer reproductive lifespan ahead of them; their mates could produce healthier offspring because of their resources; their offspring received genetic benefits from their mates. These preferences then exerted sexual selection on females to display favored traits. Those females in best condition and highest genetic quality paid fewer costs for their marginal increases in investment in these traits due to sexual selection, and the displays stabilized as honest signals of condition and quality.

In chapter 6, we discussed evidence that patterns of gynoid fat deposition and other features affected by estrogen are associated with women's reproductive health and general condition. If they are indicative of overall quality, we might also expect them to be associated with developmental instability. In fact, some evidence is consistent with this expectation. The most direct evidence comes from a study of about 200 women in Poland, in which reproductive hormones were measured throughout women's cycles. FA of finger lengths was assessed. Symmetric women had higher estradiol levels over the menstrual cycle than did asymmetric women (Jasiénska, Lipson, Ellison, Thune, & Ziomkiewicz, 2006). In this population, then, symmetric women are more fertile. They should also possess more estrogenized body and facial features, though this study did not examine those associations.

Other studies specifically examining correlations between attractive features in women and their symmetry have yielded mixed results. Grammer et al. (2002) and Schaefer et al. (2006) reported that facial and nude bodily attractiveness in women is positively associated with body and facial symmetry. Koehler et al. (2004) found that women with symmetric bodily traits have more feminine facial features than do asymmetric women. By contrast, we found no significant association between women's facial femininity and symmetry (Gangestad & Thornhill, 2003a). Women's facial symmetry predicts their attractiveness (see review in Rhodes, 2006). But reported correlations between women's facial attractiveness and body symmetry have been fairly small (e.g., Gangestad et al., 1994; Thornhill & Gangestad, 1994). Hughes, Harrison, and Gallup (2002) reported that women's symmetry predicts their vocal attractiveness. We have not, however, found women's waist-to-hip ratio (WHR) to be predicted by their symmetry (unpublished data).

One factor that complicates interpretation of this mixed pattern of results is the apparent fact that women's symmetry changes across the cycle (Scutt & Manning, 1996). These changes may be due to variations in water retention, which can affect symmetry of soft tissues and joint spaces. As result, symmetry measures in women may reflect within-individual variation that obscures meaningful between-individual variation.

Women's ornaments appear to clearly relate to reproductive health and reproductive value (see chapter 6). Evidence that they reflect variation in developmental instability and genetic quality is less conclusive.

Signals of Men's Quality

Masculinity and Testosteronization

Earlier, we discussed correlates of male symmetry. In short, symmetric men are more "masculinized" than their asymmetric counterparts. Symmetric men appear to have more masculine facial features (Gangestad & Thornhill, 2003a; cf. Koehler et al., 2004). They behave in more intrasexually competitive and confrontative ways (Simpson et al., 1999) and, in U.S. and U.K. samples, report getting in more fights (Furlow et al., 1998; Manning & Wood, 1998). Some evidence suggests that, on average, they are more muscular than asymmetric men (Gangestad & Thornhill, 1997a) and have more masculine body shapes (e.g., broader shoulders and masculine upper-body to lower-body proportions; Brown; Price, Kang, Zhau, & Yu, 2007). They have more attractive voices (Hughes et al., 2002), which may reflect their having deeper, more masculine voices (e.g., Feinberg, Jones, Little, Burt, & Perrett, 2005).

Just as researchers have hypothesized that women's estrogen-dependent features have been exaggerated as signals of condition and quality through sexual selection, researchers, including us, have argued that men's masculinized features have been under sexual selection and hence partly function as signals of condition and quality (e.g., Thornhill & Gangestad, 1993, 1999a; Penton-Voak et al., 1999). The scenario we envision is analogous to the scenario we described in chapter 6 for female estrogen-dependent features. Compared with less fit males, males who were in better condition could adaptively afford to allocate more energy and other somatic resources into mating effort and intrasexual competitiveness. A variety of "masculine" physical and behavioral traits promote mating effort. These traits covaried with male quality, then, and could serve as cues of male quality. Female mate preferences for these traits were selected. In turn, men were sexually selected to allocate greater effort into these traits and, as a result, these traits came to function partly as signals of quality.

Just as many female signals of quality are facilitated by a primary reproductive hormone in women—estrogen—many male signals of quality appear to be facilitated by the primary reproductive hormone in men—testosterone (T). T is phylogenetically old—not as old as estrogen but still quite old, originating in early vertebrates (see chapter 8). In many species, a threshold of T must be achieved for many basic reproductive traits in males (such as sex drive) to be engaged (though the T level of most males of reproductive age exceeds that threshold, and T does not appear to have strong dose-dependent effects above the threshold; for work on humans, see Bancroft, 2002). In broad strokes, it also functions to facilitate pursuit of mates, male–male competition and, thereby, access to mates. Just as the precise manifestations of estrogen have been modified within particular species or taxa, so too the precise mechanisms that regulate T have been modified in specific species since its evolutionary debut.

Following Bribiescas (2001), we conceptualize the function of testosterone in a life history framework (see also chapter 4). Specifically, we see T as a modulator of resource allocation—when resources to be allocated include energy but also time and utilization of functional structures, including neural ones. Again, from a life history perspective, organisms have finite time budgets and hence can harvest energy at a finite rate. In allocating time and energy to fitness-enhancing activities, then, they face trade-offs. Human puberty marks a time when allocation of the energy budget is shifted from growth to reproduction. In males, testosterone plays key roles in that shift.

Though production of male sperm cells is very cheap (such that, even when extremely malnourished, men still produce sperm at rates similar to those of well-nourished men;

see Ellison, 2001, 2003), reproduction is typically not cheap for males, even in species in which males do not care for offspring. Males must find and compete with other males for mates, activities that require much energy to engage in effectively. For instance, male muscle mass contributes to mating effort in many mammalian species and demands much energy to build, maintain, and use. Sexual dimorphism in musculature, particularly of the upper body, emerges in humans during adolescence. Male increases in upper body musculature are facilitated by testosterone (e.g., Basaria et al., 2002; Bhasin, 2003; Schroeder et al., 2003). Males must trade off allocation of effort to mating and somatic maintenance (e.g., immune function) and, of course, can never afford to shift all energy away from somatic maintenance; how much they can afford to allocate to mating effort depends on a variety of factors, including their condition. In species in which males do engage in biparental care, males also face a trade-off between two forms of reproductive effort, mating effort and parental effort. As we described in chapter 4, men typically experience a reduction in testosterone when mated or when fathers.

At a general conceptual level, T might be thought of as a hormone that facilitates male mating effort (Bribiescas, 2001). More T results in greater mating effort. Less T is associated with less mating effort and more somatic maintenance and/or parental investment. A variety of features facilitate mating, others facilitate parenting, yet others facilitate somatic maintenance and survival. Each feature may be modular in the sense that, for instance, muscle growth involves mechanisms separate from, say, focus on male-male competition. Endocrine hormones, such as T, are messengers in distributed communication systems that can coordinate adaptive changes in whole suites of such modular features. Selection has presumably shaped the T system—the mechanisms that regulate its release and metabolism, as well as the precise distribution of T receptors in structures—such that, based on inputs to the system, it leads to optimal allocations of effort to mating, parenting, and so on, in environments in which the mechanisms were shaped by selection. This view, though undoubtedly overly simplistic, is a reasonable working model (see Bribiescas, 2001; Ellison, 2001, 2003).

Again, though T is not necessarily expensive, its effects—those that facilitate mating effort—are potentially very expensive. Different males under different circumstances and of different quality may be able to afford different levels of these costs (when allocating effort optimally; e.g., Grafen, 1990; Getty, 2006). As we noted in chapter 4, fathers may do best by allocating proportionately fewer of their energetic resources and other resources to mating. And, all else equal, males in good condition optimally allocate more energy to mating effort than individuals in worse condition. Hence, mechanisms regulating T modulate it partly as a function of condition.

In addition to facilitating muscle growth, testosterone facilitates masculinization of the male face (see Swaddle & Reierson, 2002). In one study, women judged men's masculinity based on facial photographs. Women's judgments correlated positively with men's measured T levels (Roney, Hansen, Durante, & Maestripieri, 2006; see also Penton-Voak & Chen, 2004). Testosterone also affects men's interactions with and patterns of attention to other men (see, for instance, Ellison, 2001; Mazur & Booth, 1998). Testosterone-facilitated male dominance-seeking may be expressed, among other behaviors, in greater selective attention to angry faces (van Honk et al., 1999; van Honk et al., 2000), in less pronounced smiling (Dabbs, 1997), and in more visual attention toward interaction partners (Dabbs, Bernieri, Strong, Campo, & Milun, 2001). In nonhuman animals, T suppresses fear (perhaps particularly of a social nature); two double-blind studies on humans (though women, not men) showed that an administration of T reduced fear in response to pictures of fearful faces (van Honk, Peper, & Schutter, 2005) and a fear-potentiated startle response (Hermans, Putnam, Baas, Koppeschaar, & van Honk, 2006). Reduced potentiation of fear may lead individuals to be more likely to engage in potentially injurious conflict. Men who score higher on a test of power motivation have higher T (Schultheiss, Dargel, & Rohde, 2003). Moreover, large increases in T following a real or imagined victory in a competition are associated with power motivation in men (Schultheiss, Campbell, & McClelland, 1999; Schultheiss & Rohde, 2002). In one study, the opportunity to interact with an attractive woman led to increases in male T, particularly for men who evidenced the greatest interest in her (Roney, Mahler, & Maestripieri, 2003).

Traits such as muscularity and willingness to engage in male–male contests have real costs. Muscles require energy, and contests could result in injury. One question that arises concerns what keeps traits such as facial masculinity and vocal masculinity honest signals of underlying quality. In and of themselves, they do not appear to be expensive to achieve. One plausible answer is that these traits impose socially mediated costs. In some species of birds, badges and patches regulate male-male competition. Males with bigger patches typically win bigger or better territories. Large patches are honest signals of male ability to engage effectively in male-male competition because males who have large badges will be tested. Thus, if small-badged male Harris sparrows are artificially given large badges, they are aggressed against, typically lose competitions, and end up worse off than they would if they had simply been left with their small badges (Rohwer & Rohwer, 1978). Male facial and vocal masculinity may similarly function to regulate male-male competition (see, e.g., Mueller & Mazur, 1997). Consistent with this interpretation, Puts, Gaulin, and Verdolini (2006) found that people perceive men with deeper voice pitch to be more physically and socially dominant than men with higher voice pitch. Moreover, when a man addresses another man whom he believes is less physically dominant than he is, he speaks at a lower pitch. By contrast, when a man addresses another man whom he believes is more physically dominant than he is, he speaks at a higher pitch (Puts et al., 2006).

Preferences for Masculine Traits

Masculine male traits facilitate effective performance in contests between men. Effective performance in these contests, as well as the traits that function or relate to effective performance, function to signal quality, including genetic quality. Women, then, prefer these traits in their mates. Naturally, however, men who can win dominance competitions with other men might be expected to gain access to material benefits. Women could profit from mating with masculine, dominant men because such men better deliver material benefits (e.g., food, shelter, physical protection), relative to less masculine, less dominant men. How can we know, then, that female preferences for masculine traits evolved to (at least partly) function to obtain genetic benefits for offspring? Perhaps the function of these preferences is merely to obtain nongenetic material benefits.

We note that, although masculine, dominant men possibly *could* provide greater levels of material benefits to female mates, they also might actually *deliver* fewer material benefits. In chapter 4, we described research on trade-offs between male-delivered genetic benefits and male provisioning that face females in many socially monogamous bird species. For instance, in collared flycatchers, males with large forehead patches establish territories earlier than small-patched males (Pärt & Qvarnström, 1997). They also provide genetic benefits relative to males with small patches (Sheldon et al., 1997). But small-patched males are better providers; they feed more at the nest (Qvarnström, 1999). Large-patched males reserve effort for seeking extra-pair copulations (EPCs) in the same season or, possibly, future seasons. More generally, in species in which females engage in EPC, relatively attractive males provide fewer material benefits (through foraging) than unattractive males provide (Møller & Thornhill, 1998a).

The same could be true of masculine and symmetric men, relative to more feminine and asymmetric men. Indeed, symmetric men tend to have more EPCs (Gangestad & Thornhill, 1997b). Men with masculine bodies and faces report greater success than less masculine men at short-term mating but not in forming long-term, stable pair bonds (Rhodes et al., 2005). Muscular men, relative to less muscular men, similarly succeed at short-term, but not necessarily long-term, mating (Frederick & Haselton, 2007). As Rhodes et al. (2005) conclude, their "findings suggest that individuals of high phenotypic quality have higher mating success (and, we note, for males, particularly short-term mating success) than their lower quality counterparts" (p. 186).

In a study in which we administered questionnaires to both men and women, which they completed privately in separate rooms, we found that symmetric men, relative to their less symmetric counterparts, invested less in their romantic relationships, based on responses by both men and women to a validated measure of self- and partner investment in their relationship (Ellis, 1998). Symmetric men, compared with asymmetric men, were relatively unwilling to give their time to their partners, were dishonest with their partners, and sexualized other women more often. Though they were seen as more able to provide physical protection than asymmetric men, we found no evidence that they were willing to dedicate time to do so (see Gangestad & Thornhill, 1997b). Women, similar to females of other species, may face trade-offs between various forms of benefits—genetic benefits and nongenetic material benefits—that mates can provide (e.g., Gangestad, 1993; Penton-Voak et al., 1999).

Women's and men's perceptions of masculine and feminine men are consistent with women's facing this trade-off. Men with feminine faces are perceived to be warmer, more agreeable, and more honest than men with masculine faces (Fink & Penton-Voak, 2002). Men with masculine faces are seen to be more likely to engage in male-male competition (e.g., get into physical fights) and pursue short-term matings (e.g., sleep with a lot of women, cheat on partners), whereas men with relatively feminine faces are seen to

be more likely to be good, stable, long-term mates (e.g., be caring and emotionally supportive, be great with children; Kruger, 2006).

As we have emphasized throughout this book, the primary evidence for function and
the selective pressures that gave rise to adaptations is to be found in features' design.
According to the view that women trade off material benefits for genetic benefits to be
mated with masculine men, attraction to male masculinity should have been shaped
by selection to be *conditional*—to depend on conditions that affect (or ancestrally would
have affected) the relative value of heritable condition and paternal investment. A number of lines of evidence suggest that it is.

1. *Preference varies as a function of relationship context.* The face women find most attractive in short-term mates is more masculine than the face they find most attractive in
   long-term mates (Penton-Voak et al., 2003). In one recent study, women were asked to
   choose which male they'd prefer as a mate, a male shown to have a masculine face or a
   male shown to have a feminine face. As it happened, mating context drove female preferences. When choosing a sex or affair partner, most women (57% and 66%, respectively) choose the masculine male. When it came to choosing a marriage partner,
   however, most women (63%) preferred the feminine male. (As we previously described
   in chapter 4, women even more strongly preferred the feminine male as a son-in-law
   [73%] and similarly thought that their parents would prefer them to date the feminine
   male [71%].)

2. *Attractive women have a stronger preference for masculine faces.* Little, Burt, Penton-Voak,
   and Perrett (2001) reasoned that attractive women need not trade off male condition
   and investment as markedly as must unattractive women; masculine men should
   be more likely to invest in relationships with attractive women. In fact, attractive
   women do more strongly prefer facial masculinity (Little et al., 2001; Penton-Voak
   et al., 2003).

3. *Preference varies with culture.* Penton-Voak, Jacobson, and Trivers (2004) proposed that
   women's preference for masculinity should have been selected to be sensitive to cues
   of the relative value of condition (and genetic benefits) and investment of male mates
   in their local ecologies. In Jamaica, infectious disease is more prevalent and male
   parental investment less pronounced than in the United Kingdom. They predicted and
   found that Jamaican women show greater preference for facial masculinity than do
   British women.

As we discuss in detail in chapter 9, women also particularly prefer masculine male
traits when in estrus, and particularly so as sex partners. Changes in preferences across
the cycle, we argue, reflect female design to weight signals of heritable condition more
heavily when they are fertile, particularly when selecting a sex partner.

In sum, the design of female preferences, in concert with other evidence, is consistent
with the view that male masculine features function partly as signals of heritable quality and, hence, intrinsic genetic benefits to offspring. We know of no alternative hypothesis that can explain the variety of findings consistent with this view.

Male Intelligence and Related Attributes: Signals of Quality?

We noted earlier that FA is negatively associated with psychometric intelligence. Symmetric men score higher on standardized tests of intelligence than do asymmetric men. (The same may well hold for women [e.g., Furlow et al., 1998], though fewer data are available on women. See also Rahman et al., 2004.) Might intelligence be a cue of heritable fitness? Indeed, might intelligence have been sexually selected to signal heritable fitness?

Miller (2000) devoted most of a full-length book to developing the thesis that, indeed, human intelligence has been sexually selected to signal heritable fitness. His claim is that intelligence functions much like the peacock's tail: Because it is difficult for individuals in poor condition to develop a brain that demonstrates the complexity we recognize as human intelligence, well-developed brains (and their cognitive manifestations) advertise intrinsic good genes.

If, indeed, human intelligence advertises good genes, we suspect that its origins as a signal are more similar to the origins of female ornaments and male masculine traits than to the origin of the peacock tail. In chapter 5, we described two routes through which a trait that ultimately becomes a signal of quality first acquires an association with fitness. The first is the preferred-signal-through-sensory-bias route. In that scenario, the trait is first preferred due to the by-product of a sensory adaptation that does not function in mate choice. The trait then evolves through sexual selection to a point at which it becomes associated with quality because high-quality individuals can best afford to develop large signals. The peacock tail purportedly evolved to be a signal of quality through this route.

In the second route, a functional trait is first correlated with heritable fitness simply because individuals of higher quality can better afford it. Individuals of the other sex prefer the trait because it is a cue of quality. The value of the trait added by its being a cue of fitness leads individuals to dedicate more effort to the trait, exaggerating it as a signal. Female gynoid fat depots and male muscularity do not function *purely* as signals; they do have sexually selected signaling properties, but they have functions that were not sexually selected, as well. Similarly, human intelligence may have sexually selected signaling properties, though, in our view, it clearly has other functions as well (see Gangestad & Simpson, 2007, for a range of views about the evolution of human intelligence).

If intelligence has evolved as a signal of genetic quality, however, we might expect it to have many of the same correlates that male masculine features do. Intelligent men, similar to masculine men, should be expected to be advantaged in short-term mating and hence be more likely to engage in short-term sexual relations than less intelligent men. They should invest less in their romantic relationships. They should be particularly preferred by women as sex partners, not long-term mates. And they should be more strongly preferred where genetic quality has large effects on fitness.

Most of these expectations are not borne out. Male intelligence does not correlate positively with number of sex partners in college samples (I. Tal, unpublished data; E. White, unpublished data). We know of no evidence that intelligent men are less kind, caring, and investing in their relationships than less intelligent men. And women tend to seek intelligence in long-term mates more than they seek it in short-term mates (Gangestad et al., 2007). Gangestad et al. (2006) did report that, cross-culturally, women's preference for intelligence in a mate covaries positively with parasite prevalence, just as preferences for physical attractiveness and health do.

One possibility is that intelligence did in fact evolve as a signal of genetic quality, but one that operates in circumstances in which high quality males do not do best by exerting mating effort. Rather, just as attractive males in bird species in which few extra-pair mating opportunities are available do best by provisioning rather than seeking extra-pair mates (Møller and Thornhill, 1998a), perhaps intelligent men do best by investing in their relationships and offspring. Answers to the question of why intelligence would operate in this fashion, but masculinity does not, will require additional theoretical and empirical work.

Another possibility is that some manifestations that display intelligence do in fact operate in much the same fashion that male masculinity does. Humor and creative displays may function as mating effort, which women find attractive in sex partners (e.g., Miller, 2000). The question of why these displays would work in this way although the underlying trait they purportedly display—intelligence—does not will, once again, require additional work to answer.

Do Indicators Capture Variation in Heritable Fitness?

Earlier in this chapter, we argued that there is much heritable variation in fitness in natural populations. This heritable variation is the basis for the evolution of mate choice for intrinsic good genes. For individuals to be able to choose mates for intrinsic good genes, however, there must be available to them phenotypic features that reflect that heritable variation. Moreover, these features should capture the heritable basis of fitness but not strongly reflect the nonheritable basis of fitness. Hence, good indicators of heritable fitness should not only have much genetic variation (i.e., large CV_A values) but should also have high heritabilities (e.g., Pomiankowski & Møller, 1995; Rowe & Houle, 1996).

We have argued that a variety of masculine features, in concert with some other indicators (e.g., coordination, as might be reflected in dance; e.g., Brown et al., 2005), function as signals of genetic quality. To say so does not imply that they covary with fitness in modern environments. It does imply that they did so in ancestral environments and that female preferences for these features evolved because they covaried with fitness. At least in ancestral environments, these traits (or some linear or nonlinear combination of them, reflecting how they are utilized as cues) should have high CV_A values. They should also possess high heritabilities. Future work should address their genetic variation and heritabilities.

As we have discussed, these features and preferences for them do possess signatures of having been shaped partly in the context of a system of honest signaling of heritable fitness. In light of these signatures, we once again proceed with the working assumption that, indeed, sexual selection for displaying and choosing mates with intrinsic good genes played a role in the evolution of these features and women's preferences for them.

Summary

Potentially, individuals can choose mates for one or more of three types of genetic benefits for offspring. Intrinsic good genes are alleles that benefit offspring independent of the genotype of the mate chooser. Compatible genes are alleles that work well with the alleles of an individual mate chooser, though not with all the alleles of all mate choosers. Choice for diverse genes leads to diversification of offspring.

Adaptive mate choice for intrinsic good genes requires that there exist additive genetic variance in fitness in the population. Until the past two decades, evolutionary geneticists have typically assumed that fitness has little genetic variance because selection persistently removes it. It now appears that fitness components (e.g., longevity, fecundity) have a lot of additive genetic variance in natural populations relative to traits under stabilizing selection. A variety of processes contribute to the maintenance of additive genetic variance in fitness, despite selection. One important one is mutation-selection balance. Others are Red Queen or antagonistic coevolutionary processes, which lead to relatively rapid changes over time in which alleles are favored. Host-pathogen coevolution is one example (a process that may operate in concert with negative frequency-dependent selection). Other examples include forms of intraspecific antagonistic coevolution, such as that due to maternal-fetal conflicts of interest. One potentially important form of intraspecific antagonistic coevolution is sexually antagonistic coevolution. Though it appears that slightly deleterious mutations can maintain a substantial amount of additive genetic variance, the precise relative contributions of processes responsible for this variation in humans remains unknown.

Although the viability-indicator and Fisherian models of mate choice have traditionally been thought to represent competing views about the nature of good-genes mate choice, in fact they appear to anchor the ends of a sexual selection continuum. These ends differ not in terms of whether individuals who possess good genes (as well as their offspring) are truly high-quality individuals, as once thought. Instead, they differ in terms of how individuals who are of high quality have been selected to allocate their effort, relative to individuals of low quality. At the viability-indicator end of the continuum, high-quality individuals maintain a viability advantage over low-quality individuals. At the Fisherian end of the continuum, high-quality individuals have been shaped to strongly invest in traits that lead to mating benefits and, accordingly, may die, on average, at younger ages than low-quality individuals do.

Adaptive mate choice for compatible genes requires nonadditive genetic effects on fitness, which can be due to dominance (e.g., heterosis) or epistatic effects. Two examples in humans and many other organisms appear to be mate choice for MHC dissimilar individuals and incest avoidance.

Adaptive choice for diverse genes may be selected when individuals in the same family benefit by being different from one another. Women's preference for the scent of men heterozygous at MHC loci may be one example.

Developmental instability, as reflected in fluctuating asymmetry, is associated with fitness in many natural populations. In modern human populations, developmental instability appears to be at least moderately heritable.

We suggest that, just as a primary female reproductive hormone, estrogen, facilitates the development of sexually selected indicators of genetic quality in women, a primary male reproductive hormone, testosterone, does the same in men. A variety of "masculine" traits, known or likely to be testosterone-facilitated, are associated with developmental instability in men. Because women prefer men who possess these traits as mates, owing to the traits being indicators of intrinsic genetic quality (at least ancestrally), these men typically invest less in long-term relationships with women. As expected, then, women particularly prefer these traits in sex partners rather than in stable long-term mates; attractive women prefer these traits in long-term partners more than do unattractive women (as these men are more likely to invest heavily in relationships with attractive women); and preferences for these traits appear to vary across cultures. In cultures and ecologies in which paternal care is particularly important, women should prefer masculine men as long-term partners less than in cultures and ecologies in which paternal care is less important.

8   Estrus

The Concept of Estrus

One dictionary definition of *estrus* is "the periodic state of sexual excitement in the female of most mammals, excluding humans, that immediately precedes ovulation and during which the female is most receptive to mating" (*American Heritage Dictionary of the English Language*). In mammalian reproductive biology, the *estrous cycle* is equivalent to the ovarian cycle. The *estrous phase* refers to the phase of high fertility and ovulation in the cycle. Estrus is typically synonymous with estrous phase. Many biologists do not refer to reproductive cycles of female nonhuman Old World primates as estrous cycles. Rather, cycles in these primates are often referred to as menstrual cycles, in reference to the blood flow that occurs at approximately 30-day intervals. As we emphasize later, however, the mid-cycle phase occurring within females of these primates appears to share homologies with the estrous phase of other mammalian species (e.g., Dixson, 1998; Nelson, 1995). Other scholars reserve the term *menstrual cycle* to refer to the ovarian cycles of human females exclusively, which reflects the widespread assumption—one that, we argue in chapter 9, is clearly wrong—that, of all mammalian species, humans alone lack estrus. (This view is expressed in the dictionary definition we quoted above.) *Behavioral estrus* is typically defined as a restricted period of proceptivity and receptivity characterized by mammalian females' behavioral readiness to mate, in addition to attractiveness to males, usually, though not invariably, coinciding with relatively high probability of conception (e.g., Beach, 1976; Nelson, 1995; Symons, 1979). A synonym for behavioral estrus is *heat* (Nelson, 2000).

Estrus Is Not Restricted to Mammals

The Homology of Estrus

Though estrus has traditionally been applied solely to mammalian females, this convention is arbitrary. The estrous phase and behavioral estrus can be observed in all vertebrate taxa, and, we propose, estrus is homologous within vertebrates. Physiological machinery leads female goldfishes and garter snakes to emit, as incidental by-products, hormones or derivatives associated with egg maturation. These emissions accompany enhanced female sexual motivation and attractivity to males (Mendonca & Crews, 1996; Shine et al., 2003). These mechanisms also possess homology with the physiology of the estrous phase and behavioral estrus in the female house mouse (see later discussion). More generally, the similarity of physiological machinery typically associated with fertility across vertebrate taxa arises, in part at least, from its descent from a common female vertebrate ancestor—one that possessed estrogen-facilitated egg maturation, accompanied by enhanced sexual motivation and attractivity.

Analogies to estrus exist in nonvertebrates. Female moths secrete a hormone (restricted to arthropods and their close relatives) soon after the onset of adulthood. It stimulates ovarian development and female mating behavior (for a review, see Nation, 2002; see also Ringo, 1996). In vertebrates, a quite different hormone with a different phylogenetic origin, estrogen, plays a functionally similar role (Nelson, 2000). In vertebrates, as well as many invertebrates, fertile females emit scents highly attractive to males and are simultaneously behaviorally receptive to mating.

Ichthyologists, herpetologists, and ornithologists rarely describe the reproductive seasonality of female fish, frogs, toads, salamanders, or reptiles (including birds) in terms of estrus (for rare exceptions, see Jones et al., 1983, for a discussion of estrus in *Anolis* lizards; Aldridge & Duvall, 2002, on pit vipers). Nor do these biologists speak of fertile-phase females in these species that exhibit sexual proceptivity, receptivity, and attractivity (i.e., females characterized by the defining qualities of mammalian behavioral estrus) as estrous females. Although Whittier and Tokarz (1992) described the sexual behavior of female reptiles in terms of these qualities, for instance, they did not go so far as to refer to reptilian females exhibiting these qualities as estrous females.

Again, the reproductive cycles of all female vertebrates are regulated by physiological mechanisms, hormonal and neural, that are homologous in part. Vertebrates share a pattern of hormones that typifies high fertility within female reproductive cycles (reviews in Crews & Silver, 1985; Jones, 1978; Lange, Hartel, & Meyer, 2002; Liley & Stacey, 1983; Lombardi, 1998; Nelson, 2000; Smock, Albeck, & Stark, 1998; Whittier & Tokarz, 1992). For example, in all nonmammalian vertebrate species studied, females' estrogen levels are above a basal concentration at the time when they mate with males (Crews & Silver, 1985), precisely the pattern observed across diverse taxa of mammals (e.g., Nelson, 2000). As well, the hormones associated with ovulation appear to promote female attractivity in vertebrates in general; typically, the attractivity of fertile females is mediated by effects of estrogen (Nelson, 2000). And ovariectomy suppresses female sexual behavior within all vertebrate taxa (though, as we discuss later, nonovarian hormones may primarily affect sexual behavior in some species; Adkins-Regan, 2005; Nelson, 2000). Although studied less intensively than hormonal homology, similar, apparently homologous neurological structures appear to produce heightened female sexual motivation at peak fertility in the reproductive cycle across vertebrate taxa (e.g., Lombardi, 1998; Smock et al., 1998).

In light of these homologies, the convention of using distinct taxon-specific language to describe the sexuality of vertebrate females at peak fertility in their reproductive cycles makes little sense. It fails to recognize that important aspects of the physiology underlying the sexuality reflect homologies. Worse yet, it hinders that recognition. For this reason, we apply the term *estrus* to the fertile state of all female vertebrates in their reproductive cycles. We furthermore argue that this usage makes scientific sense, because, we propose, estrus is homologous across all vertebrates. It first appeared 400–450 million years ago in a species of fishlike animal ancestral to all vertebrates. As estrogen-facilitated female sexual motivation at high fertility in the reproductive cycle apparently characterizes all (or virtually all; see later discussion) vertebrates, the principle of parsimony supports our proposal (see chapter 2, this volume).

Phylogeny of Estrus   Figure 8.1 depicts vertebrate phylogeny, as generally accepted (e.g., Tree of Life website). We propose that significant events in the phylogeny of estrus occurred at time points A, B, and C (based on data from Thornton, 2001). At 450 million years ago (time point A in the tree), gnathostomes (jawed vertebrates) and lampreys diverged. The common ancestor of these two lineages, evidence suggests, possessed an estrogen receptor (a protein that binds with estrogen) that subsequently evolved into two estrogen receptors within gnathostomes (ERα and ERβ). The presence of an estrogen receptor, in turn, is a signature that estrogen was, in these species, physiologically functional. Estrogen in the modern-day lamprey regulates the reproductive maturation and behavior of both sexes; indeed, blood levels of estrogen are sexually monomorphic. It is reasonable to infer that estrogen levels in the common ancestor of lampreys and gnathosomes were also sexually monomorphic.

Estrogen-facilitated feminization of reproductive maturation and behavior, then, made its phylogenetic debut in the species of fish-like animal ancestral to gnathostomes (at time point B in the tree). This species had ERα and ERβ. This dimorphism was maintained in the common ancestral species of the teleost fishes and tetrapods, which diverged at time point C in the tree, approximately 400 million years ago. And it was subsequently maintained in all branches of teleosts and tetrapods (Thornton, 2001).

Estrogen appears to play a role in the regulation of reproduction in certain mollusks, branchiostomes, and echinoderms. It is the most ancient form of steroid regulation of reproduction (Thornton, Need, & Crews, 2003). Estrogen regulation of female reproduction, however, is not synonymous with estrus. When estrogen evolved the capacity to specifically regulate female reproductive maturation and simultaneously affected female sexual motivation and sire discrimination (see later discussion), estrus came to be. Again, based on the phylogeny we have sketched out, it is reasonable to assume that the reproductive behavior of both females and males of the ancestral species of the lampreys

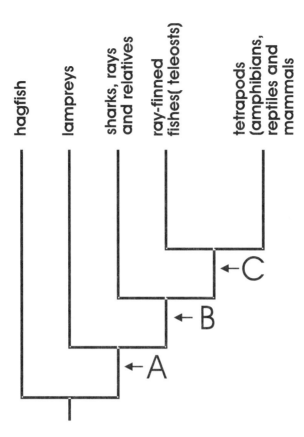

Figure 8.1 The phylogeny of the vertebrates indicating the timing of significant evolutionary events pertaining to the phylogeny of estrus. A is dated at 450 million years ago and marks the divergence of lampreys from jawed vertebrates (gnathostomes). Lampreys possess an estrogen receptor that is homologous with and ancestral to the gnathostome estrogen receptors, α and β. B is the ancestral species of all the gnathostomes (vertebrates proper). It had estrogen α and estrogen β receptors. The teleost and tetrapod lineages diverged at 400 million years ago. The ancestral species that gave rise to these two lineages (C) had estrogen α and estrogen β receptors. This phylogeny of estrogen is based on findings in Thornton (2001).

and the gnathostomes was influenced by estrogen, and possibly equally so. No clear-cut evidence of estrus in this species exists. Only with the divergence of lampreys and gnathostomes do we see female-specific estrogen effects. Hence, we propose that estrus originated in the ancestral species of the gnathostomes—at time point B. (Though no clear evidence for an earlier origin exists, it is possible, we note, that estrus first emerged in the species of Bilateria that was ancestral to the protostomes and deuterostomes;

see phylogeny of steroid receptors in Thornton et al., 2003. If so, however, estrogen-regulated reproduction was lost secondarily in the ancestral species of the Ecdysozoa [arthropods, nematodes and relatives]; Thornton et al., 2003.)

Estrus in many species involves steroid hormones other than estrogen as well. Progesterone plays an important role in some vertebrates (Nelson, 2000). Progesterone is involved in the regulation of female reproduction across nearly all vertebrate groups (e.g., Nelson, 2000; Rasmussen & Murru, 1992), though its role arose more recently than did that of estrogen. The progesterone receptor originated early in the gnathostome lineage (Thornton, 2001). Androgen also affects female sexual motivation and may facilitate estrus in many vertebrate groups (e.g., Nelson, 2000; Rasmussen & Murru, 1992). The gnathostome androgen receptor evolved very early in vertebrates, though more recently than the progesterone receptor (Thornton, 2001).

In some vertebrates, notably externally fertilizing fishes and amphibians, non-steroids such as prostaglandins also control female sexual motivation (Adkins-Regan, 2005; Argiolas, 1999; Liley & Stacey, 1983). Liley and Stacey (1983) distinguish two types of hormonal regulation of female sexual behavior in vertebrates. In externally fertilizing species, in which mating occurs at oviposition rather than at ovulation (as in internally fertilizing vertebrates), hormones that regulate oviposition, such as prostaglandins, are important proximate causes of female sexual behavior. In contrast, estrogens emitted by maturing egg-bearing follicles regulate female sexual motivation in internally fertilizing species. Estrous sexuality may also play an important role in externally fertilizing species, however, as estrogen plays a causal role in the process of egg development; in that role, it may affect female sexual psychology in ways that regulate mate choice later, when oviposition occurs. In one anuran studied in detail (the túngara frog), estrogen may affect female sexual behavior at oviposition (Lynch, Crews, Ryan, & Wilczynski, 2006).

In addition to estrogen, other hormones may be involved in regulation of estrous mate choice in specific species. Eliminating activity of a gene responsible for oxytocin in house mice (through a gene knock-out method) prevents expression of females' preference for unparasitized males (Kavaliers et al., 2005). Hence a combination of hormones (estrogen, progesterone, oxytocin, and perhaps others) plays a causal role in mice.

In summary, major hormonal regulators of estrus—estrogen, progesterone, and androgen—existed in early vertebrates; estrogen regulation arose first. Although all three hormones affect female sexual motivation in many vertebrate taxa, estrogen appears to do so in the most widespread fashion (e.g., Nelson, 2000).

The Case of the Musk Shrew   One female mammal long thought to exhibit sexual behavior not facilitated by estrogen is the musk shrew (Soricidae, Insectivora). Female shrews copulate prior to follicular development, a time when blood estrogen levels are very low. The shrew aromatizes testosterone into estrogen in the brain, however, and this estrogen controls female sexual motivation (Rissman, 1991). The musk shrew is not an exception to the general rule that estrogen facilitates female sexual motivation in vertebrates.

Functional Similarity of Estrus Across Vertebrata

We have proposed that estrus is homologous (or, perhaps more precisely, possesses certain fundamental homologous features) across vertebrates. We also propose that estrus shares a basic function across all vertebrates: to obtain sires of superior genetic quality. Indeed, we specifically propose that the phylogenetic conservation of estrus within all vertebrate lineages (even if it has been lost in rare instances; e.g., see chapter 11) has occurred because it has been maintained by selection for this common functional effect—good-genes female choice. Hence estrus in fishes and amphibians has been maintained by selection for the same function that has maintained it in reptiles and mammals.

We note that these claims are not tautological. Traits that possess common functions may be analogues, not homologues. And a trait arising in a common ancestor may be maintained in many lineages without it being an adaptation, let alone an adaptation with a common function across the lineages. Incidental effects can be universally homologous within a phylum. The bones within teleost and tetrapod vertebrates are white or nearly white in color, but this feature has not been maintained by selection for its beneficial effects. Whiteness is an incidental by-product of selection for physiology (largely, densely packed calcium phosphate) conferring structural strength. Later in this chapter, we present evidence that, unlike the whiteness of bones, estrus was retained in vertebrate lineages by direct selection for it, not because estrus covaried with another directly selected trait.

Our proposal that estrus is homologous in vertebrates does not imply that estrus is identical in all vertebrate species. Naturally, many specific vertebrate species have evolved specialized, lineage-specific estrous adaptation, which coexists with the homologous features of estrus universal among vertebrates. Hence the female house sparrow and jungle fowl solicit copulations during the fertile phase of the reproductive cycle, and this phase within the species shares some homologous behavioral, motivational, morphological, and physiological similarities. Estrus in each of the two species is also dissimilar to that of the other in particular ways that function in the particular sexual ecology of each species. Hence, for instance, estrous female house sparrows prefer conspecific males with large, melanin-based breast badges, whereas estrous hens prefer roosters with large combs. Though females of these bird species both solicit copulations by crouching, females of other species are receptive at estrus in different ways: lordosis in the female rat, neck-bending in the female *Anolis* lizard, posterior body straightening in female snakes, or, as in many species, simply standing still to permit mounting (for descriptions of these estrous behaviors, see Nelson, 2000; Whittier & Tokarz, 1992).

The features that fully characterize estrus within any particular vertebrate species, then, had multiple origins during its descent with modification by selection. Hence, consider estrus in the house mouse (*Mus musculus*). The first estrous novelty to appear in ancestors of house mice, which was selected and maintained throughout its lineage, arose, we have argued, in the ancestral fish-like species from which all vertebrates are descended, a close phylogenetic relative of hagfish and lampreys. Other novelties may have originated at more recent time points along the lineage: for example, in the species of fish ancestral to sharks and other gnathostomes; in the species of sarcopterygiian fish ancestral to all tetrapod vertebrates; in the amphibian species that gave rise to all the amniote vertebrates (reptiles and mammals); in the reptilian species ancestral to all mammals; in the mammalian species ancestral to all rodents; in the species of rodent ancestral to the genus *Mus*. Rodent estrus is facilitated by the ratio of progesterone to estrogen (Nelson, 2000), for instance, and this specific feature of estrus in rodents possibly arose only in the species ancestral to rodents but not ancestral to other mammals.

The estrus of human females, too, has had multiple origins during its descent with modification by selection for sire choice. Women's estrus shares more homologous traits with estrus of other Old World primates (Catarrhini) than with New World primates (Platyrrhini) or the primates comprising the Strepsirhini. As a result of common ancestry, women's estrus is likely more similar in many ways to the estrus of the house mouse than to the estrus of goldfish or hen.

The fact that estrus within particular taxa is characterized by features very different from those of estrus of other species does not, of course, mean that estrus within the different taxa functions in completely different ways or does not possess common origins. Birds, for instance, are an unusual group of reptiles in that olfaction appears to play little role in their sexual behavior (but see Hagelin, Jones, & Rasmussen, 2003). Naturally, however, this fact does not imply that birds lack estrus. The attractivity of estrus in birds is due to features detected by males largely or solely through visual and acoustic modalities. In other vertebrates, males typically assess female-emitted scent associated with cycle fertility through olfaction and taste (Halpern, 1992; Mason, 1992; Thornhill, 1979). At some time point after birds diverged from other reptiles, novelties in attractivity of estrous female birds arose.

Ecological settings in which adaptive estrous choice for good genes occurs will generally be more similar within than between lineages. Nonetheless, widespread exceptions may exist, reflecting similar selection on estrous females in distantly related lineages. For example, certain estrous adaptations of lekking birds and mammals (e.g, choice for males that hold central territories) may reflect convergence. Similarly, estrous females in some fishes, pinnipeds, and penguins may have independently evolved to assess male genetic quality via their ability to swim well or fast. And, possibly, convergent evolution has rendered certain features of women's estrus more similar to taxonomically distant groups more similar functionally than comparable features in close relatives. (As we later discuss, for instance, women's estrus may have been shaped by pair bonding, which may have shaped estrus in many birds as well.)

The Importance of Recognizing Estrus as Taxonomically Widespread

The concept of estrus, we have argued, should be applied to the state of selective sexual motivation and related activities of females in the fertile phase of the reproductive cycle, regardless of the vertebrate taxon to which females belong. But why should it matter? Why is it useful or meaningful to designate the fertile state in all female vertebrates as estrus?

Use of a common term to recognize a common origin and a shared function is not merely semantic. Common usage is embedded within and given meaning by a theoretical framework—a theory about the nature, historical causes, and function of fertile sexuality in all vertebrates. This theoretical framework can promote research and discovery in three ways. First, it encourages analysis of the homologous traits involved in estrus across relatively closely related vertebrate species (e.g., across lizards or birds), as well as across distantly related vertebrates (e.g., fish and mammals). Phylogenetic analyses tend to examine traits within specific orders (e.g., primates) and to be restricted to particular traits that may co-occur with estrus (sexual skins and swellings; see Dixson, 1998; Sillén-Tullburg & Møller, 1993; Strassmann, 1996b). The theoretical framework we propose encourages phylogenetic reconstruction of a much broader set of morphological, physiological, and behavioral features that characterize vertebrate estrus.

Second, recognition that estrus is a general phenomenon promotes the application of the comparative method in the study of functional design and thus in understanding of historically effective selection. The comparative method uses data on divergence in adaptation of closely related taxa and convergence of adaptation in distantly related species to identify function. If woman's estrous sexuality possesses design to obtain good genes for offspring and homologous estrous sexuality of a female house mouse and a sage grouse possess similar design, a comparative approach yields convincing evidence for a fundamental function of estrus.

Third, this theoretical perspective promotes recognition that, in species in which extended sexuality occurs, there may well exist two functionally separable (even if overlapping) sets of sexual adaptations operating during different phases: one operating during an estrous phase typically associated with ovulation and high probability of conception and a phase of female extended sexuality associated with low or no probability of conception. Sage grouse hens visit potential mates at a lek during a restricted estrus corresponding to the egg-laying period and pick sires that are healthy (resistant to malaria and lice; Boyce, 1990). Female collared flycatchers, by contrast, mate outside of the fertile phase, especially with pair-bond mates (see chapters 3 and 10). These differences, however, should not blind us to seeing that sage grouse hens and female collared flycatchers share common features: They both possess estrus, which functions to obtain good genes. A difference between sage grouse hens and female collared flycatchers is that the latter also possess extended sexuality. Similar to extended sexuality where it occurs in mammals, female extended sexuality in collared flycatchers and other passerine birds appears to function to obtain material benefits (e.g., see chapter 3). The distinction between estrous sexuality and extended sexuality within vertebrates, broadly considered, can aid comparisons across species of the function(s) of estrus and of extended sexuality—specifically, the nature of the benefits females gain by sexual motivation inside and outside the cycle phase of peak fertility, through which selection shaped forms of female sexuality.

In some species (e.g., some bats and snakes), females mate in one season, store sperm, and then ovulate and produce offspring in a later season (Aldridge & Duvall, 2002; Birkhead & Møller, 1993b; Crews & Moore, 1986). In these species with "dissociated reproduction," according to our framework, estrus does not co-occur with ovulation or fertilization. Components of estrous adaptation, we argue, likely function to secure genetically superior sires. Hence, female choice that is focused on certain male traits because they connote superior genetic quality is estrous sexuality, whether or not it is associated with ovulation and fertilization. Females in species with dissociated sexuality possess estrus. By contrast, female matings that focus on obtaining nongenetic material benefits from males and have little or no prospect of fertilization exemplify extended sexuality. Females of species with dissociated reproduction may or may not possess extended sexuality. In these species, estrogen may underlie the sexual motivation of estrous females, despite the fact that estrus does not co-occur with ovulation. In at least one bat species, female sexual motivation appears to be independent of ovarian hormones (Mendonca et al., 1996). As in the musk shrew, however, estrous sexuality may involve nonovarian estrogen. If estrogen is in fact not involved in estrous sexuality in this bat, it would be a rare exception to the general pattern observed across mammals.

Adaptive extended sexuality, in our framework, typically requires two circumstances. One is male delivery of nongenetic material resources in exchange for sexual access to females. The second is that females have significant control of paternity of offspring, which, we hypothesize, is accomplished through the functional design of estrus. As male material assistance to females did not arise with the earliest vertebrates, estrus evolved earlier than extended sexuality. Indeed, estrus may have been the sole form of female sexuality for a long period of early vertebrate history. The occurrence and distribution of extended female sexuality in the fishes, amphibians, and nonavian reptiles will remain unknown until studies assess the extent of female sexual behavior outside estrus in these taxa. Such studies require awareness that estrus and extended sexuality are distinct functional forms of female sexuality. Here again, we hope that our framework will promote exploration of female sexuality in various nonmammalian vertebrates along lines not yet widely considered.

The Design of Estrus

Theory

The term *estrus* (introduced in the late 1800s to describe the female equivalent of male rut; Wikipedia, The Free Encyclopedia) derives from the Greek word for the gadfly, *oistros*. It refers to the frenzied state of the resisting cow when a bot fly (gadfly) attempts to lay eggs on her. By analogy, the estrous-phase female mammal is purportedly in a frenzied state of madness in her desire to mate. The analogy, though perhaps appropriate in some ways, may unfortunately imply that "sex-crazy" estrous females may be driven to and be satisfied by mating with any male. Many mammalogists and other biologists share this view. Alexander and Noonan (1979), for example, explicitly referred to the "relatively uninhibited receptivity of some estrous female mammals, which may accept essentially any male" (p. 442).

This view typically arises from the belief that estrous sexuality functions to ensure conception. Females are fertile during estrus. Within that time window, and only within it, females can conceive offspring. Hence, females seek sperm to achieve conceptions. That is, according to this view, female estrous behavior functions to solicit sperm from males.

As we discussed in another context (female signaling of fertility, or lack thereof), the view that females pay large costs to solicit sperm is theoretically problematic (see chapter 5). As we detailed there, sexual selection on males typically ensures that males will find and be willing to inseminate fertile females. The problem the typical female is likely to face is not a lack of males ready and willing to copulate and offer sperm, but rather having far too many of them around.

Modern evolutionary thinking, then, leads to a very different conceptualization of estrus. At the time of estrus, females can produce offspring, which represents alloca-tion of parental investment (see chapter 5). Adaptive allocation of parental investment demands that females scrutinize ecological conditions when deciding how and when to expend this investment. A widespread, important ecological condition affecting the adaptive expenditure of female parental investment appears to be the quality of the paternal genes that offspring will receive (see chapter 7). Hence female choices of whom to mate with during estrus represent choices of how to allocate valuable and limited reproductive effort to offspring. Selection should shape female estrous adaptations to make these choices wisely. Some males are better sires for a female's offspring than are other males. Estrous adaptations should function to lead females to discriminate between males and prefer to mate with males that represent adaptive sire choice.

In many vertebrate species, and mammalian species in particular, males provide very little or no material benefits to females or their offspring. In these instances, adaptive sire choice should substantially be about choice of a mate who can deliver genetic benefits to offspring (as well as minimization of costs of superfluous, nonadaptive, and outright maladaptive mating). In species in which males do deliver material benefits (e.g., includ-ing but not exclusively pair-bonding species with biparental care), males provide these material benefits but also still deliver genes to offspring. In these species, then, females should still generally possess estrous adaptations to assess and discriminate males' abil-ity to deliver genetic benefits to offspring. At the same time, females in these species can-not ignore the implications of sire choice for the expected future flow of material benefits they receive from pair-bond partners. Hence during estrus they must gauge not only the genetic benefits they could secure from sires but also the impact of sire choice for the flow of material benefits. These aspects of estrus clearly apply to humans, and we discuss them in chapters to come (chapters 10 and 12). In this chapter, we emphasize estrous adaptations for choosing sires able to deliver good genes to offspring.

We note that adaptation for seasonal achievement of reproductive condition by females and estrous adaptation are functionally distinct, despite being expressed during the same time period in most species. Adaptation for seasonal reproduction evolves by selection for timing that coincides with the most ecologically suitable time for offspring production for mothers or offspring or both. Estrus involves female sexual adaptation for obtaining sires of superior genetic quality.

Examples of Estrous Female Choice

If, in fact, sire choice meaningfully affects offspring success, we should expect that fertile females have been shaped to possess adaptations that favor choice of some males—those who possess superior genetic quality—to sire offspring over other males, all else being equal. And, indeed, evidence points to such adaptation in a variety of nonhuman mammals. In the Asian elephant, a bull's degree of dominance (predicted by testosterone level) predicts peak-fertility females' positive response to his scent, which is stronger than females' preference for this scent outside of peak fertility (Schulte & Rasmussen, 1999). Estrous-phase American bison cows approach dominant males but run away from subordinate ones. Accordingly, dominant males obtain more matings than do subordinates (Wolff, 1998). Similar behavior characterizes female African elephants and pronghorn antelopes in estrus (Byers, Moodie, & Hall, 1994; Poole, 1989). Indeed, estrous preference of pronghorns demonstrably results in offspring with good genes for survival (Byers & Waits, 2006). Estrous topi antelopes prefer lek males over resource-holding males (Bro-Jorgensen, 2003). Red deer in estrus prefer the roars of larger males over the roars of smaller males (Charlton, Reby, & McComb, 2007). Estrous meadow voles prefer males with good spatial ability over males with poor spatial ability (Spritzer, Meikle, & Solomon, 2005). Guinea pigs in estrus prefer heavier, more vigorously courting males over other males (Hohoff, Franzin, & Sachser, 2003). Similarly, female pademelons (a species of marsupial) prefer to associate with the largest male when presented with males of different sizes, but only during estrus (Radford, Croft, & Moss, 1998). In house mice, only females in estrus show an olfactory mate preference for the scent of males with wild-type t-alleles (over males bearing a t-allele that lowers fertility and survival of offspring; Williams & Lenington, 1993). Estrous house mice also prefer the scent of males carrying MHC dissimilar alleles (Potts, Manning, & Wakeland, 1991) and males with relatively few parasites (Kavaliers & Colwell, 1995a, 1995b). Estrous snow voles prefer the scent of males with relatively high levels of hematocrit (an indicator of good nutrition and health) and developmental stability (low FA; Luque-Larena, López, & Gosálbez, 2003). Preferences by estrous females for male-produced odor stimuli have been documented in a variety of other rodents (see reviews in Gosling & Roberts, 2001; Hurst & Rich, 1999). Fallow deer time their estrus to coincide with the presence of older dominant males (Komers, Birgersson, & Ekvall, 1999). The scent of dominant males, but not that of subordinate males, induces estrus in the house mouse (Novotny, Ma, Zidek, & Daer, 1999), and only female house mice in estrus prefer the scent of dominant males (Rolland, MacDonald, de Fraipont, & Berdoy, 2004). Beach (1970) noted that his studies of sexual behavior in domestic dogs revealed that females in estrus "are not indiscriminately receptive, but accept some males much more readily and enthusiastically than others" (p. 445; see also Le Boeuf, 1967). Female choice during estrus also has been observed in tree squirrels (Koprowski, 2007), 13-lined ground squirrels, and feral domestic cats (see discussions later in this chapter). More generally, biologists increasingly recognize that female choice by estrous females in nonprimate mammals importantly determines male mating success (Ginsberg & Huck, 1989).

And the same appears to be true of nonhuman primates. Estrous pygmy lorises prefer competitive males (Fisher, Swaisgood, & Fitch-Snyder, 2003). In the laboratory, estrous-phase rhesus macaques prefer males whose faces have been experimentally manipulated to reveal exaggerated red coloration, a testosterone-facilitated male sexual ornament (Waitt et al., 2003). The copulation calls of female catarrhines may be estrous female-choice adaptations that function to promote conception with high-quality males and exclude mating with other males (see chapter 5). In female nonhuman primates with extended sexuality, copulation with dominant males often coincides with peak fertility (see chapter 5, this volume). Though the relative impacts of female choice, male-male competition, male sexual coercion, and male choice on male access to females are not always clear, female choice appears to play a major role in many nonhuman primates (Dixson, 1998; Pazol, 2003; Stumpf & Boesch, 2005; Wallen, 2000).

Consider, once again, common chimps (see also chapter 3). Males socially dominate females. Yet females show more selective mating during peak estrus (fertile phase of the estrous cycle of about 3–4 days) than during the low-to-zero fertility days of the cycle surrounding both sides of peak estrus (Stumpf & Boesch, 2005). Female rates of proceptivity (which decrease during estrus) and resistance (which increase at that time) affect male access to mating, with (in the small sample Stumpf and Boesch studied) up-and-coming dominant males having greater access. By contrast, during the extended female sexual phase, female chimps mate more promiscuously with multiple males. Peak estrous sexuality of chimps may well function largely to obtain good genes for offspring, unlike extended sexuality, which arguably secures material benefits (see chapter 3).

Because estrus has not been thought to characterize nonmammalian vertebrates, the estrous behavior of females in these species has been less well studied than that of mammals. As we previously discussed, collared flycatchers preferentially mate with extra-pair males when fertile and prefer as extra-pair males those who possess a purported indicator of good genes (see chapter 3 and also chapter 10, this volume; cf. Brommer et al., 2007). Recent research reveals that, as female túngara frogs approach the time for egg deposition (and hence egg fertilization), their ability to discriminate acoustical signals of conspecific males increases (Lynch, Rand, Ryan, & Wilczynski, 2005). Female midwife toads too possess enhanced ability to discriminate males' calls during the ovulatory phase of their reproductive cycle (Lea, Halliday, & Dyson, 2000). In an African cichlid fish, gravid females in fertile phase, but not females in infertile phase, prefer territorial males with high levels of behavioral activity (Clement, Grens, & Fernald, 2005).

Finally, in the guppy, estrus lasts a few days. The ova develop just before the birth of a litter. Female guppies are sexually receptive, respond to males, and mate within an interval of several days prior to giving birth. They are sexually unreceptive at other times (i.e., they lack extended sexuality). Males are sexually attracted to a scent that receptive females emit (Houde, 1997). Though fertile-phase guppies often mate with multiple males during a single estrus, they partly control the size of inseminates. As a result of female preference for their semen, more highly ornamented males transfer larger ejaculates (Pilastro, Evans, Sartorelli, & Bisazza, 2002), which may account for these males' greater paternity in broods with multiple sires (Pitcher, Neff, Rodd, & Rowe, 2003).

In sum, across a wide variety of mammals and other vertebrate taxa, estrous females are choosy, not indiscriminate. Furthermore, in a host of systems, evidence clearly points toward estrous female choosiness for mates of superior genetic quality.

Estrus and Constraints on Female Choice

Conception-Assurance Adaptation

The common belief that estrous females are indiscriminate and interested in mating with most any male, despite the many studies that contradict it, may partly be rooted in misleading observations made in laboratory settings in which females possess little choice. As Nelson (2000) wrote:

> Female mammals in estrus have often been portrayed as "out of control" because they appear to be indiscriminate about their mating partners, but part of this portrayal results from the laboratory testing situations used, especially for rodents. Dogs and many other mammals (especially primates) display substantial selectivity when in estrus. (p. 281)

At the same time, the view that estrous sexuality importantly reflects adaptation that functions to secure sires of high genetic quality does not preclude female sexual adaptation to assure conception. When females are conditionally constrained in their mate choices (or pay heavy search costs or costs for resisting available males), their best option may be to relax their standards of mate choice and mate with available males. In these species, however, we hypothesize that estrous adaptations that function to obtain sires with superior genes also exist. That is, we propose that, in some species, females possess conditional conception-assurance adaptation *in addition to* estrous adaptations to obtain genetic benefits for offspring. We also suspect, but cannot know for certain, that female conception-assurance adaptation is relatively uncommon across species. In most species, we suggest, sexual selection on males will have designed them to reliably find and inseminate fertile females, and females will not have been exposed to the conditions that could favor conditional conception-assurance adaptation.

When costs of assessing male genetic quality are high (e.g., due to predation risks or energy expenditure), estrous females should be naturally selected to reduce them—for instance, by limiting their movement. Some biologists assume that high costs of choice lead females to sample just a few males, thereby restricting female ability to obtain sires with good genes (e.g., Crowley et al., 1991; Pomiankowski, 1987; Real, 1990). In a review of evidence on 11 species, Gibson and Langen (1996) found that searching females sampled only an average of four males (for additional evidence on the pied flycatcher and bowerbird, see also Dale, Rinden, & Slagsvold, 1992; Uy, Patricelli, & Borgia, 2001).

At the same time as females constrain their movement to reduce search costs, however, males should be sexually selected to encounter sedentary estrous females (see chapter 5). Hence females in many species may be presented with and able to choose males of relatively high quality, even when they limit search costs. Of course, we do not argue that female choice is cost-free in terms of energy and time. Rather, we simply suggest that future studies consider the possibility that sexual selection on males typically means that suitable males are available to fertile-phase females. (Cost reduction may also take the form of relatively low-cost choice strategies, such as copying the choices of other females observed to choose mates; Pruett-Jones, 1992; Godin, Herdman, & Dugatkin, 2005.)

We similarly expect that sexual selection on males to find fertile females generally means that seasonal constraints on reproduction do not typically limit the ability of females to have their eggs fertilized by males of high genetic quality. Exceptions may exist during late season if, at that time, many males have died.

Another situation that may have been recurrent at sufficiently high rates to yield selection for conception-assurance adaptation and hence reduced discrimination is that in which females must deposit eggs (to be externally fertilized) within a short period of time following egg maturation, as in anurans and many fishes. For example, túngara frogs must lay their eggs soon after egg maturation (Lynch et al., 2005). (In contrast, female midwife toads can store mature eggs for a considerable period of time; Lea et al., 2000.) Accordingly, selection has produced a female sexual adaptation of reduced discrimination, leading female túngara frogs to mate with any males exhibiting any call type. Reduced discrimination, however, is highly conditional, characteristic only of females that are highly gravid with eggs. Moreover, this conception-assurance adaptation coexists with estrous sexuality for obtaining good genes. Indeed, egg-laden female túngara frogs retain a strong *preference* for male calls that combine the whine-chuck, which may be a preference for a sire of superior genetic quality (see Lynch et al., 2005, 2006). Only when offered no choice, then, will gravid females mate with lower quality males. In this species, males of high genetic quality may not be present, despite sexual selection on males to find egg-laden females, because of high rates of sex-specific predation or limitations imposed by breeding habitat (e.g., as in other anurans, the absence of a local, suitable body of water for mating aggregations and egg deposition).

As Nelson (2000) notes (see the preceding quotation), artificial laboratory studies may have misled researchers to the view that estrous female mammals mate indiscriminately. If an estrous female is placed in a cage with a single male, mating typically occurs. It need not follow that estrous females possess no preferences. First, in most mammalian species, males compete for access to mates, and hence females may perceive single males with whom they are caged as winners of a competition. (Otherwise, why would they have gained access to a female, with no interference from other males?) Second, female willingness to mate in these situations may reflect the same kind of adaptation that leads female túngara frogs to mate indiscriminately when highly gravid: back-up adaptation to assure conception under severe mate-choice constraints.

By no means, however, do females generally express adaptation to ensure conception when mate choices are limited. Zoos and captive breeding programs, in fact, frequently face the problem that females reject males that zookeepers provide as mates, and not merely because cage conditions are unnatural (Møller & Legendre, 2001). Indeed, the Allee effect can be observed in many species: In conditions of low population density and, accordingly, poorer opportunities for mate choice, females reproduce at lower rates, even when healthy and well nourished (for a review, see Møller & Legendre, 2001). Hence females of many species appear to possess adaptation to reduce immediate expenditure of valuable parental investment when mate choice options are restricted.

Sexually Ardent Males Constrain Female Choice

As we have emphasized, female reproductive success is generally limited by female abilities to optimize the expenditure of their parental investment, where optimization partly depends on ability to place good paternal genes in offspring through appropriate choice of sires. Ability to secure good genes by females, however, is not typically limited by male willingness or ability to deliver them by mating; males are selected to be willing and able to deliver them. Female choice may nonetheless be constrained widely by the outcomes of intense male-male competition for females. Male sexually selected adaptations that function to promote *male* reproductive success may lead to outcomes that are not in female reproductive interests. Hence males may be selected to manipulate and/or control female reproduction. Sexually antagonistic selection, which gives rise to a "battle of the sexes" (intersexual antagonistic coevolution; Arnqvist & Rowe, 2005), dates in the history of life to the appearance of anisogamy (Gowaty, 1996; Parker et al., 1972), and it continues. Males are relentlessly selected to circumvent female mate choice by control, manipulation, and coercion, and females in turn are selected to control the timing and other events surrounding adaptive parental investment, including sire quality. (See also Gowaty, 1997, on "sexual dialectics theory.") Earlier, we mentioned female resistance to males and other female traits that apparently evolved in this context (e.g., estrous female guppies' behaviors of seeking stream habitats that are too costly for low-quality males to occupy and female primate copulatory calls that reduce coercion by unwanted sires; see chapter 5).

From a theoretical perspective, it makes sense that females who possess free choice can attain greater reproductive success than females whose choice is constrained (assuming that females choose males who provide genetic benefits). Empirical studies demonstrate, in dramatic fashion, this phenomenon in Madagascar cockroaches, fruit flies, house mice, and mallard ducks (Bluhm & Gowaty, 2004a, 2004b; Drickamer, Gowaty, & Holmes, 2000; Moore, Gowaty, & Moore, 2003; Partridge, 1980). In these studies, females achieved higher reproductive success when allowed to reproduce with preferred males than when assigned a male partner by researchers or (in studies of cockroaches and ducks) when males were permitted to interfere with female mate choice through manipulation or coercion (see review in Møller & Legendre, 2001).

Estrus, Polyandry, and Multiple Paternity

Multiple Paternity in Relation to Types of Genetic Benefits

In many vertebrate species, females typically mate with more than one male during a single estrus. One might think that, if females seek genetic benefits for offspring during estrus, they should mate only with the single male who, of available mates, displays the greatest potential for good genes. In fact, however, that need not be the case.

As we discussed in chapter 7, females may seek three different kinds of good genes (Jennions & Petrie, 2000; Zeh & Zeh, 2001): intrinsic good genes, compatible or complementary genes, and diverse genes. Females may seek any one, or even all three, within the estrous phase. Indeed, in some species across diverse taxa females may produce offspring with multiple paternities within one litter or clutch (for birds, see chapter 10, this volume; for mammals, see partial review in Zeh & Zeh, 2001; for lizards, Morrison, Keogh, & Scott, 2002; for snakes, McCracken, Burghardt, & Houts, 1999; for turtles, Pearse, Janzen, & Avise, 2002; for fishes, Kelly, Godin, & Wright, 1999). Offspring with different fathers may optimize different good-genes choice. Clearly, if females choose mates for diverse genes, the point of sire choice should be to obtain offspring with different fathers. This strategy appears to characterize the choices of females of the ruff, a lekking bird without male parental care (Lank et al., 2002). But females may also seek multiple fathers if some males offer intrinsic good genes and others offer compatible genes, even in absence of selection for diverse genes. And, as we discussed in chapter 7, females may select compatible genes through postcopulatory choice mechanisms; some forms of genetic compatibility may be most readily assessed when male DNA can be scrutinized within the female reproductive tract (Zeh & Zeh, 1997).

Sexually Ardent Males and Multiple Paternity

Male sexual ardor may also partly explain multiple paternity in some species. Against the interests of females, selection may favor males that can achieve conception despite failure of females to choose these males as sires. In feral domestic cats, multiple paternity is rare, and socially dominant males sire most offspring when the density of males is moderate (Say, Devillard, Natoli, & Pontier, 2001; Say, Pontier, & Natoli, 2002), presumably due to female choice (Ishida, Yahara, Kasuya, & Yamane, 2001). By contrast, when male density is unusually high, as in many urban areas, multiple paternity in single litters is common. When males are densely distributed, single males cannot monopolize estrous females. Similarly, in the common toad, multiple paternity within single egg batches is due to high density and an excess of males, which leads to multiple males amplexing a female at oviposition (Sztatecsny, Jehle, Burke, & Hoedl, 2006).

Multiple paternity within litters in the 13-lined ground squirrel may similarly be the result of ardent males partly circumventing female preference. In these squirrels, estrus lasts a few hours, during which females typically mate with multiple (usually two) males. Males who first find a female sire 75% of her litter. Females may prefer these males because they may produce sons skilled in mate searching or refinding females after interruption of sexual interactions by later-arriving males (as first-finding males are) or may be in better condition (as first-finding males also are; see Schwagmeyer & Parker, 1987, 1990). By contrast, the paternity that late-arriving males achieve may be due largely to their ability to obtain fertilization despite female preference for skilled mate-searchers in good condition.

Perhaps the best illustration of both female choice during estrus and effort by ardent males to inseminate females contributing to multiple paternity in single broods is provided by the guppy (Kelly et al., 1999). In this species, males are attracted to the scent produced by estrous females and court them. Females prefer males with colorful ornamentation and that vigorously courtship display. When rates of predation by larger fish are high, however, males adaptively reduce their courtship effort and, instead, are more likely to attempt forced copulation without courtship. In the absence of courting males, females in turn relax their otherwise high standards for mate choice. Kelly et al. (1999) found that multiple paternity within broods varied considerably across 10 natural populations. In those populations in which level of predation was relatively high, so too was the level of multiple paternity. When the level of predation was low, the level of multiple paternity was low as well. This pattern is explained if, when female choice is adaptively relaxed because males provide less information through courtship, sexually coercive males achieve greater success in conception than when female choice fully determines paternity.

Are Estrous Females Designed to Promote Sperm Competition?

As just discussed, multiple paternity may occur despite female preference when nonpreferred males coerce matings with estrous females. Alternatively, estrous females may seek multiple paternity through polyandrous matings, as when females pursue multiple types of genetic benefits during a single estrus.

One additional hypothesis discussed widely in the literature is that females are adapted to promote sperm competition (e.g., Baker & Bellis, 1995; Birkhead & Møller, 1998; Shackelford & Pound, 2006). According to this idea, by mating with multiple males during a fertile period, females establish sperm competition. If winners come from males with highly competitive ejaculates, sons too may possess competitive ejaculates. If winners come from males more fit in general (and hence able to produce more viable sperm), sons may also be relatively fit.

The sperm competition hypothesis requires that individual females *prefer* multiple mates during a single estrus. Multiple mating solely due to coercive strategies of males is not consistent with it. Studies of polyandry have not explored in detail whether females prefer multiple mates during a single estrus. Rather, researchers typically record polyandry by individual females without regard to cycle phase or cause (e.g., female preference for multiple mating vs. avoidance of costs of resistance; see, e.g., the many studies of mammals reviewed by Wolff & Macdonald, 2004).

One exception is research on red jungle fowl by Ligon and Zwartjes (1995). Hens in laying condition (i.e., estrus) were offered a choice of two roosters. The roosters were tethered to prevent forced matings with hens. Each hen was run in multiple mating trials across several days with the same pair of roosters. Nearly all hens chose to mate with both males over successive trials, revealing that jungle fowl hens choose to obtain sperm from more than one male during the production of a single clutch of eggs. It should be noted that males in this study were isolated before the trial and prevented from engaging in male-male competition both before and during the trials. This procedure may elevate and equalize roosters' self-assessment of social dominance and hence their display of dominance to choosing hens. Hence the results show only that estrous hens prefer insemination by multiple socially dominant males. They do not demonstrate that hens prefer multiple mating with males of lower quality. The results are consistent with an explanation that hens run sperm competition races to be inseminated with good sperm. They are also consistent with female postcopulatory choice of compatible genes (e.g., Zeh & Zeh, 1997).

In chapter 10, we discuss the issue of whether women are adapted to promote sperm competition.

Estrus Will (Nearly) Always Accompany Extended Female Sexuality

In many species, females have estrus with no accompanying extended female sexuality. The converse, however, should occur rarely, if ever: Females should not possess extended sexuality without estrus. Female extended sexuality makes little adaptive sense without estrus. Female choice adaptations of estrus provide control over paternity, which allows extended sexuality to reap male nongenetic material benefits provided by an unsuitable sire at low to zero probability of fertilization by him. Extended female sexuality, in other words, will be favorably selected only when there is adaptation that functions to control sire choice in place in the female's reproductive repertoire.

Because males in species with extended female sexuality imperfectly discriminate female fertile cycle state from infertile states, it behooves them to copulate when opportunity presented by female sexual interest arises, irrespective of the female's cycle phase. Hence, when females are available for mating across their cycles, extended male sexual competition occurs. Selection then can favor females that choose mates during low to zero fertility cycle phases, but typically females will use different criteria when exercising choice at these times than during estrus. As we discuss in the next chapter, the most detailed support for different female choice criteria during extended sexuality than during estrus, is found in research on human females.

In some bird species with biparental care, females do not appear to prefer mating with males other than primary mates, even when fertile. In these species, adaptive sire choice may simply constitute choice of the social partner, regardless of his relative genetic quality, and female sexuality during the fertile phase may differ minimally from female sexuality during extended sexuality. As we mentioned earlier, implications of choosing a sire other than a pair-bonded partner for the flow of nongenetic material benefits received by that partner cannot be ignored if female choice of a sire during estrus is to be adaptive; in some cases, the implications of losing a partner's investment may outweigh any potential benefits derived from seeking another male's genes for offspring. We discuss possible examples in chapters 10 and 12. As we also discuss in these chapters, however, species in which females possess no adaptation to choose sires other than social partners are relatively rare among socially monogamous birds. Furthermore, humans do not qualify as such a species.

The Musk Shrew Probably Has Estrus

As noted earlier, the musk shrew has been widely claimed to be a mammal that lacks estrus. Females mate prior to, during, or after ovulation. Hence, unlike the typical mammalian female, female musk shrews do not limit sexual behavior to the period of ovulation (Schiml, Wersinger, & Rissman, 2000). For this reason, researchers have argued that this species lacks estrus. Female mating motivation across the reproductive cycle does not, however, imply an absence of estrus. (See chapter 9 on women.) A claim that a species lacks estrus despite extended sexuality is a claim that females of the species apply precisely the same choice criteria to mating choice within and outside of fertile states. Little is known about the sexual behavior and mating system of musk shrews in nature, though much laboratory research has been conducted on their reproductive biology (Jameson, 1988). Female musk shrews are energy limited due to a high metabolic rate and a very limited ability to store energy. They have high rates of food consumption when food is available, and food availability determines female reproductive capability (Temple, Schneider, Scott, Korutz, & Rissman, 2002). Females acquire food from territories held by males. Males are sometimes polygynous, and multiple females cohabit the territory held by a single male.

We suggest that female musk shrews do have both extended sexuality and estrus. Extended sexuality in this species functions to allow females to occupy and feed in males' territories, which, in turn, allows females to achieve an energy threshold sufficient for ovulation. That is, acquisition of material benefits may explain why female shrews routinely mate outside the ovulatory phase of the cycle. At estrus, we propose, female musk shrews exhibit mate preference for male traits associated with high genetic quality. Evidence supporting the existence of estrus in the musk shrew would be comparable to that in other mammals. If, for example, future research reveals that female shrews at the ovulatory phase of high conception probability show a sexual preference for certain males (e.g., socially dominant males) but during other cycle phases are sexually attracted to a less restricted subset of males, the appropriate conclusion is that musk shrews possess estrus. Clearly, female musk shrews have extended sexuality. Again, we propose that it only makes theoretical sense that they also possess estrus in their reproductive cycle, which functions to control the quality of their offsprings' sire(s).

Estrous Phase Is Temporally Restricted

Estrus in Nonhuman Vertebrates

Estrous behavior typically occurs during a restricted time period, whether it is coupled with conception or seasonally dissociated with ovulation. The duration of estrus is often about 12 hours to 2 days in ungulates (e.g., domestic cow, goat, pig; white-tailed deer; sable antelope; Landaeta-Hernandez et al., 2002; Romano, 1997; Steverink et al., 1999; Thompson & Monfort, 1999; White, Hosack, Warren, & Fayrerhosken, 1995; Young, Nag, & Crews, 1995). Female lemurs are in estrus about one day (Stanger, Coffman, & Izard, 1995; Wrogemann & Zimmermann, 2001). In tree squirrels (Sciuridae), females may attract males for several days before estrus, but females are typically in estrus for less than 1 day and often less than 8 hours (Koprowski, 2007). Based on data for the duration of female sexual receptivity, we surmise that estrous behavior typically lasts several days in lizards, snakes, and turtles (Whittier & Tokarz, 1992; Young et al., 1995; Weiss, 2002). Estrus in female birds appears to coincide with the period just prior to and during egg-laying and thus the reproductive cycle phase of maximum conception probability (see Birkhead & Møller, 1992). Estrus is said to last 6–9 days in African wild dogs, but mating coincides with a brief peak of estrogen levels (Monfort et al., 1997).

Only Estrous Copulation Has Maternal Investment Implications

Biologists have often assumed that copulation in general is costly to females (e.g., Symons, 1979; Trivers, 1972). We suggest that copulation is less costly to female reproductive success (particularly in species with extended female sexuality) than often assumed. The higher cost of copulation to females than to males reflects females' higher obligate investment, relative to males', necessary for offspring production. Female sexual behavior need not be costly, however, when females are not fertile. Copulation by female birds or women occurring during extended sexuality has no cost in terms of its implication for expending maternal investment on a resulting conceptus. (We note, however, that costs of infertile sex need not be equal for the two sexes, as other costs of mating may be sexually asymmetric. For instance, the cost of contracting sexually transmitted disease may be greater for females—who, for instance, are more likely to suffer infertility as a result; see, e.g., Nunn, 2003—than for males. Similarly, male seminal fluid may contain toxins or neuroactive substances designed to manipulate female behavior against her interests; see, e.g., Rice, 1996, and references therein.)

The low cost of copulations during extended sexuality to females in terms of their parental investment suggests that the benefits females obtain from extended sexuality need not be as large as assumed under the traditional view that female copulations involve heavy costs in potential parental investment. If, in fact, copulation outside of the fertile phase is not highly costly to females, male-delivered material benefits need not be particularly great for them to exceed those costs, yielding adaptive extended sexuality. Even material benefits that are seemingly minor or subtle can pay for the costs of extended female sexuality—for instance, nuptial feeding of hens by roosters; mate guarding by some male birds, reducing risk of sexual coercion for females; grooming of some nonhuman female primates by males; and males forming temporary social alliances with females that assist females to access status and its associated resources (see chapter 3).

As we discussed in chapter 4, the capacity of men's material benefits and services to promote women's reproduction is major and nonsubtle in humans living as hunter-gatherers and probably accounts for the extreme degree of extended sexuality seen in human females.

Human Estrus Was Not Lost or Extended

Our analysis of estrus is contrary to typical views in the literature about women's estrus. As noted in chapter 1, many authors have concluded that human females lost estrus during their evolutionary history or that they extended estrus through the evolution of permanent sexual ornaments that deceptively signaled high conception probability in the menstrual cycle, thereby disguising cycle-related fertility (e.g., Alexander, 1990; Alexander & Noonan, 1979; Burt, 1992; Morris, 1967; Steklis & Whiteman, 1989; Strassmann, 1981; Symons, 1979; Szalay & Costello, 1991; Turke, 1984; and many others). In general, these authors have assumed erroneously that estrus functions to signal peak conception probability in the cycle and that estrus in a primate is equivalent to the presence of female sexual swellings. Burt (1992), for example, argued that sexual swellings, and hence estrus, was lost in the hominin line, not through direct selection associated with advantages of concealed estrus but because swellings are costly and, in human females, the benefits of signaling ovulation with swellings did not pay for their costs (see also Pawlowski, 1999a).

Estrus, however, is not about signaling to males a female's fertile state. More generally, then, phylogenetic analyses of the loss of a particular type of sexual ornament of Old World primates—female sexual swellings (Sillén-Tullberg & Møller, 1993; Strassmann, 1996b)—by no means address the loss of estrus among these species. Sexual swellings were independently lost in primates about 6 (Strassmann, 1996b) or 8–11 (Sillén-Tullberg & Møller, 1993) times. We have no reason to believe, however, that estrous sexuality has ever been lost through evolution in primates.

As we have emphasized in this chapter, estrous sexual adaptations partly function to lead females to prefer as sires males that offer superior genes for offspring. In chapter 9, we lay out the evidence that supports the claim that women have not lost estrous adaptations of this sort. If correct, woman's estrus may have had its first phylogenetic origin in the fish-like ancestor of all the vertebrates, though, similar to other vertebrates, specific features of women's estrus have had multiple origins (including some features with origins since hominins diverged from their common ancestor with chimpanzees).

Again, because scholars have often linked estrus with signaling of fertility, they focus on the loss of female signaling of fertility in humans—which, we argue, is misguided because ancestral hominins never had adaptations to signal fertility. At the same time, selection on ancestral hominins may have directly favored female traits that disguise the incidental physiological and other correlates of cycle-related high fertility, as well as disguise sexual interest at peak cycle fertility in sires with good genes, except when mating with them. We discuss this evidence in chapter 11. In a very circumscribed sense, then, women may be thought to have lost *behavioral* estrus (*heat*), as women's overt proceptivity, receptivity, and attractivity, we suspect, does not greatly change across the cycle (though see later chapters). This possibility, however, should not be confused with the broader claim that women have lost estrus and estrous adaptations (see chapter 9).

Summary

Although the estrogen receptor arose earlier, estrogen-facilitated discriminative female sexual motivation at the high fertility phase of the reproductive cycle, thus estrus, had its phylogenetic debut in the fish-like ancestor of all the vertebrates. Estrus, then, is homologous across all vertebrates. The maintenance of estrus after its phylogenetic origin involved selection for its effect of good-genes sire choice. Furthermore, lineage-specific selection accounts for the different forms of good-genes preferences exhibited by females in different taxa. In many species, estrus is facilitated by steroids other than estrogen and, in externally fertilizing species, by certain nonsteroids. However, estrogen remains a fundamental proximate cause across vertebrates. The homology of estrus across the vertebrates is seen too in neurological structures affecting female sexual motivation at the fertile phase of the female reproductive cycle. Estrus occurs in the musk shrew and in species with dissociated female sexuality.

Although the term *estrus* has been applied almost exclusively to female sexuality at the fertile cycle phase in nonhuman mammals, it is important to recognize the homology of estrus across vertebrates. Such recognition promotes phylogenetic research on female sexuality, as well as the separation and functional analysis of the two types of female sexuality in species with dual female sexuality.

Studies of a variety of nonhuman mammals and other vertebrate species reveal that estrous females are choosy, not indiscriminate as often thought, preferring male traits that reflect actual or likely high genetic quality. Estrus does not function to obtain sperm but instead functions to achieve adaptive sire choice. Adaptation to choose sires that possess genes that increase offspring reproductive value is almost always, if not always, a fundamental component of estrus. Conception-assurance adaptation likely is uncommon because sexual selection on males has designed them to find and inseminate fertile females. Estrus is retained in species with conception-assurance adaptation.

Estrus may function in some species to obtain a combination of genetic benefits for offspring (diverse, compatible, and intrinsic good genes), leading to multiple paternity or sperm competition. If females have adaptation to promote sperm competition, they will exhibit a preference for multiple males during a single estrus. Multiple mating by estrous females and multiple paternity may arise commonly as a result of sexually ardent males coercing copulations.

In species with biparental care and pair bonds, adaptive sire choice must consider the impact of sire choice on the future flow of material benefits and their implications for reproductive success. Only rarely, however, do these impacts fully override potential benefits of seeking a sire other than a social partner for good genes, such that females possess no adaptation for seeking good genes during their fertile phase. Humans do not qualify as such a species (see chapters 10 and 12).

Typically in vertebrates estrus is the sole type of female sexuality. Extended female sexuality will not evolve without accompanying estrus because estrus serves to assure sire quality. Only estrous copulations have potential for maternal investment. Extended sexuality matings do not have this cost.

Loss of estrus and concealed estrus are not equivalent. Human estrus was not evolutionarily lost. Nor was it extended in the form of permanent female ornamentation that deceptively signals female cycle fertility. The evolutionary loss of sexual swellings in certain Old World primates is not equivalent to loss of estrus.

9   Women's Estrus

Women Are Expected to Possess Estrus

As we noted in our introductory chapter, over four decades ago researchers offered a definitive conclusion about women's sexuality: Over the course of the last several million years, women had lost estrous sexuality. Modern women possess no distinct fertile sexuality within their cycles. The theoretical perspective we presented in chapter 8 casts serious doubt on this conclusion. Nearly all vertebrates, we suggest, possess estrous sexuality. Many researchers, however, have misunderstood estrus. Estrous sexuality does not function to obtain sperm or ensure fertilization; these effects are by-products of estrus's function. Female reproduction is only very rarely limited by the availability of male sperm; typically, the abundant male willingness and ability to deliver sperm that sexual selection yields ensures that females have no trouble obtaining sperm. Rather, estrus functions to lead females to target as sires males who offer genetic benefits for offspring, *relative to other males*. Females in estrus should not be indiscriminate in their mate choices. Indeed, they may be even choosier during estrus than outside of it. In species with extended sexuality, what should distinguish estrous sexuality from extended sexuality are mate preferences: the features that females find most sexually attractive.

Nearly a decade ago, we discussed what has become known as the ovulatory shift hypothesis (Gangestad & Thornhill, 1998; Thornhill & Gangestad, 1999b). (We should note, however, that this hypothesis was anticipated several years earlier by Karl Grammer's [1993] pioneering work on women's scent preferences. Hence, he should be credited for formulating this hypothesis.) The hypothesis predicts that women prefer male features that would have been indicators of male ability to deliver genetic benefits to offspring, particularly when they are fertile in their cycles. That is, when fertile in their cycles, women should find purported indicators of good genes more attractive than they do outside of the fertile phase. Penton-Voak et al. (1999) offered an important addendum to this hypothesis: Preference shifts should be particularly strong when women evaluate men's "sexiness" or attractiveness as sex partners, not their attractiveness as long-term, pair-bonded mates. The traits women prefer in long-term partners should be relatively constant across their ovarian cycles. Hence, when fertile in their cycles, women should find purported indicators of good genes more sexually attractive than they do outside of the fertile phase.

In the current chapter, we review the considerable body of evidence that speaks to the ovulatory shift hypothesis, most of it accumulated in the past decade. In chapter 10, we address the implications of women's estrus for an understanding of human mating patterns considered more broadly.

Women's Estrous Phase

Women's estrus is comparable in length to that of many other vertebrates. The "fertile window" is an approximately 6-day period of time ending on the day of ovulation. Probability of conception associated with insemination rises gradually for a couple of days, then more steeply, peaking 1 to 2 days prior to ovulation, if not the day of ovulation itself (e.g., Wilcox et al., 1995; Dunson, Baird, Wilcox, & Weinberg, 1999). The day before the window and the day after ovulation are associated with very low probabilities of conception. The timing of the window in the menstrual cycle is somewhat unpredictable, even in women with regular cycles (e.g., Dunson, Weinberg, Baird, Kesner, & Wilcox, 2001; Jöchle, 1973; Wilcox, Dunson, & Baird, 2000; Wilcox, Dunson, Weinberg, Trussell, & Baird, 2001). A common pattern, however, is that shortly after menses, which typically lasts 5 days, probability of conception rises and peaks on about day 12, then sharply declines at ovulation, typically on day 14 in a 28-day cycle (Wilcox et al., 2001; see figure 9.1). In a sample of 221 women without fertility problems and attempting to conceive, Wilcox et al. (2001) found that 84% had regular cycles. Peak fertility in an individual woman's cycle, however, is brief, typically 2–3 days preceding ovulation. Presumably, this phase has the greatest homology with the estrous phase of most other vertebrates.

Bullivant et al. (2004) found that women initiated sex most frequently during the fertile window (though see our extended discussion of the long-debated issue of whether women's "sexual desire" is maximal during the fertile window in chapter 10). Based on this finding, they suggested that women's fertile window be termed "the sexual phase." As they appropriately noted, some terminology conventionally used by human reproductive biologists is problematic. For instance, the "follicular phase" spans days that differ dramatically in terms of typical types and quantities of hormones released and associated conception risk. Nonetheless, Bullivant et al.'s term for the fertile window, the *sexual phase*, is perhaps even more misleading. It implies that women are not sexual outside of this phase (i.e., do not possess extended sexuality), which is plain wrong. The term *estrus* captures the comparative homology, as well as

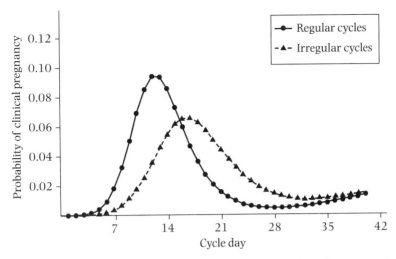

Figure 9.1 The probability of pregnancy/conception risk with one act of intercourse in relation to day of the menstrual cycle for women with regular and irregular cycles. N = 221, 16% irregular cycles. Based on figure 2 of Wilcox et al. (2001); reprinted with permission of Elsevier.

important components of evolved estrous function (see chapter 8 in conjunction with evidence presented in this chapter and the next) and hence, we argue, rightly applies to women's fertile window.

Estrous Women Particularly Prefer the Scent of Symmetric Men

Grammer (1993) assessed the ovulatory shift in women's scent preference for a particular chemical (an androgen; see section on the scent of androgens). By contrast, we first tested the hypothesis by examining whether fertile women are particularly attracted to the scents of particular men, ones who arguably possess an (ancestral) indicator of genetic quality. Specifically, we predicted that women particularly prefer the scent of symmetric men when fertile in their cycles. As we discussed in chapter 7, low fluctuating asymmetry reflects a relatively high level of developmental stability, which, in turn, may reflect low mutation load, ability to resist pathogens, and/or ability to resist the untoward effects of ontogenetic stresses in people.

In our initial study (Gangestad & Thornhill, 1998), we recruited 42 male participants. We measured each man's symmetry on 10 traits (see chapter 7). We then gave each man a clean T-shirt to wear for two consecutive nights during sleep. Men were instructed to wash their sheets in unscented soap (which we provided), not to sleep with another person, not to eat certain strong-smelling foods (e.g., spicy foods), not to drink alcohol or smoke, and to refrain from using scented deodorants or cologne. Men wore their shirts the same two nights and, on the morning of the third day, each man returned his shirt to us in a clear plastic bag identified by an arbitrary code number. We then had 28 normally ovulating women (each placed in a small room by herself) smell each shirt and rate its scent on dimensions of "pleasantness" and "sexiness." These two ratings were highly correlated, and hence we added a woman's ratings of a man's shirt to form an overall measure of how attractive she found the scent of that shirt. For each woman, we regressed her attractiveness ratings of the shirts on men's symmetry and took the slope (changes in ratings as a function of changes in symmetry) as a measure of a woman's preference for the scent of symmetry.

In this study, we estimated each woman's fertility risk based on her day of the cycle and actuarial probability of conception given by a study by Jöchle (1973). In general, fertility risk is near zero the first several days of the cycle and after day 17. It is highest in the window of days 9–13. We found a strong, statistically robust correlation between women's probability of conception and preference for the scent of symmetry (see Gangestad & Thornhill, 1998, for details).

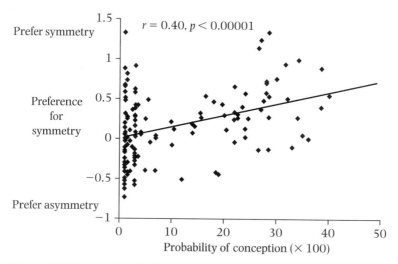

Figure 9.2 The relationship between normally ovulating women's preferences for the body scent of symmetrical men and the women's risk of conception across the menstrual cycle. Each point is the regression coefficient of a woman's scent attractiveness ratings of T-shirts worn by men regressed on men's body asymmetry. Conception risk is based on actual data from a large sample of women (Jöchle, 1973; Baker & Bellis, 1995). The line is the least squares regression line. Figured is a summary of data from three separate studies (Gangestad & Thornhill, 1998; Thornhill & Gangestad, 1999b; Thornhill et al., 2003); N = 141 women. The relationship was statistically significant in each of the three studies.

Subsequently, three other studies using similar double-blind methodologies have replicated this finding: Thornhill and Gangestad (1999b), Rikowski and Grammer (1999), and Thornhill et al. (2003). These studies have extended findings, as well. For instance, Thornhill and Gangestad (1999b) controlled for the number of showers that men took. (Men could bathe with unscented soap as often as they wished, and, in that study, number of showers covaried positively with the attractiveness of men's scent.) Partialing out this factor actually increased the association between women's fertility risk and their preference for the scent of men's symmetry. (This procedure had this effect because the factor controlled for—the number of showers men took—was unrelated to men's symmetry.) In addition, the effect remained when men who had been ill during the days the shirts were worn were removed from the analysis. Furthermore, women's preferences for the scent of symmetry do not appear to be mediated by women's greater sensitivity to men's scent; women's ratings of intensity of the scents as a function of symmetry did not change across the cycle. Nor did women's ratings of intensity of the scents change across the cycle. Figure 9.2 shows how women's preference for the scent of symmetry changes across the cycle in the three studies we conducted.

Two studies included reproductive-age women who were using hormone-based contraceptives (a contraceptive pill or Depo-Provera). These women showed no preference for the scent of men's symmetry when mid-cycle (e.g., Gangestad & Thornhill, 1998; Thornhill & Gangestad, 1999b).

Other Preferences for Male Scents Vary Across Women's Cycles

The Scent of Dominance

Havlíček, Roberts, & Flegr (2005) examined whether young women not using hormonal contraception would prefer the scent of young men high on the trait of social dominance, as assessed by a standard self-report questionnaire. Forty-eight men wore cotton pads in their armpits for 24 hours. Freshly collected pads were presented to 65 women, who rated their sexiness. Women varied in cycle phase when they made their ratings. Women in the fertile phase of their cycles, but not women in infertile phases, rated the high-dominance men as sexier smelling than the low-dominance men. Preference for the scent of dominant men was furthermore particularly strong for fertile-phase women in a pair-bond relationship. Fertile-phase women who were single did not reveal a robust preference for the scent of dominant men (figure 9.3). Havlíček et al. (2005) also had the women rate the pads for intensity of the scents. Fertile and infertile women rated the scents of nondominant men as more intense. Fertile-phase women's attraction to the scent of male dominance does not, then, appear to be mediated by greater sensitivity to men's odors.

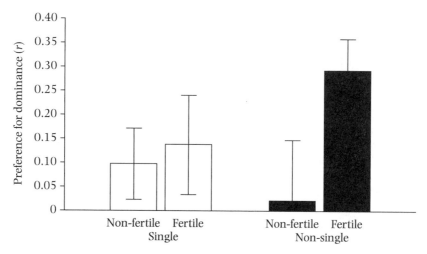

Figure 9.3 Mean (± SE) correlation coefficient between the psychological dominance score of men ($N = 35$) and the odor attractiveness rated by single (open bars) and romantically partnered women (shaded bars) in the fertile (late follicular) and infertile phases of their cycle. The difference between the two categories of single women is not statistically significant, whereas that between the two groups of partnered women is ($p < 0.001$). There were 30 female scent raters in the fertile phase and 35 female raters in other cycle phases. The raters were not using hormonal contraception. Based on figure 1 of Havliček et al. (2005). Reprinted with permission of The Royal Society.

The Scent of Androgens

No one has established, at this time, what chemicals in men's scents are responsible for women's preferences for the scent of symmetric men or their preferences for dominant men (or, indeed, whether these chemicals are the same for both symmetry and dominance). Some research hints at possibilities. The androgen androstenol, a chemical precursor of androstenone, importantly contributes to body odor. It gives scent a musky odor. Its production is furthermore sexually dimorphic, with men producing much more than women (see review by Gower & Ruperelia, 1993; see also Pause, Sojka, Krauel, Fehmwolfsdorf, & Ferstl, 1996). Women appear to be more sensitive to the scent of real or synthetic musk when fertile in their cycles (see review by Grammer, 1993; also Savic, Berglund, & Lindström, 2005).

Grammer (1993) was the first researcher to explicitly test the hypothesis that women particularly prefer a putative male marker of genetic quality when fertile. He conjectured that androstenol and androstenone may reflect testosterone levels in men, which may in turn reflect genetic quality. He found that young women not using hormonal contraception find the scent of androstenone (placed on a pad in a laboratory setting) least unpleasant during days of high fertility in the menstrual cycle (days 9–12). By contrast, women using hormone-based contraception did not show this response during the same cycle days. As Grammer (1993) recognized, these findings raised doubts about the essentially universal conclusion in the literature of human reproductive biology that estrus had been lost during the evolution of woman. Hummel, Gollisch, Wildt, and Kobal (1991) had actually found the same effect of fertility on women's preference for the scent of androstenone 2 years earlier, but they did not discuss the findings in light of function for seeking a sire of high genetic quality. We note that these researchers found that women rate the scent of androstenone "less unpleasant" when fertile. Androstenone does not generally yield a pleasing scent. By contrast, androstenol, from which it is derived, does. (Androstenol has a smell similar to that of sandalwood.) Future research may examine whether women particularly prefer the scent of androstenol when fertile.

At this time, it is not known whether symmetric or dominant men emit high levels of androstenol, such that women's preference for this androgen in fact mediates fertile women's preference for the scent of symmetric and dominant men. As we discussed in chapter 7, markers of masculinity and testosteronization may well be indicators of male robustness and genetic quality (or, more accurately, could have been such indicators ancestrally). And, as we discuss in this chapter, fertile women are particularly attracted to a variety of masculine traits in men. For these reasons, we surely would not be surprised if female preferences for the scent of androgens when fertile do indeed tell the story behind women's preferences for the scent of symmetric and dominant men. Future research should address this possibility.

At the same time, researchers have not found a robust effect of cycle phase on women's preferences for the scent of men with high concurrent testosterone levels. Based on a fairly small sample of 19 Finnish men and a sample of 36 non-pill-using women, Rantala, Eriksson, Vainikka, and Kortet (2006) found no association between scent ratings and men's T levels during any cycle phase. (They did find evidence that women prefer the scent of men with high cortisol levels, but these preferences did not vary across the cycle.) In one of our T-shirt studies mentioned earlier, we measured T levels of men. We found a trend in the predicted direction: Women tended to prefer the scent of men with high T levels more strongly when fertile than when infertile (unpublished data). (We found no evidence of a preference for the scent of men with high cortisol levels.)

Fertile Women Are Particularly Attracted to Masculine Faces

Following our first study on women's preferences for the scent of symmetry, other researchers began examining other preferences. In particular, Penton-Voak and his colleagues explored how women's preferences for masculine facial features change across the cycle. Through computerized technology, multiple faces can be digitized and combined to create average faces. Perrett et al. (1998) used this methodology to produce an image corresponding to an average male face and an image corresponding to an average female face. Each average face reflects average sex-typical features. Though many facial features are quite similar, on average, across men and women, some facial features differ across the sexes. These latter features are influenced during development

214 The Evolutionary Biology of Human Female Sexuality

by reproductive hormones—in particular, testosterone and estrogen (see chapter 7). By exaggerating the differences between the average man's and woman's faces or averaging the sex-typical features in a variety of proportions, Perrett et al. created an array of male faces that varied from relatively androgynous to hypermasculine. When asked which face they found most attractive, normally ovulating Scottish and Japanese women, on average, actually tended to choose a male face somewhat more feminine than the average male face. In other samples and in other populations, women, on average, prefer a slightly masculine face (e.g., Johnston, Hagel, Franklin, Fink, & Grammer, 2001; Penton-Voak et al., 2004; Penton-Voak & Perrett, 2000; see also chapter 7, this volume).

Penton-Voak et al. (1999) examined whether women's preferences shift across the cycle. Specifically, they hypothesized that, no matter what face women preferred on average, when fertile normally ovulating women prefer a face more masculine than the face they most prefer when infertile. This prediction was based on the hypothesis that a masculine (e.g., more testosteronized) male face is a marker of greater genetic quality (or would have been ancestrally; see chapter 7, this volume). In addition, and as noted previously, Penton-Voak et al. (1999) predicted that this shift would be pronounced when women evaluate men's attractiveness as short-term mates (e.g., sex partners) but not as long-term, stable mates.

In two different studies—one in the United Kingdom and one in Japan—Penton-Voak et al. (1999) found the predicted shift toward greater preference for facial masculinity during women's fertile phase. And this result appears to be robust. Additional studies revealed precisely the same shift: one using U.K. samples (Penton Voak & Perrett, 2000) and one using a mixed U.S./Austrian sample (Johnston et al., 2001). One additional study found mixed support for the hypothesis (Scarbrough & Johnston, 2005). Relatedly, Roney and Simmons (in press) found that estrous women, compared with women in other phases, particularly find men who possess high concurrent levels of testosterone facially attractive.

Some of these studies specifically examined changes in women's evaluation of men in short-term mating and long-term mating contexts separately. All yielded the effect only when women evaluated, in essence, men's "sexiness" and not when women evaluated men as stable, long-term partners (Penton-Voak et al., 1999; Johnston et al., 2001).

The preference shift is apparently not a by-product of a mid-cycle shift in women's generalized differences in face perception. Johnston et al. (2001) found no fertile-cycle-phase shifts in the other facial ratings they studied (such as women's judgments of female attractiveness, ratings of men's dominance based on facial features, and a variety of other ratings).

One neuropsychological study found additional evidence that women's responses to men's faces change across the cycle (Oliver-Rodriguez, Guan, & Johnston, 1999). The P300 is a brain response appearing approximately 300 milliseconds after the presentation of a stimulus. It generally covaries with the emotional salience of the stimulus. Oliver-Rodriguez et al. (1999) found that the magnitude of the P300 response of the evoked potential of normally ovulating women in estrus (late follicular phase) correlated with their rating of male facial attractiveness but not with their ratings of female facial beauty. During the infertile phase, women's responses covaried with both male and female attractiveness judgments. The greater emotional salience of attractive men's faces to women during the fertile phase of their cycles may well be related, at least in part, to enhanced response to men's facial masculinity. Additional neuropsychological studies are needed to pinpoint the nature of these effects.

The shift in women's preferences for masculine features at estrus may yield by-products in face perception. Macrae, Alnwick, Milne, and Schloerscheidt (2002) presented a series of male and female faces to normally ovulating women and asked them to categorize, as quickly as possible, each face as either male or female. They found that, when fertile, women could more quickly categorize male faces. Their ability to categorize female faces did not change across the cycle. One interpretation of this finding is that women are more attuned to masculine features when fertile. Macrae et al. (2002) hint that the shift reflects women's greater need to categorize male faces quickly when fertile. As we see little obvious advantage resulting from ability to categorize masculinity just milliseconds faster when fertile, we pose the alternative interpretation that this effect on speed of categorization is in fact not functional in and of itself but rather is a by-product of women's attunement to (and preference for) male facial masculinity.

Other Shifts in Preferences for Male Facial Features

Preferences for Dark (More Masculine) Facial Pigmentation

Frost (1994) found that normally ovulating women are more attracted to the faces of men with darker skin pigmentation when they are fertile than when they are infertile. In one sense, this effect is yet another instance in which fertile women particularly prefer masculine male faces. Within all races, skin tone is sexually dimorphic: Men have browner skin tone (reflecting greater concentrations of melanin in the skin) and ruddier skin tone (reflecting greater concentrations of hemoglobin in the skin; Frost, 1994; Jones, 1996). In turn, these sex differences appear to implicate sex hormones. In particular, testosterone tends to enhance melanin production, whereas estrogen suppresses its production in a variety of species, including humans (see review in Manning, Bundred, & Henzi, 2003). Just as fertile women particularly prefer structural facial features that are masculine, then, they also appear to prefer color and/or texture of faces that are masculine.

One might wonder why testosterone promotes production of melanin in the skin, whereas estrogen suppresses it. Melanin appears to be an antimicrobial. It defends against infection by viruses, bacteria, fungi, and possibly malaria (Mackintosh, 2001; Manning, Bundred, & Henzi, 2003). It may also possess antioxidant properties (McGraw, 2005), though under some conditions it appears to contribute to oxidative stress (Hegedus, 2000). Trade-offs between costs and benefits of internal and external deposition of melanin may be key to understanding sex differences and the cue or signal value of melanin.

Preferences for Facial Symmetry

As we have seen, women are particularly attracted to masculine faces when fertile. Moreover, male facial masculinity may be associated with male symmetry (Gangestad & Thornhill, 2003a; but see Koehler et al., 2004). Based on these associations, one might expect that estrous women will be particularly attracted to the faces of men who possess more symmetric bodies. And, indeed, a study supports precisely this prediction: Thornhill and Gangestad (2003b) found that women's conception risk predicts positively their preference for the faces of men whose bodies were measured to be relatively symmetric (figure 9.4).

Naturally, this result must be mediated by factors other than male bodily symmetry, as women never viewed men's bodies. That is, men's symmetry must have been associated, in this sample, with facial features women find particularly attractive when fertile. Though we did not measure specific facial features in this sample of men to explore which ones were responsible for the preferences, we have reason to suspect a role for male masculine facial features (see also Scheib et al., 1999).

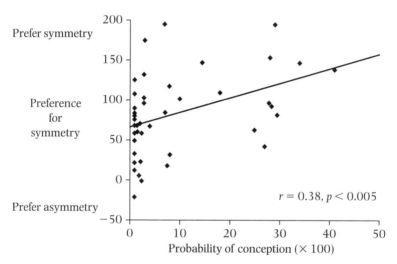

Figure 9.4 The relationship between women's facial attractiveness preference for men with symmetric bodies and the women's probability of conception across the menstrual cycle ($r = 0.38$, $p < 0.005$). Each point is the regression coefficient of a woman's physical attractiveness ratings of the men's facial pictures regressed on the men's body symmetry. Conception risk is based on actual data (Jöchle, 1973; Baker & Bellis, 1995). There were 65 men who were rated and 45 women (nonusers of hormone-based contraception) raters. The line is the least squares regression line. Data from Thornhill & Gangestad (2003b).

Another possibility, however, is that men with symmetric bodies also have more symmetric faces. And, perhaps, fertile women are particularly attracted to male faces that are symmetric (just as fertile women are particularly attracted to masculine faces). Koehler and colleagues (Koehler, Rhodes, & Simmons, 2002; Koehler, Rhodes, Simmons, & Zebrowitz, 2006) evaluated this hypothesis. They found that, in general, women prefer symmetric male faces over asymmetric male faces (as also found by other researchers; see Rhodes, 2006, for a review). Their studies, however, revealed no change across the cycle in women's preferences for symmetric faces; infertile women preferred symmetric faces just as much as fertile women preferred these faces. Unfortunately, Koehler and colleagues used menstruating/early follicular phase women as the infertile group in both studies. By contrast, most studies include women in the luteal phase. As high levels of progesterone—which is elevated only during the luteal phase—may suppress fertile phase preferences and hence importantly account for variation in preferences (e.g., Jones, Little, et al., 2005; Puts, 2006b; Garver-Apgar, Gangestad, & Thornhill, in press), studies that exclude women in the luteal phase may drastically underestimate differences across phases. Studies that have included women in the luteal phase have yielded mixed results: In two studies, Little, Jones, Burt, and Perrett (2007) did find that women particularly prefer more symmetric male faces when in the fertile phase of their cycles, especially when they are pair bonded and are evaluating men as short-term sex partners, whereas Cárdenas and Harris (2007) found no effect. (We note that some studies have found no evidence of an association between facial and body symmetry; e.g., Gangestad & Thornhill, 2003a. Partly on that basis, in fact, we have expressed concern that facial symmetry may not tap organism-wide developmental instability particularly well and, hence, may not function as a cue of developmental instability; see Gangestad & Thornhill, 2003a. More work on these issues is needed.)

Fertile Women Particularly Prefer Masculine Male Voices

Women prefer more masculine faces and, it appears, more masculine scents when fertile. Perhaps not surprisingly, then, they also prefer more masculine male voices. Voice pitch is sexually dimorphic. The fundamental frequency (pitch) of men's voices is lower than that of women's voices. And this difference holds up even when differences in body size are statistically controlled. Of course, the pitch of boys' voices lowers during puberty, and voice pitch becomes increasingly sexually dimorphic during adolescence. Testosterone and, perhaps, other androgens play roles in these changes. In general, male voices that women most prefer are of lower pitch (i.e., are more masculine) than average male voices (see review by Puts, 2005).

Puts (2005) experimentally manipulated men's voices to be of greater or lesser pitch. He then had 142 normally ovulating women rate the sexual attractiveness of men for long-term and short-term relationships based on voice alone. (Through the experimental manipulations, content and phrasing of individual men's readings could be controlled.) Women fertile in their cycles preferred lower-pitched male voices more strongly than did women infertile in their cycles. This pattern held in analyses of natural variation in men's voices, as well as analyses of experimental variations in men's voice pitch. Furthermore, this effect was particularly strong when women rated men's sexiness (attractiveness in short-term relationships) rather than their attractiveness as stable mates. Put otherwise, fertile women rated low-pitched male voices more attractive for short-term sexual relationships than for long-term or committed relationships. But the same could not be said of infertile women. (See also Puts, 2006b, who found that hormonally contracepting women do not have the same preference mid-cycle.)

Feinberg et al. (2006) further replicated and importantly extended these findings. They experimentally manipulated two sexually dimorphic vocal traits: vocal pitch and variation associated with length of the vocal tract. Women in the fertile phase of their cycles preferred voices more masculine along both dimensions (i.e., deeper voices and voices emanating from longer vocal tracts) than did women in infertile phases of their cycles. Feinberg et al. (2006) furthermore demonstrated that women's preference shifts are specific to *male* vocal traits. They found no evidence that women's preferences for the two vocal traits in *women's* voices change across the cycle. (Feinberg and colleagues, 2006, also found that women's preferences for male vocal masculinity are moderated by female estrogen levels; we discuss this finding further later in the chapter.)

Fertile Women Particularly Prefer Dominant and Intrasexually Competitive Behavioral Displays

As we discussed in chapter 7, Simpson et al. (1999) studied men's direct intrasexually competitive behavioral displays in a situation in which men competed for a potential lunch date with an attractive woman (e.g., men's explicit put-downs of another man and claims that they were "better" than their competitors). Men who exhibited more of these displays, as predicted, tended to be symmetric compared with their counterparts who refrained from making direct comparisons. In a subsequent study, Gangestad and colleagues (2004) were interested in assessing normally ovulating women's reactions to these and related displays and how their attraction to them changed across the cycle.

In Simpson et al.'s (1999) study, all 76 men had been videotaped. Gangestad et al. (2004) had these videotapes coded for a variety of specific behaviors (e.g., amount of time spent smiling, amount of time looking downward), impressions (e.g., how confident men appeared, how nervous men were), and verbal content (e.g., whether men said they were superior to their competitors, whether they emphasized that they were nice guys). Factor analysis of these ratings revealed two largely independent dimensions. First, men varied in the extent to which they displayed social presence (e.g., appeared composed vs. looked downward). Second, men varied in the extent to which they displayed intrasexual competitiveness (e.g., said they were superior to their competitors vs. reflecting the personalities of nice guys). Gangestad and colleagues (2004) then had 237 normally ovulating women view a portion of the interview (about 1 minute per male participant) of approximately half the men (with each woman rating one of two different subsets of the men) and rate each man on dimensions of attractiveness as a stable mate (long-term mate attractiveness) and attractiveness as a sex partner or affair partner (short-term mate attractiveness).

Results of the study are shown in figure 9.5. As predicted, women were particularly attracted to men who displayed both social presence and intrasexual competitiveness when they were fertile—but only when they evaluated men as short-term, and not long-term, mates. Preferences for the two traits were analyzed separately, and this pattern was found for both. Gangestad et al.'s (2004) finding that women's preferences vary as a function of fertility only when they rate men's sexiness and not their long-term attractiveness is one typically found in studies that separately evaluate both aspects of attractiveness. As already noted, short-term preferences, but not long-term preferences, for male masculine facial features, male voices, and male facial symmetry (in one study) have been reported. As we discuss later, women's short-term preferences, but not long-term preferences, for men's creative production and tallness have also been demonstrated.

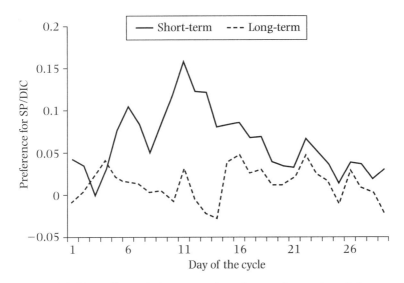

Figure 9.5 Normally ovulating women's preference for men's social presence (SP) and direct intrasexual competitiveness (DIC) as a function of the day of the cycle and long-term versus short-term mating context; points along lines are 3-day moving averages. Preference is the average regression slope of individual women's ratings regressed on men's SP and DIC, with men's physical attractiveness ratings statistically controlled. High-fertility days span from day 6 to day 14, with peak fertility at day 12. Based on figure 1 of Gangestad et al. (2004). Reprinted with permission of Blackwell.

Other Preference Shifts Across the Cycle

Preferences for Masculinity and Physical Attractiveness

Just as male faces and voice can reveal masculinization, so too can male bodies. Testosterone promotes muscle growth, often in the service of intrasexual competition and mating effort (see chapter 7). Women prefer male bodies that are muscular, particularly through the chest, arms, and back, without being clumsily overbuilt (e.g., Frederick & Haselton, 2007). If male muscularity, similar to other masculine features, is a marker of male condition and, ancestrally, may have been an indicator of good genes, might muscularity also be a trait that women find particularly attractive when fertile?

In their study of men videotaped for a potential lunch date, Gangestad, Garver-Apgar, Simpson, and Cousins (2007) had women rate all men for their degree of muscularity. Fertile women, compared with their infertile counterparts, did indeed find the men who were rated as relatively muscular particularly attractive as short-term, relative to long-term, mates. This effect held up when men's behavioral displays (social presence and intrasexual competitiveness, each found to be particularly sexy to fertile women) were statistically controlled.

In a related experimental study, Little, Jones, and Burriss (2007) found that women particularly prefer men's bodies that are masculine more when fertile than when infertile in the cycle, and this effect was stronger when women evaluated men as short-term (sex) partners than long-term relationship partners.

These results may explain findings from laboratory studies examining women's responses to male nudity. In seminal work, Slob, Bax, Hop, Rowland, and ten Bosch (1996) found that women are physiologically more aroused by and responsive to sexually explicit visual stimuli mid-cycle. Krug, Plihal, Fehm, and Born (2000; see also Krug, Pietrowsky, Fehm, & Born, 1994) found that, when they are fertile rather than infertile in their cycles, normally ovulating women exhibit a higher event-related brain potential (a late positive component occurring about 500–700 milliseconds poststimulus) after viewing nude men than after viewing various nonsexual stimuli. This brain response purportedly reflects added processing of the stimulus and hence interest. And, indeed, fertile women rated the nudes more positively than women at menses or in the luteal phase. As the nude males shown to women likely exhibited attractive musculature (as they were intended to be sexually provocative stimuli), we suspect that these results derive from a shift toward greater interest in and sexual attraction to attractive, muscular, male bodily features when women are fertile.

Indeed, in research we recently conducted, we found convergent evidence that women are more sexually attracted to and stimulated by attractive visually presented male features when fertile than when infertile in their cycles (Gangestad, Thornhill, & Garver-Apgar, 2007a). We asked a sample of approximately 50 women (all of whom were involved in committed romantic relationships) to fill out two self-report measures twice, once when fertile (as verified by a luteinizing hormone surge, which occurs about 24–48 hours prior to ovulation) and once when infertile (typically in the mid-luteal phase). First, we asked about their mate preferences for physical attractiveness (with 10 items, including "I place a very high importance on a potential mate's physical attractiveness"; "Unattractive facial features are a real turn-off to me, even if the person has other positive attributes"; and "It's hard for me to understand why some people place such high importance on a person's physical attractiveness" [reverse-keyed]). Second, we asked about women's interest in attractive bodily features (with 10 items, including "I find the thought of a very attractive body of the opposite sex very exciting"; "If I met someone I found very attractive right now, I would fantasize about what they would look like without clothes on"; and "Seeing the arm or leg muscles of an attractive opposite-sex person subtly flex would be a real turn-on right now"). Women were asked to think about how they felt *at that moment*, not how they felt in general.

As predicted, we found ovulatory shifts in both preferences. Women reported greater preference for physical attractiveness and greater interest in attractive bodily features when fertile than during the luteal phase (see also Gangestad, Garver-Apgar, Simpson, & Cousins, 2007, on fertile-phase preferences for physical attractiveness). The fertility shift in women's interest in attractive bodily features was especially strong in this study. Future research should specify more particularly the nature and significance of male bodily features and their displays that are particularly attractive to fertile women.

Perhaps relatedly, Pawlowski and Jasienska (2005) found that women more often preferred a greater degree of sexual dimorphism in stature (i.e., taller men) when they were in the fertile phase of their menstrual cycles and when the partners were chosen for short-term relationships.

Preferences for Male Creative Production

In light of Miller's (2000) hypothesis that creativity is among an array of "mental fitness indicators" in humans, Haselton and Miller (2006) asked whether normally ovulating women's preference for men's talent and creativity over their preference for men's material success (resource holdings) is enhanced when women are fertile in their cycles, and especially when women evaluate men as short-term mates. Women read a pair of vignettes. One depicted a man who displayed artistic or entrepenurial talent but had relatively little financial holdings to show for it. Another depicted a man who had financial holdings but little artistic or entreprenurial talent. When in the fertile phase of their cycles, women favored the talented man over the wealthy man when evaluating men as short-term partners. Their preferences for one sort of man over another as long-term, stable mates did not change across the cycle.

One limitation of this finding is that women may have read into the vignettes a variety of characteristics that differentiate men. These inferred features, rather than talent versus wealth per se, may have driven the effects. Thus, for instance, women may have inferred that talented but financially less endowed men were less constrained, open, and broadly competent and confident. One or another of these traits may have caused the observed effects.

Miller (2003) has suggested that fertile women may find humorous men particularly sexy. He found that fertile women rate humorous men (as described in vignettes) more attractive than they do when infertile. The critical mating context (short term vs. long term) by humor (highly humorous vs. not humorous) interaction, however, was not significantly robust. These results must be interpreted cautiously.

As we discussed in chapter 7, men's symmetry covaries with typical functional and anatomical cerebral lateralization. Furthermore, as mentioned, a number of studies demonstrate a positive association between men's IQ and their developmental stability. As we previously concluded, though it is plausible that intelligence and associated talents (e.g., creativity, possibly ability to produce humorous content) are fitness indicators, they do not possess the hallmarks of other fitness indicators (e.g., being particularly preferred in short-term mates). Though Haselton and Miller's (2006) findings suggest that these traits may be particularly preferred in short-term mates by women at mid-cycle, other evidence is inconsistent or equivocal in this regard. In their videotape-rating study, Gangestad, Garver-Apgar, Simpson, & Cousins, (2007) found no evidence that men who appeared "intelligent" were particularly preferred as short-term or long-term partners when women are fertile (for further details, see the next section). More work on shifts in women's preferences for male mental fitness indicators is required before any firm conclusions can be drawn.

Not All of Women's Preferences Are Exaggerated When They Are Fertile

Preferences for Behavioral Traits Valued in Long-Term Mates

Women are sexually attracted to masculine features and a variety of other indicators of developmental stability when fertile. But is it simply the case that all of their mate preferences strengthen when fertile? The view that we have put forward—that women should be particularly attracted to ancestral indicators of men's genetic quality during estrus—predicts no. Women have preferences for traits that are not and were not ancestrally indicators of good genes. In particular, if men engaged in parental care and men and women cooperatively raised offspring, women should have evolved preferences for male traits indicative of willingness and ability to provide care (and this is the case; see chapters 4 and 7, this volume). Indeed, as we previously noted, Buss's (1989b) cross-cultural survey of mate preferences revealed that the characteristic most valued in a mate by both men and women across cultures was "kindness and understanding." We see no reason to believe that men who are kind and understanding to their mates have higher genetic fitness. Indeed, as we also have noted, one study of college students involved in romantic relationships indicated that symmetric men invest less in their relationships (Gangestad & Thornhill, 1997a) and are less faithful to partners (Gangestad & Thornhill, 1997b) than asymmetric men. Compared with less masculine men, men with masculine faces are viewed as less trustworthy, faithful, and investing in their relationships (e.g., Penton-Voak & Perrett, 2001) and poorer long-term mate choices (e.g., Kruger, 2006; see also Scheib, 2001).

We have discussed women' preferences for male displays of social presence and intrasexual competitiveness. In additional analyses of the same videotapes of men, Gangestad,

Garver-Apgar, et al. (2007) had separate samples of women rate how men came off in terms of 10 different traits women might find attractive in a mate. They then examined which traits women found particularly attractive in a short-term mate when fertile (that is, the three-way interactions between men's trait level, women's estimated conception risk, and relationship context—short term vs. long term). When fertile, women were particularly sexually attracted to men perceived as confrontative with other men, arrogant, muscular, physically attractive, and socially respected. In contrast, no shifts in women's preference for men perceived as intelligent, kind, likely to be financially successful, or possessing qualities of a good father were detected. And men who were perceived to be faithful were actually *less* sexually attractive to fertile women than infertile women. (Put otherwise, fertile women were particularly sexually attracted to men who appeared to be unfaithful types.) The traits perceived to be particularly sexy to women during their fertile phase are traits found to be more valued in short-term partners than in long-term partners. The traits that fertile women did not find particularly sexy are ones more valued in long-term mates than short-term mates—the qualities of good, stable mates. (As we previously discussed, Haselton & Miller, 2006, found that infertile-phase women prefer uncreative men with wealth over creative but poorer men. As creativity and wealth were confounded in this study, cycle effects on preferences for each cannot be independently estimated.)

Some traits valued in long-term stable partners may be particularly attractive to women with high progesterone levels, characteristic of the infertile luteal phase and pregnancy. Jones, Little, et al. (2005) examined normally ovulating women's preferences for femininity in men's and women's computerized faces in relation to their progesterone and estrogen levels across the menstrual cycle, as estimated from published norms for cycle days. Hormone-level estimation came from published data based on the cycle day. When women's progesterone levels were high, they found femininity in both men's and women's faces more attractive than when their progesterone levels were low. Estimated estrogen levels did not predict either preference.

Both men and women with feminine faces are perceived to be more cooperative and helpful (e.g., Perrett et al., 1998). Women who may be pregnant (during the luteal phase) or are pregnant, then, may prefer investing social partners. Relatedly, DeBruine, Jones, and Perrett (2005) found that estimated progesterone levels positively predicted women's preferences for self-similar male and female faces. This effect may reflect a preference to ally with kin in preparation for or during pregnancy. Perhaps relatedly, Wedekind et al. (1995) found that women using hormonal contraception (which raises progesterone, as does pregnancy) prefer scents of MHC-similar scent donors over scents of MHC-dissimilar donors. Similarity of MHC alleles may have functioned, ancestrally, to mark kin relatedness.

During infertile phases of the cycle, women have other preferences for male traits that may connote potential for receipt of male-provided services and benefits. Estimated progesterone level in the cycle of normally ovulating women positively predicts their preference for men with less masculine voices (Puts, 2006b). Similarly, Frost (1994) reported evidence that women in the luteal phase (associated with high levels of progesterone) prefer less masculine facial skin tone. As we discussed in chapter 7, men's testosteronization appears to negatively covary with their romantic and paternal investment.

Women who are using hormone-based contraception do not appear to show shifts in preferences for indicators of genetic quality, as we have noted. Hormone-based contraception typically raises progesterone levels (Gilbert, 2000) and softens the mid-cycle peak in estrogen and hence creates a hormonal milieu more typical of a woman in the luteal phase. That women using hormone-based contraception do not show cycle effects suggests that hormones do play a role in these effects. As Puts (2006b) notes, different fertile-phase preferences may be affected by different hormones or combinations of hormones (see also Garver-Apgar, Gangestad, & Thornhill, in press; Roney & Simmons, 2008; Welling et al., 2008).

Preferences for Current Health

Perhaps also related to women's special concerns when pregnant are findings that progesterone levels are associated with increased attraction to men who appear healthy (and less attraction to men who appear unhealthy). If a woman contracts an infection when pregnant, the developing fetus may be harmed. Immune responses are modulated by progesterone in ways that facilitate successful implantation of the zygote. When progesterone is elevated, women tend to avoid eating foods that could contain harmful bacteria, especially meat (Flaxman & Sherman, 2000; Fessler, 2001, 2002). Jones and colleagues (Jones, Little, et al., 2005; Jones, Perrett, et al., 2005) manipulated the perceived current health in computerized composite faces (while controlling for face shape and hence facial masculinity and symmetry). To do so, they had people rate perceived health and then created separate composites of faces, one of healthy-looking faces and another of unhealthy-looking ones. Healthy-looking faces had a darker color, whereas unhealthy ones exhibited pallor. Women in the luteal phase of their cycles, pregnant women, and women using oral contraceptives show greater aversion to unhealthy faces (Jones, Perrett, et al., 2005). All of these conditions are associated with high progesterone, which may thus be responsible for the effects. Indeed, this research found that estimated progesterone levels predict women's aversion to unhealthy faces. This shift may reflect an adaptation similar to avoidance of foods that may carry pathogens: to promote healthy fetal development by avoiding contagion.

These findings do not necessarily contradict Frost's (1994) finding that fertile-phase women prefer masculine, darker male faces, despite the fact that healthy faces preferred by luteal-phase women appeared to have darker skin tone. Healthy and masculine skin tone may vary in a largely independent fashion, as two different dimensions. More work may be needed to identify more specifically the nature of these variations.

As we discussed in chapter 7, traits that signal superior genes do not necessarily signal current health (e.g., Getty, 2002; Kokko et al., 2002; though they may well typically covary with immunocompetence [as opposed to parasite load]; Møller, Christe, & Lux, 1999). Women's preferences for good-genes indicators and avoidance of mates with current infections may hence be partially independent, as Jones et al. discuss. Preferences for men who are disease-free may have been selected more strongly by nongenetic material benefits that result from affiliating with others who are healthy currently (and hence capable of delivering material benefits now) and will not pass on contagious infections.

Some Women May Strongly Prefer Good-Genes Indicators Even During Infertile Phases of Their Cycles

We have argued that, when fertile, women particularly prefer in mates a subset of valued traits: indicators of male genetic quality. When outside the fertile phase of their cycles, they maintain strong preferences for traits that are particularly valuable in a long-term mate. As we delve into more deeply in chapter 10, preferences for good-genes indicators when women are mid-cycle may lead some women to seek genetic benefits from men other than primary partners (i.e., extra-pair partners). Specifically, not all women can possibly have as primary partners men who possess indicators of good genes. Those women who do not may be those most likely to be attracted to men other than primary partners when fertile (see chapter 10).

Some women, however, may be able to obtain as primary partners men who possess indicators of good genes. In particular, women who have high value on the mating market may be able to attract these men as primary partners. These women, then, may benefit from pursuit of (and hence preference for) men who possess these features throughout the cycle. And, indeed, as we mentioned in chapter 7, women who are physically attractive have particularly strong preferences for male facial masculinity and facial symmetry (Little et al., 2001). Attractive women show stronger preferences than unattractive women for male facial attractiveness for long-term mateships, whereas unattractive women show stronger preferences for male attractiveness than unattractive women for short-term mateships (Little et al., 2001; Penton-Voak et al., 2003; see also Clark, 2004, and Rhodes et al., 2005). Moreover, the preferences for male facial and vocal masculinity of women who have relatively high estrogen across the cycle (and who may, then, have higher reproductive capacity than other women; see chapter 6) shift less (and hence persist at higher levels throughout the cycle) than do the preferences of women with relatively low estrogen (Feinberg et al., 2006). (For preferences for vocal masculinity, see figure 9.6.) These findings purportedly reveal strategic differences in women's preferences, dependent on their own attractiveness to men. Specifically, they support the hypothesis that attractive women can extract more materials and services from attractive men and hence reveal more consistent sexual preference for male masculine traits and symmetry across the cycle.

These findings illustrate a more general trend for biologists to increasingly recognize that female mate selection is often condition-dependent (e.g., Lynch et al., 2005). An early example is work on alternative female choice tactics in a species of scorpionfly (Thornhill, 1984b). Females of this species appear to trade off good-genes choice for material benefits delivered by males (nuptial gifts). Both female large body size (and hence high energetic cost to maintenance) and hunger independently and positively affect the willingness of females to copulate with males with small food gifts. Such males are inferior hunters and competitors with other males and thus, apparently, are

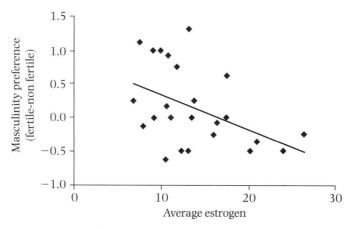

Figure 9.6 Normally ovulating women's average estrogen (mean of low and high fertility phases of the menstrual cycle) negatively predicts ($p = 0.02$) the magnitude of the cycle shift in preference for male vocal masculinity. The masculinity preference is a woman's preference at estrus minus her preference outside estrus (early follicular and luteal). Women with relatively low estrogen show the greatest cyclic shifts in the preference. The line is the least squares regression line. Based on figure 2 of Feinberg et al. (2006). Reprinted with permission of Elsevier.

of relatively low genetic quality. Thornhill (1984b) reviewed the literature of reports of other species that imply adaptive, conditional shifts in female choice. For evidence on a variety of nonhuman species supporting the hypothesis that female choice is adaptively condition-dependent, see Lynch et al. (2005).

Preferences for Compatible Genes and Diverse Genes: MHC Traits

As we discussed in chapter 7, good genes take different forms. The kind of good genes preferences discussed thus far—preferences for masculine traits and indicators of developmental stability—are purportedly preferences for intrinsic good genes. These are genes good for the offspring of any woman. Compatible or complementary good genes are genes good in a mate of an individual woman because they work well with her particular genes. MHC genes dissimilar from the woman's own genes may be instances of compatible genes.

One question that arises is whether women should, from our theoretical perspective, be expected to prefer compatible genes, as well as intrinsic good genes, when fertile. It is not clear that they should. On the one hand, compatible genes are, just like intrinsic good genes, good genes. And if estrus promotes mating with males who possess good genes, women should prefer males who provide compatible good genes at estrus. On the other hand, compatible genes may also be sought after in long-term, stable mates. Men with intrinsic good genes should be highly sought after by many women for those genes. They hence make less stable mates. Indeed, it is simply not mathematically possible for all women to end up pair bonded to men who possess high levels of intrinsic good genes. A man who, for a given woman, possesses compatible genes, by contrast, should not be highly sought after by all other women; he will not possess genes compatible with many other women. It is quite possible, in fact, for all women to be pair bonded with men who possess compatible genes. For this reason, it may make sense for women to prefer indicators of compatible genes in the infertile phases of their cycles, as well as the fertile phase.

In one of our three T-shirt studies of preferences for scent, we collected blood and typed each individual's MHC alleles at three loci, A, B, and DRβ. As discussed in chapter 7, Claus Wedekind and colleagues (Wedekind et al., 1995; Wedekind & Füri, 1997) found that individuals of both sexes have a sexual preference for the scent of others who possess dissimilar MHC alleles at these loci (see also Santos et al., 2005). Studies prior to ours did not examine variation in preferences across the cycle. (In fact, Wedekind selected women who were in the fertile phase only.) In our study with colleagues, we did so (Thornhill et al., 2003). We found no evidence for it; women's fertility risk was almost totally unrelated to their preference for the scent of MHC dissimilarity ($r = 0.03$). (In fact, as discussed in chapter 7, we found no overall preference for MHC dissimilarity in women, though we did find one in men). The lack of cycle-related variation, again, is not inconsistent with the view that estrous sexuality includes adaptation that functions to lead women to seek men with good genes. (Indeed, as we discuss in the next chapter, MHC dissimilarity to a mate does have implications consistent with our view of estrus.)

One female preference that we did detect was a preference for the scent of men who are heterozygous at MHC loci. As we described in chapter 7, this preference was very robust. As we noted there, a main benefit of having a heterozygotic mate is that he could produce a family that is more diverse at MHC alleles than could a homozygotic mate. Ancestrally, heterozygotes may have had longer survival and greater health and hence have been better investors in offspring (Mitton, 2000). Either benefit is most valuable in a long-term partner with whom a female will have multiple offspring. The former benefit, in particular, may matter little if a female has just one offspring with the chosen male (see chapter 7).

Interestingly, then, we found little variation across the cycle in women's preference for heterozygotic males (indeed, we found a nonsignificant trend for women to particularly prefer the scent of heterozygotic males *outside* of the fertile period; see figure 9.7). Moreover, the pattern of preference for heterozygosity across the cycle significantly differed from the pattern of preference for the scent of men's symmetry in the same study (which peaks, of course, at mid-cycle). Possibly, during the fertile phase of the menstrual cycle, preferences for indicators of good genes in sex partners increase, with corresponding dampening of the strength of preferences for traits particularly valuable in long-term partners.

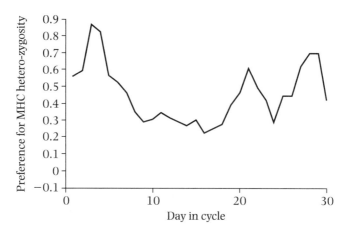

Figure 9.7 Women's preferences for the scents associated with men's MHC heterozygosity across the menstrual cycle. Points along line represent 5-day moving averages. Total number of women raters was 65. Based on data from Thornhill et al. (2003).

Women's Resistance to Poor Mating Options When Fertile

Disgust Toward Maladaptive Sex

In addition to preferring indicators of male genetic quality when fertile in their cycles, such women should also be expected to possess adaptations that lead them to avoid poor mating options. One set of poor mating options are those that would produce mal-adapted offspring. Consistent with theory, then, during the fertile phase of the cycles, women exhibit greater disgust toward maladaptive sex (e.g., incest and bestiality) than they do during infertile phases. As should also be expected, they do not exhibit greater disgust toward nonsexual objects (e.g., excretory products, rotting foods; Fessler & Navarrete, 2003).

Potential for Sexual Coercion

Mid-cycle, women may also act to avoid sexual coercion even more than they do when infertile (e.g., Bröder & Hohmann, 2003; Chavanne & Gallup, 1998; Petralia & Gallup, 2002). Chavanne and Gallup's antirape theory states that: (1) ancestrally, men some-times circumvented female choice and forced copulation with females; (2) selection led to female antirape adaptations (e.g., wariness of going alone to places where women risk rape); (3) although rape was always costly to female victims (e.g., owing to physical harm and psychological trauma), women paid higher costs of rape during the fertile phase, for at that time they paid additional costs of possibly having a child sired by a father they did not choose and who may not invest in the child; and (4) because efforts to avoid rape entail costs (e.g., by limiting movement), the effort that women put into antirape tactics has been adaptively shaped to be most intense when the benefits of risk reduction were the greatest—that is, when the costs of rape were most devastating. Women do claim that they would avoid situations perceived to be those associated with higher risk of sexual coercion when fertile more so than when infertile in their cycles (Bröder & Hohmann, 2003; Chavanne & Gallup, 1998; see also chapter 12, this volume.)

Women's Estrus Redux

Fertile Women Prefer Purported Indicators of Genetic Quality

A considerable body of research reveals that the value women attribute to putative markers of male genetic quality increases during their fertile phase of the menstrual cycle. Fertile women particularly prefer the scents associated with male symmetry, social dominance, and particular androgens; male facial masculinity; male vocal masculinity; male behavioral social presence and intrasexual competitiveness; male bodily muscularity and sexual attractiveness; and, possibly, facial symmetry (though findings are mixed). Fertile women experience greater disgust in response to maladaptive sex, such as incest and bestiality, and tend to avoid situations perceived to be associated with risk of sexual coercion.

The theory that urged researchers to explore changes in women's preferences across the cycle, leading to the discovery of these effects, states that women have been designed by past selection to manifest preferences for sires with superior genetic quality for offspring. These preferences should be maximal during the periovulatory phase because that phase coincides with maximum risk of conception and, hence, is the time at which female choice has the greatest effect on offspring genetic quality. That pursuit of high genetic quality is the fundamental design of these fertile-phase preference shifts is consistent with the robust finding that the effects are specific to women's attraction to men's "sexiness" and do not apply to women's preferences for men as long-term, stable partners.

We propose that these effects reflect adaptations, ones functional in the ancestral past. They are adaptations shaped and maintained by genetic benefits ancestral females obtained for their offspring by mating with men with superior genetic quality. Some effects demonstrated in the literature are mere by-products of these adaptations (e.g., the effect on women's ability to discriminate male faces quickly, the effect on women's interest in nude men). The shifts in preferences for specific male features at estrus in particular reflect adaptation.

Alternative Hypotheses for Women's Preference Shifts

Might women's preference shifts at estrus have been shaped by benefits other than genetic benefits for offspring? Two possibilities are that women prefer masculine men mid-cycle to obtain benefits of physical protection and a fertile ejaculate.

It is surely plausible that masculine men could better protect women from unwanted advances by other men. Though we found that symmetric men invest less into their relationships overall, they perceive themselves and are perceived by their romantic partners as better able to physically protect those partners (Gangestad & Thornhill, 1997a). Protection is a form of male investment in a relationship that perhaps minimally interferes with a man's efforts to attract partners other than his pair-bond mate. Male protective ability may be attractive to women in general and, compared with investment in the form of time, honesty and sexual exclusivity, men's investment in protection may compete with their pursuit of additional partners to a lesser degree. Quite possibly, female preference for male protection occurs in addition to choice for superior genes. At the same time, it is not obvious that, if females choose men for physical protection alone, they should do so specifically when fertile.

Women's preferences for symmetric and masculine men conceivably reflect adaptation for obtaining a fertile ejaculate. Manning and colleagues (Manning, Scutt, & Lewis-Jones, 1998; Manning, Scutt, Wilson, & Lewis-Jones 1998; see also Baker, 1997; Firman, Simmons, Cummins, & Matson, 2003) found that men's body symmetry positively predicts ejaculate size and sperm quality. Soler et al. (2003) found a positive association between men's facial attractiveness and semen quality. The benefits of obtaining good sperm need not be alternatives to obtaining superior genes. Men's ejaculate quality may be importantly affected by individual variation in parasite resistance and immune system quality. Skau and Folstad (2003) argue that the ability to bear the cost of the immunosuppression necessary for producing high-quality ejaculates is possessed primarily by males of high genetic quality.

Fertile Phase Sexuality Is Not Merely Ramped Up Infertile Sexuality

Women's fertile sexuality, we have argued, is not merely heightened infertile sexuality. It is not merely the case that, at mid-cycle, women prefer more of all they prefer when they are infertile. Furthermore, contrary to Buller's (2005) claim, fertile women are not simply more "sexual" than infertile women. Women's preferences shift in a more fine-tuned way. And some preferences are heightened during the infertile phase, relative to the fertile phase.

As we discussed, women prefer male features valuable in long-term, stable mates just as much as or, in some instances, even more during infertile periods compared with fertile periods. Women's preferences for kind, faithful men who would make good fathers do not shift or are more pronounced during the luteal phase. So too may be women's preferences for men who possess genes that could diversify a family's MHC alleles.

Furthermore, women do not possess uncontrolled and indiscriminate female proceptivity and receptivity during the fertile phase of their cycles. Rather, their libido is highly selective. As we argued in chapter 8, estrus in general has been misinterpreted by many biologists. In general across vertebrate species, estrous sexuality should be designed to lead females to be selective, except in response to rare circumstances.

The body of evidence we have reviewed is profoundly inconsistent with the common view that woman has no estrus. It is also inconsistent with the related view that woman's sexuality has been liberated from the influences of ovarian-cycle hormones. That normally ovulating women experience cyclic shifts in sexual preference for putative markers of male genetic quality but users of hormone-based contraception do not establishes that reproductive hormones play a role in women's estrus.

Tarín and Gómez-Piquer (2002) noted the similarity between estrous sexual motivation in female mammals in general and women's mid-cycle sexual motivation implied by their high cycle-fertility preferences of the scent of men's symmetry and masculinity. Their analysis is correct, we suggest, in claiming that women possess estrus. At the same time, it is incomplete because it does not recognize the functional significance of mammalian estrus, much less woman's eroticism during their fertile phase. Woman's estrous sexuality, specifically its proceptivity and receptivity, reflects preference for phenotypic markers of potential male genetic quality. We propose that this eroticism evolved because it led to the production of offspring of high genetic quality. Again, as we emphasized in chapter 8, women's specialized sexuality during the fertile phase of the menstrual cycle is apparently homologous with and reflects adaptations functionally similar to adaptations of estrus in other mammals and female vertebrates in general.

Women, we argue, possess two functionally distinct forms of sexuality and associated adaptations for mate preference, specialized so as to be manifested at different times of the menstrual cycle (Gangestad & Thornhill, 1998; Little, Jones, Penton-Voak, Burt, & Perrett, 2002; Thornhill & Gangestad, 1999a; also see Penton-Voak et al., 1999; Penton-Voak & Perrett, 2001). The evidence we have reviewed indicates that one set of adaptations—which together constitute what we refer to as extended sexuality—functions at low- or zero-fertility menstrual-cycle phases. Its function is to increase access to nongenetic material benefits. The other set of adaptations—which together constitute what we refer to as estrous sexuality—functions during fertile phases to lead women to particularly prefer male indicators of genetic benefits for offspring. When pregnant or using hormonal contraception, women also manifest sexuality similar in design to that at infertile cycle phases. These two sets of sexual psychological adaptations of women show specialized function (Symons, 1987; Thornhill, 1990, 1997). To say that these two sets of adaptations have different functions is to say that each was ultimately forged by forms of past selection that were not causal in forging the other adaptation.

That is not to say that no adaptations overlap fertile and infertile phases. Given that women's estrus functions in the context of a species in which females rely on material benefits delivered through stable pair bonds, women cannot completely ignore the costs they may pay for estrous sexuality in the currency of lost material benefits. Hence, though the forms of estrous sexuality we have discussed in this chapter have, we argue, been shaped largely by benefits in the form of good genes for offspring, women must also possess adaptations that operate even when women are fertile to avoid loss of a primary partner or his investment. In chapters 10 and 12, we discuss how women's estrous sexuality operates in the context of pair-bonded relationships.

At the same time, when women do evaluate men as long-term partners, whether during fertile or infertile phases, they must concern themselves with characteristics that relate to *both* a flow of nongenetic material benefits that they can receive from a partner (including allocation of time to invest in a relationship or offspring) *and* genetic benefits for offspring. As expected, then, we see little evidence that adaptations for women's long-term mate choices are expressed differently during estrus and during extended sexuality.

Ovulation Frequency Is Not an Evolutionary Constraint

Before moving on, we discuss one final topic. It has been suggested to us that woman cannot possibly have female choice adaptation that functions only around ovulation, because ovulation is too uncommon an event to provide a context for effective selection for specialized ovulation-related adaptation. In natural-fertility human populations, such as the Dogon of Africa, the modal pattern of ovulatory cycles for a young woman is about two per 2 years (Strassmann, 1996a). In such populations, low-frequency ovulation in women stems from relatively frequent pregnancy, lengthy lactation, and associated infertility, as well as amenorrhea stemming from periodic paucity of nutrients and low levels of fat storage.

The effectiveness of selection in shaping adaptation, however, is not determined by the frequency of a problem that causes selection. Recurrence and significant selection pressure across generations are necessary for evolution by selection. Infrequent events coupled with large fitness consequences can generate major phenotypic evolution (e.g., Buss, 2005; Williams, 1992). (After all, individuals are born but once, but that surely does not imply that selection has not shaped adaptations specific to gestation and the birth process.) Sire choice based on genetic quality, we have argued, is an event with large fitness consequences. That the selection for preference in fertile-cycle-phase women for mates and sires of high genetic quality has been effective in leading to adaptations for such preference is indicated by the evidence, reviewed previously, that high-fertility-phase women have this form of special-purpose female choice. Said differently, fertile-phase women's sexual preferences appear to show functional design for obtaining good genes for offspring and therefore the history of effective selection for such design. More generally, however, organisms commonly adapt to even rarer events than the infrequent ovulation in hunter-gatherer women. For example, the once-in-a-lifetime reproductive effort of semelparous plants and animals is characterized by special design shaped by selection. (In chapter 12, we address a related concern that ovulation by women not nursing offspring is a rare event after the birth of first child in traditional forager populations.)

Summary

Contrary to conventional wisdom, abundant evidence demonstrates that women possess estrus. In important respects, women's estrus is functionally similar to, as well as homologous with, estrus is other mammals, and even vertebrates in general. The design of women's estrus is to motivate sexual interest in men who are relatively masculine in scent, facial and bodily features, muscularity, voice, and skin tone; who have scent and appearance associated with bilateral symmetry; and who, perhaps, are relatively high in creative intelligence. These preferences are reasonably interpreted as sexual preferences for male features than connote good phenotypic and genetic quality (or did ancestrally, when the preferences evolved; see chapter 7). Hence the function of these adaptations of women's estrus is to obtain through mating genes that enhance the reproductive value of offspring. Estrous women reveal additional specialized mate choice through aversion to maladaptive sexual behavior and avoidance of sexual coercion. (Aversion, avoidance, and rejection of males are expressions of female mate choice no less than mate preference is; e.g., Kokko et al., 2003.)

The pattern of mate choice demonstrated by women at the fertile phase of their cycles is not merely an exaggeration of preferences they possess during infertile phases of the cycle. During the infertile luteal phase, women may prefer relatively feminized men and men who appear healthy currently. The preferences of women who are pregnant or using hormonal contraception are similar to those of women in the luteal phase. All of these conditions are associated with high levels of progesterone. Women in infertile phases do not ignore indicators of good genes, as they are relevant to choice of a long-term, investing mate. Attractive women, who can garner long-term investment from a male of high quality, in particular value purported indicators of good genes when infertile. In general, however, preferences for long-term mates do not change much across the menstrual cycle.

10 Women's Estrus, Pair Bonding, and Extra-Pair Sex

Women's Estrus Exists in the Context of Pair Bonding

In the preceding chapter, we argued that women possess estrus, just as all (or nearly all) mammalian and, indeed, vertebrate females possess estrus. Estrous sexuality is marked by adaptation that functions to obtain from sires genetic benefits for offspring. Hence females should generally find relatively fit males more sexually attractive when fertile in their cycles. As the result of research accumulated just over the past decade, women's fertile sexual interests, relative to their interests when infertile in their cycles, are understood perhaps better than those of any other vertebrate female. And the research to date points to a definite conclusion: When fertile, women are sexually attracted by purported (ancestral) markers of genetic fitness in men. Women possess estrous sexuality, as we conceptualize it.

Earlier in this book, we discussed views of human mating patterns. And, though debate exists, we generally sided with those who have argued that humans have evolved to care for offspring biparentally. Throughout much of human history, men and women have paired up, more often than not monogamously, and cooperatively raised offspring. This pattern, we noted, is very unusual for a mammal. Women's estrus must be understood within this context. Before considering women's estrus within it, we reflect on fertile-phase sexuality within a taxon of species in which similar patterns of biparental care is very common: passerine birds. The human sexual selection system exhibits important analogy with those of many of these birds.

Pair Bonding, Estrous Sexuality, and Extra-Pair Paternity Within Birds

Extra-Pair Copulation in Birds

Biparental care characterizes the majority of bird species, some rodents, and a few primates, among other species. In these species, both sexes may be choosy, with each sex's mate preferences exerting sexual selection on the other sex's features. As we discussed in chapter 4, recent modeling suggests that complementarity of each sex's parental investment may be an important factor favoring biparental care (Kokko & Johnstone, 2002); that is, biparental care may evolve when the sum total beneficial effect of the sexes' efforts exceeds the sum of the beneficial effects of males and females investing in offspring individually.

It was once thought that biparental bird species were sexually, as well as socially, monogamous. As behavioral ecologists well know, spectacular findings accumulated in the late 1980s and 1990s showed that, in fact, both males and females in many species frequently copulate with individuals other than their social partners (i.e., engage in extra-pair copulation, or EPC). And, indeed, on average across species, 10–15% of offspring are sired by males other than a social father, with rates ranging from close to zero to about 70% (e.g., Birkhead & Møller, 1995; Petrie & Kempenaers, 1998).

The ultimate reasons for which male birds are motivated to engage in EPC are not difficult to understand: Males can increase their reproductive success if they sire offspring raised in nests other than their own. Indeed, in collared flycatchers on the island of Gotland (Sweden), EPC accounts for a substantial amount of variation in reproductive success across males (Sheldon & Ellegren, 1999). Less obvious are the reasons that females would engage in EPC. Whether females are inseminated by their own social partners or by extra-pair males, their eggs end up in their own nests. They and their social partner care for any resulting offspring. Indeed, if social partners can discriminate between their own offspring and those of extra-pair males, partners could invest less in offspring of extra-pair males, thereby imposing a cost on female EPC. How, then, do females increase reproductive success by being inseminated by males other than their social partners?

Two major classes of benefits to female EPC have been proposed: direct benefits and genetic benefits. Direct benefits include food and protection. In some species, it is not obvious how extra-pair males deliver those benefits. Hence, genetic benefits have received much attention recently (e.g., Jennions & Petrie, 2000). As females purportedly choose mates on the basis of genetic benefits when males do not provide care, it is perhaps not surprising that their choices could be influenced by consideration of genetic benefits even when males do provide care. Having male assistance in raising offspring does not eliminate selection on females to obtain good genes.

As we discussed in chapter 7, three kinds of genetic benefits exist: intrinsic good genes, compatible genes, and diverse genes. The three types are not mutually exclusive.

Within individual species, any one or all may operate to affect female choice. A review of the evidence finds support for the importance for all three, though to varying degrees across species (Jennions & Petrie, 2000).

In a number of species, evidence consistent with a role for intrinsic good genes exists. Behavioral or direct DNA fingerprinting data indicate that males who are solicited as extra-pair mates or who are responsible for extra-pair fertilizations possess features that distinguish them from other males in a number of species: attractive zebra finch males (Houtman, 1992); dusky warblers who sing for longer periods of time at high amplitude (Forstmeier, Kempenaers, Meyer, & Leisler, 2002); black-capped chickadees who are more socially dominant than in-pair males (Otter, Ratcliffe, Michaud, & Boag, 1998); great reed warblers with a more extensive song repertoire (Hasselquist, Bensch, & von Schantz, 1996); collared flycatchers who have a more extensively developed condition-dependent secondary sexual character (a larger forehead patch; Sheldon et al., 1997); common yellowthroats who have a large sexually selected black facial mask (Thusius, Peterson, Dunn, & Whittingham, 2001); barn swallows who have a higher song rate (Møller, Saino, Taramino, Galeotti, & Ferrario, 1998). In some instances, researchers have linked preferred traits to fitness components: Extensive singing at high frequencies is associated with male longevity in dusky warblers (Forstmeier et al., 2002); song repertoire of a father predicts the postfledging survival rate of his offspring in great reed warblers (Hasselquist, 1998; Hasselquist et al., 1996); offspring sired by male collared flycatchers with a larger forehead patch are in better condition than their half-siblings within the same brood (Sheldon et al., 1997). Females in some species (e.g., Bullock's orioles; Richardson & Burke, 1999; western bluebirds; Dickinson, 2001) favor older males as EPC partners, perhaps because age itself is a marker of viability (see Brooks & Kemp, 2001).

At the same time, studies in a variety of species in which extra-pair mating is common have yielded negative or equivocal evidence for intrinsic genetic benefits: for example, razorbills (Wagner, 1992); hooded warblers (Stutchbury et al., 1997); sedge warblers (Buchanan & Catchpole, 2000; but see Langefors, Hasselquist, & von Schantz, 1998). Some researchers have argued that genetic benefits achieved in particular species are better understood in terms of compatible genes (e.g., in bluethroats; Johnsen, Andersen, Sunding, & Lifjeld, 2000; Johnsen, Lifjeld, Andersson, Ornborg, & Amundsen, 2001; in pied flycatchers; Rätti, Hovi, Lundberg, Telegstrom, & Alatalo, 1995) or diverse genes (e.g., in great tits; Otter et al., 2001; see also Lubjuhn, Strohbach, Brun, Gerken, & Epplen, 1999) rather than intrinsic good genes. For instance, a study of the savannah sparrow found that females paired to MHC-similar males, compared with females paired to MHC-dissimilar males, engage in more EPC, as measured by extra-pair paternity (EPP) of young (Freeman-Gallant et al., 2003). Females of this species use EPC (during estrus at least) to seek compatible genes.

Akçay and Roughgarden (2007) recently performed a meta-analysis of 121 studies testing claims that female birds use EPC to obtain genetic benefits. (Typically, studies compared traits of extra-pair males with those of in-pair males.) They found that, whether findings were sorted by studies or by species, about 3 in 7 cases provided support. Furthermore, this was true whether studies assessed benefits of intrinsic good genes or whether compatible gene hypotheses were assessed. The strongest mean effect sizes were found for secondary sexual characteristics and age (each, when expressed as a correlation coefficient, being approximately 0.3, with more ornamented and older males favored for EPC). Larger males were also favored, on average, as EPC partners. As the authors noted, these findings can be interpreted in multiple ways. They favored the interpretation that studies to date do not strongly support good-genes hypotheses for avian EPC. We, by contrast, view these results as being very consistent with the possibility that female EPC in birds often, even if not always, functions to obtain genetic benefits. If the average study had 50% statistical power to detect a true meaningful effect, for instance, one could expect only 50% of the studies to find support, when all true effects were meaningful. If average power was 65% to find effects, which existed in 65% of species, one could expect only 42% of studies to yield supportive evidence. The overall pattern of results appears to us much closer to what would be expected if females commonly seek genetic benefits through EPC than if females never or rarely do. Nonetheless, more work is needed to more precisely characterize and explain the pattern of positive and negative findings.

Not all biparental bird populations have high EPP rates. Many species of raptors, seabirds, and island birds, for example, have rates less than 5%, and some are close to 0%. As we discuss (see later in this chapter and chapter 12), low rates are not inconsistent with females possessing adaptations for engaging in EPCs for genetic benefits. Males in many species evolve counteradaptations to reduce the risk of investing in offspring not their own, and these counteradaptations suppress the EPC rate.

Estrous Sexuality in Birds

Female birds store sperm in sperm storage organs, which then can be released to fertilize eggs days or weeks following initial storage. Females producing a clutch of multiple offspring typically lay eggs over a period of days. Sperm loss over time is variable but in wild birds can be quite high (Sax, Hoi, & Birkhead, 1998). As a result, the fertile period is typically graded, with copulations some days (e.g., in superb fairy wrens and collared flycatchers, a day to a few days prior to the laying of the first egg; Double & Cockburn, 2000; Michl et al., 2002) more likely to be responsible for fertilizations than copulations earlier in the period (Birkhead & Møller, 1995). Females can produce broods of mixed parentage; in fact, even in species with high EPP rates, broods typically include at least one offspring sired by the main pair-bond partner (Birkhead & Møller, 1995).

In many socially monogamous bird species, pairs mate outside of the fertile period (e.g., prior to their seasonal fertile period or after laying is completed) or on days early in the fertile period unlikely to result in fertilization; they have extended sexuality. Extended sexuality should typically function to obtain material benefits from social partners (see chapter 3). (In some species, it has been argued to also serve the function of mate assessment; e.g., Mougeot, 2000. But in such cases, it is reasonable to expect that female sexual motivation functions to secure material benefits from males, as assessment need not involve mating.) Components of sexuality at peak fertility, by contrast, should serve to increase the chances that a male with good genes will sire offspring.

This period, again, is homologous with mammalian estrus and hence is appropriately defined as estrus (see chapter 8).

In species in which females obtain genetic benefits through extra-pair mating, females at peak fertility should (1) not mate indiscriminately; and (2) relative to their mating during less fertile periods, prefer as sires males who possess indicators of genetic benefits. These predictions are those we offered for females in estrus in general (see chapter 8). Additional predictions, however, follow for females in species in which males and females form socially monogamous pairs. Some females will have paired with mates of superior genetic quality (intrinsically good and perhaps also compatible genes). These females should typically have little reason to look past the paired male for good genes for offspring. Not all females can pair with a male who has better than average genetic quality, however. Hence, (3) overall, female EPC rates should increase during the fertile period, and (4) females paired with males who possess lower than average genetic quality should have the highest likelihood of engaging in EPC.

In a variety of species these patterns are found:

Currie, Krupa, Burke, and Thompson (1999) studied wheatear (members of the fly-catcher family). They experimentally removed resident males when females were either fertile or after eggs were laid. The EPC rate increased when males were not present during females' fertile phase, but not during incubation. Fertile females, however, were choosy. They could rebuff male copulation attempts. Only males who were in better body condition than the removed primary partner succeeded in obtaining EPCs. Hence the proportion of offspring sired by extra-pair males within a brood negatively covaried with body condition of the pair-bond male.

Dickinson (1997) similarly detained resident male western bluebirds when females were fertile. Females were choosy about with whom to engage in EPC. Neighboring males typically approached females within 15 minutes of male detention. But only half of females approached by males accepted even a single EPC. Females prefer EPCs with older males (Dickinson, 2001).

In common yellowthroats, each male possesses a black facial mask, which functions as a sexually selected ornament. Masks vary in size by a factor of about two. Males with larger masks have higher reproductive success than those with smaller masks (Thusius et al., 2001), outcompete other males for territories (Tarof, Dunn, & Whittingham, 2005), and are preferred by females in controlled choice experiments (Tarof et al., 2005). Pedersen, Dunn, and Whittingham (2006) found that females make forays into male neighboring territories only during their fertile periods; none were observed outside of females' fertile period. Furthermore, females preferentially foray into territories of males with masks larger than that of the primary partner. (A male is more likely to foray into the territories of neighboring females that are fertile, and especially if the resident male has a smaller mask than his own.) As a result, males with large masks have more EPCs.

Extra-territorial forays into neighboring male territories made by fertile females have been documented in other species as well. In superb fairy wrens (which have the highest EPP rate documented to date), females make predawn forays into neighboring male territories when fertile. Furthermore, of five females tracked in this study, all EPP was attributed to males they visited on these forays (Double & Cockburn, 2000). Similarly, a study of hooded warblers found that 80% of females made extra-territorial forays when fertile, compared to 0% when not fertile (Neudorf, Stutchbury, & Piper, 1997). In a review of work on 19 species of raptors, Mougeot (2004) found that EPCs mostly occurred during the fertile period.

Perhaps female control of paternity through estrous sexuality has been most thoroughly documented in a population of collared flycatcher on the island of Gotland in the Baltic Sea. As we discussed in chapter 3, although male and female pairs form socially monogamous unions in this population, an average of 15% of eggs are sired by extra-pair males. When sexually mature, males develop a patch of white feathers on their foreheads. Males with large patches outcompete other males for early breeding. They also account for a disproportionate number of the extra-pair fertilizations, partly due to female preference (Sheldon & Ellegren, 1999). In support of the hypothesis that large patched males are selected as extra-pair mates for their good genes, a controlled study revealed that offspring of males with large forehead patches tend to be in better condition (as measured by standard body weight assessments) than their half-siblings sired by small-patched males (Sheldon et al., 1997). Females whose social mates have relatively small forehead patches, moreover, are particularly likely to engage in EPCs (Michl et al., 2002; Sheldon, Davidson, & Lindgren, 1999). Furthermore, females preferentially engage in EPCs in the middle of the fertile period, at least 2 days after their last in-pair insemination. Hence, through the number of extra-pair insemination events was only 1.33 per cuckolding female in this study, the ratio of sperm from extra-pair and in-pair inseminations was at least 5:1 (Michl et al., 2002). (An interesting additional finding in this species is that the offspring of males with large forehead patches tend to be male, the sex that most benefits from having such a sire [Ellegren et al., 1996], which suggests that flycatchers adaptively adjust the sex-ratio of offspring depending on their own qualities or the qualities of their mates.)

In sum, it seems fair to conclude that, in a variety of species, females possess adaptations characteristic of estrous sexuality that are not operative during extended sexuality and that function to obtain genetic benefits through highly discriminating partner choice and EPC.

Why Don't Females in All Socially Monogamous Birds Solicit EPC?

Exceptions to these patterns exist. For instance, Low (2005a,b) found that most EPCs in New Zealand stitchbirds are forced, not female initiated or accepted. Moreover, acceptance of attempted EPCs does not appear to be predicted by male quality. In the great skua, males, when they forage, leave females in the breeding territory . Despite ample opportunity for EPCs, females rarely solicit EPCs, including during the fertile period (Catry & Furness, 1997). And, as we noted already, in many raptors, seabirds, and island birds the EPP rate is low. If females can gain good genes through EPC, why don't they do so in all socially monogamous species?

The likely reason in the majority of instances is that investing males pay a cost for female EPC—which has led them to evolve tactics to reduce the costs they pay for female

EPC. In socially monogamous species in which pairs have neighboring males, an inevitable conflict of interest between the sexes arises. All else being equal, females mated to males not possessing the best genes could benefit by getting genes from someone else. At the same time, selection operates on investing males to prevent cuckoldry. For instance, males can engage in selective mate guarding. They can attempt to increase rates of copulation. They can detect offspring that are not their own and refuse to invest in them. If males evolve abilities to prevent or impose costs on female EPC, then "all else" is not equal; for females to benefit through EPC, the benefits of the good genes obtained through EPC must be greater than the costs incurred by EPC or EPC attempts. If males can impose costs that exceed the benefits of female EPC, females may engage in EPC only very rarely. As we described in chapter 8, estrous sexuality should ultimately function to adaptively control sire choice. Obtaining good genes for offspring is an important part of adaptive sire choice in most species, and it is the major relevant consideration for females of some species. But females in species in which males and females form pair bonds typically receive important nongenetic material benefits from social partners, and a relevant consideration of adaptive sire choice for them is the implications of EPC for the future flow of those benefits.

Consistent with this interpretation, Møller (2000) found that, in pair-bonding birds in which male parental care is very critical to offspring survival (assessed through experimental removal of resident males), EPP rates are lower than in species in which male parental care is less critical (see also Griffith, Owens, & Thuman, 2002). As Albrecht, Kreisinger, and Piálek (2006) reported, EPP rates furthermore decline with estimated costs of direct selection imposed by males withdrawing care (but see Griffith, 2007, for a critique of the method of arriving at those estimates, drawn from Arnqvist & Kirkpatrick, 2005.) (Another factor affecting EPP rates, we note, is the variable value of genetic benefits across species. When genetic variance in male fitness is low, female EPC for genetic benefits is less likely to be adaptive. Petrie, Doums, and Møller (1998) found that, across species and populations, genetic variation positively predicts EPP rates in socially monogamous birds.)

We stress, however, that a low rate of EPC need not imply that females lack adaptations that function to obtain good genes at estrus. When females could potentially gain through EPC due to genetic benefits, selection operates on each sex against the interests of the other sex; "sexually antagonistic adaptations" evolve (see chapter 7). Female adaptations to engage in EPC selectively evolve. In response, male counteradaptations to suppress female EPC evolve. Counter-counteradaptations may be selected in females, and so on; the ensuing arms race may have no final stable solution. Depending on which sex evolves more effective adaptations (which may depend on ecological factors affecting the ease with which males mate-guard, the relative value of good genes and male assistance for females, etc.), the actual extra-pair sex rate may be high (more than 20%) or low (5% or less). Even when it is low, however, the conflict of reproductive interest exists. Accordingly, an extensive array of sexually antagonistic adaptations may have evolved. (For similar reasons, the current rate of infectious disease need not be high for both hosts and pathogens to have evolved many adaptations that function in host-pathogen conflicts, and the failure rate of pregnancies need not be high for both mothers and fetuses to possess batteries of adaptations and counteradaptations that function to give an edge over control of flow of maternal resources to fetuses. See Rice & Holland, 1997.)

We can make this point another way. For male adaptations that impose the costs that prevent female EPC (mate guarding, paternity detection, etc.) to be selected, females must engage in EPC. In the absence of female EPC, there is no benefit to offset the costs males pay for adaptations preventing female EPC. Female EPC may occur despite there being no adaptations for female EPC; EPCs may be forced on females by males. Again, however, in absence of any male counteradaptation, females could benefit through EPC when males vary in genetic quality. And in that light, male adaptation functioning to prevent female EPCs has likely been selected, at least in part, to counter female adaptation for EPC (see game theoretic modeling, for instance, by Fishman, Stone, & Lotem, 2003). Hence, male adaptations that effectively function to keep the occurrence of EPC of female partners low often evolved at least partly because females possess adaptation to selectively engage in EPC.

Females could pay costs for EPC for reasons other than male counterstrategies. Perhaps most notably, when a female's offspring are sired by multiple fathers, sibling-sibling conflict increases. This conflict imposes costs on siblings and their reproductive success. It may also impose costs on female fecundity directly through maternal-fetal conflict in mammalian species (e.g., Haig, 1993).

In a recent paper, Arnqvist and Kirkpatrick (2005) attempted to estimate the genetic benefits (based on within-clutch fitness of extra-pair vs. in-pair young) and direct costs (in the currency of lost male investment) of female EPC, based on data aggregated across a number of pair-bonding species, and argued that the costs exceed the benefits. They hence proposed that female EPC adaptation is absent and argued instead that males manipulate females into EPC, which are not in female reproductive interests (see, for instance, Holland & Rice, 1998, on "chase-away sexual selection"). Yet, in light of findings on female choosiness, selective forays, partner choice, and/or timing of choice (i.e., at peak fertility), too many predictions made by a model of EPC for good genes but not a model of male manipulation have been confirmed in some species to seriously entertain the latter for them (e.g., see preceding findings on collared flycatchers and common yellowthroats). Some EPC, indeed, is almost certainly due to male manipulation and against female interests. Moreover, in some species, EPC no doubt serves functions other than acquisition of genetic benefits for offspring (see chapter 3). As the aggregate meta-analyses performed by Arnqvist and Kirkpatrick (2005) did not strictly identify EPC motivated by choice of sires of superior genetic quality, the genetic benefits of those particular EPCs may have been diluted by other EPCs and hence underestimated. Their conclusion could be flawed for that reason. A full critique of Arnqvist and Kirkpatrick's (2005) methods and finding is provided by Griffith (2007).

Estrous Sexuality, Pair-Bonding, and Extra-Pair Mating in Humans

Just as estrous sexuality in many bird species exists in the context of pair formation, social monogamy, and extensive biparental care, so too must human estrus be situated in this same context. In the previous chapter, we documented the many ways in which women's preferences for men change across the menstrual cycle. As we emphasized, the preferences for masculine traits and indicators of developmental stability enhanced during estrus pertain primarily if not exclusively to evaluation of men as sexual partners. In general, preferences for long-term partners have not been found to change across the cycle. If, through recent hominin history, most females of reproductive age had primary mates, then these preference shifts largely functioned to affect selection of a sexual partner when fertile. And, if the partner found most attractive by a female when she was fertile was not the primary partner, that partner was an extra-pair male. Generally speaking, the primary context in which estrous preference shifts have been maintained by selection in human evolutionary history is the potential benefits to obtain genetic benefits through EPC.

Estrous sexuality within a pair-bonded species with genetic benefits through EPC has a number of implications. First, preference shifts should be observed in women with primary partners. In a number of cases, researchers have found that women in relationships experience preference shifts more strongly than unmated women. For instance, Penton-Voak et al. (1999) found that women in relationships only or mainly accounted for the effect of fertility status on the shift in preferences for masculine faces. And Havliček et al. (2005) found that only mated women particularly preferred the scent of dominant men when fertile. Not all studies, we note, have yielded similar effects. (Most studies in the literature, in fact, have not included relationship status as a variable.) At the same time, no study to date has reported a preference shift that occurs selectively in unmated women.

Second, if women with primary partners experience preference shifts that affect their attraction to real men in their lives, then women with primary partners should report more frequent or stronger attraction to men other than primary partners when they are fertile. On average, women may not report any greater attraction to their own primary partners.

Third, some women are mated to men who possess features found most attractive during women's fertile periods. As these women could have lost investment from males through EPC, they should not show increased attraction to men other than primary partners when they are fertile. Indeed, these women may experience increased attraction to their primary partners during estrus. Put otherwise, the expected increase in attraction to men other than primary partners during estrus should be accounted for by women with partners lacking features particularly favored by women when fertile.

Fourth, compared with the period of extended sexuality, during estrus women should, on average, feel less committed to their partners and more willing to engage in sex with men other than primary partners.

Fifth, just as males in bird species are expected to possess counteradaptations that suppress the likelihood of their primary partners' EPC, so too men are expected to possess paternity assurance and anticuckoldry adaptations.

Sixth, in light of male paternity assurance adaptations that impose costs on female EPC, women's estrous sexuality should be shaped to be responsive not solely to the benefits of potential EPC; it should also be shaped to be responsive to the costs of EPC imposed by males. Hence, women's estrous sexuality should be sensitive to factors that affect the likelihood or size of the cost of male efforts to assure paternity and prevent investment in offspring not their own.

We have already discussed the first implication. The remainder of this chapter considers the second, third and fourth implications, which concern patterns of women's sexual interests and attitudes across the menstrual cycle. We also discuss women's arousal, orgasm, and potential cryptic mechanisms of choice. In chapter 12, we turn to the fifth and sixth implications, which pertain to men's counteradaptations and their evolutionary effect on women's estrous psychology.

Women's Sexual Motivation and Interests Across the Cycle

Do Women Experience Greater Sexual Desire Near Ovulation?

As we discussed in our introductory chapter, following the discovery of estrogen and its changing levels across the reproductive cycle, researchers naturally turned to ask whether women experience an increase in sexual desire mid-cycle, when estrogen levels are highest and fertility is maximal—precisely when (it was assumed) females of other mammalian species experience a surge in libido that functions to obtain sperm. Do women initiate sexual activity with partners more when fertile than during other phases of the cycle? Do couples accordingly have sex more frequently near ovulation than otherwise? (The latter effect could be due to greater male sexual motivation mid-cycle, which we discuss in chapter 12.) Over the years, many studies have been conducted and many reviews written. Gray and Wolfe (1983) offered an early review. They concluded that the literature yielded mixed results across studies. Overall, however, research suggested a weak but discernable elevation in sexual behavior in couples, including female-initiated sex, at mid-cycle. Manson's (1986) and Hill's (1988) expanded reviews offered the same conclusions and appealed for still more research. Updated reviews of the literature emphasized the many mixed findings and drew the conclusion that no reliable pattern in women and men's sexual behavior across the menstrual cycle has been documented (Meuwissen & Over, 1992; Steklis & Whiteman, 1989). Still later overviews asserted contradictory conclusions: Dixson (1998) concluded that no reliable pattern exists, whereas Hrdy (1997) and Pawlowski (1999a) discerned from much the same literature a meaningful mid-cycle peak in sexual motivation. Strassmann's (1999) critique of Pawlowski's review raises questions about his conclusion. And inconsistency continues. A number of recent studies and analyses suggest that some women do, after all, experience a mid-cycle increase in sexual motivation and initiate sex more at that time (Bullivant et al., 2004; also Burleson, Trevathan, & Gregory, 2002, in part; Hedricks, Piccinino, Udry, & Chimbira, 1987; Regan, 1996; Wilcox et al., 1995). One well-conducted study that used hormonal measures to detect ovulation on a sample of 68 women who used highly reliable methods of nonhormonal contraception (IUD or tubal ligation), and who therefore had no reason to avoid sex when fertile, found a 24% increase in sexual activity on the 6 most fertile days compared with the remaining nonbleeding days (Wilcox et al.,

2004). By contrast, an analysis of reports from over 20,000 women in 13 developing countries around the world (gathered as part of a 1998 demographic and health survey) detected no change in frequency of sexual intercourse across the cycle, aside from lower frequency during menses (Brewis & Meyer, 2005).

The contradictory conclusions suggest to us that the unreliable findings are not merely due to sampling variability. That is, against a background of null findings in large samples (e.g., Brewis & Meyer, 2005), positive results are too frequent and, in some cases, strong (e.g., Wilcox et al., 2004) to be due to chance alone. Something about women's sexual desires appears to be changing across the cycle. We suggest that the persistent problem that researchers and scholars have encountered when trying to answer the question of whether women's sexual desires change across the cycle is simply that the question has been framed in the wrong way. Researchers have typically assumed that the change to be expected is a change in overall sexual drive and have generally studied intercourse or female-initiated intercourse with an established partner to assay that drive. That is, researchers have examined whether women want to have sex during mid-cycle based on the assumption that, should they have a sexual need or urge, naturally they will look first to their primary partners to fulfill it. But, once again, estrous sexuality is not merely heightened sexual drive. Estrous sexuality is discriminating sexual interest. Depending on whether women's partners fulfill or do not fulfill their discriminating sexual appetite when fertile, women may or may not express greater interest in having sex with their partner. Regan's (1996) conclusion that *some* women experience enhanced sexual desire at mid-cycle fits this view. But direct evaluation of this alternative view requires answers to a different set of questions.

Women's Attraction to In-Pair and Extra-Pair Men

If women do possess adaptation selected for genetic benefits for offspring achieved through EPC, at estrus women should find men who possess indicator traits of male fitness (at least under ancestral conditions) sexually attractive. Indicator traits of male fitness are expected to be highly variable across males (Gangestad, 1993; Pomiankowski & Møller, 1995; Rowe & Houle, 1996; see chapter 7, this volume). And, in a population in which most females are mated monogamously, many of them will not be mated with men who possess these traits. Hence, estrous pair-bonded females should often be sexually attracted to men other than their primary partners.

In collaboration with Christine Garver (now Garver-Apgar), we examined this prediction in a sample of romantically involved college students (Gangestad, Thornhill, & Garver, 2002). Young women not using hormone-based contraceptives served as research participants. They filled out questionnaires about their sexual attractions, interests, and fantasies experienced over the past 2-day period in two separate sessions. One session was conducted within 5 days before a luteinizing hormone (LH) surge, the period of high fertility, with the surge corresponding roughly to peak fertility (Bullivant et al., 2004; Wilcox et al., 1995). Another session was conducted during the luteal phase, with attempts made to target a day about 1 week after ovulation and to avoid days just preceding menstruation. Questions addressed attraction to the primary partner ("[I] felt strong sexual attraction toward my primary current partner," "fantasized about sex with a current partner") as well as attraction to someone other than a primary partner ([I] "felt strong sexual attraction toward someone other than a current partner," "fantasized about sex with a stranger or acquaintance/past partner"). We found that, on average, women's sexual attraction to and fantasy about primary partners differed minimally and statistically insignificantly between the fertile and luteal phase. By contrast, the same women's attraction to and fantasy about "extra-pair men" (which could have included former boyfriends, strangers, friends, and acquaintances) was substantially and significantly higher during the fertile phase of the cycle than during the luteal phase. Furthermore, analyses showed that the increase in attraction to extra-pair men during the fertile phase was significantly greater than the minimal difference in attraction to primary partners. This study also yielded a significant increase in strength of the effect of phase on attraction to and fantasy about extra-pair men as women's high fertility session approached the day of the LH surge, as expected if women's extra-pair interest rises with the degree of cycle-related fertility and associated hormonal influence (figure 10.1).

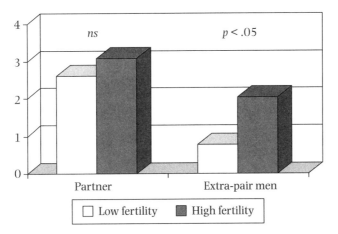

Figure 10.1 Women's sexual attraction to their primary partners and men other than their partners (Y-axis) during the fertile and luteal phases. White bars represent luteal phase. Gray bars represent fertile phase. Women's attraction to extra-pair men was significantly greater during the fertile phase. No difference in women's attraction to primary partners across phases was detected. The interaction is significant. Based on data from Gangestad et al. (2002).

In a replication (and extension; see later discussion) on a sample of women who, on average, were fairly committed to their relationships, Garver-Apgar and we found the same pattern: greater attraction to extra-pair men during the fertile phase than during the luteal phase, no significant difference in attraction to own primary partners across phases, and a significantly greater increase in attraction to extra-pair men compared with primary partners during the fertile phase (Gangestad et al., 2005a).

Pillsworth, Haselton, and Buss (2004) did not find this pattern of results. Instead, they found that normally ovulating, pair-bonded women reported greater in-pair sexual attraction and no change in interest in extra-pair men. The authors noted that the relationships of women in their study were generally new and satisfying; many may have been characterized by a period of infatuation with partners and enhanced male material benefit delivery to females. Women in these relationships may have estimated a relatively high cost/low benefit for EPC. Indeed, they found that, as relationship length increased, so too did women's extra-pair attraction during the fertile phase. No doubt, a variety of factors may moderate fertile women's sexual attraction to primary partners and extra-pair men. As we have already mentioned and as we discuss shortly, these factors should include characteristics of women's primary partners. (We hold off discussion of two additional studies that have examined changes in women's extra-pair interests across the cycle as a function of partner characteristics—studies by Haselton and Gangestad, 2006, and Pillsworth and Haselton, 2006—until turning to that topic.)

The studies discussed thus far concerned only women's in-pair and extra-pair *sexual attraction*. They did not examine women's actual in-pair and extra-pair sexual activity. Because the perceived costs of EPC may typically overwhelm the potential genetic benefits women detect (not consciously, but as reflected in evolved adaptations), women can perhaps be expected to actualize their extra-pair sexual interests in only a very small proportion of cycles (and, as we discuss in chapter 12, contingent on factors that affect male-imposed costs). One study did examine changes in frequency in actual EPC across the cycle. Data were collected through solicitation of responses to a questionnaire in a national magazine (*Company*) in the United Kingdom. Normally ovulating women with primary pair-bond partners were asked to report whether their last copulation occurred with the in-pair partner or an extra-pair partner; 6% were with extra-pair partners. Whereas in-pair sex tended to occur no more during the fertile phase than during the luteal phase, the frequency of extra-pair sex tended to be relatively high during the fertile phase (Bellis & Baker, 1990). In sheer numbers, the peak rate of EPCs (occurring about the 10th day of the cycle) was about 2.5 times the minimal rate of EPCs (occurring the last week of the cycle; Baker & Bellis, 1995).

Baker and Bellis (1995) also found a relatively high rate of EPCs during menses. Accordingly, Gomendio et al. (1998) argued that these data imply that EPC is not adaptive. As women may engage in EPC for functions other than obtaining a quality sire for offspring, this conclusion is premature. EPC during extended sexuality may largely function to obtain nongenetic material benefits from males. At the same time, additional research on occurrence and timing of EPC (as well as moderators of timing and occurrence, including in-pair and extra-pair partner characteristics) is needed.

Male Partners' Phenotypic Indicators of Intrinsic Genetic Quality Should Moderate Women's Sexual Interests

As we saw with some species of birds, females should not indiscriminately engage in EPC with high-quality males when fertile. Both its benefits and costs can vary as a function of conditions that vary across females, and selection should favor adaptation in females to assess those conditions. One major factor affecting the benefit of engaging in EPC to obtain genetic benefits for offspring is, obviously, the quality of the primary mate. When paired with males of relatively low genetic fitness, females can garner genetic benefits for offspring through EPC with a high-quality male. When paired with males of high quality, however, the potential benefits females can obtain through EPC are minimal.

Earlier, we discussed a study we performed with Christine Garver-Apgar (Gangestad et al., 2005a) replicating our prior finding that women report greater sexual attraction to extra-pair men during the fertile phase than during the luteal phase. As we also noted, that research extended the previous study, as well. Once again, we asked normally ovulating women in romantic relationships to report on their sexual attraction to and fantasy about their primary partners, as well as extra-pair men, on two separate occasions: once when fertile in their cycle (as documented by an LH surge) and once during the infertile, luteal phase. In addition, we asked women to bring their primary partners to the sessions. As part of the assessment, we measured men's asymmetry on 10 bilateral traits, composited to form an overall measure of fluctuating asymmetry (see chapter 7). As already noted, we found that women reported greater sexual interest in extra-pair men when fertile than when infertile. And, as expected, overall women with relatively asymmetric partners reported greater sexual interest in men other than primary partners.

Did we also find, however, that women with asymmetric men largely accounted for the effect of fertility status? Yes. A significant fertility status (fertile vs. luteal) × partner asymmetry interaction on extra-pair sexual interest emerged. This interaction is depicted in figure 10.2. As can be seen in the figure, when in their luteal phase, women generally experience relatively low levels of attraction to men other than partners. And this level does not vary as a function of primary partners' asymmetry. The overall increase in sexual attraction to extra-pair men by women during the fertile phase is experienced only for women with relatively asymmetric men. Women with symmetric men experience the same low levels of attraction to extra-pair men. In contrast, women with the most asymmetric men report, on average, a greater than twofold increase in sexual attraction to men other than primary partners.

Women's sexual interest in their own partners revealed an interaction in the opposite direction, as illustrated in figure 10.3. Once again, during the luteal phase women's attraction to primary partners did not vary as a function of partners' asymmetry. When fertile, however, women paired with symmetrical men reported greater attraction to their partners than women paired with asymmetrical men (Gangestad et al., 2005a).

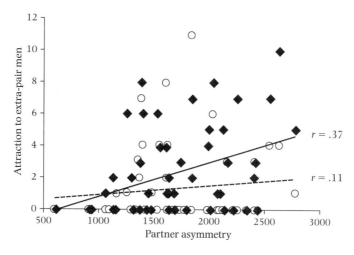

Figure 10.2 Scatterplots and regressions of women's sexual attraction to extra-pair men as a function of their primary partner's fluctuating asymmetry. Solid diamonds and solid regression line are for women during their estrous phase ($r = 0.37$, $p = 0.006$). Open circles and dashed regression line are for the women during their luteal phase ($r = 0.11$, ns). Partner asymmetry is the sum of the asymmetries across the 10 bodily traits measured times 10,000 (to eliminate decimal places). A value of 2,000 is equivalent to a mean of 2% asymmetry (relative to mean trait's size in the sample) across the 10 traits. The lines are least squares regression lines. Based on figure 1 of Gangestad et al. (2005a).

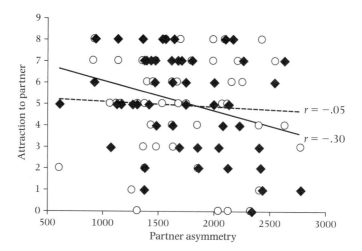

Figure 10.3 Scatterplots and regressions of women's sexual attraction to the pair-bond partner as a function of his fluctuating asymmetry. Solid diamonds and solid regression line: estrous phase; open circles and dashed regression line: luteal phase ($r = -0.30$, $p = 0.027$, and -0.05 [ns], respectively). The lines are least squares regression lines. Partner asymmetry as in figure 9.2. Based on figure 2 of Gangestad et al. (2005a).

In this study we also measured another variable that importantly affects women's interest in extra-pair men: their relationship satisfaction (see, e.g., Thompson, 1983). It makes sense that relationship satisfaction affects women's interest in extra-pair men: Satisfaction may partly gauge women's sense of the overall level of benefits they are obtaining in their relationships relative to what they are putting into them. Hence women's satisfaction with their relationships partly reflects their sense of what they would lose if they no longer had their partners. When we also included relationship satisfaction in the analysis of women's attraction to extra-pair men, we found that it did have a substantial effect. Critically, however, it did not diminish the crucial interaction between phase of the cycle and partner asymmetry. It is interesting to compare the relative effects of relationship satisfaction and partner asymmetry on attraction to extra-pair men resulting from a regression analysis performed separately on reports from the luteal phase and reports from the fertile phase. As shown in figure 10.4, during the luteal phase, relationship satisfaction had a very large effect. It also had a substantial effect during the fertile phase—but the effect of men's asymmetry was even larger (though not significantly so). Hence, even independent of relationship satisfaction, partner asymmetry had a large effect. We did not find that satisfaction moderated the effect of partner asymmetry. Thus we found no evidence that women who are very satisfied in their relationships are not attracted to extra-pair men when paired with asymmetrical men.

Two studies tested related predictions. Haselton and Gangestad (2006) asked normally ovulating women to keep daily diaries for 1 month. A high fertility window was estimated from cycle length using a backward counting method. (This method assumes

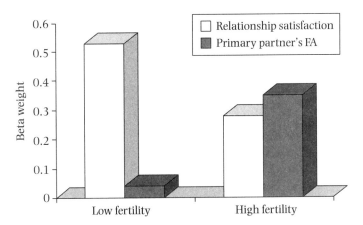

Figure 10.4 Effect sizes (β) for women's relationship satisfaction (open bars) and male primary partners' fluctuating asymmetry (shaded bars) in predicting women's sexual attraction to extra-pair men during the luteal and fertile phases. Note that β weights for relationship satisfaction are negative, but their signs are ignored in this figure for purposes of comparing predictive power. Based on data from Gangestad et al. (2005a).

a fairly constant day of ovulation 14 days before the first day of menstrual flow. The length of the luteal phase is indeed less variable than is the length of the follicular phase; e.g., Baird et al., 1995.) For women with pair-bonded partners, diary reports of women's attraction to and flirtation with men other than primary partners during this window were contrasted with the same reports made during the luteal phase (excluding the last few days prior to menstruation). No overall effect of fertility status was detected (though it was in the predicted direction, with fertile phase > luteal phase). Women were also asked to report their partners' attractiveness as short-term mates and long-term mates, however. The difference between the ratings reflected the extent to which a partner was viewed as sexually attractive but not particularly investing. As predicted, men's sexual attractiveness moderated the effect of fertility status on women's extra-pair interest: Women who were paired with men whose asset was as good long-term but not sexually attractive partners reported greater sexual attraction to and flirtation with men other than their partners on high-fertility days. Women who were paired with sexy but not particularly good long-term partners showed no such pattern (figure 10.5). Pillsworth and Haselton (2006) used an LH test to verify the fertile window. They too found that men's sexual attractiveness moderated the effect of cycle phase on women's attraction to

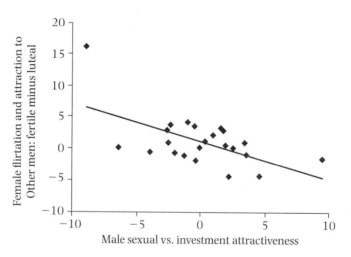

Figure 10.5 Scatterplot of shift in women's flirtation and attraction to men other than a primary partner (estrous phase minus luteal phase) as a function of men's relative sexual-minus-investment attractiveness. These values are residuals (centered around the sample mean), with potential confounders (male overall mate value, female physical attractiveness, and breakup status) partialled out (see Haselton & Gangestad, 2006). The line is the least squares regression line (partial correlation = −0.58, p = 0.004). From figure 1 of Haselton and Gangestad (2006).

men other than primary partners: Women with partners they perceive to be relatively unattractive reported greater interest in extra-pair men when fertile.

In chapter 9, we detailed the variety of shifts from infertile to fertile phases in what women find sexually appealing. Those shifts document a number of adaptations that function during women's estrus to obtain genetic benefits for offspring from sires of high fitness—that is, that evolved historically because of benefits derived through sire choice. The findings pertaining to women's sexual attraction to men other than women's primary partners strongly suggest another set of estrous adaptations. In theory, women partnered with sexually attractive and women partnered with sexually unattractive men encounter, in their lives, men who possess features that women typically find attractive when fertile. Yet women with sexually attractive and symmetric men do not claim to experience enhanced attraction to men other than partners when fertile. These results suggest that women's assessment of their mates modulates their sexual attraction to extra-pair men with appealing features. This adaptation (or set of adaptations, given assessment of multiple features) makes sense for females with primary investing mates but who can potentially and contingently garner genetic benefits for offspring through EPC.

Commonality of Women's MHC Alleles With Male Partners' MHC Alleles Moderates Women's Sexual Interests

Thus far, we have focused on male partners' intrinsic good genes in discussions of both avian and women's extra-pair sexual interests. But male partners' good genes in terms of their MHC complementarity may moderate women's estrous attraction to men other than partners, as well (see, e.g., the study on savannah sparrows discussed earlier; Freeman-Gallant et al., 2003). Indeed, our colleagues and we have found that, just as male partners' asymmetry and attractiveness moderate the impact of the estrous phase on women's attraction to extra-pair men, so too does male partners' sharing of MHC alleles with women (Garver-Apgar et al., 2006). MHC alleles were assayed at three highly polymorphic loci in 48 couples, and the proportion of alleles shared by partners (ranging from 0.00 to 0.67) was calculated. Once again, women (all normally ovulating) reported attraction to extra-pair men and partners over the preceding 2 days twice: once when fertile, once during the luteal phase. During the luteal phase, women reported low levels, independent of MHC sharing. During the fertile phase, by contrast, whereas women with men sharing no alleles reported low levels in line with those reported during the luteal phase, the same was not true of women with men sharing alleles. Women's sexual attraction to men other than partners during estrus significantly correlated with proportion of alleles shared with partners. The study also revealed that MHC sharing between coupled men and women negatively predicts women's sexual responsivity to and sexual satisfaction with partners. It positively predicts women's sexual infidelity in the relationship (though not sexual infidelity in previous relationships, as expected if EPCs stem from lack of genetic complementarity between partners; see figures 10.6 and 10.7).

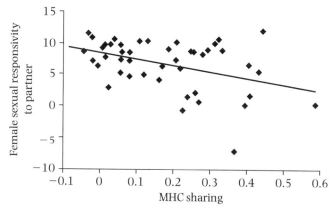

Figures 10.6 Women's sexual responsivity to their primary partners as a function of proportion of shared MHC alleles. Women's sexual responsivity is an average of a self-report and a partner report (which correlated 0.57). Women's age and relationship length were partialled out. (With these variables controlled, some residuals of MHC sharing were negative, resulting in some values slightly less than zero.) The correlation between MHC sharing and women's sexual responsivity is −0.40. The line is the least squares regression line. Based on data from Garver-Apgar et al. (2006).

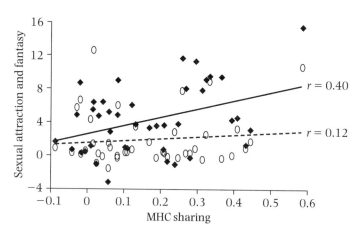

Figure 10.7 Scatterplots and regressions of women's sexual attraction to extra-pair men as a function of the proportion of MHC alleles shared with their primary partners. Solid diamonds and solid regression line represent women during their estrous phase ($r = 0.40$, $p = 0.007$). Open circles and dashed regression line represent the women during their luteal phase ($r = 0.12$, ns). Women's age and relationship length are partialed out. (With these variables controlled, some residuals of MHC sharing were negative, resulting in some values slightly less than zero.) The lines are the least squares regression lines. Based on data from Garver-Apgar et al. (2006)

As we reported in chapter 9, in a previous study with colleagues, we did not find that women's preferences for the scent of MHC-dissimilar men change across the cycle. (In that study, we did not find any preference in women for MHC dissimilarity, though other studies have.) Is there a contradiction between that finding and the preceding findings? Possibly not, at an underlying theoretical level. Preferences for indicators of intrinsic good genes should be particularly enhanced during estrus because not all women can possibly pair-bond with high-quality males, though all women can potentially engage in an EPC with one. The same expectation need not hold for preferences for complementary genes in a pair-bonding species. Different women will desire different men for their complementary genes. Hence (nearly) all women, in theory, could pair up with males possessing complementary genes. Preference for MHC-dissimilar scents, then, may function to adaptively guide mate selection both during extended sexuality (when women may be evaluating men as primary partners) and during estrus (when women may be evaluating men as sires, as well).

At the same time, for women who do end up with primary partners who do not possess complementary genes (e.g., because they were appealing on other dimensions), adaptations that contingently shift attention to men other than primary partners during estrus could be selected. This is the pattern documented to date.

This pattern requires differential evaluation of information about MHC similarity depending on the decision to be informed. It need not differentially affect attraction across the cycle. But it should nonetheless differentially inform decisions about whether to be open to attraction to men other than primary partners across the cycle. At mid-cycle, women partnered with men who share MHC alleles need not be looking outside the relationship for MHC dissimilarity; they could also simply be looking for an overall better set of genes from a sire, including intrinsic good genes.

Changes in Women's Sexual Attitudes Across the Cycle

Yet another feature is expected if estrous sexuality has components selected through genetic benefits garnered through EPC: Women should feel more willing to engage in sex outside of a committed pair bond during estrus. They should feel more willing to engage in sex with someone with whom they do not feel close. With Garver-Apgar, we examined this prediction in the same sample of coupled women we discussed earlier. At fertile phase and luteal phase sessions, women filled out a questionnaire concerning their sexual attitudes and were asked to respond as they felt at that moment, not how they felt in general. One dimension identified in this set of attitudes we referred to as sexual opportunism. Items measuring it included "I believe in taking my sexual pleasures where I find them," "The thought of an illicit sex affair excites me," "Sometimes I'd rather have sex with someone I didn't care about," and "If an attractive person (of my preferred sex) approached me sexually, it would be hard to resist, no matter how well I knew him/her." Women scored significantly higher on the aggregate measure when fertile than during the infertile phase. At estrus, they endorsed approximately 25% more items than they did when in the luteal phase (Gangestad, Thornhill, & Garver-Apgar, 2007a).

Similarly, Sheldon, Cooper, Geary, Hoard, and DeSoto (2006) asked women their motives for sex at different points in the cycle and found that women report less interest in sex for intimacy when fertile. They found suggestive but less conclusive evidence that interest in sex for partner approval, self-enhancement, or self-affirmation decline at mid-cycle.

Perhaps relatedly, Jones, Little, et al. (2005) found that partnered women expressed less commitment to their relationships when progesterone levels are low, as during estrus. This effect perhaps reflects greater willingness to risk loss of the relationship during the fertile window. Alternatively, the higher levels of commitment observed when progesterone levels are high (during the luteal phase) may be a product of selection for increased pursuit of long-term investment from the partner during pregnancy.

Women's Sexual Arousal During Estrus

Sexual Satisfaction and Arousal Is Important to Women for EPC

Greiling and Buss (2000) asked women what they want or would want in a one-night stand outside of their main sexual relationship. The women rated a variety of features on a 1–9 scale, with 9 being most desirable. Compared with what women want from their main partner (various forms of male investment; Buss & Schmitt, 1993), women desired from their extra-pair partner sexiness, sensuality, physical attractiveness, and high desirability to the opposite sex; indeed, each of these items received mean ratings above 8.0. In addition, women focused on sexual gratification. In particular, women's copulatory orgasms appear to be important, as women rate them as more of a potential asset in extra-pair sex than merely experiencing sexual gratification (Greiling & Buss, 2000).

These desires to experience during a brief extra-pair affair climactic zenith with a physically attractive, sexy, sensuous man who is highly desired by many women is fully consistent with desiring an EPC to obtain a sire with intrinsic high genetic quality. We suggest that, more generally, it represents the sexual desires of women during the estrous phase.

Greiling and Buss (2000) found that women desire other features in a one-night stand partner suggestive of other functions for EPC (e.g., obtaining a male protector or backup long-term mate, testing the potential for a new long-term relationship, and evaluation of personal mate value). Studies showing that women report greater sexual attraction to and fantasy about extra-pair men during estrus, particularly when they are paired with a relatively unattractive primary partner (e.g., Gangestad et al., 2005a; Haselton & Gangestad, 2006; Pillsworth & Haselton, 2006), suggest that desire for sexual gratification operates more often during estrus than extended sexuality. Alternative desires, we suggest, operate more often during extended sexuality than estrous sexuality. Future research can address this prediction.

Is There a Sex Difference in Sexual Arousal?

The evidence that woman has estrus perhaps provides some insight into answers to a question that frequently pops up in the literature on human sexuality: Which sex has the greatest sexual motivation (see Baumeister, Catanese, & Vohs, 2001; Hrdy, 1997; Oliver & Hyde, 1993; Symons, 1979)? It has been debated in discussion of both gender equality and evolved sex differences. The usual answer is that men, by far, have more libido. Indeed, many studies reveal that men, compared with women, masturbate more, fantasize more frequently about explicit sexual encounters, complain more often about the mate's sexual withholding, have sexual wishes that account for essentially all infractions of sexual norms and laws, do not require partner's sexual interest for sexual arousal, and so on (Baumeister et al., 2001; Buss, 1994; Oliver & Hyde, 1993; Symons, 1979; Thornhill & Palmer, 2000). There is no doubt that these research findings are reliable. Earlier, we argued that researchers traditionally asked a question about general libido that less than enlightened our understanding of women's fertile-phase sexuality. Here, the question of sex differences in general libido tells us something, but perhaps less about libido than about the conditions under which libido is stimulated. Men's sexual desires, relative to women's, are much less discriminating. This sex difference is expected by parental investment theory (Trivers, 1972).

At the same time, this analysis does not address a key question raised by research on women's estrus: Is there a sex difference in peak sexual arousal? We have argued that women's estrous sexuality is designed to be both highly selective and, under historically adaptive circumstances, manifested with great lust toward a mating partner with superior genes. When the sex-specific design of sexual arousal in humans is taken into account, it is reasonable to suggest that there is no sex difference in the level of peak sexual arousal (see Hrdy, 1997, and Okami, 2004, for related discussion).

Additional Estrous Adaptations: Eating, Drinking, and Walking

Fessler (2003) has argued that the periovulatory nadir in food, salt, and water consumption by women reflects adaptation to focus effort on mating rather than ingestive behavior when conception is the most critical goal. The nadir, at least in feeding, is observed in estrous females of other mammalian species (see Fessler, 2003). We suggest that in women and other species in which females exchange sex for food during extended sexuality, the feeding nadir at estrus may be designed to solve the problem of hunger leading to mating with males who offer food but are of inferior genetic quality. If so, the nadir is a design feature of woman's estrus that ultimately serves mate and sire choice.

Another cycle-related change that may have been directly selected concerns ambulatory activity. Women walk and move at mid-cycle more than they do during infertile phases (see review in Fessler, 2003). The same estrus peak in movement has been documented in a variety of other female mammals. Heightened activity of domestic cats or dogs in heat provide familiar illustrations. Particularly as women do not seek greater access to food or water during estrus, Fessler (2003) suggests that the feature in women is adaptation designed for seeking exposure to high-quality mates at the time of peak fertility.

Consistent with his interpretation, Haselton and Gangestad (2006) found that women in estrus expressed greater interest in attending social gatherings at which they might meet men than the same women did during the luteal phase. Grammer, Renninger, and Fischer (2004) reported a similar effect. At the same time, women's increased ambulatory activity at estrus is not manifest as increased levels of random walking. Indeed, during estrus women avoid certain situations likely associated with increased probability of sexual coercion (e.g., walking along unlit streets alone at night; Bröder & Hohmann, 2003; Chavanne & Gallup, 1998). Estrous women increase their activity, but selectively, in ways that maintain female control of mating. The adaptation of increased movement may be, we suggest, functionally fine-tuned to enhance likelihood of encountering men of superior genetic quality while reducing the likelihood of exposure to nonpreferred, sexually coercive males. Estrous women's enhanced walking motivation, then, may be the functional analog (and hence independently evolved) or functional homolog of the estrous female guppy's directed movements to stream microhabitats that contain high-quality males, but not low-quality ones (see chapter 5) or the forays that estrous female common yellowthroats make into territories of desirable sires with large facial masks (see earlier in this chapter).

During estrus, women's interpretations of emotional expressions of others differ from their interpretations in the luteal phase. One study found that women interpret facial expressions—whether of men or women—when progesterone levels are high (as in the luteal phase; Derntl, Kryspin-Exner, Fernbach, Moser, & Habel, 2008). In particular, women in the luteal phase tended to overinterpret negative expressions as anger and disgust (see also Conway et al., 2007). Both of these reactions represent threats (whether from the target perceived or pathogen-relevant stimuli to which the target reacts). Possibly, changes in women's interpretations of others' expressions reflect adaptations of avoidance of threat or contagion in conceiving or pregnant women.

Is Female Orgasm a Cryptic Choice Mechanism?

Baker and Bellis (1995) proposed that women's sexual arousal—specifically, their orgasm—functions cryptically in good-genes mate and sire choice (see also Thornhill et al., 1995; Thornhill & Gangestad, 1996). Cryptic female choice involves female adaptations that are manifested during and after mating and are designed to bias fertilization in favor of certain mates over others (Thornhill, 1983; Thornhill & Alcock, 1983; see Eberhard, 1996, for review across taxa; for nonhuman primates, especially Reeder, 2003, also Dixson, 2002). Cryptic female choice is expected to be favored by selection in species in which females mate with multiple mates, especially when individual females have a dual mating preference of sires with superior genetic quality and males willing to deliver material benefits (Thornhill, 1983, 1984b). Cryptic female choice in this case allows females to control paternity of their offspring and bias it toward males of superior genetic quality while simultaneously garnering male-provided material benefits by extended sexuality.

Orgasm may bias retention of sperm in circumstances in which women have mated with multiple men within a fertile period, thereby favoring the male with superior genes for offspring (Baker & Bellis, 1995). These claims have been controversial (see, e.g., Birkhead, 1995; Levin, 2002); additional studies are needed. Female copulatory orgasm may be an adaptation of cryptic female choice in some nonhuman primates and other mammal species. Troisi and Carosi (1998) reported that female Japanese macaques experience orgasm more frequently when mating with dominant males than when mating with subordinate males. (See also Eberhard, 1996, for other examples in nonhuman species of females' differential treatment of ejaculates from different mates.)

Orgasm need not function as a cryptic means of mate choice for it to bias paternity in the context of mating. It may lead women to engage in selective mating, and perhaps particularly so during estrus.

Several studies have demonstrated that women are more likely to experience copulatory orgasm with some males than with others. Thornhill et al. (1995) found that women whose partners were more symmetric and more physically attractive experienced orgasm during a greater proportion of copulations than women with less symmetric and less physically attractive men. Møller, Gangestad, and Thornhill (1999) found the same association with men's symmetry in a partially overlapping sample. Shackelford et al. (2000) replicated the effect of men's physical attractiveness. Montgomerie and Bullock (1999) reported a failure to replicate the association between copulatory orgasm and men's symmetry.

Recently, Lloyd (2005) concluded, in a book-length treatment, that women's sexual orgasm is a by-product, not an adaptation of female choice. The by-product hypothesis, proposed by Symons (1979), predicts no pattern in female orgasmic response in relation to women's mating partners' traits. In our view, Lloyd's (2005) conclusion is premature. A key problem with her treatment is her choice of criteria for identifying adaptation. She argued that woman's orgasm is not adaptation because evidence that women's orgasms promote their current reproductive success is lacking. But, as we discussed in chapter 2, female orgasmic behavior in relation to offspring production is not fundamental to identifying female orgasm as an evolved adaptation. The key evidence resides in functional design. Puts and Dawood (2006) provide a detailed critique of Lloyd's central arguments and conclusions.

At the same time, in light of the evidence, it is also premature to conclude that women's orgasm functions in sire choice. One limitation of existing research is that it has rarely separated woman's sexuality into its two apparently functionally distinct components: estrus and extended sexuality. One possibility yet to be deeply explored is that estrous copulatory orgasm in women functions to obtain sires with superior genetic quality for offspring, whereas copulatory orgasm in women during extended sexuality (as well as faked female sexual arousal, whether during estrus or extended sexuality) functions to increase material benefits from men. Women in marital relationships with investing men, for instance, have orgasm during mating with their partners more frequently than do women married to less investing men (see review in Thornhill & Furlow, 1998). One might expect that these orgasms tend to occur during extended sexuality.

In this regard, it is interesting to note two different potential functions of female orgasm: selective sperm uptake and social bonding (Thornhill et al., 1995; Thornhill & Gangestad, 1996). The latter occurs, perhaps, through oxytocin release. One speculation worth investigating is that oxytocin-based physiology functions primarily during the extended aspect of woman's sexuality, leading to bonding with primary partners delivering material benefits. Selective sperm uptake may function during estrus. Possibly, women release relatively little oxytocin during estrus.

One study has examined orgasm during estrus and during extended sexuality separately. Garver-Apgar et al. (2006) asked women to report on the frequency of orgasm over a 2-day period twice, once in the fertile phase and once in the luteal phase. MHC allele sharing (i.e., MHC incompatibility) significantly and negatively predicted frequency of female orgasm during the fertile phase, but not during the luteal phase. Unreported analyses found nonsignificant interaction effects with phase (albeit in predicted directions) for men's symmetry and physical attractiveness. Unfortunately, this study did not ask women to separately report copulatory and noncopulatory orgasm. Much work remains to be done.

Do Women Mate With Multiple Partners to Run Sperm Competition Races?

In some species, females are thought to mate with multiple males within a single estrous phase to set up sperm competition races (see, e.g., Birkhead & Møller, 1992). The idea is that males may vary with respect to how well their sperm can win such competitions, that this male quality may be an important component of male fitness, and that the best way for females to assess this quality is to test male sperm directly by, effectively, "running" races within their reproductive tract. The "winner" of such a race achieves a conception. Baker and Bellis (1995) suggested that women have been selected to double-mate during their fertile phase (with both an in-pair partner and an extra-pair partner) to run sperm competition races. We suggest that women's multiple mating is unlikely to function to promote sperm competition. Though women, we have argued, do value indicators of intrinsic good genes and compatible genes when in estrus, as those genes clearly provide benefits to offspring, we are unconvinced that the benefits of promoting sperm competition as a means of evaluating sperm viability itself provided benefits that outweighed costs ancestrally.

If women were designed to promote sperm competition, one might expect them to possess specialized adaptations that facilitate sperm competition. In birds, females possess organs designed to store sperm. Women possess no sperm storage organs or other devices that are candidate sperm competition facilitators. One might also expect that women's reproductive tracts would permit sperm to live following insemination for a long period of time. But most conceptions occur within a 2–3 day window in humans (e.g., Wilcox et al., 1995). Hence, ejaculates from different men can compete for fertilization only when a mature egg is imminent during estrus (for further discussion, see Thornhill, 2006).

The Functions of Estrous Mate Preferences and Sexual Desires Revisited

Buller's Proposal

We have argued that mate preferences and patterns of sexual desire that distinguish estrous women from women during phases of extended sexuality are adaptations forged by the benefits afforded by seeking sires of high genetic quality. Buller (2005), by contrast, has suggested that women's desires when fertile are by-products of other adaptation rather than mate- and sire-choice adaptations. Specifically, he claimed that ovulatory cycle effects can be explained by appeal to three adaptations that did not evolve in the context of extra-pair mating: (1) the "sex drive," a desire for regular and fulfilling sex, together with efforts to satisfy that desire; (2) a peak in sexual desire during the fertile phase of the ovulatory cycle, resulting in greater female-initiated sexual activity; and (3) a preference for symmetrical males, particularly when women seek partners who will satisfy sexual desires. These three adaptations could result in the empirical patterns observed as follows. When a woman is sexually dissatisfied in her relationship, she will be more likely to have extra-pair sex should an opportunity to do so present itself. In selecting an extra-pair partner, she will use the same criteria that she uses to choose a long-term mate but, because her desire for sexual satisfaction is heightened, she will weight factors related to sexual satisfaction, including symmetry (and, by extension, related features), most highly. Women hence are expected to initiate extra-pair sex with symmetrical men more often than expected at random. Moreover, because the sex they initiate will tend to be directed toward their extra-pair partners (given that sexual dissatisfaction led to the affairs in the first place), their sex with their extra-pair partners will tend to occur when women are most likely to initiate sex—at mid-cycle. Patterns of sex with and sexual interest in extra-pair partners, Buller argued, are therefore merely by-products of adaptations that evolved in the context of female mating more generally. Hence they constitute no evidence of female adaptation that specifically evolved in the context of extra-pair mating.

Surely, alternative explanations for ovulatory cycle variations in women's preferences and sexual interests are welcomed. They also, however, should be subjected to the same level of close scrutiny as the hypothesis that these variations evolved in the context of extra-pair mating. Buller's (2005) theory, in our view, leaves so many questions unanswered and findings unexplained that it can be seriously questioned, if not rejected outright.

First, a crucial piece of argument in Buller's theory is that female sexual desire is enhanced mid-cycle. We have already detailed the gist of the extensive literature that speaks to that claim: The evidence is very mixed.

Second, Buller's argument states that sexual desire evolved to be at a peak when women are fertile because it was designed for "reproduction." This claim is the familiar claim that women (similar to females in general) seek sperm when fertile. Again, we have already detailed the theoretical and empirical problems with that view. Estrus is not about getting sperm per se in general, and not so in women specifically.

Third and relatedly, Buller does not explain why women find some desired features "sexually" attractive, whereas others are found to be good in a mate but not sexually attractive. Indeed, he appears to acknowledge that symmetry may well be found particularly sexually attractive because it was ancestrally an indicator of good genes. Ability or willingness to invest in a mate are important to mate choice, he notes, but are not found sexually attractive. But why would women have evolved to find indicators of good genes, but not indicators of investment, particularly sexually attractive? He provides no explanation but rather takes this "fact" as a starting point. Our explanation actually does not propose that women do not find indicators of investment sexually attractive. Some may well do so, particularly in particular contexts (e.g., feminine male faces may be attractive to women because they connote warmth; see chapter 7). The face they find most attractive when mid-cycle, however, is more masculine than the one they find most attractive during the luteal phase (e.g., Penton-Voak et al., 1999). Buller's account is unable to explain these findings.

Fourth, Buller's argument is not able to explain some of the specific findings of ovulatory cycle shifts. Though not explaining precisely why women find particular features sexually attractive (as just noted), his theory does suggest that whatever women find sexually attractive in general, they will find particularly appealing mid-cycle, when they can conceive (and hence desire sex more). As we have emphasized, however, that is simply not the case (see chapter 9).

Fifth, Buller's argument appears to be unable to account for recent findings about the role of sexual satisfaction in accounting for cycle shifts. Gangestad et al. (2005a), once again, found that women paired with asymmetrical men are particularly likely to experience heightened attraction to men other than their partners mid-cycle and experience less attraction to their partners mid-cycle, relative to women paired with symmetrical men. According to Buller's account, these associations should be mediated by sexual satisfaction. Controlling for sexual satisfaction, then, should eliminate the effects. In fact, however, controlling for relationship satisfaction did not reduce the size of the effect of partner asymmetry on changes in female sexual interests at all (Gangestad et al., 2005a). Additional analyses of these data show that controlling for *sexual* satisfaction specifically also does not reduce the size of the effect. Moreover, though sexual satisfaction not surprisingly predicts sexual attraction to partners relative to attraction to other men, it did not have differential effects across the cycle in this study; Buller's theory suggests it should. Similarly, Haselton and Gangestad (2006) found that women who report that they find their partners less sexually attractive (relative to their attraction as investing mates) are particularly likely to be attracted to and flirt with men other than partners mid-cycle. Controlling for a measure of sexual satisfaction in the relationship did not eliminate this effect.

In sum, Buller's (2005) alternative by-product account of changes in female sexual interests across the cycle cannot explain a number of key findings and, furthermore, leaves a variety of important questions unanswered. (For a related but somewhat different by-product hypothesis, see Roney, 2005.)

Gray and Wolfe's Theory

Gray and Wolfe (1983) proposed yet another explanation for changes in women's sexual interests across the cycle. Sexual activity in monogamous species peaks early in the pair bond and subsequently declines (see review in Kleiman, 1977; also, in humans, Buss, 1994, 2003a,b). Gray and Wolfe hypothesized that women become interested in sex mid-cycle to ensure conception with a long-term male partner who becomes sexually unmotivated as time passes after pair bonding. This hypothesis is challenged by the theoretical problem faced by proposals that females are designed to signal fertility: Males are sexually selected to pursue copulations with females in the absence of signals. And, indeed, Klusmann (2002, 2006) has found that declines in the frequency of sexual activity across time in human pair bonds are due to declines in female sexual interest, not male sexual interest; as expected based on theory, the latter changes minimally over time. Of course, this idea is also inconsistent with many findings we report in the last two chapters.

Enquist, Rosenberg, and Temrin's Theory of Female Extra-Pair Copulation

Enquist, Rosenberg, and Temrin (1998) hypothesized that female sexual infidelity functions to obtain assistance from primary pair-bond mates rather than to secure superior genes. Females achieve this effect, they argued, by displaying sexual receptivity to extra-pair males during the fertile phase. Though offered as an explanation of the evolution of female EPC in pair-bonding birds, the hypothesis could in principle be applied to humans. It fails, however, to account for key features of women's estrus: estrous preferences for men with putative phenotypic indicators of intrinsic genetic quality, estrous interests in men other than partners conditional on their partners lacking those same indicators of genetic quality (and independent of female relationship satisfaction), estrous interests in men other than partners conditional on their partners lacking compatible MHC alleles.

This is not say that women's interest in extra-pair men exclusively revolves around their interests in acquiring good genes for offspring. They do so for a variety of reasons (e.g., Greiling & Buss, 2000). And, as noted by Buss (2003b), women may indeed use sexual interest in extra-pair males to manipulate primary partners in various ways (e.g., to demand their attention). We simply make the point that Enquist et al.'s hypothesis does not explain the nature of women's estrous sexuality. (See also Lumpkin, 1983, for a discussion of female manipulation of males, based on the males' desire to avoid cuckoldry, in birds.)

Summary

We argue that the evidence reviewed in the past two chapters strongly points to functional design for good-genes choice in women's fertile phase sexuality (in contrast to extended sexuality). In addition to being supported by evidence for design, however, the interpretation we favor is strongly supported by the phylogenetic and functional comparative evidence. Women's fertile-phase sexuality possesses apparent homology with and functional similarity to the fertile-phase sexuality of other mammalian females and vertebrates in general (chapter 8). These phylogenetic considerations pose serious challenges to incidental-effect interpretations of women's fertile-phase sexuality, whether Buller's (2005) or others. As we have stressed throughout this book, the evolutionary history of woman's sexuality is importantly informed not only by design considerations within humans but also by phylogenetic and functional comparative evidence.

The Roles of Ovarian Hormones

As we discussed earlier in this chapter, researchers have debated since the 1930s whether the ovarian cycle of women influences their sexual motivation (see Bullivant et al., 2004; Wallen, 2000). A central issue in this debate concerns the role of ovarian cycle hormonal changes. Until fairly recently researchers commonly assumed that the sexual behavior of female Old World primates is liberated from the influence of female reproductive cycle steroids. No longer is that assumption viable for many nonhuman Old World primates; though females of many of them exhibit extended sexuality, their sexual choices may change during peak estrus (e.g., Stumpf & Boesch, 2005), and hormones may well play a role (Dixson, 1998; Wallen, 2000; Pazol, 2003). Research we have reviewed provides good reason to believe that women's sexual behavior, as well, is importantly under endocrine regulation.

In nonprimate mammals, female proceptivity and receptivity typically co-occur with surges of estrogen. Correlational, as well as experimental, studies demonstrate a causal role of estrogen (see review by Nelson, 2000). In many of these species, estrogen (and in rodents, progesterone as well; Nelson, 2000) is a precondition of females' ability to copulate. In Old World monkeys and nonhuman apes, ability to copulate is not linked critically to estrogen. Many possess extended sexuality (see Wallen, 2000). In these catarrhines, it is nonetheless widely thought that females' motivation to copulate remains affected by the estrous peak in estrogen (see reviews of evidence in Bancroft, 1987, 2002; Dixson, 1998; Nelson, 2000; Wallen, 2000).

Pazol's (2003) work on blue monkeys, however, paints a more complicated picture. Female blue monkeys exhibit fairly dramatic forms of extended sexuality. At times, they mate across the entire menstrual cycle, during pregnancy, and as subfertile adults, prior to the point at which conceptive cycling begins. In this species, elevated estrogen levels characterize periods of sexual activity outside the fertile phase of the cycle, including prior to reproductive cycling and during pregnancy; in some cases, elevations equal those associated with estrus. Pazol also found elevated progesterone levels prior to cycling. She hypothesized that female blue monkeys have evolved adaptation for extending sexuality in the cycle through a combination of elevated estrogen as a hormone that motivates mating and elevated progesterone that suppresses ovulation.

Women's ability and motivation to copulate also clearly does not singularly depend on cycle-related estrogen peaks. Interestingly, however, the rise in progesterone during the mid-luteal phase is accompanied by a second (though more gradual and less dramatically peaked) rise in estrogen (e.g., Ferin et al., 1993). Possibly, as in blue monkeys, human females possess endocrine adaptation that supports interest in sex throughout the cycle.

Clearly, however, women's estrous sexual interests differ from their sexual interests during extended sexuality. Hormone-based adaptations that support women's dual sexuality, while maintaining sexual interest through the cycle, may have evolved in women. Estrogen and perhaps testosterone (which rises somewhat mid-cycle; see van Anders, Hamilton, Schmidt, & Watson, 2007) may facilitate women's estrous sexual interests. Progesterone, which rises and remains high during the luteal phase, may suppress estrous sexual interest. As we noted in chapter 9, a number of studies have attempted to map cyclic changes in estrous sexual preferences onto hormonal variations (using day-to-day actuarial data on typical changes). These preferences typically map onto cyclic changes of no one hormone (cf. DeBruine et al., 2005), suggesting that multiple hormones play roles. Interestingly, Puts (2006a), Jones, Little, et al. (2005), and Garver-Apgar et al. (2007a) each found that an estrous preference—attraction to masculine voices, masculine faces, and the scent of symmetry, respectively—was negatively associated with estimated progesterone levels, consistent with the interpretation that progesterone suppresses estrous sexuality. Garver-Apgar et al. (2007a) also found evidence for an independent positive association with estrogen, whereas Puts (2006b) found provisional evidence of a negative effect of prolactin. Based on measured hormones, Welling et al. (2007) found a positive association between women's testosterone levels and their attraction to masculine male faces. By contrast, Roney and Simmons (2008) detected an association between estrogen levels and women's preference for the faces of men with high testosterone. A number of hormones (estrogen, testosterone, LH, FSH) rise in close connection mid-cycle (see Bullivant et al., 2004), however, perhaps accounting for some of the inconsistency in findings to date. It suffices to say that some estrous cycle hormone(s) likely contribute(s) to women's estrous sexuality, and probably multiple ones do so. Though suppressing women's estrous sexuality, progesterone could very well enhance forms or motives underlying extended sexuality (e.g., DeBruine et al., 2005; Jones, Little, et al., 2005). Whether combinations of hormones affect changes in estrous sexual interests in close phylogenetic relatives (see Stumpf & Boesch, 2005) in a homologous fashion is unknown at this time.

At a number of points in this chapter and the preceding one, we have noted that women using hormonal contraceptives do not typically exhibit estrous sexuality. Instead, they tend to display sexual preferences and interests similar to those of women who are in the luteal phase. That is, these contraceptives bring about a hormonal state of extended sexuality across the entire cycle. Depo-provera and Norplant effectively induce extended states of elevated progesterone. Some contraceptive pills do, as well. More generally, contraceptive pills suppress the large spike in estrogen and tend to elevate progesterone. Their elimination or attenuation of estrus is fully consistent with the idea that progesterone suppresses estrus and enhances extended sexuality.

Research on hormone replacement therapy to treat women who possess little sexual interest—whether postmenopausal or not—demonstrates important roles of estrogen, progesterone, and testosterone in women's sexual motivation. Though each of these hormones administered separately often enhances female libido, combinations of all three have larger effects (as in tibilone therapy; e.g., Castelo-Branco et al., 2000; Davis, 2002; Egarter, Topcuoglu, Vogl, & Sator, 2002; Floter, Nathorst-Boos, Carlstrom, & von Schoultz, 2002; also Rako, 2000).

That ovarian hormones affect women's sexuality, of course, does not imply that women's sexuality is not under the control of exogenous or environmental factors (see review of this issue in Wallen, 2000). Women track the quality of the environment. Women's sexuality and other aspects of reproduction are linked to (ancestral) cues of reproductive opportunities (Thiessen, 1994). For example, when women lack gynoid fat stores (have low energy load; Ellison, 2001), they do not cycle (Frisch, 1990). Their lack of cycling naturally, then, affects the nature of their sexual interests. Similarly, features of women's primary mateship affect their sexuality within the mateship (Gangestad et al., 2002; Gangestad et al., 2005a; Garver-Apgar et al., 2006; Thiessen, 1994; Thornhill et al., 1995; Thornhill & Furlow, 1998). More generally, factors that affect the adaptive value of female expenditure of their limited parental investment should affect female sexuality in other vertebrates, as well as invertebrates.

Summary

Biparental, socially monogamous birds have a mating system relatively similar to that of humans. Pair-bonded female birds commonly copulate outside the pair bond, and most frequently during estrus. In many cases, female EPC preferences focus on male traits that potentially connote male genetic quality and that in some cases actually secure good genes for offspring. Estrous female choice adaptation that functions to get good genes is distinct from female extended sexuality adaptation. In some of these species there is low to no extra-pair copulation because costs of the behavior do not exceed benefits.

Evidence supports the hypothesis that women's estrous mate preferences function to an important extent to get good genes for offspring, including from an extra-pair partner. First, women in romantic relationships show estrous preferences more strongly than unmated women. Second, estrous women with primary partners are more strongly attracted to other men than to their partners. Third, women's estrous interest in extra-pair men is greater when the partners have low genetic quality than when they have high genetic quality. Fourth, estrous women feel less committed to their relationships and more sexually motivated to mate with other men. These findings generally are in marked contrast to findings about female sexual preferences and interests outside estrous during extended sexuality.

Two additional predictions of the hypothesis that women's estrous preferences include seeking good genes from extra-pair partners are: (1) men will possess counteradaptation that functions to provide paternity given women's estrous interest in high-quality extra-pair men, and (2) women will possess counteradaptation to the counteradaptation of men. Support for these two predictions is discussed later in the book.

Women's high desire for orgasm and sexual satisfaction during short-term mating may reflect estrous motivation to obtain intrinsic good genes by extra-pair copulation. There may not be a sex difference in degree of peak sexual arousal when women's estrous sexual zenith is taken into consideration. Cryptic female choice may play a role in women's good-genes preferences during estrus.

Estrous women's nadir in food intake and increase in ambulation and interest in socializing with men may be adaptations that function to promote good-genes mate choice. The feeding nadir may avoid mating with food-providing men of low genetic quality. The increase in ambulatory and socializing activity may be analogous or homologous to estrous adaptation to improve good-genes mate choice in certain other female vertebrates.

Overall, research findings indicate that women's estrous sexuality is not merely a by-product of generalized heightened sexual motivation to ensure conception regardless of male quality. Instead, it is adaptation that functions to achieve conception by a male of high genetic quality, including contingently through extra-pair mating.

11   Concealed Fertility

What Is Concealed Fertility?

According to most authors, some female primates, often thought to be only women, possess concealed, cycle-related fertility, whereas most mammalian females do not (e.g., Alexander & Noonan, 1979; Andelman, 1987; Baker & Bellis, 1995; Beach, 1976; Burley, 1979; Campbell, 2002; Cartwright, 2000; Dixson, 1998; Marlowe, 2004; Strassmann, 1981; Symons, 1979; Turke, 1984). In its fullest form, female concealment of cycle-phase fertility has typically been thought to require three features. First, a concealing female does not "know," that is, perceive, discriminate, or respond to, her own peak fertility in the estrous/menstrual cycle. Hence, she exhibits no variation in sexual motivation across the cycle. Second, and relatedly, female sexual behavior has become independent of variations in female reproductive-cycle hormonal influences. That is, endocrine changes across the cycle do not significantly produce variation in female sexual motivation. Third, female morphological, behavioral, or olfactory signals of peak fertility in the estrous/menstrual cycle are greatly reduced or altogether absent. Hence, males cannot discriminate female peak fertility in the cycle.

In fact, females of most nonhuman primate species claimed to be "concealed ovulators" (e.g., vervet monkeys, tamarins, and marmosets) do experience changes in sexual activity through the ovarian cycle and exhibit peak sexual interest and attractiveness to males when fertile. As their estrus makes available to others estrous-phase-specific olfactory and visual (e.g., behavioral) stimuli, it may not be concealed, even if human observers cannot detect their estrus due to absence of sexual swelling (see Dixson, 1998, for a review; also Carnegie, Fedigan, & Ziegler, 2005; DeVleeschouwer et al., 2000; Digby, 1999; Ferris et al., 2004).

As we have discussed, women too appear to have estrus. Hence, the first and second features typically thought to be components of concealed ovulation do not describe women's biology. Women do act in ways at their peak fertility different from how they act when not at peak fertility. Their mate preferences and sexual dispositions change across the cycle. And hormonal variations appear to play key roles, even if those roles are not completely understood at this time.

With regard to the third feature, we have argued that estrus is not an adaptation to signal cycle-related fertility and that female animals only rarely advertise their fertility in the reproductive cycle. Instead, males are designed through sexual selection to focus attention on cycle-related fertility cues that arise incidentally from female cycle changes in fertility status.

As we argue in this chapter, if women do possess concealed fertility in any meaningful sense, it is with regard to these incidental cues. Possibly, women have been designed to suppress these cues, leaving men scant information about their fertility status, as it varies across the cycle (though, as we also discuss, men do not lack cues altogether.) If so, women perhaps possess adaptation meaningfully referred to as concealed fertility.

Prior to discussing the theories that purportedly explain this human adaptation and findings pertinent to evaluating claims that women possess concealed fertility, we must explicate, at a conceptual level, several crucial distinctions.

Incidental Nondisclosure of Fertility Status Versus Concealed Fertility

The accuracy with which males can identify females at peak fertility in their cycle varies across species in relation to the conception-related validity of the by-products of reproductive-status changes cycling females emit, incidentally resulting in some degree of nondisclosure of fertility in some species. That is, information that males can possibly glean about females' cycle-related fertility is typically imperfect, as the informative cues arise incidentally; females are not selected to signal their fertility in their reproductive cycle. In most species, males are selected to detect females' peak fertility and pursue females so detected. But the accuracy with which males can discriminate fertility is limited by the informativeness of by-product cues that females emit. In some vertebrate species, maximum cycle-related fertility in females may not coincide perfectly with peak sexual motivation among males as a result. Some variation in male ability to detect female fertility status, then, arises for reasons having nothing to do with female adaptation to conceal fertility cues. It arises, quite simply, from the fact that females are rarely if ever selected to signal fertility, and hence the cues that males must use to infer female fertility status are incidental cues.

Chacma baboons illustrate the unreliability of by-product cues emitted by females in estrus. Male chacma baboons detect and are sexually motivated by by-products produced by females in estrus. In part, males pick up cues in scent associated with high levels of estrogen. At the same time, however, males require several months of experience with an individual female before they can distinguish her conceptive from nonconceptive estrous cycles. As a result, high-ranking males in this species consort with estrous females less than expected under the assumption that all estrous females are equally valuable for male reproductive success (Weingrill et al., 2003). The by-product cues that sexual selection has designed male chacma baboons to be attuned to are ambiguous, it seems, and can best be discerned through learning each female's cue profile.

And so it may be with most vertebrates. Females may not be designed to "fool" males. Selection simply has not designed females to advertise to males their fertility status. Males do the best job they can do, given incidental information available to them.

An interesting alternative, however, is the possibility that direct selection in females to hide cycle fertility has led to concealed-fertility adaptation in the chacma baboon and other species.

Two distinct phenomena, then, must be distinguished. The word "concealment" implies an active "hiding" or suppression of information. When we refer to *concealed cycle fertility* or *concealed estrus*, we refer to an evolved outcome due to direct selection on females to suppress information related to cycle fertility, leading to concealed-fertility adaptation. We distinguish concealed cycle fertility or concealed estrus from *undisclosed cycle fertility* or *undisclosed estrus*. The word "undisclosed" does not imply an active process of disguise. Similarly, the term *disclosed cycle fertility* does not imply active advertisement; that is, disclosed cycle fertility does not imply signaling of cycle fertility. Undisclosed cycle fertility applies to instances in which males are not able to perfectly discriminate female cycle fertility, but not because females have been under direct selection to disguise cycle fertility. Put otherwise, undisclosed female fertility refers to cases in which the unreliability of information occurs incidentally. Concealed cycle fertility refers to cases in which the unreliability of information occurs by design. Again, disclosed cycle fertility and undisclosed cycle fertility, we suspect, characterize the vast majority of vertebrate and invertebrate taxa.

Concealed Fertility Versus Extended Sexuality

The concepts of concealed fertility and extended sexuality are sometimes conflated. Extended sexuality exists when females are sexually receptive or proceptive outside of the fertile period (see chapter 3). We have argued that it typically involves adaptation to obtain material benefits from males. It may be argued that concealed fertility typically evolves only in species with extended sexuality. If females are sexually receptive only during the fertile phase, there is no point to concealing fertility in other ways. But the converse is not true: Extended sexuality need not—and perhaps does not typically—involve concealed fertility. Males need not be "fooled" by females to copulate with them outside of their fertile window. Rather, males will copulate with infertile females whenever information about females' fertility status is imperfect and copulation, for males, is relatively cheap. Hence, even when fertility is merely undisclosed, not actively concealed, males will copulate with females when those females are in phases of extended sexuality.

Consider, for example, extended sexuality in some pregnant langurs. Females exhibit sexual motivation during pregnancy, which leads males that usurp a harem to copulate with them and, as a result, refrain from aggression toward impending newborns. This behavior has been referred to as "sham estrus" (Hrdy, 1981; also see chapter 3, this volume), deceptive signaling of fertility outside the fertile phase of the cycle. We suggest that, in fact, female langurs need not deceptively signal cycle fertility. Estrus, we have argued, is not a signal of cycle fertility. Male langurs, as well as males of other species, with or without extended female sexuality, possess imperfect information about female cycle fertility. Female sexual interest at any time is often sexually stimulating to males as a result of sexual selection designing males to be sexually motivated as if they know that females must mate in order to conceive. The pregnant female langur's sexual motivation is extended female sexuality. By design, her motivation is timed to coincide with a threat to her unborn baby by a take-over by a new male to obtain material benefits from him, and this timing of extended sexuality to obtain material benefits is, we have argued, an integral part of extended sexuality adaptation. But she need not "fool" him into thinking she is in estrus. He presumably does not have perfect knowledge of when her estrus occurs, based on incidental cues. Her behavior (and similar behavior of pregnant female mammals of various species; see chapter 3) involves extended sexuality but need not involve concealed fertility.

Various sexual behaviors of infertile-cycle-phase females in other species with extended sexuality similarly need not involve deception about cycle fertility. Rather, males are attracted to sexually motivated females because they invite sexual access and because males lack perfect information about fertility status. Men and males of other species with extended female sexuality are designed to home in on these cues. Naturally, the cues can be deceptive—but they do not deceive males about fertility status. A woman may flirt with a man, with no genuine sexual interest, to obtain benefits from him. She deceives him about promise of sexual access (e.g., Buss, 2003b). But he is not deceived by her flirtations into thinking that she is fertile. Rather, men imperfectly know when women are fertile, which means that, historically, copulations even with infertile women were often worth pursuing, and therefore men have been selected to pursue them. (As we discuss, human females do conceal fertility. Here, we simply refer to the fact that flirtation itself is not a deceptive signal of fertility.) Similarly, males of other species may pursue copulations with sexually interested but infertile females, but not because females are actively deceitful about their fertility status. Rather, males have imperfect knowledge of fertility based on incidental cues, rendering copulations with no chance of resulting in conception nonetheless worth pursuing from the male's point of view.

Based on the same reasoning, the sexual ornaments of female adolescents in some primate species are not deceptive signals of cycle fertility (see chapter 5). We have argued that these ornaments honestly signal future reproductive value. Adolescent ornamental exaggeration is typically accompanied by adolescent female sexual motivation (a form of extended sexuality). Both function to gain access to male-delivered material benefits but not through deception about fertility.

Concealment of Cycle Fertility Versus of Ovulation

In the literature on women's sexuality, concealed *peak fertility* and concealed *ovulation* are often equated. In fact, however, peak cycle fertility precedes ovulation by 1 to 2 days, coinciding with the LH surge, which stimulates ovulation within 12–48 hours (Santoro et al., 2003). Immediately following ovulation, the probability that an insemination will result in conception plummets (e.g., Wilcox et al., 1995). The major hypotheses in the literature for concealment of ovulation in fact attempt to explain the apparent lack of highly reliable cues of peak fertility. They do not explain concealment of ovulation per se. We hence focus on concealment of high fertility in the menstrual cycle, not of ovulation, by women and therefore speak of "concealed fertility" (even when discussing purported explanations of "concealed ovulation").

A topic in discussions of women's concealed fertility is *mittelschmurtz*—the pain or characteristic sensations occurring with the release of an egg. Purportedly, *mittleschmurtz* constitutes awareness of ovulation. If so, it is not awareness of peak fertility. As just noted, fertility plummets immediately following ovulation. Hence, knowledge of ovulation is of little or no use to women desiring foresight of peak fertility. Possibly, *mittelschmurtz* is an adaptation signaling to women that an ovulatory cycle without conception has occurred (Baker & Bellis, 1995), but another possibility is that it does not reflect adaptation whatsoever. *Mittelschmurtz* has been thought to speak to the issue of concealed fertility because peak fertility and ovulation have been conflated erroneously (e.g., Alexander & Noonan, 1979; Burley, 1979; Gray & Wolfe, 1983; Manson, 1986).

What Is Concealed in Concealed Fertility?

Concealed fertility, we have argued, does not consist of presenting signals of fertility outside the fertile phase. Again, females rarely signal fertility in any case. Neither, we have noted, does concealed fertility consist of merely being sexually receptive outside of the fertile phase (even if, once again, females that conceal fertility do typically if not always engage in extended sexuality).

Concealed fertility, we suggest, should be thought of as *concealed estrus*. Two aspects of estrus are suppressed by direct selection in concealed fertility. First, females emit various incidental effects of estrous adaptations (involving, e.g., increased estrogen levels), which males use to detect fertility status. Concealed fertility results from direct selection on females to suppress or reduce these incidental effects. Second, females in estrus manifest sexual motivation with potential sires able to benefit offspring with good genes. Females who conceal fertility suppress the manifestations of estrous sexuality outside the contexts of these sexual interactions and hence conceal its expression conditionally. Here, it is not so much that females conceal estrous sexuality by merely being sexual outside of the fertile phase (extended sexuality). Rather, when females conceal fertility they suppress their estrous proceptivity toward particular males outside of contexts involving only those males, as well as incidental effects associated with estrous adaptations (e.g., cyclic variations in hormones).

The concept of concealed fertility implies a target from whom fertility is concealed. Fertility could be concealed from women themselves, from other women, from men, or from both sexes. Concealment of fertility could entail disguise of estrous sexual adaptation or other aspects of peak fertility or both. In this chapter and the next, we discuss these various forms of concealment of fertility in relation to woman's estrous sexuality. (In some species, estrous cues could also be used by predators to detect prey. We do not consider concealment due to interspecific conflicts of interest here.)

Theories of Women's Concealed Fertility

Undisclosed Fertility as a By-Product

A number of theories argue that women possess adaptations that conceal fertility. Other theories, by contrast, argue that women possess undisclosed fertility as an incidental by-product. We first discuss theories of undisclosed fertility.

Undisclosed Fertility Is a By-Product of Large Adrenal Glands   At least two hypotheses argue that women possess undisclosed cycle fertility, not concealed fertility (table 11.1). Spuhler (1979) proposed that women do not disclose fertility as a by-product of large adrenal glands. Enlargement of these glands, in turn, is possibly adaptation associated with distance walking. They may incidentally affect androgen levels

Table 11.1 Hypothesis for the Evolution of Women's Concealed Fertility.

I. Concealed fertility as individual-level adaptation
 A. Concealed fertility functions to secure paternal investment (Alexander & Noonan, 1979; Lovejoy, 1981; Miller, 1996; Strassmann, 1981; Turke, 1984).
 B. Concealed fertility functions to secure food in exchange for sex (Hill, 1982; Symons, 1979).
 C. Concealed fertility functions to protect female's offspring from infanticide by males (Hdry, 1979; 1981; Schroder, 1993).
 D. Concealed fertility functions to obtain an extra-pair sire for offspring with better genes than possessed by the in-pair mate (Benshoof & Thornhill, 1979; Schroder, 1993; Symons, 1979).
 E. Concealed fertility functions to assure conception, which would be avoided if females perceived their peak in fertility in the cycle (Burley, 1979).
II. Concealed fertility as an incidental effect of other adaptation
 A. Concealed fertility as a by-product of adaptive large adrenal glands, which function in distance walking (Spuhler, 1979).
 B. Concealed fertility as a by-product of the directly selected loss of sexual swellings, which are assumed to signal peak fertility in the menstrual cycle (Burt, 1992).

and, hence, sexual motivation. This idea purports to explain extended sexuality more than undisclosed fertility (see preceding discussion), but, as these traits are conflated, the literature entertains it as an explanation for concealed fertility (e.g., Gray & Wolfe, 1983; Turke, 1984).

Undisclosed Fertility Is a By-Product of Selection Against Sexual Swellings   Burt (1992) hypothesized that undisclosed fertility evolved because sexual swellings were selected against in female humans. Sexual swellings, he argued, did not provide sufficient benefits derived from signaling of fertility to outweigh their costs. In this scenario, undisclosed fertility was indirectly selected—a by-product of direct selection against swellings (also see Pawlowski, 1999a).

Evaluation   Spuhler's (1979) hypothesis, again, is more accurately portrayed as an explanation of extended sexuality as by-product. It is challenged by data that extended sexuality is characterized by adaptive design (see chapters 3 and 9).

Burt erroneously assumes that female sexual swellings advertise fertility and that otherwise female fertility is undisclosed. As already discussed, many mammalian species without sexual swellings exhibit disclosed cycle fertility (see chapter 5). Burt also assumes that hominin female ancestors had swellings; they apparently did not (Sillén-Tullburg & Møller, 1993).

Despite the inadequacies of these particular accounts, we note, it is possible that women do possess (partially) undisclosed fertility and not concealed fertility. Concealed fertility, again, requires adaptation for suppression of fertility cues. Later, we review evidence pertaining to concealed fertility.

Concealed Fertility as Adaptation

A number of theories purportedly explain concealed fertility in humans in terms of adaptation for concealment.

Concealed Fertility Promotes the Pair Bond   An old literature claims that concealed fertility promotes the human male-female pair bond or family organization. It fails to explain, however, why concealed fertility is adaptive for individual women and thereby would evolve as female adaptation (for further discussion of this literature and references, see Benshoof & Thornhill, 1979; Gray & Wolfe, 1983; Manson, 1986). Typically, women's extended sexuality is conflated with concealed fertility in this literature and claimed to promote pair bonding but, again, with no detailed explanation of benefits to women driving evolution of this trait. In some cases, even in relatively recent literature (Daniels, 1983), advantage at the group level, not the individual level, is claimed to have driven the evolution of concealed fertility.

Other hypotheses about women's concealed fertility do argue that suppression of fertility cues was directly selected because of a net individual reproductive advantage to females (see table 11.1).

Concealed Fertility Enhances Ability to Obtain Male-Delivered Material Benefits
One set of hypotheses proposes that concealed fertility allowed women to obtain material benefits from males in the form of paternal investment (Alexander & Noonan, 1979; Lovejoy, 1981; Miller, 1996; Strassmann, 1981; Turke, 1984) or food in exchange for sex (Hill, 1982; Symons, 1979). The argument is that concealed fertility (1) leads a mate and/or other males to provide material benefits throughout the menstrual cycle, rather than only when females signal fertility through estrus, and (2) provides a mate with high paternity reliability, as other males do not sexually compete for females not signaling fertility through estrus. In this scenario, furthermore, (2) reinforces (1).

Concealed Fertility Confuses Paternity and Thereby Protects Against Infanticide
Another hypothesis proposes that concealed fertility protects offspring from infanticide by males. Specifically, concealment, coupled with multiple mating, confuses reliability of paternity for males, rendering male infanticide—an adaptation to bring females with dependent offspring into reproductive condition earlier—maladaptive (Hrdy, 1979, 1981; Schroder, 1993, in part). Whereas the paternal investment and food-for-sex hypotheses view concealed fertility as a means of increasing reliability of male parentage, the infanticide-avoidance hypothesis may seem to claim that it reduces confidence of paternity. In fact, however, a better construal is that female multiple mating gives each male enough paternity confidence to render his maltreatment of her offspring maladaptive. A female's strategy, then, is not to confuse paternity but, instead, to give credibility to the possibility of each male being the father (see also Hrdy, 1981). As we discussed in chapter 3, Hrdy's paternity-confusion hypothesis is a form of the hypothesis that females possess extended sexuality to obtain male-delivered material benefits.

Concealed Fertility Assures Copulation in the Face of the Pain of Childbirth   Nancy Burley (1979) argued that women's large brains allowed them to appreciate the connections between mating, conception, childbirth, and time-consuming child care. Women who recognized that sex during fertile periods could lead to childbirth might avoid mating when fertile and thereby leave few descendants. Accordingly, fertility concealed from the self and mates evolved.

Concealed Fertility Facilitates Cuckoldry   This hypothesis argues that concealed fertility was selected directly because it allowed adaptive extra-pair copulation (EPC) with males of higher genetic quality than a pair-bond mate. Concealed fertility functions to reduce effective mate guarding by a pair-bond mate and thereby facilitates pursuit of extra-pair males of superior genetic quality without the pair-bond mate's interference or loss of a pair-bond mate's investment (Benshoof & Thornhill, 1979; Schroder, 1993, in part; Symons, 1979, in part).

Evaluation   As we argued in chapter 3, extended sexuality typically functions to allow females to obtain material benefits from males. As we also noted then, for females to acquire material benefits through extended sexuality, males must not be able to detect female cycle-varying fertility perfectly. In theory, then, suppression of cues of fertility could facilitate a flow of male-delivered benefits in currencies of parental investment, food for sex, or infanticide reduction. That is, concealed fertility could serve the function of enhancing the ability of females to obtain benefits achieved through extended sexuality.

In fact, however, as typically construed, these theories are incomplete explanations of extended sexuality. Typically, they do not discriminate between the adaptations of extended sexuality and concealed ovulation. Instead, they argue that concealed ovulation is achieved through extended sexuality: By being sexually receptive beyond the period immediately preceding ovulation, females conceal their fertility status. Hence these theories typically are meant to imply that concealed ovulation is tantamount to a loss of a distinct estrous sexuality. As recently as 1990, Richard Alexander argued that the paternal-investment hypothesis predicts that neither men nor women have knowledge of cycle-related fertility, as sexuality is "continuous" across the cycle (Alexander, 1990). Indeed, he argued that this hypothesis best explained the data on the basis of this prediction; in 1990, it seemed reasonable to assert that women's sexuality is fairly continuous across the cycle. These theories assumed that estrous sexuality simply functioned to facilitate conception (see chapter 8). Its loss, then, presumably functioned to conceal the fertile period at little cost to conception probability. Burley's (1979) hypothesis similarly implies a loss of estrus (see Gray & Wolfe, 1983).

As we documented in the preceding two chapters, however, women's estrus has not been lost. Women's preferences and sexual interests change across the cycle. Women, then, can discriminate between different phases of their cycles (even if they do not "consciously" recognize that estrous sexuality is fertile sexuality). As we will see later in this chapter, men, too, can discriminate women's fertile sexuality at better than chance levels. The existence of estrus is simply inconsistent with these theories as they have been cast. Extended sexuality does not merely "tack on" an extended period of estrous sexuality. It functions differently from the way estrus functions. Given differences between estrus and extended sexuality, extended sexuality in no way "conceals" estrus and, hence, the fertile phase.

What is required as an adequate explanation, then, is an explanation of why estrous is retained while cues of the estrous period are suppressed. The cuckoldry hypothesis is one such explanation. It offers a number of predictions.

First, it predicts that women's fertile sexuality is distinct from their extended sexuality. At estrus, women should be particularly attracted to men who exhibit purported indicators of good genes. As we saw in chapter 9, women's estrus is indeed characterized by a variety of enhanced preferences for indicators of developmental stability and masculinity.

Second, it predicts that, during estrus, women's attraction to men other than primary partners should be enhanced, but selectively so. Specifically, they should be attracted to particular men other than primary partners only if their primary partners lack indicators of good genes for offspring. As we saw in chapter 10, the evidence to date is consistent with this prediction.

Third, the cuckoldry hypothesis for concealed fertility goes beyond these findings and predicts that women's estrus will be concealed in the two ways we outlined earlier.

First, incidental effects associated with estrous adaptations should be reduced. Second, some behavioral effects of estrus—selectively greater interest in men other than primary partners—should be concealed from primary partners and, indeed, most anyone other than the men to whom fertile women are attracted. Women need not conceal their estrus from men they seek; there are no benefits to doing so. They do benefit by concealing their estrus from a primary partner, should he not be one of those men.

The cuckoldry hypothesis, then, expects that estrus will be concealed selectively. Women's estrus is not concealed from themselves. And it is not concealed from men to whom they are particularly attracted during estrus.

Although the cuckoldry hypothesis may appear to be about acquisition of good genes, in fact it is just as much about acquisition of material benefits. Again, a precondition of concealed estrus, according to this view, is that pair-bonded males deliver material benefits. Hence, in some sense the cuckoldry hypothesis is not opposed to hypotheses that argue that concealed fertility functions to enhance male delivery of material benefits. Rather, it argues that, although these hypotheses specify some of the benefits of concealed fertility, they do not fully specify the benefits. Concealed estrus fosters a flow of material benefits from males outside of estrus while simultaneously enhancing effective sire choice during estrus. Toward the end of the chapter, we suggest that the cuckoldry hypothesis for concealed estrus, while adequate to account for concealed estrus, must be broadened to account for all of the circumstances under which concealed fertility could be selected. As we argue then, females could, in principle, benefit through concealed estrus whenever they receive material benefits during extended sexuality from males other than those preferred as sires, whether cuckoldry is involved or not.

The Nature of Fertility Concealment in Women

Women Possess Typical Mammalian Estrus

Before we discuss evidence that women have been selected to conceal their fertility in various ways, a few additional words about ways women's fertility is not concealed are in order. Cycle-related peak fertility, once again, is not hidden from women themselves. As their preferences, patterns of attraction, and experience change across the cycle, women themselves surely discriminate their fertile periods from their infertile periods. Naturally, there is no presumption that they consciously associate estrous sexuality with "fertility"—but, of course, there is no presumption that females of any other species not exhibiting concealed fertility do so, either. Rather, it is generally presumed that females of nonhuman species experience the world differently when fertile and hence discriminate fertile from nonfertile periods. The same, we argue, is true of women.

This claim contrasts with the opinion typically expressed in the literature. Many scholars throughout the history of the study of human sexuality have claimed that women possess virtually no knowledge whatsoever of their peak fertility in the menstrual cycle (e.g., Alexander, 1990; Burley, 1979; Strassmann, 1981; Turke, 1984; and many others; but see Benshoof & Thornhill, 1979; Hrdy, 1997; Small, 1996; Wallen, 2000). As Diamond (1997) put it, "it's especially paradoxical that a female as smart and aware as a human should be unconscious of her own ovulation, when female animals as dumb as cows are aware of it" (p. 5). One basis for the claim is that, as Burley (1979) and Strassmann (1996a) point out, even the medical profession did not realize that peak fertility occurs at mid-cycle until near 1930. Before that time, it was commonly thought that conception was most likely near or during menses (see discussion in Burley, 1979; Marlowe, 2004; Strassmann, 1996a). Indeed, as recently as 1923, a form of contraception medically recommended to women was abstinence from sex during menstruation (Walker, 1997). Similarly, traditional people do not typically appreciate a link between estrous sexuality and conception. Marlowe (2004) asked men and women of the Hadza of Tanzania what leads to pregnancy. The Hadza understood that sexual intercourse leads to pregnancy, but the vast majority claimed that conception occurs immediately after menstruation ends. Clearly, people would not be so ignorant about when the fertile period exists if women were similar to female dogs or cats, who, when confined to a home without a mate, are conspicuously active and exhibit dedicated, intense effort to break out to join males attracted to their estrous scents. As scholars have noted, there is indeed something different about women's manifestation of fertile sexuality.

What is different, however, is not loss of estrus. Again, the evidence amassed for the existence of women's distinct fertile sexuality cannot be denied. How is it that women possess this distinct fertile sexuality yet we do not see that people everywhere recognize it? We suggest that it is because women have been designed to conceal it, except very selectively—when it is adaptive to manifest it in the event of copulation with a partner of superior genetic quality. Women are different, as scholars traditionally have claimed. They are not different in lacking estrus, however. They are different in the extent to which they conceal estrus. The behavioral changes women experience at estrus are much more covert than what is typically observed in other species, as the cuckoldry hypothesis expects.

Concealment From Main Partner

The cuckoldry hypothesis for women's concealed fertility predicts that selection acting on females has favored estrous-phase concealment of sexual interest in nonpartner men, as well as of physiological side effects and emotional by-products occurring at estrus, from primary pair-bond partners. The benefits of doing so are preventative. If male primary partners know when their mates are fertile, they could take additional steps to prevent females' estrous pursuit of mates with superior genetic quality. Furthermore, detection of a woman's interest in other men by a primary partner could lead the partner to divest himself of her and/or her offspring.

As we have argued, men's investment in offspring is a critical component of the human adaptive complex. Women's fitness depends critically on male investment in offspring. At the same time, women's ability to acquire good genes for offspring critically affects her fitness, as well. Selection has retained estrus in women, which functions to obtain good genes from sires. But, as we argued in the preceding chapter, women's estrus exists within the context of a species with intensive biparental care. And, indeed, the reason that selection has acted on women to conceal estrus is because of the tantamount importance of biparental care (see also Benshoof & Thornhill, 1979). Naturally, selection has operated on men to detect fertility and to protect their own reproductive interests against those of their partners (see review on the link between paternity reliability and investment in Clutton-Brock, 1991; also, for birds, Møller & Cuervo, 2000). We describe outcomes of that selection process in the next chapter.

In our study of couples, colleagues and we examined whether women do engage in attempts to resist men's efforts to track their behavior and whereabouts more frequently when fertile than in the infertile luteal phase. Alita Cousins (unpublished) developed a measure of women's resistance to men's efforts to guard them. Sample items include, "I avoided situations where my partner might be able to check up on me" and "I hid stuff from my partner so that my partner wouldn't be able to find it." Women filled out this measure twice, once when fertile and once during the luteal phase, each time reporting on frequencies of behavior over the preceding 2 days. As predicted, women engaged in efforts to resist their primary partners' vigilance more frequently during the fertile phase than during the luteal phase (Garver-Apgar, Cousins, Gangestad, & Thornhill, 2007).

Estrous Emotional Withholding and Regulation

Information concerning the emotional states of others can constitute useful knowledge for predicting their behavior. Humans possess adaptations designed to glean information about others' mental states and intentions from facial and other bodily expression (e.g., Goel, Grafman, Sadato, & Hallett, 1995; Leslie, 1987). Often, it pays individuals to signal to others one's own internal states. For example, an expression of fear in response to danger can simultaneously protect one's allies and recruit their assistance in thwarting the danger. But it does not always pay to express emotional states and, hence, selection has shaped adaptations that contingently regulate the expression of emotion or other cues of internal states. In context-specific ways, people suppress fear, anger, pain, anxiety, sadness, happiness, and other emotions. (See Cosmides & Tooby, 2000, for a general discussion of the evolution of emotional regulation.)

In a similar fashion, women's sexual interests and motivations are conditionally expressed. Women's sexual emotional regulation during estrus may involve adaptations to disguise cycle-related emotional by-products at peak fertility, as well as adaptations to disguise estrous sexual interest in nonpartner men from main partners.

Women's emotions appear to be influenced, in part incidentally, by adaptive schedules of reproductive hormones, especially estrogen, and other physiological systems across the cycle (e.g., Berlanga & Huerta, 2000; Freeman & Halbeich, 1998; Singh, Berman, Simpson, & Annechild, 1998; VanGoozen, Frijda, Wiegant, Endert, & VanderPoll, 1996; Walker, 1997). We propose that expression of emotions incidentally varying across the cycle, which could allow others to discriminate peak fertility in the cycle, have been suppressed by selection acting on females.

As well, women's actual estrous sexual passion (which occurs by design, not incidentally) should be manifested primarily or only in mating with a partner of superior genetic quality. Moreover, conditionally expressed deceptive sexual arousal in estrous women may play a role. Pair-bonded men's and women's independent reports of the woman's copulatory orgasm frequency positively covary, which suggests that men assess information about partner's erotic response during mating (Thornhill et al., 1995). In theory, a woman may strategically affect a man's perception of women's assessment of his mate value through manipulation of her erotic response to partners (which assumes no deliberate, conscious manipulation on women's part). Naturally, women who possess partners of high genetic quality are expected to express estrous sexual passion toward them. These topics have received little attention from researchers thus far.

Emotional regulation may suppress estrous emotional cues in the presence of other individuals, as well, including other men (including men who are potentially sexually coercive) and relatives.

Though women are more emotionally expressive than men, women, interestingly, appear to also possess greater ability to control facial and body expressions of emotions (see Bjorklund & Kipp, 1996). They may also be better at suppressing unwanted thoughts (e.g., selectively avoiding thoughts about potential mates). Bjorklund and Kipp propose that these emotional features may be adaptation for EPC, along the lines we sketched earlier. If so, then women's abilities to suppress emotion may vary across the cycle and be particularly keen when women are at peak fertility. Indeed, this ability may even vary as a function of women's partners' characteristics (e.g., developmental stability). As Cosmides and Tooby (2000) note, the literature on human emotions has focused almost exclusively on expression rather than suppression (for exceptions, see Gross, 1998, 2002). A potentially fruitful avenue of research would examine how women's emotional regulatory abilities vary as a function of their potential to garner genetic benefits from EPC.

Concealment From Oneself

Women are exposed to cues that permit them to discriminate their own estrous sexuality from extended sexuality; the male features to which they are attracted change across the cycle, as do their sexual interests. At the same time, women may also benefit from selective self-deception. Specifically, women may be benefited by believing that, in fact, they have no motivation to copulate outside of a pair bond (see also Benshoof & Thornhill, 1979). As a number of scholars have speculated, one's ability to deceive another may be enhanced by self-deception about one's own interests and motives (Alexander, 1979; Benshoof & Thornhill, 1979; Freeman & Wong, 1995; Trivers, 1985). Estrous women may be aware of their attraction to nonpartner men, yet simultaneously deny, to themselves as well as others, any true interest in sex outside of their pair bonds.

Reduction of Physiological Incidental By-Products

Thus far, we have focused on concealment of the behavioral manifestations of estrus. As we emphasized in chapter 5, however, males of many species do not discriminate females simply on the basis of behavioral cues. They do so on the basis of physiological by-products of estrous adaptations. Prior to egg-laying, for instance, many birds gain considerable weight from the developing eggs. Males can discriminate female fertility status on the basis of weight gain (e.g., Low, 2004). In many primate species, males identify fertile females on the basis of scent. As female reproductive hormones change, so too does the chemical composition of females' secretions as an incidental by-product. Males are selected to attune to variations in scent that validly discriminate fertile from infertile states. Women's hormonal states likely also affect the chemical composition of their secretions. And they undergo a variety of other physiological changes during estrus, as well (see chapter 9).

From the perspective of women, the optimum, all else equal, may be a complete suppression of incidental effects that men or others could use to discriminate their fertile states. At the same time, not all else is equal. The suppression of incidental effects may be costly, for it may interfere with the functional expression of the adaptations giving rise to the by-products. Thus, for instance, a way to reduce any possibility that by-products of estrogen metabolism are secreted in greater quantities during estrus would be to eliminate the rise of estrogen at mid-cycle. But elimination of estrogen mid-cycle is not possible without disruption of a multitude of reproductive (including estrous) adaptations that depend on cyclic variations in estrogen. Selection is expected to maximize net benefits—the benefits of suppression minus any costs that result from suppression. The point at which benefits are maximized need not be constant across evolutionary time. As males become more sensitively attuned to incidental effects that have been suppressed, females may benefit from increased suppression, leading selection to suppress fertility cues further. Due to continuing coevolution of such adaptations in each sex to counter those in the opposite sex, neither sex is expected to be fully adapted to the adaptations of the other. Imperfect adaptation in coevolving antagonists may characterize the design of both sexes in the context of peak-fertility detection by men and peak-fertility disguise by women.

Residual By-Products of Estrus

Cues of Women's Fertility

In the preceding section, we argued that women have been under selection to suppress or reduce incidental effects or other signs of estrus. That is, women do have concealed fertility. As we just discussed, however, concealed fertility need not imply a complete absence of cues associated with fertility. Selection can operate on women to suppress fertility cues (i.e., conceal fertility) without eliminating cues.

Women do exhibit a variety of by-products of estrus at mid-cycle, which remain not fully suppressed. Women have a smaller waist-to-hip ratio when fertile than during some other phases of the cycle (Kirchengast & Gartner, 2002; Singh, 2002a,b), likely due to retention of fluid during the premenstrual and menstrual phases. The asymmetry of women's breasts or other soft tissues also changes in subtle ways across the cycle (Manning, Scutt, Whitehouse, Leinster, & Walton, 1996; Scutt & Manning, 1996), peaking about 2 days prior to ovulation, then declining sharply prior to ovulation. Once again, these effects may reflect variations in fluid retention. Furthermore, the pitch of ovulating women's voices changes across the menstrual cycle and is highest at peak fertility. The voices of fertile-phase normally ovulating women are rated more attractive, too, whereas the attractiveness of voices of women using hormonal contraception does not vary across the cycle (Bryant & Haselton, 2007; Pipitone & Gallup, 2007). We suggest, however, that this variation is mere by-product. Though men may find women's bodies and voices more attractive when they are in estrus, men's attraction does not imply that women signal that they are in estrus by enhancing their attractiveness.

Roberts et al. (2004) claimed to find that women's faces are more attractive during the fertile phase than during the luteal phase. In fact, their conclusion may be premature, as their statistical analysis did not permit generalization across women, but only generalization across raters of a very specific set of women's photos. Once again, however, if women's faces are indeed more attractive when they are fertile, it is not because they signal their fertility by enhancing their attractiveness. Rather, estrogen may enhance their attractiveness incidentally. Men have evolved to detect the subtle incidental effects on women's facial features (e.g., the smoothness of their skin) that result.

Women's gait may change across the cycle as well (Grammer, Fieder, & Filova, 1997; Grammer, Keki, et al., 2003; Provost, Quinsey, & Troje, 2007). Again, these effects are probably incidental effects of hormonal factors during estrus. Possibly, they are incidental to estrous women's greater ambulatory activity (Fessler, 2003; see chapter 10, this volume).

Women report that they feel more attractive when fertile than during the luteal phase (Haselton & Gangestad, 2006; Singh, 2002). As a result of their perception that they are more attractive, women may dress in ways that others perceive as sexier or more attractive when they are fertile. Recent studies by Haselton, Mortezaie, Pillsworth, Bleske-Recheck, and Frederick (2007) and Durante, Li, and Haselton (2007) found precisely that effect. And Grammer et al. (2004) reported that normally ovulating women attending nightclubs and university classes wear clothing that reveals more skin at mid-cycle. Furthermore, salivary estradiol predicted the amount of skin women exposed. Perhaps as a result of their own enhanced attractiveness, fertile women rate other women's, but not men's, faces as less attractive than do infertile women (M. L. Fisher, 2004).

One impressive study examined the tips that women performing as gentlemen's clubs' lap dancers receive as a function of their time in the cycle (Miller, Tybur, & Jordan, 2008). When in the fertile phase of their cycles, normally ovulating women earned tips given by men nearly double the amount they earned when in the luteal phase or menstrual phase. Women using hormonal contraceptives did not show a similar effect. Moreover, the advantage in earning power of normally ovulating women over women using hormonal contraception was greatest during the fertile phase. The precise features (physical, olfactory, behavioral) of dancers responsible for their greater earnings mid-cycle are unknown but appear to be affected by hormonal changes.

Once again, we interpret these effects as incidental by-products of changes in women's physiology. They may also be by-products of women's enhanced but discriminating sexual motivation when fertile. They do not imply, however, that women seek to compete for insemination or even for insemination by a genetically high-quality male. Women's change in attire itself is incidental to increased estrogen during estrus or, possibly, incidental to enhanced sexual motivation to obtain a high-quality sire. Similarly, we do not expect that women compete with other women more when fertile than when infertile.

The various by-products of women's estrus we discuss here may well have been acted on by direct selection for concealment during the evolution of woman's concealed fertility/concealed estrus, but their elimination may be constrained by the costs of suppression due to disrupted estrogen regulation.

Scent

Women smell better to men when they are fertile than when they are infertile. Poran (1994) studied seven pair-bonded couples in which females were not using hormonal contraception. Men rated the odor of their pair-bond mates in mid-cycle as significantly more desirable than their odor when they were infertile in the menstrual cycle. Four larger studies confirmed this effect. In each, men rated the scent of normally ovulating women who were strangers to them (using a worn T-shirt methodology; Havliček, Dvorakova, Bartos, & Flegr, 2006; Kuukasjärvi et al., 2004; Singh & Bronstad, 2001; Thornhill et al., 2003). One additional study failed to replicate the effect (Thornhill & Gangestad, 1999b); in light of the multiple successful replications, that result may be a Type II error. Doty, Ford, Preti, and Huggins (1975) asked men and women to rate the scents of tampons worn by four normally ovulating women at different points in their cycles. They found that vaginal scents are most pleasant during the ovulatory phase but, given the weak effect sizes, nonetheless concluded that "it is unlikely that humans can use vaginal odors reliably to determine the general time of ovulation" (p. 1317). (These authors did not separately report effects on men's and women's ratings.)

Kuukasjärvi et al. (2004) also asked men to rate the scent of a sample of women using hormone-based contraception. They found that only the scent of normally ovulating women became more sexually attractive to men at mid-cycle. This result strongly suggests that the estrous cue that men respond to derives from reproductive hormonal variations. The estrogen peak occurring just before ovulation is a likely candidate. (Maximal scent attractiveness in this study was at 12.7 days, close to the expected estrogen peak.) If estrogen produces the scent to which men are attracted, woman's scent of cycle fertility is homologous with that in the many species of mammals and other vertebrates (see chapter 5; figure 11.1a and 11.1b.)

Kuukasjärvi et al. (2004) also asked normally ovulating women to rate the sexual attractiveness of women's body scents (figure 11.1c and 11.1d). There was an effect approaching statistical significance ($p = 0.07$) for the scent of normally ovulating women to be more attractive at mid-cycle. If reliable, this effect may be due to (1) a female ability that is an incidental effect of men's ability to detect estrous scent or (2) a female adaptation, possibly playing a role in sexual competition.

Marlowe (2004) notes that men smelled T-shirts at very close range in these studies. They may hence not speak to whether men can usefully discriminate women's fertility

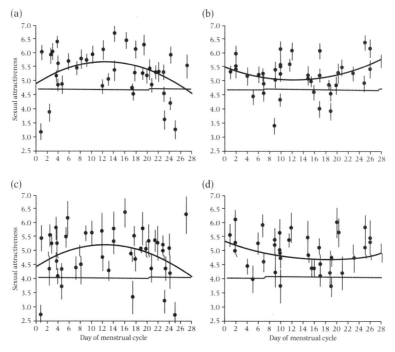

Figure 11.1 Sexual attractiveness of women's body scent (on T-shirts) across the menstrual cycle. The points are means (± SE). Sexual attractiveness ratings on T-shirts on a 1–10 scale (10 = most attractive) rated by 31 men (a and b) and 12 women (c and d). Dotted lines show the average ratings for clean T-shirts. (a) Normally ovulating women rated by men; (b) women using hormone-based contraception rated by men; (c) as in (a), but rated by normally ovulating women; and (d) as in (b), but rated by normally ovulating women. From figure 1 of Kuukasjärvi et al. (2004). Reprinted with permission of Oxford University Press.

status. Of course, the maximum distance over which men can evaluate women's scent is unknown. Even if detecting fertility through scent requires intimate interaction, however, male partners could benefit from being able to discriminate women's fertility status.

If men are not particularly good at detecting the scent of women's estrus at a distance—as we suspect is the case—it would be fully consistent with the hypothesis that, in fact, women have been selected to suppress their scent. In most mammals and many other vertebrates, estrous olfactory emissions attract males over considerable distances. That men are not attracted to women over similar distances is almost surely not simply due to weakened ability of men (or humans in general) to detect scents. We are comparatively very good at discriminating even small concentrations of the scent of many fitness-relevant stimuli (e.g., food; for comparative data on detection of aliphatic esters [found in fruits], see Laska & Freyer, 1997). Rather, men's relative inability to discriminate women's estrus on the basis of scent is probably due to selection on females to reduce the production or emission of chemicals that serve as cues of their fertility status.

In the chacma baboon, males can discern conceptive menstrual cycles from nonconceptive menstrual cycles based on olfaction, at least in part (Weingrill et al., 2003). Males have the ability to discern conceptive from infertile cycles in several nonhuman Old World primates (see review in Alberts et al., 2006). As in women (see chapter 3), nonconceptive cycles are frequent in these species (about 70% in nulliparous savanna baboons; Alberts et al., 2006). No study has examined similar abilities in men.

In the T-shirt studies we have described, nonvaginal body odors were operative. The scent of vaginal copulins may also provide cues of peak female fertility. Grammer et al. (1997) asked men to smell copulins, a behaviorally active chemical fraction of the vaginal secretions in macaques and present in women's vaginal fluids. Their testosterone levels rose, relative to controls who smelled water, and particularly so if men were exposed to copulins collected from women at mid-cycle. Additional research on copulins, other chemical cues to fertility, and men's hormonal responses to scents is needed.

Cue Versus Signal

Some scholars have argued that the subtle changes in women's physiology across the cycle, leading them to be more attractive to men in some regards, suggests that women do in fact signal fertility status. Scutt and Manning (1996), for instance, argued that the subtle changes in asymmetry function to advertise women's fertility to primary partners, thereby securing their investment. We have argued that, instead, women do not signal to men their fertility status through enhanced facial attractiveness, bodily attractiveness, scent attractiveness, or sexy dress. Indeed, we argue that women emit these cues despite selection against their expression, not because women have been selected to emit them. A variety of considerations argue against these features being signals. First, females rarely signal fertility in general (chapter 5). Second, no evidence exists that women possess complex design for emitting these signals; indeed, changes in physiology (e.g., fluid retention affecting waist and symmetry, skin texture, and chemical composition) are generally understandable as by-products of changes in women's hormonal status or other estrous adaptations rather than as outputs of design for signaling. Third, men should be expected to be attracted to those features that are enhanced when women are fertile, even subtly, as a result of strong sexual selection on them to detect female fertility alone. That men find women at estrus more attractive than infertile women need not be explained by a system in which females signal, by design, to males.

Fourth, changes in women's attractiveness across the cycle—whether in the form of bodily attractiveness, facial attractiveness, or scent attractiveness—are very subtle. By contrast, fertile females in many other primate, mammalian, and vertebrate species are highly attractive to males—and typically in absence of any signaling system whatsoever. Indeed, the changes in women's physiology permitting men to detect female estrus at better than chance rates are so subtle that, as we emphasized earlier in this chapter, throughout the history of the study of women's sexuality, the overwhelming consensus is that, not only do women not signal estrus, but also women's fertility has evolved to be, for all intents and purposes, completely concealed. Now, in fact, that claim is overstated. Men can detect women's fertility status and, as we see in the next chapter, men act on that information for their own benefit. But at the same time, the changes in women's physiology across the cycle that men can potentially detect are, within comparative perspective, subtle indeed.

Finally, there is simply no evidence that women benefit through male detection of their fertility status. Indeed, as we see in the next chapter, women appear to pay costs as a result of emitting incidental by-products of fertility cues that men detect. Male primary partners pick up on these cues and act on them for their own benefit and, at least at times, against the interests of female partners.

As we stressed toward the beginning of this chapter, undisclosed fertility must be distinguished from concealed fertility. Concealed fertility implies that direct selection has operated on females to suppress or reduce the cues associated with estrus. Concealed fertility, however, does not imply a complete absence of fertility cues. Women emit a variety of fertility cues. Do we know for an absolute fact that women have adaptation for concealed fertility—that is, that these cues are residual cues following a history of selection to suppress them? In fact, we cannot offer that conclusion with 100% certainty. Important comparative work on female production and emission of chemicals, as well as male ability to detect products of estrogens or other chemicals that reliably predicts fertility within the cycle, remains to be done. We are confident that, in the end, the evidence will show that women do have adaptation for concealed fertility. Again, in broad comparative perspective, men appear to be poor at detecting female estrus.

Men Cannot Simply Count Days

Some people have wondered why men do not simply note menstruation and then count cycle days to target the fertile window in the menstrual cycle. Naturally, if men do not know when fertility occurs (and, indeed, in the modern west, it was known only within the last hundred years), they would have no idea how many days to count. Men can rely only on residual cues of fertility status.

Adaptation for Concealed Fertility in Nonhuman Species

Socially Monogamous Birds

Benshoof and Thornhill (1979) hypothesized that the cuckoldry hypothesis for concealed cycle fertility may apply to species other than *Homo sapiens*—that is, that there has been convergent evolution of the concealment of estrus. Colonial socially monogamous birds (which show variable extra-pair paternity across species) are especially good candidates for the possession of concealed-fertility female adaptation. In general, when biparentally investing pairs live in close proximity, some costs of infidelity are reduced.

Colonial socially monogamous birds nest in close proximity, and hence aspects of their social lives are similar to those of humans. As we discussed in chapter 10, EPC and associated extra-pair paternity are common in socially monogamous bird species, including some colonial birds (Barash & Lipton, 2001; Birkhead & Møller, 1992; Westneat & Sherman, 1997).

Socially monogamous birds often mate within pairs at substantial rates over extended periods of time, beginning up to about 3 weeks before estrus (Barash & Lipton, 2001; Birkhead, 1979; Birkhead & Møller, 1992; Low, 2004; Wysocki & Halupka, 2004). The typical fertile period in female birds, that is, estrus, is from about 5 days before egg-laying begins until the day of the penultimate egg (see Birkhead & Møller, 1992). Much in-pair mating in these species hence occurs when females are not at peak fertility (see references just cited). Females in such species, then, have extended sexuality that may function to increase females' access to material benefits and services from the pair-bond partner and sometimes from other males in the population (see chapters 3 and 10). Males purportedly copulate with infertile female partners because their information about female fertility status is incomplete.

As we discuss in the next chapter, males of many socially monogamous bird species possess adaptation to enhance their chances of paternity in the absence of certainty about when female partners are fertile. A question remains unanswered: Is female estrus in these birds simply undisclosed, leading males to be uncertain about the fertility status of females? Or has selection operated on females to suppress cues of fertility?

Olfactory cues emitted by females are used by males in many vertebrates to find females in the high-fertility phase of their reproductive cycle. The general absence of well-developed olfactory ability in birds may limit male ability to detect peak female fertility. All else being equal, females may have an edge in the coevolutionary race between males to detect estrus and females to conceal it.

Female birds in estrus, however, gain much weight, due to both the enlargement of the oviducal tissue and developing eggs. Their weight gain presents challenges to females to conceal fertility. Female weight increases yield secondary changes in flight and ambulation (e.g., Low, 2004). Concealment of estrus may hence be most effective early in the estrous phase.

Stitchbirds may partially conceal estrus by hiding in the nest eggs laid early in a clutch (Low, 2004). Stitchbirds engage in male-female pair bonding and social monogamy with EPC. Female EPCs are often aggressively forced by males (see Low, 2005a). Egg hiding hence may function to conceal fertility from extra-pair males that might force matings on fertile females. Male stitchbirds commonly investigate nests in other birds' territories and could gain information about a female's fertility by observing an egg in the nest. The female stitchbird's egg-hiding behavior is unlikely to function in reducing egg predation (see Low, 2004).

In pair-bonding birds, then, behavioral crypsis of estrus could include egg hiding. Other estrus-concealing adaptations, especially early in estrus before weight gain, might also exist. Future research may investigate whether socially monogamous female birds hide estrus from pair-bond partners for the function of cuckoldry and from extra-pair males for rape avoidance.

Other Pair-Bonding or Consortship Species

Other male–female pair bonding or consortship species, in vertebrates or invertebrates, are candidates for the presence of female adaptation that functions to conceal fertility in the service of extra-pair mating to obtain superior genes for offspring. Some scholars have suggested that nonhuman primates possess female adaptation for concealing fertility. But any male-female pair bonding or consorting species may exhibit such adaptation. Candidates include species characterized by a variety of mating systems in which a male and female associate beyond the duration involved in courtship and mating: systems involving biparental care, heterosexual friendships and other heterosexual alliances in primates, and male-female associations due to male mate guarding (for nonhuman primates, see Hawkes, 2004; Smuts, 1985). Any time a fertile-phase female and a male associate for an extended time beyond mating, whether the extended association is coerced by the male or not, female extra-consortship mating for genetic quality may be adaptive and, if so, may lead to direct selection on females to conceal estrus. Adaptive female extra-consortship copulation to obtain superior genes is particularly likely when the pair-bond or consort male provides nongenetic material benefits to females and/or their offspring (e.g., food, protection from sexual or other coercion) and when males' willingness to deliver material benefits negatively covaries with their genetic quality.

In addition to socially monogamous birds, a number of species may be particularly good candidates for study of concealed fertility. As we noted, chacma baboon males must associate socially with a female for several months to be able to reliably discriminate her conceptive from her nonconceptive cycles (Weingrill et al., 2003). Do these females possess adaptation for concealed fertility? Additional research is needed to assess whether the ambiguity associated with fertility is the result of concealed-fertility adaptation or incidental undisclosed fertility. Other intriguing candidates for further investigation include male-female pair-bonding fish (reviewed by Whiteman & Cote, 2004), nonprimate mammals (reviewed by Kleiman, 1977), and male—female consorting lizards such as *Sceloporus virgatus* (Weiss, 2002) and *Crotophytus collaris* (Baird, 2004).

Is Extra-Pair Mating Necessary? A Broadening of the Cuckoldry Hypothesis

In non-pair-bonding and nonconsortship species, female extra-pair-bond or extra-consortship mating is absent by definition. Hence, in such species, the cuckoldry hypothesis of concealed estrus provides no reason to expect direct selection on females to conceal estrus. But might females be selected to conceal fertility nonetheless? We suspect so. In fact, these considerations lead us to propose that the key elements necessary for the evolution of concealed fertility do not include extra-pair mating (cuckoldry) per se.

Earlier in this chapter, we noted that females could potentially benefit from concealed estrus whenever they obtain material benefits during extended sexuality from males other than those preferred as sires. The cuckoldry hypothesis pertains to one such circumstance: that in which females receive material benefits from a pair-bond male while men other than pair-bond mates can possibly offer better complements of genes to offspring. But circumstances other than this one may also fit the general requirement.

Consider the theory of infanticide reduction. Females reduce the risk of infanticide by engaging in multiple mating with all males in a group throughout a period of extended sexuality. They do so during a phase of extended sexuality, rather than during the fertile period, so that they gain the benefits of infanticide reduction while simultaneously maintaining control of sire choice by estrous adaptations. Suppose, however, the hypothetical situation in which female fertility in such a species was completely disclosed to males; through incidental cues, males clearly knew when females were fertile. As a result, females could not effectively give all males paternity confidence through extended sexuality. Accordingly, selection could operate on females to suppress cues of fertility. It would appear, then, that infanticide reduction can explain concealed fertility, as well as extended sexuality. Perhaps the food-for-sex hypothesis could generate a similar argument.

Let us ask, however, what elements within the infanticide-reduction hypothesis account for concealed fertility in this example. On the one hand, concealed fertility enhances females' ability to execute a strategy of reducing infanticide through extended sexuality. This is the element stressed by the infanticide-reduction theory. On the other hand, a crucial element appears to be that concealed fertility allows a female control over which male sires her offspring while she simultaneously obtains nongenetic material benefits from other males. If females were not selected to care which male sired their offspring, they could reduce infanticide by mating with all males when fertile and not conceal fertility. The ability to choose a sire and yet obtain direct benefits from other males is, of course, the core of the cuckoldry hypothesis. Yet, in this instance, cuckoldry is not involved.

We propose a broadening of the cuckoldry hypothesis for concealed fertility. The core elements that drive the evolution of concealed fertility possibly do not require cuckoldry per se. Rather, as noted earlier, the fundamental conditions that may lead to selection of concealed fertility are that (1) females are selected to maintain control of sire choice and (2) females are also selected to obtain nongenetic material benefits from other males, partly through extended sexuality. Although selection on females to cuckold males engaging in parental care satisfies these conditions, so, too, may many other conditions—potentially, selection on females to distribute paternity confidence across many males while maintaining control of her offspring's sire, or selection on females to obtain food items from nonfathers during extended sexuality.

Now, in fact, these are conditions that, in theory, could lead to selection for concealed fertility. The precise conditions that actually have led to the evolution of concealed fertility are unknown. Selection for cuckoldry may be a particularly potent force. In that instance, males providing nongenetic material benefits pay large costs to do so. Those costs mean that intense efforts to assure paternity may pay. And those intense efforts fuel the coevolutionary race between males to detect the fertility status of female partners and females to conceal it. When males pay smaller costs for the direct benefits they provide to females (e.g., not aggressing against infants), the costs they are willing to pay to deliver those benefits in the absence of high paternity certainty may be relatively low. If males are quite uncertain about female fertility status based on incidental cues, selection may not be sufficiently strong to have any appreciable effect on female concealment. Possibly, then, selection on females to conceal fertility has never been sufficiently strong in any extant species in which females multiply mate to reduce infanticide to actually lead to effective concealed fertility. Perhaps additional cost-benefit modeling of the precise circumstances in which concealed fertility could be favored would be useful.

At this time, we cannot rule out the possibility that selection on females to suppress cues of fertility is widespread across species. Ironically, it may apply even to females who possess sexual swellings. Sexual swellings, of course, have been claimed to be signals of fertility. We have argued that they are not but, rather, are likely signals of condition or quality. In theory, female primates who exhibit sexual swellings have been under selection to suppress incidental cues of fertility so as to control paternity of their offspring while simultaneously obtaining nongenetic material benefits from other males through extended sexuality.

Summary

Three widely held assumptions about human sexuality that led scholars to the conclusion that women have lost estrus are not true of women. Instead, women have estrus, which is mediated by ovarian hormones, and hominin female ancestors never had sex swellings that signal cycle fertility. Nonetheless, women likely possess concealed estrus, in the sense that incidental cues associated with estrus are suppressed.

Concealed estrus is adaptation directly selected for the purpose of concealing peak cycle fertility. It is distinct from undisclosed estrus, which is cryptic incidentally as a result of absence of selection for signaling estrus. Similarly, disclosed estrus is incidental to hormonal and other changes across the cycle. In all likelihood, females of the vast majority of taxa possess estrus that is disclosed or (to varying degrees) undisclosed. Female sexual motivation outside estrus is extended sexuality, not deceptive estrus. Women's concealed estrus involves direct selection for hiding hormonal, emotional, behavioral, and other features that correspond to peak cycle fertility. The suppression of some cues, particularly behavioral ones, should be conditional. Under some circumstances, at least, they may particularly be concealed from pair-bond partners.

The various published hypotheses about women's concealed fertility are critically reviewed. Of these, only the cuckoldry hypothesis is consistent with research findings. As stated and typically interpreted, the others are challenged by the existence of women's estrus and men's ability to detect it. The cuckoldry hypothesis proposes that women's concealment of cycle fertility functions to allow contingently adaptive extra-pair copulation for good genes while retaining material benefits from a pair-bond partner. Males are selected to detect estrus despite selection on females to conceal it. A sexually antagonistic coevolutionary race between females who evolve better concealment and males who evolve counteradaptation for detection ensues.

The cost of concealment of estrus, in particular estrus's reliance on reproductive cycle hormones affecting fertility, constrains how far selection can go to eliminate the disclosure of estrus by women. The cycle-related cues of estrus that women exhibit during estrus reflect this limitation. These changes are by-products, not signals of estrus.

Concealed estrus adaptation may occur in certain nonhuman species, notably in those with pair-bonding or male-female consortship accompanied by female extra-pair copulation for good genes. In these species the cuckoldry hypothesis may apply.

We broaden the cuckoldry hypothesis for concealed estrus. The basic conditions under which concealed estrus may be selected are that females are selected to maintain control over sire choice during estrus but obtain material benefits from males, including, at times, males other than potential sires, during extended sexuality. These conditions do not require the formation of pair bonds per se and hence do not require cuckoldry (or extra-pair sex in the usual sense). Whether, absent cuckoldry, selection operates on females to conceal estrus, in light of males' abilities to detect undisclosed estrus, is unknown at this time.

12   Coevolutionary Processes
Men's Counterstrategies and Women's Responses to Them

Sexually Antagonistic Coevolution

Coevolutionary Arms Races

As we discussed in chapter 7, biologists have long recognized that species may coevolve with other species in an antagonistic fashion; for example, the coevolution of predator–prey, host–pathogen, or competitors for the same food source. Through antagonistic coevolution, new adaptations in one species (e.g., traits in predators that increase their ability to capture prey) evoke selection on the other species (e.g., prey) to evolve counteradaptations (e.g., defenses)—which may then produce selective pressures on the first species to counter those counteradaptations, and so on. Antagonistic coevolution of adaptation and counteradaptation can continue through long stretches of evolutionary time, resulting in persistent evolutionary change in both species. As we also discussed, just as genes within two species' genomes can antagonistically coevolve in response to their interaction, genes within a single species can coevolve antagonistically. Sexually antagonistic coevolution is a prime example. (For a general discussion of intersexual conflict theory and evidence in nonhuman animals, see, in particular, Arnqvist & Rowe, 2005; Burt & Trivers, 2006; Chapman, Arnqvist, Bangham, & Rowe, 2003; Hammerstein & Parker, 1987; Holland & Rice, 1999; Parker, 1979a; Rice, 1996; Rice & Holland, 1997.)

Intrasexual Antagonistic Coevolution

Through sexual reproduction, two individuals' genes are passed onto an offspring they jointly conceive, and hence the offspring is a vehicle through which each individual's genes can be propagated. Nonetheless, of course, reproduction is by no means a purely cooperative enterprise between mates. Selection will favor individuals' treating their mates' outcomes as just as important as their own when each individual can reproduce only with that particular mate. In such a case, the death of the mate ends the individual's reproductive career just as surely as does the individual's own death. By creating living groups of just two individuals—one member of each sex—experimental biologists have created these circumstances in laboratory populations (e.g., Holland & Rice, 1999). In nearly all natural populations, however, they rarely if ever exist. Instead, the events that optimize one partner's reproductive outcomes do not perfectly match those that optimize the other's. Mismatches reflect reproductive conflicts of interest between the sexes within mateships, which can generate selection for features that promote the fitness of one sex at the expense of the fitness of the other sex. As we discussed in another context in chapter 7, the outcome of such selection is referred to as *sexually antagonistic adaptation* (for a review, see Arnqvist & Rowe, 2005).

Sexually antagonistic coevolution exists in species without paternal care. Through ingenious artificial selection procedures, Rice (1996) allowed *Drosophila melanogaster* males to evolve in one experimental lineage while preventing females from evolving counteradaptations in a separate control line (see chapter 7). Tests performed after 30 generations clearly demonstrated male adaptation to target females. Wild *Drosophila melanogaster* typically mate promiscuously, and males make frequent attempts to induce remating on the part of females. Males in the experimental line had increased capacity for remating with females who had previously mated with competitor males taken from the control line. At the same time, competitor males were less able to remate with females previously mated with experimental males and to displace sperm inseminated by experimental males, even when experimental males were no longer present. In mixed groups, experimental males outreproduced control competitor males.

Male adaptation, furthermore, evolved at the expense of female fitness. Females mated to experimental males had higher mortality than those mated to control males, with no compensating increase in fecundity. Proteins in male *Drosophila melanogaster* seminal fluid are low-level toxins to females (e.g., Fowler & Partridge, 1989). The increased female mortality rate was likely mediated by a greater exposure to (due to an increased remating rate) and enhanced toxicity of male seminal proteins. The toxicity of male seminal fluids to females is probably an incidental by-product of beneficial effects on male reproductive success. The proteins can harm other males' sperm (e.g., Harshman & Prout, 1994). Some may enter the female's circulatory system and influence her neuroendocrine system in ways that benefit the male (e.g., by reducing her remating rate; e.g., Aigaki, Fleischmann, Chen, & Kubli, 1991). The costly effects on females are thus sexually antagonistic outcomes of male adaptation.

In species with biparental care delivered by pair-bonded males and females, particular forms of sexual conflicts of interest arise—conflicts over each party's efforts to engage in extra-pair copulation (EPC). They arise partly because, though both individuals of a pair benefit equally from the caring that either parent does for the pair's shared genetic offspring, only the individual actually performing the care pays the costs. Hence members of a pair have conflicts of interest over the way that each individual in the pair allocates time and energy. Conflicts also arise due to the fact that males can be cuckolded—induced to invest time and energy into caring for young not their own genetic offspring.

Conflict of Interest Over Male Extra-Pair Copulation Mating Effort   There often exists a conflict surrounding male effort to obtain extra-pair matings. Even if both sexes have historically been interested in extra-pair matings, males have probably paid a large bulk of mating effort costs (e.g., signaling costs, costs of intrasexual competition, courtship costs). These efforts do not benefit female partners and detract from male parental effort, which does benefit the female partner.

Conflict of Interest over Female EPC   Conflicts of interest over female extra-pair sex also clearly exist. We have already detailed the ways in which females in biparentally investing pairs can benefit through EPC with males possessing genetic quality superior to that of their mates, while maintaining the material benefits delivered by mates. Naturally, these EPCs are not in the reproductive interests of male primary partners. Selection should hence shape counteradaptations in males to discourage or prevent partners from engaging in EPC, should restrict the possibility that offspring produced by female mates are sired by EPC partners, and should limit costly investment in offspring sired by EPC males. In turn, sexually antagonistic selection may favor adaptations in females to be better able to engage in EPC without incurring costs, or counteradaptations in males to reduce female ability to do so, and so on.

Male Strategies and Counterstrategies to Female EPC

Investing male partners can engage in a number of strategies designed to increase their fitness in response to females who could potentially benefit through EPC.

Mate Guarding   Males can attend to females closely or track their whereabouts.

Frequent In-Pair Copulation   Males can attempt to copulate with their in-pair partners frequently. If a female partner has engaged in EPC, the in-pair partner increases his chances of siring offspring if he has viable sperm in the female reproductive tract at all potentially fertile times. Moreover, the greater the number of his own sperm in the female reproductive tract when ovulation occurs, relative to numbers of sperm inseminated by rival males, the greater the chances that one of his own sperm will "win."

Fertility Detection   Both of the preceding strategies, mate-guarding and frequent in-pair copulation, can be performed more effectively and efficiently if male partners know when females are fertile in their cycles. Males are hence selected to be able to detect female fertility within the cycle.

Paternity Detection   If males cannot prevent female EPC and extra-pair paternity through mate guarding and frequent copulation, they can protect their reproductive interests through detection of nonpaternity of offspring based on phenotypic cues, such as physical similarity of self to offspring or scent-based cues of MHC genes. If males are sufficiently good at detecting nonpaternity, and if paternal care is highly important to offspring success, females can perhaps be dissuaded from engaging in EPC even in the absence of male mate guarding and frequent in-pair copulation.

Female Counterresponses—and So On

In response to male adaptation to female EPC, females may evolve counteradaptation to increase abilities to engage in EPC. As we emphasized in the preceding chapter, for instance, females may frustrate male attempts to mate-guard through concealment of fertility in the cycle. Adaptation to conceal estrus from main partners in human ancestral females, in turn, may have selected, for male counteradaptation, mechanisms that function to detect female peak fertility despite female adaptation to disguise fertile states. The evolution of male mechanisms, then, selects for more sophisticated female disguise adaptation, which selects for male counteradaptation, and so on, leading to an unending coevolutionary race between males and females.

Males and females may each become involved in other, related coevolutionary races. For instance, male and female adaptation to control events in the female reproductive tract affecting conception may coevolve.

Both sexes should furthermore be expected to evolve ever more fine-tuned and sophisticated conditional strategies. For instance, male mate guarding may require costly expenditures of time and energy. Males should be expected to expend those costs under circumstances in which they have ancestrally paid off and to refrain from intensive mate guarding when, ancestrally, costs exceeded benefits. Females similarly should have evolved to judge the costs of EPC, as well as the potential benefits. Naturally, we need not expect that males and females calculate costs and benefits either consciously or unconsciously. Rather, selection will have shaped psychological adaptations to respond differentially (e.g., mate guard vs. do not mate guard) as a function of cues that, ancestrally, were sensitive to relative costs and benefits of different tactics.

Equilibria?

Coevolution does not bring the reproductive interests of parties out of conflict and "resolve" them. Coevolutionary races hence do not have stable equilibria so long as either party can potentially introduce new strategies to undermine recently evolved adaptations of the other party. Hosts and pathogens will always have conflicting reproductive interests and therefore never stop evolving in response to one another. The same holds true of males and females with conflicts of interest. Rice's (1996) experimental manipulations that "halted" the coevolution of females testify to the consequences of one sex not continually evolving in response to the other sex: The other sex evolves adaptations that give it an edge in the conflict.

Nonetheless, outcomes of a coevolutionary race at individual points of time may be fairly stable in certain respects over time. Long-lived organisms are susceptible to disease and must constantly struggle in coevolutionary races with pathogens. But organisms that have evolved a long-lived life history (and whose fitness hence depends heavily on longevity) may consistently allocate substantial effort to resist pathogens (Robson & Kaplan, 2003). Despite their constant coevolutionary race with pathogens, long-lived hosts stably tend to live longer (e.g., through greater allocation of effort to immune function) than organisms that have evolved short-lived life histories.

In species characterized by biparental care, we should expect an analogous situation. Among all bird species with biparental care, male parental investment affects offspring quality more strongly in some species than in others. Species vary in how much time and energy males put into parental care or, put otherwise, the extent to which individuals' fitness depends on paternal investment. In species in which males invest much care and female fitness depends much on male care, one expects that males will have evolved more effective strategies for protecting that investment and that females will take fewer risks of losing paternal investment. Compared with other biparentally investing species, then, the rate at which females engage in EPC may, as a result, be stably lower in those species highly dependent on biparental care. But by no means need that imply that, in these species, the sexes have stopped coevolving in response to one another, just as a stable long life of a species (and fairly effective resistance to infection) does not imply that it no longer coevolves with pathogens (Møller, 2000). Rather, the characteristic outcomes of the conflict between males and females over female EPC in some species will generally lead to fewer EPCs in some species than in others.

An implication of this observation is that the extra-pair paternity rate within a species is not an accurate barometer of how much coevolutionary conflict exists between males and females over female EPC. Again, consider coevolutionary conflicts between hosts and pathogens. Specifically, consider a species whose age-specific mortality rates due to infectious disease are half of those in another species. In no way can we infer that the host–pathogen coevolutionary struggle is less intense in the species with lower rates of death due to infectious disease than in the other species. It may be less intense—but it need not be. The marginal value to fitness of longevity may have historically been greater in the species with a lower mortality rate due to infectious disease, hence leading individuals in that species to invest more heavily in mortality reduction adaptations that hosts implement to struggle against pathogens.

Similarly, a species with a relatively low extra-pair paternity rate (say, 5%) need not have been subject to less intense sexually antagonistic conflict over female EPC than a species characterized by a much higher rate (say, 25%). Through historical coevolution, each sex of both species may have accumulated an extensive array of "armaments" that function within the conflict. The species' extra-pair paternity rates may differ because the marginal utility of male paternal care has historically been greater in one species than in the other, leading to stable differences in the characteristic outcome of the conflict. That is, when males extensively invest in offspring, the marginal utility for them to engage in increasingly costly tactics to guard paternity diminishes less rapidly than does the marginal utility for males who invest in offspring less intensively, yielding lower rates of female extra-pair paternity in the former species than in the latter one (see also Arnqvist & Rowe, 2005, e.g., p. 223).

Paternity-Enhancement and Anticuckoldry Adaptations in Birds

Mate Guarding

In a variety of bird species, males have evolved adaptations for mate guarding. Mate guarding in some species is particularly intense when female partners are fertile in the reproductive cycle. In house martins, male mate guarding increases in frequency about 7 days before commencement of egg laying (Riley, Bryant, Carter, & Parkin, 1995). And the distance at which male northern mockingbirds position themselves from females during the fertile period is about half the typical distance separating social mates during infertile phases (Bodily & Neudorf, 2004). Similar patterns have been documented in European barn swallows (Saino, Primmer, Ellegren, & Møller, 1999), bearded vultures (Bertran & Margalida, 1999), Seychelles warblers (Komdeur, Kraaijeveld-Smit, Kraaijeveld, & Edelaar, 1999), and New Zealand stitchbirds (Low, 2005b).

In some species, males guard mates extensively despite low female EPC rates. Korpimäki et al. (1996), for instance, found that male kestrels guard mates despite very low rates of female EPC (1% of total copulations). Male Australian magpies increase their mate guarding when females are fertile; in the same population, the extra-pair paternity rate has been observed to be very low (just 3%). These instances illustrate the point we just made: A low rate of extra-pair paternity need not imply that males have not been under strong selection to prevent cuckoldry. Indeed, the low rates of EPC in these species may exist precisely because males have been under strong selection pressures to prevent the possibility that their substantial paternal investment will be squandered in the face of potential fitness benefits females can accrue from EPC. (Male raptors, for instance, provide effectively all food for females prior to egg laying, during incubation, and for much of the nestling period.)

Mate guarding can arise for reasons other than adaptation to female-initiated EPC. Mate guarding can also function to prevent sexual coercion, in which cases it typically serves the interests of female mates, as well as the pair-bond males. In New Zealand stitchbirds, females appear to rarely if ever initiate EPCs. Most EPCs are forced by males, and whether a female accepts an EPC is not predicted by the extra-pair male's quality (Low, 2005a). Wysocki and Halupka (2004) argued that, through a prolonged sexual period, female blackbirds may induce males to mate-guard, thereby reducing female risk of sexual coercion.

In a variety of species, however, male mate guarding clearly does function to serve male interests in sexual conflicts over paternity of offspring. In the brown thornbill, mated pairs spend more time together during the fertile period, largely due to male efforts to maintain close proximity to female mates (Green, Peters, & Cockburn, 2002).

The extra-pair paternity rate in this population is fairly low (about 6%), as males are fairly successful at preventing female EPC.

In some species, a male's mate guarding is conditional not only on fertility status of his female partner but also on male traits that affect the value of mate guarding relative to competing activities. The feathers of male bluethroats reflect ultraviolet light to variable degrees. Females prefer to mate with males with feathers that reflect relatively much UV light. Johnsen, Andersson, Ornborg, and Lifjeld (1998) experimentally manipulated ultraviolet reflectance of males. Males whose reflectance was reduced through manipulation guarded their mates more frequently as a result, purportedly because their partners were accordingly more likely to seek EPC. More generally, male allocation of effort to mate guarding is sensitive to its opportunity costs, as well as potential risks of EPC. In Montagu harriers (a semicolonial raptor), presentation of a male decoy led males to guard mates more frequently and in-pair copulation rates to rise (Mougeot, Arroyo, & Bretagnolle, 2001). In red kites, as the density of neighboring males increases, so too does the percentage of time males spend in their own territories (Mougeot, 2000). Black-throated blue warbler males that have opportunities for EPC themselves engage in less mate guarding than males with few EPC opportunities (Chuang-Dobbs, Webster, & Holmes, 2001). Male yellow warblers (Yezerinac & Weatherhead, 1997) and western bluebirds (Dickinson, 2001) make fewer forays into neighboring territories when their own mates are fertile (cf. Dickinson, 1997). Osorio-Beristain and Drummond (1998) found that, in blue-footed boobies, most copulations between males and extra-pair females occurred when the males' own social partners were infertile. In a related vein, when Møller (1987) experimentally removed male barn swallows when their partners were fertile, neighboring males with fertile female partners increased the intensity of their male guarding, whereas neighboring males with infertile female partners did not.

In many species, male singing appears to function to advertise cues of the singer's own quality to neighboring females, thereby enhancing opportunities for high-quality males to EPC. In willow warblers and a variety of other species, males sing very little when their own mates are fertile, presumably because of a trade-off between effort to seek EPCs and efforts to guard mates (for a review, see Gil, Webster, & Holmes, 1999; see also Hall & Magrath, 2000). In contrast, male European robins (Tobias & Seddon, 2000), blue grosbeaks (Ballentine, Badyaev, & Hill, 2003), and some other passerine species (Møller, 1991) sing more frequently or sing more song variants when social mates are fertile than when mates are infertile, perhaps because singing functions to attract or retain a mate. In some species, males' singing when female mates are fertile ("fertility announcement") may honestly advertise male quality (Møller, 1991), deter male rivals from foraying into the male singer's territory, and attract extra-pair females.

Mate guarding may be accompanied by threat of retaliation for female EPC. Valera, Hoi, and Kristin (2003) removed female lesser gray shrikes from their territories during their fertile period. After they were returned to their territories, their male partners retaliated physically against them. Extra-pair paternity in this species is rare.

In-Pair Copulation and Sperm Competition

In some species, males cannot effectively guard females and simultaneously engage in foraging. For instance, in many species of raptors, males must separate from social mates to forage for prey. Typically, females and young are highly dependent on foraging by males, who are the main providers from a time prior to egg laying until the late nestling period. A strategy of paternity enhancement by paired males that is an alternative to or that complements mate guarding has evolved in many of these species. Through frequent in-pair copulation, resident males increase the likelihood that offspring are sired by them and not by extra-pair partners. Hence, in many species of raptors, the in-pair copulation rate is exceptionally high—up to several hundred times per clutch (Mougeot, 2004). Korpimäki et al. (1996), for instance, observed the in-pair copulation rate of Eurasian kestrels to be one copulation every 1.3 hours during vole season. Across 3 years, the extra-pair paternity rate was only 5%, 0%, and 0%, respectively.

In Montagu's harrier, Arroyo (1999) found that in-pair copulation rates increased during the fertile period and were maximal 1 day before the laying of the first egg, close to maximal female cycle fertility. Extra-pair copulations nonetheless occurred. About 5% of copulations were EPCs, all occurring during females' fertile period. As the number of neighbors increased, so too did the in-pair copulation rate, most notably at peak EPC risk, during the laying period.

More generally, Mougeot (2004) reviewed the literature on copulation rates during the fertile period in 49 species of raptors and found similar patterns. In most species in which EPCs have been studied (13 of 19), females were observed to engage in EPC (and especially when fertile). The proportion of females that engaged in EPC increased with breeding density. Similarly, the rate of in-pair copulation was associated positively with breeding density. And, indeed, the proportion of females who engaged in EPC predicted rates of in-pair copulations across species, just as expected if high rates of in-pair copulation arise from an in-pair male strategy that functions to reduce the likelihood of cuckoldry. (By contrast, rates of copulation prior to the fertile period are not predicted by the EPC rate; those copulations have functions other than increasing paternity rate. See Mougeot, 2004.) Males of frequently copulating species have larger testes, indicating that raptors have adaptations in response to sperm competition with extra-pair males that are both behavioral and physiological.

In general, rates of extra-pair paternity are low in raptors. Across 10 species studied, they range from 0 to 5% and average barely over 1% (Mougeot, 2004). Once again, however, these low rates do not imply the absence of historical conflict between males and females within pairs over control of paternity. Indeed, to the contrary, raptors reveal a variety of clear footprints of historical antagonistic coevolution: Females do occasionally engage in EPC (particularly when fertile in their cycles), males invest very heavily in paternity enhancement through spectacular rates of copulation, copulation during the fertile period functions differently from copulation prior to the fertile phase, and males respond to factors that enhance the risk of extra-pair paternity by increasing the rate of copulation during the fertile period. As we previously noted, males in these species may well have evolved to be effective at guarding paternity because they invest very heavily in offspring and hence have much to lose if cuckolded (Mougeot, 2004).

In many colonial birds, as well, in-pair copulation rates are very high (e.g., Birkhead & Møller, 1998), which may be explained by the same evolutionary dynamics we have discussed for raptors.

Males may experience physiological changes as a function of their mates' fertility status in response to increased demands to engage in copulation, sperm competition, or territory defense. The testosterone levels and sperm storage capacity of Seychelles warbler males, for instance, increase during their mates' fertile phases (Van de Crommenacker, 2004).

Detection of Nonpaternity and Divestment

If a male's mate does produce offspring not his own, he can reduce his costs resulting from her infidelity by discriminating his own offspring from offspring not his own and, accordingly, withdrawing investment in those not his own. All else being equal, a male does better by preventing infidelity than by divesting in offspring produced through EPC after the fact. Divestment after the fact entails new mate search costs. In a species that breeds seasonally, a male whose mate bore only offspring of another male has missed the opportunity to reproduce that season. At the same time, females may pay considerable costs for male withdrawal of investment. Particularly when male contributions to parental investment crucially affect offspring success (and hence female mates' fitness), males who merely threaten divestment when lack of paternity is detected may powerfully deter female EPC.

In some bird species, extra-pair paternity rates are remarkably low despite no evidence that males engage in costly mate guarding or in-pair copulation tactics. Purple sandpipers are relatively long-lived arctic shorebirds. Pierce and Lifjeld (1998) found an extra-pair paternity rate of 1% in this species. Yet they found no evidence that males engaged in mate guarding or attempted to copulate at high rates. Males alone care for offspring from hatching to fledging.

Another illustration is provided by the Capricorn silvereye, an island bird of the Great Barrier Reef. Males and females form lifetime pairs soon after their fledging. Robertson, Degnan, Kikkawa, and Moritz (2001) found no instances of extra-pair paternity in 122 young. Yet males do not mate guard, and in-pair copulation rates are low—despite the fact that pairs breed in high density. Michalek and Winkler (2001) similarly observed low rates of in-pair copulation in two species of woodpeckers (the great spotted and the middle spotted), despite no cases of extra-pair paternity found in over 200 offspring. Females did not engage in EPC, yet males were not efficient paternity guards. Once again, the authors proposed that the importance of male care (and threat of its withdrawal) constrains females to be faithful. (See also Stanback, Richardson, Boix-Hinzen, & Mendelsohn, 2002, on Monteiro's hornbill, a species that stores sperm for long periods yet does not engage in EPC.)

One reason that males in these particular species need not engage in highly expensive tactics to deter EPC relates to a topic we discussed earlier (see chapter 10): In many if not all of these species, genetic variability between individuals is low due to bottlenecks and high levels of relatedness among individuals in the same group. Island bird populations, for instance, tend to have high band-sharing coefficients (indices of genetic similarity and, indirectly, relatedness between pairs of individuals). So do great spotted and middle spotted woodpeckers (nonmigratory species). Relatedness within bands increases genetic similarity, reducing the potential benefits of EPC for good genes (Møller, 2003; see also Petrie et al., 1998; Robertson et al., 2001). When the potential benefits of female EPC are low, the costs males must pay to engage in tactics that effectively deter female EPC is similarly low.

If, in these species, females do not seek genetic benefits through EPC when fertile, despite lack of male counterstrategies to directly prevent or (through frequent in-pair copulation) undermine female extra-pair mating, they may be true exceptions to the general rule that vertebrate females exhibit estrous sexuality (or, perhaps better stated, differentiated fertile and extended sexualities; see chapter 8). In these species, distinct estrous sexuality may have been lost. Phylogenetic analyses may shed light on the questions of how many times estrous sexuality has been lost within birds and whether, once lost in a lineage, estrus has ever reemerged.

Men's Paternity-Enhancement and Anticuckoldry Adaptations

Mate Guarding

Men have been observed to engage in a rich diversity of mate-guarding tactics: vigilance of partners' whereabouts and activities, monopolization of a mate's time, attempts to please a mate, and threats to abuse an unfaithful mate (see, e.g., Buss, 1994, 2000). Wilson and Daly (1992) refer to the male psychological adaptation or complex of adaptations responsible for condition-dependent mate guarding as sexual proprietariness. Proprietariness functions to exclude other males from sexual access to one's mate, as well as to deter a mate from engaging in EPC. Though both sexes express concern about a mate's infidelity, across a variety of cultures men appear to be relatively more concerned about sexual infidelity as opposed to "emotional" infidelity (strong feelings for an extra-pair person) than are women (Buss, 2000; Buss, Larsen, Westen, & Semmelroth, 1992; Buss et al., 1999; Shackelford & Pound, 2006). For instance, Crocker (1984, 1990) studied the Canela, a small South Amerindian group that traditionally engaged in ritual extramarital sex. Both sexes experienced sexual jealousy, which was the leading cause of marital dissolution. But men appear to have experienced more. In ethnographies of the Canela, Crocker (1984, 1990) refers to sexual jealousy in 13 different paragraphs. In one place, he refers merely to "jealousy between spouses" (1999, Section III.A.3.C.(3).(b)). One refers to jealousy in both sexes, but "particularly" males (1990, Section III.F.e.(2).(b)). All remaining 11 instances refer exclusively to male sexual jealousy. Similarly, Hill and Hurtado (1996, p. 230) claim that male jealousy is a leading cause of wife beating in the Ache of Paraguay. As Flinn (1987) documented in a traditional Caribbean setting and others have shown in studies in the United States, men guard mates who are young or not pregnant more intensely than they guard mates of low-fertility status (e.g., postmenopausal or pregnant mates; for reviews, see Buss, 2000; Shackelford & Pound, 2006). (Harris, 2003, has argued that, in fact, sex differences in sexual jealousy are weak or nonexistent. She does not deny, however, that sexual jealousy can be intense.)

As well, a number of scholars have argued that the psychological bases of male mate guarding by men importantly contributes to cultural practices that encourage female sexual modesty, in addition to female circumcision and claustration (Buss, 2000; Dickemann, 1981; Lancaster, 1997; Thornhill, 1991; Thornhill & Thornhill, 1987; Wilson & Daly, 1992).

The functional design of men's sexual proprietariness constitutes evidence that female EPC was of sufficient importance to males in the deep-time past to have generated direct selection that created male adaptation that functions to hinder female EPC and cuckoldry. Were there no recurrent negative effect of female EPC on male reproductive success in human ancestral settings, men would not have evolved often costly, functionally specialized arrays of emotional responses as deterrents (e.g., Buss, 2000).

By themselves, however, men's sexual proprietariness adaptations do not demonstrate that women possess adaptation that functions to gain benefits through EPC. Again, female EPC may be forced or unforced. Male proprietariness could evolve in response to EPC that is forced (or manipulated, and against female interests; see Thornhill & Palmer, 2000). As we have argued in chapters 9–11, independent evidence suggests that women have ancestrally benefited from targeted and conditional EPC.

Mate Guarding During Estrus

In chapter 11, we described evidence that indicates that women's estrus is not completely concealed. Men can potentially detect female fertility status through scent, appearance, or behavioral cues. If, based on these cues, male partners can in fact detect female estrus, they might be expected to guard mates more intensely when women are fertile in their cycles, analogous to the ways that males of some bird species increase their mate guarding when female partners are fertile. Multiple studies demonstrate that, indeed, the intensity with which men guard mates varies across their partners' cycles.

With Christine Garver-Apgar (Gangestad et al., 2002), we asked normally ovulating women to report the occurrence of a variety of male mate-retention tactics over the preceding 2 days on two different occasions: once when women were fertile in their cycles and once during the mid-luteal phase. Our questionnaire tapped two different dimensions of mate guarding recognized in the literature. Proprietariness reflects vigilance of a mate's activities (e.g., frequently checking up on a partner), as well as negative inducements to stray (e.g., threats or expressions of anger in response to a partner's contact with other men). Attentiveness reflects attempts to monopolize a partner's time (e.g., insisting on spending time with a partner) and some positive inducements not to stray (e.g., doing favors for a mate). Women reported that their partners engaged in both proprietary and attentive tactics more frequently when they were fertile than during the infertile luteal phase (Gangestad et al., 2002). Our broad measures consisted of 11 different subscales. The subscale on which we observed the largest effect of fertility status was vigilance: Relative to when their mates were infertile, men monitored a mate's whereabouts and activities more when their mates were fertile.

Additional evidence strongly indicates that these effects arise from male efforts to engage in mate guarding and are not solicited by female partners. First, the effects were stronger for primary partners of women who did not claim that the relationship was sexually exclusive than for women who claimed it was sexually exclusive. Second, and relatedly, we were able to examine the association between changes in male vigilance across the cycle and changes in female extra-pair sexual interest across the cycle. Men whose mate guarding increased most dramatically during their partners' estrus tended to be with partners who expressed greater sexual attraction to and fantasy about other men during estrus. By contrast, changes in women's attraction to their own partners did not predict increases in male vigilance at estrus.

Haselton and Gangestad (2006) asked normally ovulating women to keep daily diaries for a month. Women reported on male partners' jealousy and proprietariness each day. As predicted, women claimed that their partners were more jealous and proprietary on days just prior to ovulation than during the luteal phase. As Gangestad et al. (2002) found, increases in male partners' jealousy and proprietariness during estrus were predicted by increases in women's attraction to men other than primary partners. Women rated their mates' sexual attractiveness and attractiveness as long-term stable mates. Men seen by their partners as good long-term mates but not sexually attractive mates were particularly likely to increase their proprietariness during their partners' estrus (figure 12.1). In a separate study, Pillsworth and Haselton (2006) did not replicate this effect but did find a related pattern: Men rated by their partners to be sexually unattractive were particularly likely during their partner's estrus to engage in positive inducements to mate retention (e.g., by being attentive and doing favors for them).

In all of these studies, women reported on men's behavior (though independent work finds that partners' reports of male behaviors of this sort do correlate highly; see Gangestad et al., 2002). In another study with Garver-Apgar, we asked both women and their male partners (who filled out questionnaires in separate rooms) about men's mate guarding at two different sessions—one just prior to the female partner's ovulation and one during her luteal phase. This study once again revealed that men are more proprietary during estrus than during the luteal phase, as evidenced by both women's and men's independent reports (Garver-Apgar, Cousins, et al., 2007). In that study, we were also able to examine whether asymmetrical men were particularly likely to ramp up their mate-guarding efforts when partners were fertile. We did not find that male primary partners' fluctuating asymmetry (FA) moderated the effect of their female partners' phase of the cycle on mate guarding.

In a rural village on the Caribbean island of Dominica, colleagues and we interviewed normally ovulating women and their male partners on 4 different days: 2 consecutive days just prior to ovulation (the day of an LH surge and the next day) and 2 consecutive days during the mid-luteal phase. Men in this village do not work 9-to-5 jobs. They have

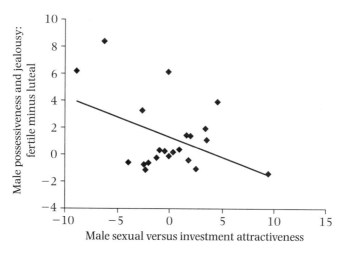

Figure 12.1 Scatterplot of shift in men's possessiveness and jealousy across the cycle (estrus-luteal phase) as a function of men's relative sexual-versus-investment attractiveness. These values are residuals (centered around the sample mean value), with male overall mate value, female physical attractiveness, and breakup status controlled. The partial correlation is −0.42, $p = 0.045$. The line is the least squares regression line. From figure 2 of Haselton and Gangestad (2006).

much leeway to decide how they spend their days: for example, working in their gardens, cultivating bay leaves (from which they extract and sell bay oil), chopping wood, fishing, or spending time with partners. We asked women and men how much time men spent with their partners each day. Of 12 men in our sample, all 12 spent more time with their partners on days their partner was in estrus than on days during their partners' luteal phase—on average, about 3 hours longer a day (Thornhill, Gangestad, Falcon, Dane, & Flinn, unpublished data).

This study also explored whether men's salivary testosterone levels change with their partners' fertility status across the cycle. On average, men's testosterone levels dropped when their partners were fertile, relative to levels during the luteal phase. This endocrinological change may be a proximate mediator of men's greater investment in their mates, including mate guarding, during their estrus. High testosterone levels tend to facilitate men's mating effort, including extra-pair mating effort. Low testosterone levels may facilitate attentiveness to and investment in partners and offspring (see chapter 4). Male mate guarding corresponds with *increased* testosterone levels in certain pair-bonding bird species (e.g., Saino & Møller, 1995). The possibility that men's testosterone differentially predicts different forms of male inducements to partner fidelity (e.g., predicting attentiveness to partners negatively but use of intrasexual threats positively) has yet to be addressed empirically.

Additional evidence that men not only perceive their partners' estrus but also act strategically on that knowledge comes from a recent study of pair-bonded men's reaction to other men's faces that varied in the dominance of their bearers (as perceived by raters; Burriss & Little, 2006). As the authors predicted, the male partners of normally ovulating women rated dominant-appearing faces (but not submissive ones) as even more dominant when their partners were fertile in their cycles. No similar shift across the cycle was observed in men partnered to women using hormonal contraceptives. The authors suggested that, as a male counterstrategy to risk of lost paternity as a result of fertile-phase women's EPC for good genes, men become more sensitive to cues of other men's dominance (as well as, perhaps, other features that women are particularly attracted to during estrus) when their partners are fertile.

Copulation Frequency

Human couples copulate far more than is necessary for reproduction per se. As many scholars have noted, much copulation occurs during women's extended sexuality (see chapter 10). We have argued that, from women's standpoint, sex during extended sexuality functions to enhance the flow of male-delivered nongenetic material benefits. In particular, we argued that women's extended sexuality enhances investment by male primary partners. It does so because of complementary adaptations that men possess: From a functional perspective, men are motivated to copulate with receptive females because, historically, copulations on average had a nonzero chance of resulting in conception. Motivation to copulate, even during women's infertile phases (which men cannot discriminate from the fertile phase with 100% accuracy), functions for men to increase reproductive success directly through conception.

Just as in-pair copulation in many bird species functions as a means for males to enhance paternity in the face of potential female EPC, however, so too does in-pair copulation for men. Todd Shackelford and his colleagues (Shackelford, Goetz, Guta, & Schmitt, 2006) tested and found support for the hypothesis that in-pair copulation and mate guarding function as concurrent male anticuckoldry tactics. In each of two large samples of romantically involved couples, they found a positive correlation between the intensity of male mate-guarding and frequency of copulation (whether reported by male or female partners). The associations were independent of male or female age, length of the relationships, or time that the couple spent together. Relatedly, Goetz and Shackelford (2006) found an association between men's mate-guarding tactics and their attempts to coerce sex from their relationship partners. Baker and Bellis (1995) proposed that men are motivated to "top up" partners regularly through "routine sex"—every few days, if not more frequently—to ensure that partners' reproductive tracts contain their own live sperm at all times aside from days of menstruation (see also Parker, 1984; Smith,1984b). Patterns of sexual intercourse within couples are, perhaps not surprisingly, characterized by periodicity: the longer the time since last intercourse, the higher the likelihood of sex on any given day (James, 1980).

Men typically desire to have more frequent in-pair sex than do women. Men, much more than women, complain that their partners resist having sex that they attempt to initiate (see review in Buss, 2000). Moreover, the course of sexual desire over the duration of a relationship differs across the sexes. In a sample of 30- to 45-year-olds, Klusmann (2006) found that men's sexual motivation remains fairly constant across the duration of the relationship; even after many years in a relationship, men desire to have sex about as often as they desired to copulate at the beginning of the relationship. In contrast, whereas women's desire for sex nearly matches men's in the early years of a relationship, it plummets as the relationship grows older. Klusmann argues that constancy of male sexual desire over time speaks to the function of sex for men to prevent cuckoldry (see also Klusmann, 2002).

If copulation partly functions to guard against cuckoldry, men should be motivated to engage in sex with partners following periods of separation, during which females may have had opportunity to engage in EPC. Shackelford et al. (2002) tested a variety of specific hypotheses that follow from this expectation. They found that, with time since previous copulation and relationship satisfaction statistically controlled, the proportion of time spent apart since the previous copulation predicts men's assessment of the attractiveness of their partners, how attractive men think their partners are to other men, and men's interest in copulating with their partners. (See also Goetz et al., 2005, on possible adaptations of men that function to displace extra-pair men's semen during copulation.) Men may also adjust the size of their ejaculates as a function of time spent apart from a partner, independent of the effects of total time since their last ejaculation (Baker & Bellis, 1995).

Men might also be expected to initiate sex with partners more frequently or more intensely when partners are in estrus. Just as researchers have long investigated whether women initiate sex more frequently when they are mid-cycle, so, too, attempts to address whether men initiate sex with partners during estrus more frequently have a long history. Though findings have been mixed, there is in fact no strong evidence that men initiate sex more frequently with partners when they are fertile than during the luteal phase (e.g., see reviews by Dixson, 1998; Gray & Wolfe, 1983; Hill, 1988; Manson, 1986; Meuwissen & Over, 1992; Pawlowski, 1999a; Steklis & Whiteman, 1989). The possibility that men's specific motives for sex (and factors that arouse men's sexual interest) nonetheless change as a function of their partner's cycle phase may be explored in future research. Perhaps, for instance, men are particularly motivated to have sex following a period of time spent apart when their partners are in estrus.

Other sexual practices could be affected by men's detection of partners' estrus. Male mammals commonly sniff and lick females' genitals. This behavior functions to detect fertility status, and it may serve similar functions in nonavian reptiles and other non-mammalian vertebrates (e.g., Mason, 1992). Cunnilingus in humans could potentially serve such a function for men. Whether it serves that function or not, patterns of cunnilingus across the menstrual cycle (e.g., whether it is more common during women's estrus) may speak to whether women have been selected to effectively conceal chemical cues of cycle-related fertility status in their vaginal fluids.

Detection of Nonpaternity

A number of investigators have devised ingenious experimental designs to probe whether men possess adaptations for discerning paternity and engaging in discriminative paternal care based on these assessments (DeBruine, 2004; Platek, Burch, Panyavin, Wasserman, & Gallup, 2002; Platek et al., 2003, 2004). In one design, a digital photograph is taken of a participant. The participant's own face or, alternatively, the face of another participant is digitally combined with the face of a small child to create two composite images of child faces—one that is "self-resembling" and one that is not. In a variety of ways, participants are asked which of the two children they would be more likely to invest in (e.g., which child they would like to spend time with, which one they would spend $50 on, which one they would rather adopt). Men prefer to invest in the self-resembling "child," an effect not due to participants being able to consciously recognize self-resemblance per se (Platek et al., 2002; Platek et al., 2003). The evidence that men's preference for a self-resembling child is stronger than women's is mixed: whereas, in a seminal study, Platek et al. (2002) found a sex difference, DeBruine (2004), who used a modified procedure, did not replicate it. Platek et al. (2004), however, reported not only a behavioral sex difference in response to self-resembling versus non-self-resembling child faces; men responded to self-resembling child faces with more overall brain activation (assessed through functional magnetic resonance imaging) than women, despite women generally exhibiting stronger brain responses to presentations of children's faces in general (see also Platek, Keenan, & Mohamed, 2005). These results are consistent with the view that men possess adaptation to discriminate and differentially invest in offspring as a function of phenotypic paternity cues. (One might ask how individuals could possibly assess self-resemblance ancestrally, prior to the invention of mirrors. Even lacking mirrors, however, individuals could assess familial resemblance.)

Potentially, males and infants have been subject to an antagonistic coevolutionary race, with infants selected to hide their paternity and males selected to discern it. Pagel (1997), for instance, speculated that babies may tend to look indistinctive as a result of adaptation to conceal cues of paternity (see also Alvergne, Faurie, & Raymond, 2007).

Men could potentially use other phenotypic cues to discriminate paternity. Prime candidates are scents associated with MHC alleles. Just as individuals discriminate individuals' attractiveness on the basis of MHC matching to themselves, so too may men discriminate likely paternity on the basis of MHC matching. Research has not yet assessed this possibility.

As we discussed earlier (chapter 4), men apparently adjust their investment in offspring as a function of their paternity certainty (see Anderson, 2006; Anderson, Kaplan, & Lancaster, 1999, 2007). More work is needed to identify precisely what cues they use to assess paternity, both in Western settings and in more traditional human groups.

Women's Counterresponses

Resistance to Mate Guarding and Fertility Detection

In response to enhanced levels of male mate guarding when female partners are in estrus, women may be expected to resist male mate guarding more intensively when they are fertile. And, indeed, as we described in chapter 11, during estrus women do indeed appear to engage more frequently in behaviors that undermine the efforts of male mates to mate-guard, remain vigilant to partners' activities, and attend female partners (Garver-Apgar, Cousins, et al., 2007).

More generally, in the preceding chapter we discussed female concealment of ovulation—active suppression of by-products of women's fertile state (e.g., by-products of reproductive hormone fluctuations)—and adaptive effects of women's estrus (attraction to men exhibiting particular traits, including extra-pair men with those traits). Adaptations to conceal fertility itself, to the extent that they truly have evolved in women (see chapter 11), should be thought of as a counterresponse to male paternity-enhancement and anticuckoldry adaptations. If men did not exert effort to protect paternity, as well as refuse to invest in offspring not their own, there would perhaps be no selective pressure on women to conceal estrus. As we also discussed in chapter 11, due to continuing coevolution of adaptations in each sex to counter those in the opposite sex, neither sex is expected to be fully adapted to the adaptations of the other. Imperfect adaptation in coevolving antagonists may characterize the design of men to detect fertility and women to disguise it.

Based on the presumption that men can detect the scent of a fertile-phase woman only at close distances, Marlowe (2004) concluded that women have "won" the coevolutionary race concerning detection or concealment of fertility. Though we agree that weak cues imply a coevolutionary struggle yielding adaptation, counteradaptation, counter-counteradaptation, and so on, it is not at all clear to us that women have "won." Men do apparently detect estrous scent, and, based on that cue or other valid cues, accordingly adjust the intensity of their mate guarding. Each sex appears to possess adaptations resulting from antagonistic coevolution, but neither has "won" the ongoing race to control conception. (To be precise, we should say that no one tactic on the part of one sex has trumped all opposing tactics in the other sex. As males and females have equal mean fitness when the sex ratio is 1:1, in principle no one sex can win a conflict of reproductive interests. Specific tactics [and alleles promoting them], however, can "win." See Arnqvist & Rowe, 2005.)

Resistance to Copulation

In species in which fertilization occurs within females' reproductive tracts, females can control which sperm reside within their tracts at any point in time via two precopulatory behavioral strategies. First, they can proceptively pursue copulation with desired males. We have argued that women may selectively do so, and particularly during estrus. Second, they can be differentially receptive to males' pursuits of copulation with them. Females of many species, including women, clearly resist many—probably the grand majority of—male attempts to woo them. In some instances, women resist their primary partners' efforts to copulate with them. (Females can, of course, also control retention of sperm via postcopulatory means of ejecting sperm.)

Women may be expected to resist male partners' desires to copulate with them contingently—as a function of their fertility status and, conjointly, as a function of their partners' indicators of quality. Specifically, women may selectively resist partners' attempts to copulate with them during the fertile period, with those paired to men possessing indicators of quality accepting their partners' sexual invitations with greater likelihood than those paired with other men. By contrast, women may less selectively resist partners' come-ons when infertile. In our study of couples with Garver-Apgar, we find mixed support for this conjecture. Garver-Apgar et al. (2006) did report that women were more likely to resist the sexual advances of their partners if their partners had similar (incompatible) MHC genes, and particularly so during estrus. Women paired with MHC-similar partners were also more likely than women paired with MHC-dissimilar men to have compliant sex with their partners—sex agreed to only after continual argument or threats of relationship dissolution. This effect, however, was not moderated by cycle phase. In unpublished analyses, we similarly found that women paired with asymmetrical men were, as expected, significantly more likely to resist their sexual advances and have compliant sex with them than women paired with symmetrical men, but the effect of symmetry was nearly equal during periovulatory and luteal phases. As our sample size was fewer than 100 women, additional research is needed before much can be said about the factors that differentially affect women's resistance to partner sexual advances during estrus and extended sexuality.

At the same time, we note that the theoretical expectations about changes in rejection rates across the cycle may not be as straightforwardly derived as we have suggested. Female rates of compliance and rejection of male sexual advances, if differing as a function of fertility and partner quality, could evolve to become cues of fertility status. That is, if females with partners of low quality do not resist sexual advances during extended sexuality but do resist them when in estrus, male partners could use partners' receptivity to their advances to infer partners' fertility status. Accordingly, we should perhaps expect women to maintain fairly regular rates of acceptance and rejection of male advances across the cycle. If they do so, women may adaptively vary rates of rejection as a function of male ability to deliver genetic benefits but maintain them at constant rates across the cycle. It is perhaps enlightening, then, that in our study both MHC similarity and male assymmetry had *main* effects on the number of male sexual advances female partners both rejected and complied with only after being pressured with arguments or threats.

Women's Assessments of Costs of Loss of Male Investment

The Costs of EPC Detection to Women and Implications for Additional Estrous Adaptations

We have characterized estrous sexuality in terms of adaptations designed to garner good genes for offspring. Much evidence speaks to the existence of those adaptations in women (see chapters 9 and 10). The fact that detection of EPC or extra-pair paternity by partners is often very costly to women, however, means that adaptive estrous sexuality should also reflect adaptations that take into account the costs of EPC. Specifically, women's attraction to men other than partners (or their willingness to act on attraction to men other than partners) should be sensitive to a variety of considerations: the risk of EPC detection, the risk of extra-pair paternity detection, the potential benefits of EPC, and the potential costs if EPC is detected. Put otherwise, estrous sexuality should generally function to enhance adaptive sire choice by females. One component of adaptive sire choice is choice of a partner who can deliver genetic benefits to offspring. But in pair-bonded species, in many instances the best sire for a woman's offspring is in fact the pair-bond mate, and not merely in instances in which the mate has good genes; the primary partner delivers nongenetic material benefits in a variety of currencies (see chapter 4) and loss of those benefits could have a drastic negative impact on a female's fitness.

In chapter 10, we discussed women's assessment of the potential benefits of EPC. Women's ability to garner genetic benefits through EPC depends on the relative value of genetic benefits offered by their primary partners, as reflected by their phenotypic indicators of intrinsic and compatible good genes. As we detailed in chapter 10, women's attraction to men other than primary partners indeed does appear to be sensitive to the indicators of intrinsic and compatible good genes that women's primary partners possess.

Women may possibly assess the benefits of EPC—or, more broadly, good genes—in other ways too. For instance, where parasites are highly prevalent (as assessed by rates of disease), the expression of genetic variability affecting broadly defined health may be particularly great, such that men's genetic quality varies substantially (e.g., Low, 1990). In more benign environments, the expression of fitness-related genetic variability may be weaker and the differences between men's genetic quality less substantial.

In addition to assessing benefits of EPC, however, women should assess costs. What do they have to lose if their partners detect EPC? Hassebrauck (2003) conjectured that, when fertile, women engage in detailed assessments of their relationships with primary partners—what he refers to as relationship "scrutiny"—more intensively than they do when infertile in their cycles. He administered a questionnaire that assessed scrutiny to normally ovulating women during the fertile phase and during the luteal phase. As predicted, he found that women scrutinize their relationships more intently when fertile.

Assessment of the Value of Men's Investment

Historically, the costs of EPC to women had much to do with what they had to lose if their partners refused to invest in the relationship or offspring. Women's willingness to engage in EPC should hence be sensitive to factors that affect loss of investment.

Relationship Satisfaction   A woman's satisfaction with her relationship partly reflects the extent to which she believes that her partner and the way he treats her give her a reasonably good "deal" given what she offers on the market. (In theory, women can be satisfied with relationships and still believe that alternatives provide them with even better outcomes; e.g., Thibaut & Kelley, 1959; empirically, however, relationship satisfaction reflects desire to maintain the relationship.) Predictably, then, the literature on extramarital affairs features relationship satisfaction as one of the best predictors of women's fidelity (see Thompson, 1983).

As we discussed in chapter 10, women's relationship satisfaction (as assessed by a standard measure) strongly and negatively predicted women's attraction to men other than primary partners in our study of couples (Gangestad et al., 2005a). It also predicted positively (albeit more moderately) women's attraction to their own partners. These relationships, however, were not moderated by phase of the cycle. Hence women who are satisfied with their relationships show *as much* of an increase in attraction to men other than partners when fertile as do estrous women dissatisfied with their relationships. During *both* phases, however, dissatisfied women report that they experience attraction to extra-pair men more frequently than do satisfied women. In stands to reason, then, that satisfied women would be less likely to actually stray.

That dissatisfied women are more likely to stray may be taken to mean that their EPCs do not have the function of seeking sires with good genes; rather, they could be looking to replace their current mates through EPC (e.g., Greiling & Buss, 2000). We do not doubt that women's EPC may serve a variety of different functions, including mate replacement. But it is a mistake to dismiss the possibility that many dissatisfied women's EPCs are nonetheless outcomes of adaptations that function to seek sires with good genes. A woman need not be looking to replace a mate by having an EPC when dissatisfied. She may simply feel that she has relatively little to lose if, in fact, her EPC is detected and she loses her current mate.

Familial Support

The cost of loss of a mate to a woman may partly depend on the availability of alternative sources of support. A woman who has a strong local familial system of support (e.g., in a society with matrifocal residence patterns) perhaps suffers less from loss of a mate than a woman who has no local support (e.g., in a society with patrifocal residence patterns).

Punishment for Infidelity   Female infidelity is punished more severely in some cultures than in others (Buss, 2000). Punishment can be severe—indeed, even death (Buss, 2000). In strong patrilineal societies in particular, males go to great lengths to assure paternity (Thornhill, 1991; Thornhill & Thornhill, 1987). (Not coincidentally, these cultures tend to have patrilocal residence patterns as well; as noted previously, patrilocality should encourage female fidelity.) No study examining changes in women's mate preferences or sexual interests across the cycle has been performed in a strongly patrilineal society in which female infidelity has extreme punitive consequences. In such cultures, do women not exhibit changes in sexual interests across the cycle? Or, alternatively, do they experience the typical changes in patterns of sexual attraction but inhibit any impulse to act on extra-pair attraction?

Parity   Women who are childless probably pay lower costs for loss of a mate than do women who are parous. In ancestral environments, a parous woman who was deserted by a mate potentially lost all (or nearly all) investment in her children by their father. And, though she could possibly replace her mate with another, her market value may have been diminished by her having children; quite simply, a man could not assume that her parental investment would be directed entirely to his children with her. Moreover, her new mate would not have invested in her existing children as much as their father would have (for studies of relatively low investment by stepfathers, see Daly & Wilson, 1988; Marlowe, 1999). Possibly, then, parous women could rarely if ever gain net benefits by engaging in EPC.

At the same time, the impact of sharing children with a female partner on the male partner's decision to divest must also be considered. Possibly, relative to a childless man, an ancestral man who shared children with a partner required a lower level of confidence that a newborn was his child to render desertion the optimal decision. His decision to desert would affect the well-being of his existing putative offspring, unlike the decision of a childless man to desert. From a game-theoretic standpoint, women's adaptations affecting their decisions to engage in EPC should have been affected not only by the costs to them should their partners desert but also by the probabilities that their partners *would* desert. And, from this perspective, it is not clear how selection should have shaped women's decisions to be contingent on parity.

Importantly, however, relatively little work on changes in women's mate preferences or patterns of sexual interests across the cycle has focused on parous women. In studies we have performed involving college students, we typically have only a few parous women. Our work in Dominica on male mate guarding across partners' cycles, by contrast, focused almost entirely on parous women.

We recently did additional statistical analysis to examine whether a variety of relationship features moderate the effects of cycle fertility on women's changing sexual interests across the cycle that we have documented in our studies with women: effects of estrus on women's attraction to extra-pair men, attitudes toward "opportunistic sex," and attraction to attractive male bodily features. Though most of the women in our studies using college samples are single and childless, our samples are heterogeneous. About a quarter of the men and women in our couples' study, for instance, cohabit, with about 15% being married. Women's self-reported levels of commitment to and investment in their relationships vary substantially. Some couples have children together. In fact, we find no evidence that any of the effects of estrus on women's sexual interests are weakened in women who cohabit with their partners, are married to their partners, are particularly committed to their relationships, invest substantially in their relationships, or have children with their partners (Gangestad et al., 2007b).

This analysis, however, does not address a concern recently expressed by Lancaster and Kaplan (in press). In traditional societies, women with existing offspring typically conceive the next offspring while nursing the previous one. (Lactation serves as an effective contraception the first year but not subsequently; on average, women in traditional populations lactate about 2.5 years; Sellen & Smay, 2001.) When women are nursing yet ovulating, they experience a richer hormonal environment than when they ovulate while not nursing. Estrogen and progesterone production support fertility, whereas prolactin and oxytocin support lactation. Possibly, the latter hormones affect the expression of estrous sexuality; indeed, possibly they suppress physiological mechanisms, of sexuality altogether. The possibility that they do suppress estrous sexuality at this point is conjectural. (For a view that the social bonding and sexual arousal systems are, in fact, quite separate and hence not under control of the same physiological mechanisms, see H. Fisher, 2004.) As no research has examined changes in women's sexuality across the ovulatory cycle of nursing yet cycling women, at present we simply do not know the effects of these hormones on estrus. (Based on normative cyclic variations in hormone levels, not measured hormones, Puts, 2006b, found some evidence that prolactin partly suppresses estrous preferences for men's vocal masculinity; by contrast, Garver-Apgar et al., 2008, detected no effect of prolactin on estrous preferences for male scents associated with body symmetry. These studies, however, did not involve nursing women.)

More generally, we still know very little about how women evaluate the nature of their relationships and the potential loss of their partners during the estrous phase. As, once again, women's fitness could have been deeply affected by loss of partner-delivered material resources of various sorts (e.g., provisioning, direct child care, protection), any complete understanding of women's estrous psychology, in terms of both proximate causation and function, must speak to this topic. Yet the topic has barely been addressed.

Human Sexually Antagonistic Coevolution in Modern Perspective

What Is the Extra-Pair Paternity Rate in Human Groups?

We have not yet addressed a question that our approach perhaps demands an answer to: Just what rate of extra-pair paternity has characterized human groups ancestrally? That is, how often have women engaged in EPC? And how often have EPCs conceived offspring?

Naturally, we cannot step back in time and measure the extra-pair paternity rate in ancestral human groups using sophisticated modern DNA techniques. To the extent that we can infer the extra-pair paternity rate in ancestral human groups, we must do so on the basis of the extra-pair paternity rate in modern groups. Anderson (2006) offers a comprehensive review of the literature. As he notes, few samples yield entirely unbiased estimates. Rates of extra-pair paternity in samples of men whose paternity is tested in laboratories because they challenge paternity clearly exceed the overall population rate. By contrast, samples of men who know paternity may be tested may be self-selected for high paternity confidence and hence yield underestimates of the overall population extra-pair paternity rate. Anderson (2006) separately treated estimates of nonpaternity (of the social father) in three categories: samples in which paternity confidence is relatively high, samples in which paternity was tested in a paternity testing lab and hence paternity confidence is low, and samples in which paternity confidence is unknown. As Anderson (2006) notes, the true nonpaternity rate in any population is probably some weighted average of high and low paternity confidence groups, probably much closer to that of the high-paternity-confidence group (as most men may well be fairly confident of paternity). (See also Simmons, Firman, Rhodes, & Peters, 2004.)

In most modern Western samples, the nonpaternity rates in high-confidence samples are typically 1–4%. It may be reasonable to guess that the true population-wide non-paternity rates are slightly higher—perhaps 2–5%. These rates are quite low, though not zero. Large surveys of married women in the United States estimate that between 15% (Laumann, Gagnon, Michael, & Michaels, 1994) and 70% (Hite, 1987) have had extramarital sex, with a median estimate (perhaps biased on the high side) being about 30% (for a review, see Thompson, 1983; also Kinsey, Pomeroy, Martin, & Gebbard, 1953; see Hansen, 1986, on extra-dyadic sex). The true level no doubt depends on the cohort selected, the number of years married (with risk probably increasing with relationship duration), and perhaps the subculture represented.

At the same time, extra-pair paternity rates are variable across populations. In some traditional or third-world samples, rates are estimated to be fairly high. For instance, in Monterrey, Mexico, the nonpaternity rate was estimated to be about 12%, and 20% within a low socioeconomic status subgroup (Cerda-Flores, Barton, Marty-Gonzalez, Rivas, & Chakraborty, 1999). Neel and Weiss (1975) estimated it to be 9% in the Yanamamo of Venezuela. (See also Hill & Hurtado, 1996, and Beckerman et al., 1998, on high rates of "secondary fathers" in the Ache of Paraguay and the Barí of Venezuela, respectively—men whom women name as nonprimary mates with whom they have had sex during pregnancy, traditionally recognized as contributors to paternity in a number of South Amerindian groups.) By contrast, Henry Harpending estimated the extra-pair paternity rate in the !Kung to be just 2% (as cited in Anderson, 2006.) Claims of high rates of nonpaternity (>10%) in India and Africa —albeit not well documented—have also been made (Anderson, 2006).

Our own reflections are the following. First, it is difficult to say precisely what the extra-pair paternity rate has been, on average, in ancestral hominin groups. We can be fairly confident that the rates were not zero (as zero rates have never been observed in any large human sample); on average, they probably totaled at least a few percent. But it is difficult to say much more about mean or median rates. Quite possibly, the rates have been quite low (< 5%). Second, the rates have probably varied across different groups faced with varying ecologies and socioecologies. Even if rates have centered across human groups at < 5%, it is not difficult to accept, based on extant data on modern groups, that they sometimes ranged upward of 10%. Indeed, some traditional groups have been reported by ethnographers to have relatively high rates of EPC (e.g., a number of societies in the Pacific islands, the Tiwi of Australia, South Amerindian groups such as the Canela; see, e.g., Crocker, 1984, 1990; Hart et al., 1987; see also Hartung, 1985). We have no estimates of extra-pair paternity rates in these particular societies.

Here, however, is a critical question: Does it matter to the claims about the role of EPC in selecting male and female adaptations we have made in this book? That is, does the extra-pair paternity rate in human groups, now and ancestrally, speak to whether women possess adaptation for EPC to obtain genetic benefits for offspring? Does it speak to whether men have counteradaptation in response to female EPC? Does it speak to whether the potential benefits of (for females) and threat of (for mates) female EPC have fueled sexually antagonistic coevolution across deep evolutionary time? In response, we reiterate a point we made early in this chapter: The extra-pair paternity rate within a species need not be a barometer of how intensively each sex has been selected to possess sexually antagonistic adaptations that function within the conflict over control of paternity of offspring. As we emphasized earlier, a species with a relatively low extra-pair paternity rate (say, 5%) need not have been characterized historically by less intense sexually antagonistic conflict over female EPC than a species characterized by a much higher rate (say, 25%). Through historical coevolution, each sex of both species may have accumulated an extensive array of "armaments" that function within the conflict, even if EPC rates have traditionally been low. That pattern we see in raptors: low rates of extra-pair paternity (typically, almost certainly lower than in human groups), yet extraordinarily high rates of copulation, with patterns that reveal signatures of design that the rates function, for males, to ensure paternity in the face of potential female EPC. If, indeed, the extra-pair paternity rate in human groups has been fairly low—5% or less—that fact may reflect the fact that the marginal utility of male paternal care has historically been relatively great in humans, leading to a stably low level of extra-pair paternity. But it does not imply weak sexually antagonistic selection (see also Arnqvist & Rowe, 2005; Møller, 2000).

As we emphasized in chapter 2, it is also important to recognize that current adaptiveness does not identify adaptation—only functional design demonstrates historically effective selection. Hence the questions of whether men possess anticuckoldry adaptation and whether women possess EPC adaptation are not answered by evidence of associated current reproductive success. They are answered by close and critical scrutiny of the signatures of historical selection found in design. That form of evidence speaks loudly in favor of both types of adaptation.

Can Genetic Monogamy Be Enforced by Paternity Detection and Divestment Threatened by Males Alone?

Earlier, we noted that some bird species are characterized by extraordinarily low extra-pair paternity rates, despite no evidence that male social partners ensure paternity through mate guarding or frequent copulation. In those species, male detection of paternity and threat to divest themselves of offspring not their own (especially in combination with low levels of genetic variability) may impose costs on female EPC sufficient to deter it and maintain genetic monogamy. And as a result, we noted, females in these species may have lost particular components of estrous sexuality. Male investment in offspring may well have been very important to offspring success in humans, too (e.g., Hill & Hurtado, 1996; Kaplan et al., 2000; see also chapter 4, this volume). Has male detection and threat to divest themselves of offspring not their own been sufficient to sustain low rates of extra-pair paternity in humans? That is, are humans similar in this regard to bird species characterized by genetic monogamy despite lack of costly male paternity guards?

Based on what we know to date, the answer is an emphatic no. Again, telltale signs of historical selection are to be found in the nature of the organisms selection has shaped.

And humans do not look like those bird species. First, females in those species appear to lack characteristic features of estrus. By contrast, human estrous sexuality is distinct from extended sexuality in many ways. During estrus, women's attraction to a host of male features is enhanced, women report greater attraction to extra-pair partners, their attraction to extra-pair partners is moderated by whether their partners possess favored features, and they report greater willingness to engage in opportunistic sex. Second, men do engage in costly paternity assurance tactics. Men mate-guard. They do so more intensively when partners are in estrus. They are interested in frequent copulation. And their motivation to copulate with their partners is affected by whether female partners had opportunity to engage in EPC since their last copulation. Third, females engage in behaviors that appear to resist male mate guarding and paternity assurance behaviors.

Could human pairs more effectively and efficiently produce successful offspring if, in fact, they did not have the sexually antagonistic adaptations we claim men and women possess? We have little doubt that they could. Male mate guarding imposes costs on both themselves (in currencies of time and energetic effort) and women (in a multitude of currencies, including risk of injury, but also constraints on independent movement and activity). If effort men expend to ensure paternity could be channeled into paternal care, they and their partners could more effectively and efficiently produce high-quality offspring. Once males and females follow a path of sexually antagonistic coevolution over deep evolutionary time, however, very often neither sex can afford to eliminate their sexually antagonistic adaptations. They pay fitness costs for doing so. And, hence, selection maintains those adaptations (or, very often, elaborates them).

That is not to say that humans could not and did not evolve to contingently express little in the way of sexually antagonistic behaviors under some specific circumstances. The evidence concerning women's estrus and male paternity guards speaks to averages; it cannot rule out contingent but meaningful and adaptive exceptions. In some situations, for instance, male paternity detection and threat of divestment of offspring could be sufficient to induce females to remain completely faithful in absence of paternity guards. We do not know, at this point, whether those circumstances exist and, if so, precisely what they are.

Intrauterine Sexually Antagonistic Coevolution

As we have noted already (chapters 5, 7, and 10), overt mate choice is not the final opportunity for females to exert influence over choice of a sire. Adaptations for cryptic choice are means by which females can, at least partly, control the fate of inseminations or conceptuses prior to committing large outlays of precious maternal investment (e.g., Eberhard, 1996; Thornhill, 1983). In chapter 7, for instance, we discussed the benefits of "inspecting" sperm for genetic compatibility postcopulation, as some aspects of compatibility will not be evident without allowing sperm to interact with female tissue in the reproductive tract. In chapter 10, we expressed skepticism that human females have been selected to mate with multiple males for the express purposes of ensuring ability to choose among sperm for compatible genetic quality, as may occur in some species (Zeh & Zeh, 2001). Nonetheless, it is quite possible—perhaps probable—that women do possess mechanisms that operate in utero to disfavor sperm or conceptuses containing either incompatible or intrinsically poor genes. Sperm or conceptuses that women themselves would do well to divest themselves of, however, may do well to strategically defend themselves. Hence the female reproductive tract is not just a cryptic arena for female choice; it is also yet another arena for sexually antagonistic coevolution.

Though conventional wisdom suggests that the maternal immune system presents a danger to foreign sperm (and, indeed, it clearly does), it now appears that a maternal immune response to sperm can facilitate favorable reproductive outcomes. Prior exposure to a man's sperm in the reproductive tract facilitates proper implantation of the zygote in the uterine wall (see Robertson et al., 2003). Maternal immune system recognition of paternal MHC alleles (and possibly other proteins) in sperm may lead to tolerance of them in a conceptus. Consistent with this interpretation, MHC allele sharing between partners is associated with lower couple-specific fertility (e.g., Black & Hedrick, 1997). The facilitating effect of maternal immune recognition of foreign MHC may be an adaptation to choose cryptically a compatible mate. This adaptation may operate not only in humans but also in close and phylogenetically distant relatives, such as mice (see Robertson, 2005). Other choice mechanisms may similarly operate to select sperm. In addition, selection in favor of immunologically familiar sperm may favor in-pair partners (over, e.g., coercive men), who would be more likely to paternally invest in offspring (see Davis & Gallup, 2006).

Once this cryptic choice system was in place, however, it seems unlikely that it would be without intersexual conflict. Even if, from the male's perspective, it would be better if one of his compatible sperm entered the egg than if one of his incompatible sperm did, there would exist some conditions in which male fitness would be hurt by biases against his sperm that were adaptive from the female's perspective. One might expect, then, male counteradaptations that undermine female attempts to disfavor sperm.

Seminal fluid contains a rich variety of immunomodulatory factors that apparently function to induce female tolerance of male antigens (e.g., Poiani, 2006). Broadly, the active immune system has two components: the cell-mediated system, which attacks viruses and bacteria that get inside cells, and the humoral system, which attacks extracellular bacteria, parasites, toxins, and other foreign bodies. Allocation of effort to one system detracts from the other. Some richly represented components of seminal fluid (e.g., TGF-ß, prostaglandins, cortisol) are not immunosuppressive so much as they are immunomodulatory. They bias expression of immunity within the female reproductive tract away from one arm of the female immune system and toward the other. Specifically, they enhance humoral immunity (e.g., Denison, Grant, Calder, & Kelly, 1999; Robertson et al., 2003) while suppressing cell-mediated activity. This bias has two effects. First, it increases immune system recognition of foreign bodies. Second, however, by suppressing destructive cell-mediated activity, it creates an environment conducive to sperm. Just as fetal and maternal interactions are, in some ways, coordinated in a way that regulates a pregnancy in favor of both, some seminal constituents may coordinate with female tissue to achieve effects that are beneficial overall to each. Conception and proper implantation is, in a grand sense, necessary for both male and female reproduction. At the same time, males benefit from greater recognition of antigens and simultaneous suppression of cell-mediated activity than is in the interest of females. It stands to reason that male seminal fluid has been selected to achieve greater effects than are in the interest of females, such that female counterresponses may be expected to evolve. That is, small differences in genetic interests between the sexes could drive a coevolutionary process through which male seminal products are selected for their powerful immunoregulatory effects on female immune function and the female reproductive tract is selected to counter these effects. Some males' seminal products (and, possibly, proteins on the sperm acrosome) may more effectively lead to conception and be sexually selected, whereas some females may be better able to control activities in their reproductive tracts in their own interests, leading to selection for resistance.

Interestingly, the female reproductive tract does appear to be biased away from humoral immunoactivity just prior to ovulation—precisely the time at which the male and female reproductive conflict of interest is in play (e.g., Franklin & Kutteh, 1999; Gravitt et al., 2003). This push-pull dynamic (males pushing humoral immunoactivity, females pulling away from it) is, potentially, a signature of antagonistic coevolution (see, e.g., Haig, 1993).

The situation we have described reflects chase-away sexual selection (e.g., Holland & Rice, 1999). Chase-away sexual selection occurs when one sex—in this case, males—are selected to produce an effect that increases their own fitness (here, through increasing access to valued female investment) but that compromises female fitness. In this instance, males suppress female immunoreactivity against their sperm. Females, in turn, are selected to reduce their sensitivity to the male-induced effect, in this case by biasing their own immunoreactivity in a direction counter to male interests. This selects for greater male efforts to achieve the effect, which selects for greater female counterresponse, and so on.

As Kokko et al. (2003) noted, however, over time chase-away sexual selection can transform into sexual selection for good genes. If male responses become exaggerated and costly to the point that male abilities to achieve them depend on their overall condition and quality, males of highest quality will be best able to achieve them. Hence the sexual selection system has become one that leads females to cryptically select as sires males with intrinsic good genes (see our related discussion of the evolution of signaling systems in chapter 5; see also Gangestad, Thornhill, & Garver-Apgar, 2005b). Possibly, female resistance to the immunosuppressive effects of men may ultimately be maintained by the fact that men whose semen is best able to overcome the resistance offer genetic benefits. If so, it functions as a cryptic choice mechanism for good genes.

We know of no strong tests of this proposal. We sketch this scenario in part to stimulate research. We also, however, wish to stimulate adaptationist thinking about sexual selection and sexually antagonistic coevolution revolving around events in women's reproductive tracts. Very little work has been done on these matters to date. (See also, however, Burch & Gallup, 2006.)

Male Induction of Ovulation

Jöchle (1973) argued that females may have adaptation to ovulate in response to sex with certain males (as in some other species; e.g., Smith, 1992). Recently, Preti, Wysocki, Barnhart, Sondheimer, and Seyden (2003) found that men's axillary sweat affects patterns of LH pulses of normally ovulating women, which reflect hypothalamic responses regulating fertility and ovulation. Potentially, selection could favor females whose ovulation is induced by male cues as a means of biasing paternity. This view requires that females are sensitive to signals that differentiate men (e.g., chemical cues associated with quality). Future research should explore what male features (if any) influence women's ovulation.

Males may benefit from affecting females' psychological responses as well. Male *Drosophila*, for instance, may benefit by suppressing their female mates' rate of remating (see chapter 7, this volume, and the first section in this chapter) and may be able to do so via neuroactive constituents of semen. Gallup, Burch, and Platek (2002) report that sexually active women who do not use condoms report less depression and more positive moods than their counterparts who use condoms. Future research may address the question of whether these associations are due to psychoactive effects of substances in male seminal fluid, as Gallup et al. (2002) suggest, and, if so, address the nature of the adaptations or by-products that give rise to these effects (see also Burch & Gallup, 2006).

Other Antagonistic Coevolutionary Races

Coevolution of Women and Coercive Men

Aside from the coevolutionary race between male and female partners over control of conception, other coevolutionary races emerge from women's estrus and its conceal-ment. One may be a race between fertile-phase females to avoid rape and males to focus sexual coercion toward females of peak cycle fertility. Women's mating effort during estrus may increase their risk of rape. Males should be under selection to use coercive sex when reproductive benefits exceed costs (or ancestrally did) and benefits are maximized when the victim is fertile. Men target women who are fertile based on *age* as victims of sexual coercion (see review in Thornhill & Palmer, 2000). The conception rate associated with rape has been estimated to be about 8%, compared with 2–4% for consensual unprotected matings. This high rate could be a by-product of women's activities (which expose women to greater risk of rape when fertile), but male adaptation to target fertile females for sexual coercion cannot be ruled out (Gottschall & Gottschall, 2003). (See Thornhill & Palmer, 2000, for a review of literature speaking to whether men possess adaptation for rape. As they state, no compelling evidence for the existence of such adaptations exists. But adaptation for coercive sex cannot be ruled out at this point, either.)

As mentioned earlier, women not taking hormonal contraception and in the high-fertility phase of the cycle engage in fewer behaviors that are considered to pose risks for rape than do other women (Bröder & Hohmann, 2003; Chavanne & Gallup, 1998). Miller (2003) found that estrous women are more wary of sexual coercion than women in the luteal phase. And Garver-Apgar, Gangestad, and Thornhill (2007) asked normally ovulating women to rate the sexual coerciveness of men interviewed for a potential lunch date, shown in videotapes. Relative to women in the luteal phase, women at estrus rated men as more sexually coercive overall (possibly reflecting a tendency to minimize false negative errors; e.g., Haselton & Nettle, 2006). Their ratings also tended to converge on the prototypic rating of women overall, which may be due to their more reliably tuning into consensually agreed-on cues of sexual coerciveness gleaned from men's demeanor.

Petralia and Gallup (2002) asked women to read a scenario and tested their hand-grip strength. Fertile-phase women who read a scenario with sexual assault content had greater strength than they exhibited at baseline. Fertile-phase women who read scenarios with other content did not. Neither did normally ovulating women in infertile phases or women using hormone-based contraception, regardless of content of the scenario they read. Possibly, women's increased testosterone at mid-cycle mediates this effect. In the Atlantic stingray, fertile females manifest a peak in both androgen level and hyperaggressivity toward males. Tricas, Maruska, and Rasmussen (2000) speculate that females in this species possess adaptation to control mate choice via increases in testosterone production and associated aggression. Rises in testosterone levels during the fertile phase are commonly observed in female vertebrates (e.g., Nelson, 2000). Whether these effects reflect estrous adaptation for mate choice in other species or humans is an open question begging to be answered (see, e.g., Møller, Garamszegi, Gil, Hurtrez-Boussès, & Eens, 2005).

More generally, concealment of women's fertility status could partly have been selected through antagonistic coevolution with coercive men, as well coevolution with primary partners.

Coevolution of Primary Partners and Other Men

Primary pair-bonded partners of women may antagonistically coevolve with other men. Male partners may be advantaged if they, but not other men, can detect their mate's estrus. The cultural conventions of women's modesty and related formal or informal rules of social conduct (e.g., rules that permit women to socially interact freely with men only if they are close genetic relatives), claustration, prescribed clothing to hide female ornamentation (e.g., veiling), and chaperoning may partly be outcomes of men's attempts to conceal from alliance partners cues of estrus in their mates (as well as, more generally, limiting opportunities for men to attract their female partners; for discussion, see Dickemann, 1981; Flinn, 1988; Hrdy, 1997; Lancaster, 1997; Low, 2001; Thornhill, 1991; Thornhill & Thornhill, 1987; Wilson & Daly, 1992).

Male or female relatives may also gain by controlling women's reproduction and hence also benefit from tracking women's fertility status. Features of these relatives may antagonistically coevolve with features of women, as well as features of men whose reproductive interests conflict with those of the relatives. Fathers guard reproductive-age daughters more than pre- and postreproductive daughters (e.g., Flinn, 1988). An open empirical question is whether they guard daughters most intently during their estrus.

Summary

Antagonistic coevolution between the sexes is one important and widespread form of coevolutionary arms race; it has produced sexually antagonistic adaptations that function to cope, although only temporarily and partially, with counteradaptation in the opposite sex. Extra-pair copulation leads to intersexual conflicts of interest in pair-bonding species such as passerine birds and *Homo sapiens*. Sexual conflicts of interest in these species arise from the cost of parental care in each sex and from cuckolded males investing maladaptively in other males' offspring. Male counteradaptations to extra-pair mating by a female partner include at least four important categories: mate guarding, frequent in-pair copulation, detection of female's fertile cycle phase, and detection of paternity of offspring.

An important feature affecting the sexually antagonistic race in species with biparental care is the extent to which male and female fitnesses depend on male parental investment. This feature varies across these species, and when paternal investment contributes importantly to offspring fitness, males should generally be expected to exhibit effective strategies for securing their investment in offspring, partly because females are expected to take few risks of losing male investment, partly because selection on males to ensure paternity should be relatively strong. Rates of female extra-pair copulation should tend to be lower in these species than in species in which male parental investment is less important to female fitness; patterns of extra-pair paternity in birds fits this expectation. Nonetheless, conflicts of reproductive interests do not evaporate; sexually antagonistic coevolution fueled by conflict over female extra-pair copulation persists. Hence, though extra-pair copulation and paternity rates vary across biparental investing species in systematic and meaningful ways, these rates do not speak directly to the intensity of past coevolutionary conflict between the sexes over female extra-pair mating. Similarly, they do not speak to the presence or absence of sexually antagonistic adaptations that, over deep-time evolutionary history, have been selected as a result of conflicts of interest. Hence, even in some bird species with very low extra-pair paternity rates, males exhibit design for paternity assurance (e.g., in the fantastically high rates of in-pair copulation in raptors).

Estrus may have been lost evolutionarily in certain species of pair-bonding birds in which evidence indicates that females do not engage in EPC for genetic benefits, despite a lack of male mate guarding and other paternity-assurance adaptations. If so, these species are exceptions to the general rule across the vertebrata that females possess estrous adaptations that function to acquire genetic benefits for offspring.

Similar to many male birds, men possess condition-dependent psychological adaptations for mate guarding. Men engage in mate guarding more frequently when their mates are in estrus than when their mates are infertile in their cycles. Men's interest in and manner of copulating are suggestive of anticuckoldry adaptations. Men, furthermore, appear to discriminate paternity based on physical resemblance to offspring and express greater interest in investing in children who exhibit cues of resemblance than in those who do not.

Women possess counteradaptations to male adaptations to prevent cuckoldry. Women engage in behaviors that undermine their partner's mate guarding more so when fertile than when infertile in the cycle. They appear to resist copulation with primary partners contingently and adaptively, though more studies on these phenomena are needed. As we discussed in chapter 11, women's concealed estrus may be an important counteradaptation. Finally, women may possess design to conditionally show interest in extra-pair men based on cues related to costs of lost male partner investment and likelihood of partner detection. This last category of counteradaptations may yield much insight into the nature of female dual sexuality, but, as yet, it has hardly been investigated.

Current extra-pair paternity and women's extra-pair mating rates are variable across human cultural settings. Whether observed in pair-bonding birds or humans, low rates are fully consistent with a deep-time history of intersexual arms races and associated effective selection for sexually antagonistic adaptations. Researchers should not be misled into inferring from low rates of extra-pair paternity that sexually antagonistic coevolution has been unimportant historically and antagonistic adaptation largely absent.

A variety of coevolutionary arms races pertaining to conflicts over female control of sire choice likely characterized human evolutionary history, including conflicts involving women and sexually coercive men, male partners and extra-pair men, and women and their relatives.

13   Reflections

What can we conclude about the evolution of human female sexuality? In this chapter, we reflect on this question and address it in broad strokes. All of our answers, we emphasize, are provisional. Their function is not to declare firm truths (even if, indeed, we would bet on many ultimately proving to be true). Rather, they offer frameworks that we hope can generate fruitful future research.

The Need for Broad Comparative Frameworks

At its most general level, our book is not merely about women's sexuality. Because women and men's sexualities coevolve, we have also addressed men's sexuality. But even more generally, the book addresses vertebrate reproductive biology and behavior. Indeed, more than a third of the book discusses reproductive biology, with no explicit reference to humans. Of course, most work on the biology of human sexuality applies general evolutionary principles and models. If our book is of any value, however, we hope that it illustrates the scientific utility of viewing humans in a comparative perspective even broader than is typically applied to generate research and frame questions about the evolutionary history of human behavior and psychology. If not understood in light of vertebrate estrus, for instance, women's mid-cycle sexuality consists of a set of intriguing findings, but without phylogenetic roots and comparisons that allow strong inference about ultimate causation. When a strong, functional comparative framework is applied, women's estrus, its concealment, and the residual cues that men use to detect and respond to estrus can be identified. Women and men's ornamentation is similarly illuminated when viewed under a broad comparative lens.

In fact, however, the comparative perspective we applied was not generated completely independently of findings on women's sexuality. Indeed, some of the arguments about vertebrate reproductive biology were informed by our reflecting on the conceptual implications of findings on women's sexuality. Humans are an interesting evolved case, and much can be learned about nonhuman sexuality by understanding human sexuality. As we noted in chapter 9, perhaps more is now known about the nature of the changes in sexual interests and preferences across the reproductive cycle in women than in most any other vertebrate female. Our provisional conclusions—again, working hypotheses—do not pertain only to women; we offer novel frameworks for understanding the broad nature of vertebrate reproductive biology. Again, perhaps the prime illustration is our perspective on the function of sexual motivation during the fertile phase of vertebrate females to obtain sires of high genetic quality.

Throughout our treatment, we recognized two distinct, ultimate causal categories for the existence of a given trait in an organism: first, phylogenetic origin of a novel phenotype and, second, maintenance after origin. Phylogenetic origin is caused by the developmental modification of a preexisting trait. Maintenance after origin is caused by evolutionary agents. The method of adaptationism is the method that identifies adaptations and by-products; it cannot identify phylogenetic origin on the Tree of Life. At the same time, causes of phylogenetic origin cannot explain the phylogenetic persistence of traits.

The two evolutionary causal categories are not just distinct; they are complementary. Naturally, a complete evolutionary understanding of any phenotypic trait requires knowledge of both origin and maintenance. But, moreover, valid phylogenetic inference about common ancestry requires evidence of homology, which itself entails discriminating homology from analogy. And the nature of trait similarity that distinguishes homology from convergence can be discovered only through detailed adaptationist analysis of phenotypic similarity. Hence, just as we need to use in tandem both phylogenetic and adaptationist methods to understand any evolved trait fully and conclusively, so too will both, in conjunction, offer the most productive avenues toward an understanding of human features.

Estrus

By 1960, scholars concluded that estrus had been evolutionarily lost in the lineage now referred to as Tribe Hominini. As we argued, women's estrus has not been lost and never was—except in the confusion of what researchers thought estrus truly is. For many years, a common dominant view of mammalian estrus has been that it functions to obtain sperm and, relatedly, to signal to males the need for sperm—and that notion remains prevalent. After the discovery of estrogen-dependent ovulation and its relation to estrus in nonhuman mammals, researchers found that, at mid-cycle, women do not exhibit greatly enhanced sexual motivation toward their partners or in general. Moreover, male partners do not act anything similar to sexually ardent bulls or squirrels that detect estrous females. From these observations and the then-standard understanding of estrus, women were declared to be estrus-free. Lost estrus, in turn, was claimed to be, observationally and functionally, equivalent with concealed ovulation. Lost estrus hence led scholars to try to understand why women concealed their ovulation.

Only in the past two decades has good-genes mate choice been considered valid theoretically and empirically supported across a variety of species. When the function of estrus is understood to be obtaining not merely any sperm but sperm containing DNA that will enhance the fitness of offspring, new predictions about women's fertile-phase sexuality arise. During this phase, women should not experience generalized or indiscriminate increases in libido. Instead, they should sexually prefer male markers of putative high genetic quality and compatible genes. They should be particularly attracted to men possessing these markers when their primary partner lacks them. And they should be more willing to engage in noncommittal, "opportunistic" sex. Finally, they should be more avoidant and resistant to matings with males of putative low genetic quality. Women outside the fertile phase of their cycles and women using hormonal contraception should possess these features to lesser degrees. When combined with intersexual conflict theory, this idea furthermore predicts counteradaptation in male partners to detect estrus and respond in ways that enhance paternity. All of these predictions are supported by numerous studies, virtually all of which were conducted in just the past decade.

Provisionally, then, we conclude that women do possess estrus, and it is functionally organized (in part) to obtain good genes for offspring, including contingently by extra-pair copulation. Women's psychological and behavioral shifts during estrus cannot be explained as by-products of generally enhanced sexual motivation at the fertile cycle phase or as by-products of women's preferences at infertile cycle times.

Estrus, we emphasize, is not designed for obtaining sperm from just any male. Neither does estrus function to signal fertility to males, who then provide sperm. The ability to obtain sperm only rarely limits female reproduction. Sexual selection on males produces male adaptations that function to use by-products of female fertility status to encounter and inseminate females at the fertile phase of their reproductive cycles. In general, across organisms, evolved sexual ardor in males eliminates selection on females for signaling fertility in the cycle. The still-widespread view that females commonly signal their cycle fertility (that is, by design) stems from two errors. First, by-products that males are designed to use to detect female fertility have been uncritically labeled female signals. Second, when females possess signals surrounding cycle fertility, the signals have uncritically been thought to be fertility signals, when in fact they are typically signals that function as honest signals of phenotypic and genetic quality. Both errors, we suggest, can be avoided if criteria that distinguish adaptations from by-products and identify the function of adaptations are critically applied (in the case of the second error, with aid from modern signaling theory).

Estrus, we have argued, is homologous across mammals. We know of no convincing evidence of its loss in any mammal species. Fertile-phase female sexual behavior has the same fundamental physiological basis throughout the Mammalia. This homology is accompanied by similar functional design of estrus across Mammalia: good-genes sire choice. In fact, however, these features are not homologous across Mammalia alone; they first appeared in a very early vertebrate and characterize species throughout Vertebrata. After estrus's origin, it has been widely maintained by selection for its effect of good-genes sire choice. Lineage-specific selection in the context of sire choice shaped the lineage-specific design of estrous sexuality across mammal species and other vertebrate taxa.

Conceivably, estrus has been lost evolutionarily in some taxa when its benefits are exceeded by costs. Certain pair-bonding bird species in which extra-pair copulation and male mate guarding are very uncommon may lack estrus because of its cost to females of lost male parental care and/or because of the absence of adaptive preference for good genes.

Within lineages, estrus may also be absent conditionally—present in some generations or populations and not in others. Given the costs of estrus, selection shaped its behavioral expression to be conditional and to be present only when benefits exceed costs. The design of concealed estrus reflects conditional expression of behavioral estrus of a sort. But conditionality may also be reflected in cases occurring on an ecological time scale on which female sexual interest in good genes is maladaptive temporarily or male parental care is paramount.

Extended Sexuality

Buss (1989) and Symons (1979) showed that women possess functional design (in the form of mate preferences and sexual motivation) for gaining nongenetic material benefits and services from men (through psychological features of mate preferences and sexual motivation). Hrdy (1979, 1981) demonstrated that female sexuality to obtain material benefits is phylogenetically ancient in Old World primates. We label the female sexual adaptations identified by these three scholars and studied by many others "extended female sexuality." It is female sexual proceptivity and/or receptivity outside the fertile phase of the estrous cycle and during all other times at which fertility is low or zero. Extended female sexuality is distinct from polyandry, but females may be polyandrous during this phase. Women have two functionally distinct sexualities. Estrus functions to identify good sires (which may or may not be the primary partner). Extended sexuality functions to secure nongenetic material benefits from males. Women's extended-sexuality mate preferences possess design features different from estrous mate preferences. Although women's preferences for stable pair-bond partners are stable across the cycle (and function to favor both genetic and material benefits), women's preferences for sex partners during estrus focus on traits that connote possession of good genes.

As many scholars have noted, women are continuously sexually receptive across the menstrual cycle. Contrary to some writings, however, continuity does not reflect loss of human estrus. Women possess dual sexuality: two functionally distinct sexualities.

As predicted, extended female sexuality occurs largely in taxa in which males deliver material benefits to females. It is common in Old World primates and by convergent evolution in pair-bonding biparental birds. Extended female sexuality, in those species that possess it, coexists with estrus, except in the rare species that may have lost estrus. This point is illustrated in studies of primates and birds with extended sexuality. Estrus, however, does not require the presence of extended sexuality. Indeed, most species of female vertebrates have only estrous sexuality.

Women's degree of extension of sexuality exceeds that of females of any other known species. A rich array of nonconceptive female sexual adaptation in humans has coevolved with the importance of male-provided material benefits subsidizing female reproductive activity, which ancestrally permitted females to dedicate more energy and time to gestation, lactation, and other forms of maternal care and to therefore accelerate the birth rate and/or reduce infant mortality. Though male efforts to subsidize female reproductive activity in traditional populations (notably, hunting) function as paternal care, they also probably function as mating effort.

Signals

Women's facial and bodily ornaments (typically, exaggerations of otherwise functional components of anatomy), we argue, are honest signals of phenotypic and genetic quality. That female ornaments in nonhuman Old World primates function as signals of fertility in the cycle is a widespread view. It has led to claims that women's ornaments are deceptive, permanent signals of cycle fertility, which (like continuous sexuality) functions to conceal estrus. We argue, along with some other scholars, that female sex skins and swellings of many Old World primates are signals, but not of cycle fertility; instead, they are signals of quality (whether immediate or long term). These signals function to obtain male-provided nongenetic benefits (though they may also result in better sires). Though they possess overlapping functions, women's ornaments and extended sexuality reflect distinct adaptations. The permanence of women's ornaments, from their maximum exaggeration during the low-fertility life-history phase of adolescence through the years of offspring production, testifies to the importance of long-term pair bonding and of male material-benefit delivery to females in human evolutionary history. Compared with other female primates with signals of quality during adolescence, adolescent human females are more sexually motivated and attractive to males, features that also constitute telltale signs of the importance of male material subsidy for enhancing the reproduction of ancestral hominin females.

Women's ornaments are ultimately the product of sexual selection acting on females. Both male mate choice and female-female competition for male-provided material benefits shaped women's ornaments. Aspects of feminine behavior, including voice, function similarly and reflect, in the functional sense of the word, ornamentation. Women's ornaments in general appear to be facilitated by estrogen during their development and may be dependent on estrogen for their maintenance. Estrogen regulates the trade-off between female reproductive effort and somatic effort. Hence degree of female ornamentation honestly reflects the amount of available energy reserve and other bodily resources allocated to and available for current reproduction—reproductive capacity. Estrogenized fat depots of women function to store fat for gestation and lactation. Because of their information value to males, however, these markers became involved in a competitive female signaling system based on honest phenotypic quality, which led to their elaboration. Women's ornaments appear to be largely redundant signals of reproductive capacity.

Women's phenotypic effects that are associated with high estrogen during estrus are not signals of cycle fertility. Many if not all of these effects are by-products of estrogen. Men have evolved to use them to identify fertility within the cycle. Possibly, as well, women's enhanced attractiveness to men during estrus may constitute incidental effects of men's greater attraction to women's signals of reproductive capacity.

One traditional interpretation of women's loss of sexual swellings is that it reflects women's loss of estrus. That interpretation is not correct. In fact, women do have estrus. Moreover, the common ancestor of chimpanzees and hominins did not have elaborate sex swellings; these features evolved in the chimpanzee lineage after its divergence from Hominini. Finally, female sex skins and swellings, in our view, function to signal reproductive capacity, not cycle fertility.

Men's ornamentation similarly reflects a history of sexual selection and honest signaling of personal phenotypic and genetic quality. It appears to be largely dependent on testosterone, which regulates the trade-off between men's mating effort and other forms of effort. Men's various bodily and behavioral ornaments are covarying signals of phenotypic and genetic quality. Women are more sexually attracted to them during estrus than during extended sexuality.

The additive genetic variance in major fitness components, and hence in overall fitness, appears to be significant and widespread across many species, including humans. Developmental stability often has significant additive genetic variance. The maintenance of the genetic variation in fitness is a result of a combination of processes. The maintenance of the considerable variation among men and women in their ornamentation partly reflects additive genetic variation in fitness due to mutation-selection balance and a variety of coevolutionary antagonistic races. Nonadditive genetic variation is important in mate choice by both sexes. Compatible good genes play a significant role. Women may possess adaptation that leads them to favor mates that diversify genetic makeup in particular ways within a family, as well.

One area in need of further research concerns the extent to which heritable fitness variation covaries negatively across the sexes as a function of sexually antagonistic loci. If males who posess genetic complements favor sons at a cost of disfavoring daughters or vice versa, sexual selection for indicators of male intrinsic good genes can be limited. Studies suggest negative intersexual correlations in heritable fitness in several species (though these estimates vary widely). As evidence on humans reveals telltale signatures of sexual selection for indicators of intrinsic good genes, we provisionally assume that the intersexual negative covariation in heritable fitness across the sexes in our species is not sufficient to significantly suppress the strength of this selection. Still, definitive research into the nature and extent of this covariation is needed.

Extra-Pair Copulation and Sexually Antagonistic Coevolution

Pair-bonding biparental birds' mating systems are similar to that of humans in many respects. In particular, estrus occurs in the context of resource flow to offspring arising through parental efforts of a male partner that, potentially, is of low genetic quality. Estrous female birds in some species appear to engage in extra-pair copulation with males of higher genetic quality. This pattern reflects female adaptation for obtaining good genes for offspring through extra-pair copulation. Reproductive conflicts of interest arise when females engage in extra-pair copulation for good genes because of the cost of cuckoldry to males. These conflicts fuel sexually antagonistic coevolutionary races. Because a coevolutionary race may result in extra-pair paternity rates that are high or low, by no means is a low (but nonzero) rate of extra-pair paternity a signature of the absence of an historical coevolutionary race. Even intense intersexual conflicts (ones that produce antagonistic adaptation in both sexes) may result in low rates. Instead, low rates often tend to reflect the costs to males of delivering material benefits to females (generating relatively effective strategies for enhancing paternity) and the benefits to females of receiving those benefits. In humans, extra-pair paternity rates tend to be low but variable across groups and places.

In some pair-bonding birds, male primary partners' quality affects extra-pair copulation by females. Estrous females paired to low-quality males are more likely than other females to copulate outside the pair bond and with a male of higher quality than the main partner. Evidence on humans suggests a similar pattern with respect to female attraction to extra-pair males. Estrous women partnered to men who are physically unattractive, who evidence developmental instability, or who possess genetic incompatibilities (at major histocompatibility complex loci) are more strongly attracted to extra-pair men compared with estrous women partnered to men without these features.

These patterns in estrous women are consistent with the view that women possess estrous adaptations that were shaped, ancestrally, in the context of extra-pair copulation. Other evidence for these adaptations include the fact that women's mate preferences for putative signals of quality are enhanced at estrus in the context of women evaluating sex partners but not long-term mates, weakened commitment by estrous women to their primary partners, and greater openness to opportunistic sex with attractive men during estrus.

Concealed Estrus

Women possess estrus, but they also possess adaptation to conceal it. Contrary to traditional views, concealed estrus (typically referred to as *concealed ovulation* in the literature) is not the loss of estrus. Concealed estrus is not equivalent to extended female sexuality (though concealed estrus may not be possible without extended sexuality). The permanently displayed estrogen-facilitated sexual ornamentation of women or the evolutionary loss of sexual swellings does not constitute concealed estrus. Finally, concealed estrus is not concealed ovulation. Ovulation marks the beginning of the infertile luteal phase. The estrous phase precedes ovulation.

Concealed estrus must also be distinguished from undisclosed estrus. Concealment of estrus involves direct selection for traits that hide estrus. Undisclosed estrus is the relative absence of cues of cycle fertility, not reflective of direct selection to hide estrus. Disclosed estrus is the presence of cues of the fertile phase. Almost never does disclosure reflect signaling of cycle fertility. Rather, it reflects detectable by-products of female reproductive machinery that mark the fertile phase (e.g., increases in estrogen). The extent to which estrus is disclosed or undisclosed varies along a continuum in vertebrate species.

As is widely acknowledged, women at the fertile phase of their cycles are different from other female mammals when fertile. The differences, however, do not reflect features due to the presence versus absence of estrus. Many differences derive from the extent to which estrus is concealed. Differences between men's interest in women when fertile and male chimps, opossums, or coyotes when a conspecific female is fertile similarly reflect the extent to which estrus is concealed. Women appear to possess adaptation to conceal estrus. That is, ancestral direct selection favored hominin females who hid the hormonal, scent, behavioral, and emotional correlates of high fertility in the cycle. The concealment of behavioral estrus is conditional on the audience; it is less concealed when the audience is a partner who possesses putative markers of quality.

Of the numerous hypotheses that have been offered to explain women's concealment of cycle fertility, only the cuckoldry hypothesis appears consistent with recent research results. Others are seriously questioned by the presence of estrus in women, men's detection and responses to it, and women's adaptations for extra-pair copulation to obtain sires of high genetic quality. The cuckoldry hypothesis states that concealment evolved in the context of adaptive female extra-pair copulation with mates of high genetic quality to reduce costs of male efforts to ensure paternity and to maintain the primary partner's investment.

Concealment of estrus does not imply the complete absence of cues that males can use to ascertain fertility status. As noted, females typically emit by-products of the physiological machinery leading to ovulation. In many species, males are selected to detect these cues. Concealment need not lead to a complete suppression of fertility cues. Indeed, because of the information value of those cues to males, selection on females to suppress them may be accompanied by selection on males to be increasingly sensitive to them. A sexually antagonistic coevolutionary race may ensue. Multiple studies demonstrate that men are more attracted to women during the fertile phase of their cycles. Again, however, that males can detect estrus does not reflect female signaling of fertility status. Indeed, it does not imply that females have not been selected to conceal those cues. The extent to which selection on females can suppress cues of cycle fertility is constrained by the costs that concealment entails by interfering with the machinery responsible for cycle fertility (e.g., the effects of estrogen).

Men not only detect fertility in the cycle but also possess an arsenal of counteradaptations in response to threat of extra-pair copulation by estrous partners. The intensity of men's mate-guarding efforts is enhanced when primary partners are in estrus. Some evidence suggests that men's testosterone levels are diminished during this time, perhaps to facilitate attention to their primary partners. Women, in turn, appear to engage in greater efforts to resist their partners' mate-guarding efforts during estrus.

One area in need of future research concerns female estrous adaptations to assess not only the benefits of extra-pair copulation (e.g., partners' abilities to deliver genetic benefits to offspring) but also its costs. Because men may withdraw delivery of material benefits when they perceive cues of partners' infidelity, women should possess adaptation to assess those costs. More generally, estrous women should possess adaptation to identify the sire who will best benefit them. In circumstances in which loss of delivery of material benefits from partners is very costly, the best sire may be a woman's own partner even when that partner does not possess the ability to deliver genetic benefits. In some species, females may have lost estrus for this reason. Women do possess estrus, but it may be conditional. The little evidence that is available suggests that women's estrus is not diminished as a function of the length of, their commitment to, or satisfaction with their relationships or even by their already having had children with their partners. But much more data, particularly pertaining to the effects of caring for children with partners, are needed before any definitive conclusions can be drawn.

Concealed estrous adaptation may exist in some nonhuman species with pair bonding or simply male-female consort formation around estrus as a result of direct selection of concealment to promote adaptive extra-pair copulation for good genes. In some nonhuman species, concealed estrus potentially may be favored by direct selection in the absence of cuckoldry when males are especially astute at detecting undisclosed estrus, when males deliver important material benefits to females, and when females exhibit extended sexuality to obtain male-provided material benefits. In these circumstances, concealed estrus may facilitate the ability of females to maintain sire choice through estrus and yet maintain the flow of male-delivered material benefits through extended sexuality.

REFERENCES

Abitbol, J., Abitbol, P., & Abitbol, B. (1999). Sex hormones and the female voice. *Journal of Voice* 13: 424–446.

Acton, W. (1857). *Functions and Disorders of the Reproductive Organs in Youth, in Adult Age, and in Advanced Life.* Churchill, London.

Adkins-Regan, E. (2005). *Hormones and Animal Social Behavior.* Princeton University Press, Princeton, NJ.

Aeschliman, P. B., Haberli, M. A., Reusch, T. B. H., Boehm, T. & Milinski, M. (2003). Female sticklebacks *Gasterosteus aculeatus* use self-reference to optimize MHC allele number during mate selection. *Behavioral Ecology and Sociobiology* 54: 119–126.

Agostoni, C., Marangoni, F., Bernardo, L., Lammardo, A. M., Galli, C., & Riva, E. (1999). Long-chain polyunsaturated fatty acids in human milk. *Acta Paediatrica* 88: 68–71.

Aigaki, T., Fleischmann, I., Chen, P. S., & Kubli, E. (1991). Ectopic expression of sex peptide alters reproductive behavior of female *Drosophila melanogaster. Neuron* 7: 557–563.

Akçay, E., & Roughgarden, J. (2007). Extra-pair paternity in birds: Review of the genetic benefits. *Evolutionary Ecology Research* 9: 855–868.

Alberts, S. C., Altmann, J., & Wilson, M. L. (1996). Mate guarding constrains foraging activity of male baboons. *Animal Behaviour* 51: 1269–1277.

Alberts, S. C., Buchan, J. C., & Altmann, J. (2006). Sexual selection in wild baboons: From mating opportunities to paternity success. *Animal Behaviour* 71: 1177–1196.

Albrecht, T., Kreisinger, J., & Piálek, J. (2006). The strength of direct selection against female promiscuity is associated with rates of extrapair fertilizations in socially monogamous songbirds. *American Naturalist* 167: 739–744.

Alcock, J. (2001). *The Triumph of Sociobiology.* Oxford University Press, New York.

Aldridge, R. D., & Duvall, D. (2002). Evolution of the mating season in the pit vipers of North America. *Herpetological Monographs* 16: 1–25.

Alexander, R. D. (1979). *Darwinism and Human Affairs*. University of Washington Press, Seattle, WA.

Alexander, R. D. (1990). *How Did Humans Evolve? Reflections on the Uniquely Unique Species*. Special Publication No. 1. Museum of Zoology, The University of Michigan, Ann Arbor.

Alexander, R. D., & Borgia, G. (1979). On the origin and basis of the male-female phenomenon. In *Sexual Selection and Reproductive Competition in Insects* (eds. M. S. Blum & N. A. Blum), pp. 417–439. Academic Press, New York.

Alexander, R. D., & Noonan, K. M. (1979). Concealment of ovulation, parental care, and human social evolution. In *Evolutionary Biology and Human Social Behavior: An Anthropological Perspective* (eds. N. A. Chagnon & W. G. Irons), pp. 436–453. North Duxbury Press, Scituate, MA.

Altmann, J., Altmann, S. A., Hausfater, G., & McCuskey, S. A. (1977). Life history of yellow baboons: Physical development, reproductive parameters, and infant mortality. *Primates* 18: 315–330.

Alvergne, A., Faurie, C., & Raymond, M. (2007). Differential facial resemblance of young children to their parents: Who do children look like more? *Evolution and Human Behavior* 28: 135–144.

American Heritage Dictionary of the English Language, 4th edition. (2000). Houghton Mifflin, New York.

Amundsen, T. (2000). Why are female birds ornamented? *Trends in Ecology and Ecology* 15: 149–155.

Amundsen, T., & Forsgren, E. (2001). Male mate choice selects for female coloration in a fish. *Proceedings of the National Academy of Sciences USA* 98: 13155–13160.

Amundsen, T., Forsgren, E., & Hansen, L. T. T. (1997). On the function of female ornaments: Male bluethroats prefer colourful females. *Proceedings of the Royal Society of London B* 264: 1579–1586.

Amundsen, T., & Pärn, H. (2006). Female coloration in birds: A review of functional and nonfunctional hypotheses. In *Bird Coloration*, Vol. 1: *Mechanisms and Measurements* (eds. G. E. Hill & K. J. McGraw). Harvard University Press, Cambridge, MA.

Andelman, S. J. (1987). Evolution of concealed ovulation in vervet monkeys (*Cercopithecus aethiops*). *American Naturalist* 129: 785–799.

Anderson, C. M., & Bielert, C. (1994). Adolescent exaggeration in female catarrhine primates. *Primates* 35: 283–300.

Anderson, J. L. (1988). Breasts, hips and buttocks revisited. *Ethology and Sociobiology* 9: 319–324.

Anderson, K. G. (2006). How well does paternity confidence match actual paternity? Evidence from worldwide nonpaternity rates. *Current Anthropology* 47: 513–520.

Anderson, K. G., Kaplan, H., & Lancaster, J. (1999). Paternal care by genetic fathers and stepfathers. I: Reports from Albuquerque men. *Evolution and Human Behavior* 20: 405–431.

Anderson, K. G., Kaplan, H., & Lancaster, J. (2007). Confidence of paternity, divorce, and investment in children by Albuquerque men. *Evolution and Human Behavior* 28: 1–10.

Andersson, M. (1994). *Sexual Selection*. Princeton University Press, Princeton, NJ.

Andrews, P. W., Gangestad, S. W., & Matthews, D. (2003). Adaptationism: How to carry out an exaptationist program. *Behavioral and Brain Sciences* 25: 489–504.

Argiolas, A. (1999). Neuropeptides and sexual behaviour. *Neuroscience and Biobehavioral Reviews* 23: 1127–1142.

Arkush, K. D., Geise, A. R., Mendonca, H. L., McBride, A. M., Marty, G. D., & Hedrick, P. W. (2002). Resistance to three pathogens in the endangered winter-run chinook salmon (*Oncorhynchus tshawytscha*): Effects of inbreeding and major histocompatibility genotypes. *Canadian Journal of Fisheries and Aquatic Sciences* 59: 966–975.

Arnqvist, G., & Kirkpatrick, M. (2005). The evolution of infidelity in socially monogamous passerines: The strength of direct and indirect selection on extrapair copulation behavior in females. *American Naturalist* 165: S26–S37.

Arnqvist, G., & Nilsson, T. (2000). The evolution of polyandry: Multiple matings and female fitness in insects. *Animal Behaviour* 60: 145–164.

Arnqvist, G., & Rowe, L. (2005). *Sexual Conflict.* Princeton University Press, Princeton, NJ.

Arroyo, B. E. (1999). Copulatory behavior of semi-colonial Montagu's Harriers. *Condor* 101: 340–346.

Babbitt, G. A. (2006). Inbreeding reduces power-law scaling in the distribution of fluctuating asymmetry: An explanation of the basis of developmental instability. *Heredity* 97: 258–268.

Bagley, K. R., Goodwin, T. E., Rasmussen, L. E. L., & Schulte, B. A. (2006). Male African elephants, *Loxodonta africana,* can distinguish oestrous status via urinary signals. *Animal Behaviour* 71: 1439–1445.

Baird, D. D., McConnaughey, R., Weinberg, C. R., Musey, P. I., Collins, D. C., Kesner, J. S., Knecht, E. A., & Wilcox, A. J. (1995). Application of a method for estimating day of ovulation using urinary estrogen and progesterone metabolites. *Epidemiology* 6: 547–550.

Baird, T. (2004). Reproductive coloration in female collared lizards, *Crotophytus collaris,* stimulates courtship by males. *Herpetologica* 60: 337–348.

Bajema, C. J., ed. (1984). Evolution by sexual selection theory: Prior to 1900. *Benchmark Papers in Systematic and Evolutionary Biology,* Vol. 6. Van Nostrand Reinhold, New York.

Baker, A. J., Dietz, J. M., & Kleiman, D. G. (1993). Behavioural evidence for monopolization of paternity in multi-male groups of golden lion tamarins. *Animal Behaviour* 46: 1091–1103.

Baker, R. R. (1997). Copulation, masturbation and infidelity: State-of-the-art. In *New Aspects of Human Ethology* (eds. A. Schmitt, K. Atzwanger, K. Grammer, & K. Schafer), pp. 163–188. Plenum Press, New York.

Baker, R. R., & Bellis, M. A. (1995). *Human Sperm Competition: Copulation, Masturbation and Infidelity.* Chapman & Hall, London, UK.

Ballentine, B., Badyaev, A., & Hill, G. E. (2003). Changes in song complexity correspond to periods of female fertility in blue grosbeaks (*Guiraca caerulea*). *Ethology* 109: 55–66.

Bancroft, J. (1987). Hormones, sexuality and fertility in women. *Journal of Zoology* 213: 445–454.

Bancroft, J. (2002). Biological factors in human sexuality. *Journal of Sex Research* 39: 15–21.

Barash, D. P., & Lipton, J. E. (2001). *The Myth of Monogamy: Fidelity and Infidelity in Animals and People.* W.H. Freeman, New York.

Barber, N. (1995). The evolutionary psychology of physical attractiveness: Sexual selection and human morphology. *Ethology and Sociobiology* 16: 395–424.

Bartosch-Härlid, A., Berlin, S., Smith, N. G. C., Møller, A. P., & Ellegren, H. (2003). Life history and the male mutation bias. *Evolution* 57: 2398–2406.

Basaria, S., Leib, J., Tang, A. M., DeWeese, T., Carducci, M., Eisenberger, M., & Dobs, A. S. (2002). Long-term effects of androgen deprivation therapy in prostate cancer patients. *Clinical Endocrinology* 56: 779–786.

Bateman, A. J. (1948). Intrasexual selection in *Drosphilia*. *Heredity* 2: 349–368.

Bates, T. C. (2007). Fluctuating asymmetry and intelligence. *Intelligence* 35: 41–46.

Baumeister, R. F., Catanese, K. R., & Vohs, K. D. (2001). Is there a gender difference in the strength of sex drive? Theoretical views, conceptual distinctions, and a review of relevant evidence. *Personality and Social Psychology Review* 5: 242–273.

Beach, F. A. (1970). Hormonal effects on socio-sexual behavior in dogs. In *Mammalian Reproduction* (eds. H. Gibian & E. J. Plotz), pp. 437–465. Springer-Verlag, Berlin, Germany.

Beach, F. A. (1976). Sexual attractivity, proceptivity and receptivity in female mammals. *Hormones and Behavior* 7: 105–138.

Beckerman, S., Lizarralde, R., Ballew, C., Schroeder, S., Fingelton, E., Garrison, A., & Smith, H. (1998). The Bari partible paternity project: Preliminary results. *Current Anthropology* 39: 164–167.

Beehner, J. C., Onderdonk, D. A., Alberts, S. C., & Altmann, J. (2006). The ecology of conception and pregnancy failure in wild baboons. *Behavioral Ecology* 17: 741–750.

Bell, G. (1985). On the function of flowers. *Proceedings of the Royal Society of London B* 224: 223–265.

Bellis, M. A., & Baker, R. R. (1990). Do females promote sperm competition? *Animal Behaviour* 40: 997–999.

Benshoof, L., & Thornhill, R. (1979). The evolution of monogamy and concealed ovulation in humans. *Journal of Social and Biological Structure,* 2: 95–106.

Bercovitch, F. B. (2001). Reproductive ecology of Old World monkeys. In *Reproductive Ecology and Human Evolution* (ed. P. T. Ellison), pp. 369–396. Aldine de Gruyter, New York.

Berg, S. J., & Wynne-Edwards, K. E. (2001). Changes in testosterone, cortisol, and estradiol levels in men becoming fathers. *Mayo Clinic Proceedings* 76: 582–592.

Berglund, A., & Rosenqvist, G. (2001). Male pipefish prefer ornamented females. *Animal Behaviour* 61: 345–350.

Berglund, A., Rosenqvist, G., & Bernet, P. (1997). Ornamentation predicts reproductive success in female pipefish. *Behavioral Ecology and Sociobiology* 40: 145–150.

Berglund, A., Widemo, M. S., & Rosenqvist, G. (2005). Sex-role reversal revisited: Choosy females and ornamented, competitive males in a pipefish. *Behavioral Ecology* 16: 649–655.

Berlanga, C., & Huerta, R. (2000). Gonadal esteroids and affectivity: The role of sexual hormones in the etiology and treatment of affective disorders. *Salud Mental* 23: 10–21.

Bertran, J., & Margalida, A. (1999). Copulatory behavior of the bearded vulture. *Condor* 101: 164–168.

Betzig, L. (1986). *Despotism and Differential Reproduction: A Darwinian View of History.* Aldine de Gruyter, Hawthorne, NY.

Bielert, C., & Anderson, C. M. (1985). Baboon sexual swellings and male response: A possible operational mammalian supernormal stimulus and response interaction. *International Journal of Primatology* 6: 377–393.

Bhasin, S. (2003). Regulation of body composition by androgens. *Journal of Endocrinological Investigation* 26: 814–822.

Birkhead, T. R. (1979). Mate guarding in the magpie *Pica pica. Animal Behaviour* 33: 608–619.

Birkhead, T. R. (1995). Human sperm competition: Copulation, masturbation, and infidelity. *Animal Behaviour* 50: 1141–1142.

Birkhead, T. R., & Møller, A. P. (1992). *Sperm Competition in Birds: Evolutionary Causes and Consequences.* Academic Press, New York.

Birkhead, T. R., & Møller, A. P. (1993a). Why do male birds stop copulating while their partners are still fertile? *Animal Behaviour* 45: 105–118.

Birkhead, T. R., & Møller, A. P. (1993b). Sexual selection and the temporal separation of reproductive events: Sperm storage data from reptiles, birds and mammals. *Biological Journal of the Linnean Society* 50: 295–311.

Birkhead, T. R., & Møller, A. P. (1995). Extra-pair copulations and extra-pair paternity in birds. *Animal Behaviour* 49: 843–848.

Birkhead, T. R., & Møller, A. P., eds. (1998). *Sperm Competition and Sexual Selection.* Academic Press, London, UK.

Bjorklund, D. F., & Kipp, K. (1996). Parental investment theory and gender differences in the evolution of inhibition mechanisms. *Psychological Bulletin* 120: 163–188.

Black, F. L., & Hedrick, P. W. (1997). Strong balancing selection at HLA loci: Evidence from segregation in South American families. *Proceedings of the National Academy of Sciences* USA 94: 12452–12456.

Blackwell, E. (1902). *Essays in medical sociology.* London: Bell.

Bluhm, C. K., & Gowaty, P. A. (2004a). Social constraints on female mate preferences in mallards, *Anas platyrhynchos,* decrease offspring viability and mother productivity. *Animal Behaviour* 68: 977–983.

Bluhm, C. K., & Gowaty, P. A. (2004b). Reproductive compensation for offspring viability deficits by female mallards, *Anas platyrhynchos. Animal Behaviour* 68: 985–992.

Bodily, R. Y., & Neudorf, D. L. H. (2004). Mate guarding in northern mockingbirds (*Mimus polyglottos*). *Texas Journal of Science* 56: 207–214.

Bondurianksy, R. (2001). The evolution of male mate choice in insects: A synthesis of ideas and evidence. *Biological Reviews* 76: 305–339.

Booth, A., & Dabbs, J. M. (1993). Testosterone and mens' marriages. *Social Forces* 72: 463–477.

Borgerhoff Mulder, M. (1988). Kipsigis bride wealth payments. In *Human Reproductive Behavior* (eds. L. L. Betzig, M. Borgerhoff Mulder, & P. Turke), pp. 65–82. Cambridge University Press, New York.

Borgerhoff Mulder, M. (1990). Kipsigis women's preferences for wealthy men: Evidence for female choice in mammals? *Behavioral Ecology and Sociobiology* 27: 255–264.

Borries, C., Koenig, A., & Winkler, P. (2001). Variation of life history traits and mating patterns in female langur monkeys (*Semnopithecus entellus*). *Behavioral Ecology and Sociobiology* 50: 391–402.

Boyce, M. (1990). The red queen visits sage grouse leks. *American Zoologist* 30: 263–270.

Bradbury, J. W., & Vehrencamp, S. L. (1998). *Principles of Animal Communication.* Sinauer Sunderland, MA.

Brewis, A., & Meyer, M. (2005). Demographic evidence that human ovulation is undetectable (at least in pair bonds). *Current Anthropology* 46: 465–471.

Bribiescas, R. G. (2001). Reproductive ecology and life history of the human male. *Yearbook of Physical Anthropology* 44: 148–176.

Bro-Jorgensen, J. (2003). No peace for estrous topi cows on leks. *Behavioral Ecology* 14: 521–525.

Bröder, A., & Hohmann, N. (2003). Variations in risk taking behavior over the menstrual cycle: An improved replication. *Evolution and Human Behavior* 24: 391–398.

Brommer, J. E., Kirkpatrick, M., Qvarnström, A., & Gustafsson, L. (2007). The intersexual correlation between lifetime fitness in the wild and its implications for sexual selection. *PloS ONE* 2: e744. doi: 10.1371/journal/pone.0000744

Brooks, R., & Kemp, D. J. (2001). Can older males deliver the good genes? *Trends in Ecology and Evolution* 16: 308–313.

Brown, W. M., Cronk, L., Grochow, K., Jacobson, A., Liu, C. K., Popovic, Z., & Trivers, R. (2005). Dance revealed symmetry especially in young men. *Nature* 238: 1148–1150.

Brown, W. M., Price, M. E., Kang, J., Zhau, Y., & Yu, H. (2007). Male body build reveals developmental stability. Paper presented at the annual meeting of the Human Behavior and Evolution Society, Williamsburg, VA.

Bryant, G. A., & Haselton, M. G. (2007). Changes in women's vocal behavior across the ovulatory cycle. Paper presented at the annual meeting of the Human Behavior and Evolution Society, Williamsburg, VA.

Buchan, J. C., Alberts, S. C., Silk, J. B., & Altmann, J. (2003). True paternal care in multi-male primate society. *Nature* 425: 179–181.

Buchanan, K. L., & Catchpole, C. K. (2000). Extrapair paternity in the Sedge warbler *Acrosephalus schoenobaenus* as veiled by multilocus DNA fingerprinting. *Ibis* 142: 12–20.

Buller, D. J. (2005). *Adapting Minds: Evolutionary Psychology and the Persistent Quest for Human Nature.* MIT Press, Cambridge, MA.

Bullivant, S. B., Sellergren, S. A., Stern, K., Spencer, N. A., Jacob, S., Mennella, J. A., & McClintock, M. K. (2004). Women's sexual experience during the menstrual cycle: Identification of the sexual phase by noninvasive measurement of luteinizing hormone. *Journal of Sex Research* 41: 82–93.

Burch, R. L., & Gallup, G. L., Jr. (2006). The psychobiology of human semen. In *Female Infidelity and Paternal Uncertainty* (eds. T. K. Shackelford & S. M. Platek), pp.141–172. Cambridge University Press, Cambridge, UK.

Burghardt, G. M. (1970). Defining "communication." In *Communication by Chemical Signals,* Vol. 1 of *Advances in Chemoreception* (eds. J. W. Johnston, D. G. Moulton, & A. Turk), pp. 5–18. Appleton-Century-Crofts, New York.

Burleson, M. H., Gregory, W. L., & Trevathan, W. R. (1995). Heterosexual activity: Relationship with ovarian function. *Psychoneuroendocrinology* 20: 405–421.

Burleson, M. H., Trevathan, W. R., & Gregory, W. L. (2002). Sexual behavior in lesbian and heterosexual women: Relations with menstrual cycle phase and partner availability. *Psychoneuroendocrinology* 27: 489–503.

Burley, N. (1979). The evolution of concealed ovulation. *American Naturalist* 114: 835–858.

Burley, N. (1986). Sex-ratio manipulation in color-banded populations of zebra finches. *Evolution* 40: 1191–1206.

Burnham, J. C., Chapman, J. F., Gray, P. B., McIntyre, M. H., Lipson, S. F., & Ellison, P. T. (2003). Men in committed, romantic relationships have lower testosterone. *Hormones and Behavior* 44: 119–122.

Burriss, R. P., & Little, A. C. (2006). Effects of partner conception risk phase on male perception of dominance in faces. *Evolution and Human Behavior* 27: 297–305.

Burt, A. (1992). Concealed ovulation and sexual signals in primates. *Folia Primatologica* 58: 1–6.

Burt, A. (1995). Perspective: The evolution of fitness. *Evolution* 49: 1–8.

Burt, A., & Trivers, R. (2006). *Genes in conflict: The biology of selfish genetic elements.* Belknap Press of Harvard University Press, Cambridge, MA.

Buss, D. M. (1989a). Conflict between the sexes: Strategic interference and the evocation of anger and upset. *Journal of Personality and Social Psychology* 56: 735–747.

?4Buss, D. M. (1989b). Sex differences in human mate preferences: Evolutionary hypotheses tested in 37 cultures. *Behavioral and Brain Sciences* 12: 1–49.

Buss, D. M. (1994). *The Evolution of Desire: Strategies of Human Mating.* Basic Books, New York.

Buss, D. M. (2000). *Dangerous Passions.* Free Press, New York.

Buss, D. M. (2003a). Sexual strategies: A journey into controversy. *Psychological Inquiry* 14: 219–226.

Buss, D. M. (2003b). *The Evolution of Desire: Strategies of Human Mating,* 2nd ed. Basic Books, New York.

Buss, D. M. (2005). *The murderer next door: Why the mind is designed to kill.* Penguin Press, New York.

Buss, D. M., Larsen, R. J., Westen, D., & Semmelroth, J. (1992). Sex differences in jealousy: Evolution, physiology and psychology. *Psychological Science* 3: 251–255.

Buss, D. M., & Schmitt, D. P. (1993). Sexual strategies theory: An evolutionary perspective on human mating. *Psychological Review* 100: 204–232.

Buss, D. M., Shackelford, T. K., Kirkpatrick, L. A., Choe, J. C., Lim, H. K., Hasegawa, M., Hasegawa, T., & Bennett, K. (1999). Jealousy and the nature of beliefs about infidelity: Tests of competing hypotheses about sex differences in the United States, Korea and Japan. *Personal Relationships* 6: 125–150.

Bussière, L. F. (2002). A model of the interaction between "good genes" and direct benefits in courtship-feeding animals: When do males of high genetic quality invest less? *Philosophical Transactions of the Royal Society of London B* 357: 309–317.

Buston, P. M., & Emlen, S. T. (2003). Cognitive processes underlying human mate choice: The relationship between self-perception and mate preference in Western society. *Proceedings of the National Academy of Sciences USA* 100: 8805–8810.

Buunk, A. P., Park, J. H., & Dubbs, S. L. (2008). Parent-offspring conflict in mate preferences. *Review of General Psychology* 12: 47–62.

Byers, J. A., Moodie, J. D., & Hall, N. (1994). Pronghorn females choose vigorous mates. *Animal Behaviour* 47: 33–43.

Byers, J. A. & Waits, L. (2006). Good genes sexual selection in nature. *Proceedings of the National Academy of Sciences* 103: 16343–16345.

Byrne, P. G., & Rice, W. R. (2005). Remating in *Drosophila melanogaster:* An examination of the trading-up and intrinsic male-quality hypotheses. *Journal of Evolutionary Biology* 18: 1324–1331.

Byrne, P. G., & Rice, W. R. (2006). Evidence for adaptive male mate choice in the fruit fly *Drosophila melanogaster. Proceedings of the Royal Society of London* B 273: 917–922.

Campbell, A. (2002). *A Mind of Her Own: The Evolutionary Psychology of Women.* Oxford University Press, Oxford, UK.

Candolin, U. (2003). The use of multiple cues in mate choice. *Biological Reviews* 78: 575–595.

Cant, J. (1981). Hypotheses for the evolution of human breasts and buttocks. *American Naturalist* 117: 199–204.

Cárdenas, R. A., & Harris, L. J. (2007). Do women's preferences for symmetry change across the cycle? *Evolution and Human Behavior* 28: 96–105.

Carnegie, S. D., Fedigan, L. M., & Ziegler, T. E. (2005). Behavioral indicators of ovarian phase in white-faced capuchins (*Cebus capucinus*). *American Journal of Primatology* 67: 51–68.

Caro, T. M., & Sellen, D. W. (1990). The reproductive advantages of fat in women. *Ethology and Sociobiology* 11: 51–66.

Cartwright, J. (2000). *Evolution and Human Behavior: Darwinian Perspectives on Human Nature.* MIT Press, Cambridge, MA.

Castelo-Branco, C., Vicente, J. J., Figueras, F., Sanjuan, A., de Osaba, M. J. M., Casals, E., Pons, F., Balasch, J., & Vanrell, J. A. (2000). Comparative effects of estrogens plus androgens and tibolone on bone, lipid pattern and sexuality in postmenopausal women. *Maturitas* 34: 161–168.

Catry, P., & Furness, R. W. (1997). Territorial intrusions and copulation behaviour in the great skua, *Catharacta skua. Animal Behaviour* 54: 1265–1272.

Cerda-Flores, R. M., Barton, S. A., Marty-Gonzalez, L. F., Rivas, F., & Chakraborty, R. (1999). Estimation of nonpaternity in the Mexican population of Nuevo Leon: A validation study with blood group markers. *American Journal of Physical Anthropology* 109: 281–293.

Cerda-Molina, A. L., Hernandez-Lopez, L., Chavira, R., Cardenas, M., Paez-Ponce, D., Cervantes-De la Luz, H., & Mondragon-Ceballos, R. (2006). Endocrine changes in male stumptailed macaques (*Macaca arctoides*) as a response to odor stimulation with vaginal secretions. *Hormones and Behavior* 49: 81–87.

Cerda-Molina, A. L., Hernandez-Lopez, L., Rojas-Maya, S., Murcia-Mejia, C., & Mondragon-Ceballos, R. (2006). Male-induced sociosexual behavior by vaginal secretions in *Macaca arctoides. International Journal of Primatology* 27: 791–807.

Chagnon, N. A. (1992). *Yanomamo: The last days of Eden.* Harvest Books, New York.

Chapman, T., Arnqvist, G., Bangham, J., & Rowe, L. (2003). Sexual conflict. *Trends in Ecology and Evolution* 18: 41–47.

Charlesworth, B. (1987). The heritability of fitness. In *Sexual Selection: Testing the Alternatives* (eds. J. W. Bradbury & M. B. Andersson), pp. 21–40. Wiley, New York.

Charlesworth, B. (1990). Mutation-selection balance and the evolutionary advantage of sex and recombination. *Genetical Research* 55: 199–221.

Charlesworth, B., & Hughes, K. A. (2000). The maintenance of genetic variation in life history traits. In *Evolutionary Genetics: From Molecules to Morphology* (eds. R. S. Singh & C. B. Krimbas) pp. 369–391 Cambridge University Press, Cambridge, UK.

Charlton, B. D., Reby, D., & McComb, K. (2007). Female red deer prefer the roars of larger males. *Biology Letters* 3: 382–385.

Charnov, E. L. (1979). Simultaneous hermaphroditism and sexual selection. *Proceedings of the National Academy of Sciences USA,* 76: 2480–2484.

Charnov, E. L. (1982). *The Theory of Sex Allocation.* Princeton University Press, Princeton, NJ.

Charnov, E. L., & Berrigan, D. (1993). Why do female primates have such long life spans and so few babies? Or, life in the slow lane. *Evolutionary Anthropology* 1: 191–194.

Chavanne, T. J., & Gallup, G. G., Jr. (1998). Variation in risk taking behavior among female college students as a function of the menstrual cycle. *Evolution and Human Behavior* 19: 27–32.

Chippendale, A. K., Gibson, J. R., & Rice, W. R. (2001). Negative genetic correlation for adult fitness between sexes reveals ontogenetic conflict in *Drosophila. Proceedings of the National Academy of Sciences USA* 98: 1671–1675.

Chrastil, E. R., Getz, W. M, Euler, H. A., & Starks, P. T. (2006). Paternity uncertainty overrides sex chromosome selection for preferential grandparenting. *Evolution and Human Behavior* 27: 206–223.

Chuang-Dobbs, H. C., Webster, M. S., & Holmes, R. T. (2001). The effectiveness of mate guarding by male black-throated blue warblers. *Behavioral Ecology* 12: 541–546.

Clark, A. P. (2004). Self-perceived attractiveness and masculinization predict women's sociosexuality. *Evolution and Human Behavior* 25: 113–124.

Clement, T. S., Grens, K. E., & Fernald, R. D. (2005). Female affiliative preference depends on reproductive state in the African cichlid fish, *Astatotilapia burtoni. Behavioral Ecology* 16: 83–88.

Clutton-Brock, T. H. (1991). *The Evolution of Parental Care.* Princeton University Press, Princeton, NJ.

Clutton-Brock, T. H., & Harvey, P. H. (1976). Evolutionary rules and primate societies. In *Growing Points in Ethology* (eds. P. P. G. Bateson & R. A. Hinde), pp. 195–237. Cambridge University Press, Cambridge, UK.

Clutton-Brock, T. H., & Parker, G. A. (1992). Potential reproductive rates and the operation of sexual selection. *Quarterly Review of Biology* 67: 437–456.

Clutton-Brock, T. H., & Parker, G. A. (1995). Sexual coercion in animal societies. *Animal Behaviour* 49: 1345–1365.

Conway, C. A., Jones, B. C., DeBruine, L. M., Welling, L. L. M., Law Smith, M. J., Perrett, D. I., Sharp, L. A., & Al-Dujaili, E. A. S. (2007). Salience of emotional displays of danger and contagion is enhanced when progesterone levels are enhanced. *Hormones and Behavior* 51: 202–206.

Collins, S. A., & Missing, C. (2003). Vocal and visual attractiveness are related in women. *Animal Behaviour* 65: 997–1004.

Corner, G. (1942). *The Hormones in Human Reproduction.* Princeton University Press, Princeton, NJ.

Cosmides, L., & Tooby, J. (2000). Evolutionary psychology and the emotions. In *Handbook of Emotions* (eds. M. Lewis & J. M. Haviland-Jones), pp. 91–115. Guilford Press, New York.

Cotton, S., Fowler, K., & Pomiankowski, A. (2004). Do sexual ornaments demonstrate heightened condition-dependent expression as predicted by the handicap hypothesis? *Proceeding of the Royal Society of London B* 271: 771–783.

Cowen, R. (1990). *History of life.* Blackwell Scientific, Boston, MA.

Cox, C. R., & Le Boeuf, B. J. (1977). Female incitation of male competition: A mechanism of sexual selection. *American Naturalist* 111: 317–335.

Coyne, J. A. (2000). Of vice and men: The fairy tales of evolutionary psychology. *New Republic* 147: 27–34.

Crews, D., & Moore, M. C. (1986). Evolution of mechanisms controlling mating behavior. *Science* 231: 121–125.

Crews, D., & Silver, R. (1985). Reproductive physiology and behavior interactions in non-mammalian vertebrates. In *Handbook of Behavioral Neurobiology* (eds. N. T. Adler, D. Pfaff, & R. W. Goy), pp. 101–182. Plenum Press, New York.

Crocker, W. H. (1984). Canela marriage: Factors in change. In *Marriage practices in lowland South America* (ed. K. M. Kensinger) pp. 63–98. Urbana and Chicago, IL: University of Illinois Press.

Crocker, W. H. (1990). The Canela (Eastern Timbira). I: An ethnographic introduction. *Smithsonian Contributions to Anthropology*, No. 33. Smithsonian Institution Press, Washington, DC.

Crow, J. F. (1997). The high spontaneous mutation rate: Is it a health risk? *Proceedings of the National Academy of Sciences USA* 94: 8380–8386.

Crowley, P. H., Travers, S. E., Linton, M. C., Cohn, S. L., Sih, A. S., & Sargent, C. R. (1991). Mate density, predation risk and the seasonal sequence of mate choices: A dynamic game. *American Naturalist* 137: 567–596.

Cuervo, J. J., & Møller, A. P. (2000). Sex-limited expression of ornamental feathers in birds. *Behavioral Ecology* 11: 246–259.

Cuervo, J. J., de Lope, F., & Møller, A. P. (1996). The function of long tails in female barn swallows (*Hirundo rustica*): An experimental study. *Behavioral Ecology* 7: 132–136.

Cuervo, J. J., Møller, A. P., & de Lope, F. (2003). Experimental manipulation of tail length of female barn swallows (*Hirundo rustica*) affects their future reproductive success. *Behavioral Ecology* 14: 451–456.

Cunningham, E. J. A. (2003). Female mate preferences and subsequent resistance to copulation in the mallard. *Behavioral Ecology* 14: 326–333.

Currie, D., Krupa, A. P., Burke, T., & Thompson, D. B. A. (1999). The effect of experimental male removals on extra-pair paternity in the wheatear, *Oenanthe oenanthe*. *Animal Behaviour* 57: 145–152.

Dabbs, J. M. (1997). Testosterone, smiling, and facial appearance. *Journal of Nonverbal Behavior* 21: 45–55.

Dabbs, J. M., Bernieri, F. J., Strong, R. K., Campo, R., & Milun, R. (2001). Going on stage: Testosterone in greetings and meetings. *Journal of Research in Personality* 35: 27–40.

Dahl, J. F., & Nadler, R. D. (1992). The external genitalia of female gibbons, *Hylobates (H.) lar. Anatomical Record* 232: 572–578.

Dale, S., Rinden, H., & Slagsvold, T. (1992). Competition for a mate restricts mate search of female pied flycatchers. *Behavioral Ecology and Sociobiology* 30: 165–176.

Daly, M., & Wilson, M. (1988). *Homicide.* Aldine de Gruyter, New York.

Daly, M., & Wilson, M. (1999). Human evolutionary psychology and animal behaviour. *Animal Behaviour* 57: 509–519.

Daniels, D. (1983). The evolution of concealed ovulation and self-deception. *Ethology and Sociobiology* 4: 69–87.

Darwin, C. (1859). *On the Origin of Species.* John Murray, London, UK.

Darwin, C. (1871). *The Descent of Man and Selection in Relation to Sex.* John Murray, London, UK.

Daunt, F., Monaghan, P., Wanless, S., & Harris, M. P. (2003). Sexual ornament size and breeding performance in female and male European Shags *Phalacrocorax aristolelis*. *Ibis* 145: 54–60.

Davies, N. B. (1992). *Dunnock Behaviour and Social Evolution*. Oxford University Press, Oxford, UK.

Davis, J. A., & Gallup, G. G., Jr. (2006). Preeclampsia and other pregnancy complications as an adaptive response to unfamiliar semen. In *Female infidelity and paternal uncertainty* (eds. T. K. Shackelford & S. M. Platek), pp. 191–204. Cambridge University Press, Cambridge, UK.

Davis, S. R. (2002). The effects of tibolone on mood and libido. *Menopause: The Journal of the North American Menopause Society* 9: 162–170.

DeBruine, L.M. (2004). Resemblance to self increases the appeal of child faces to both men and women. *Evolution and Human Behavior* 25: 142–154.

DeBruine, L. M., Jones, B. C., & Perrett, D. I. (2005). Women's attractiveness judgments of self-resembling faces change across the menstrual cycle. *Hormones and Behavior* 47: 379–383.

Denison, F. C., Grant, V. E., Calder, A. A., & Kelly, R. W. (1999). Seminal plasma components stimulate interleukin-8 and interleukin-10 release. *Molecular Human Reproduction* 5: 220–226.

Derntl, B., Kryspin-Exner, I., Fernbach, E., Moser, E., & Habel, U. (2008). Emotion recognition accuracy in healthy young females is associated with cycle phase. *Hormones and Behavior* 53: 90–95.

Deschner, T., Heistermann, M., Hodges, K., & Boesch, C. (2003). Timing and probability of ovulation in relation to sex skin swelling in Wild West African chimpanzees, *Pan troglodytes verus*. *Animal Behavior* 66: 551–560.

Deschner, T., Heistermann, M., Hodges, K., & Boesch, C. (2004). Female sexual swelling size, timing of ovulation, and male behavior in wild West African chimpanzees. *Hormones and Behavior* 46: 204–215.

DeVleeschouwer, K., Heistermann, M., VanElsacker, L., & Verheven, R. F. (2000). Signaling of reproductive status in captive female golden-headed lion tamarins (*Leontopithecus chryomelas*). *International Journal of Primatology* 21: 445–465.

de Waal, Frans F. M. (2001). *Tree of Origin: What Primate Behavior Can Tell Us about Human Social Evolution*. Harvard University Press, Cambridge, MA.

Diamond, J. (1997). *Why is Sex Fun? The Evolution of Human Sexuality*. Basic Books, New York.

Dickemann, M. (1981). Paternal confidence and dowry competition: A biocultural analysis of purdah. In *Natural Selection and Social Behavior* (eds. R. D. Alexander & D. W. Tinkle), pp. 417–438. Chiron Press, New York.

Dickinson, J. L. (1997). Male detention affects extra-pair copulation frequency and pair behaviour in western bluebirds. *Animal Behaviour* 53: 561–571.

Dickinson, J. L. (2001). Extrapair copulations in western bluebirds (*Sialia mexicana*): Female receptivity favors older males. *Behavioral Ecology and Sociobiology* 50: 423–429.

Digby, L. J. (1999). Sexual behavior and extragroup copulation in a wild population of common marmosets (*Callithrix jacchus*). *Folia Primatologica* 70: 136–145.

Ding, Y. C., Chi, H. C., Grady, D. L., Morishima, A., Kidd, J. R., Kidd, K. K., Flodman, P., Spence, M. A., Schuck, S., Swanson, J. M., Zyang, Y. P., & Moyzis, R. K. (2002). Evidence of positive selection acting at human dopamine receptor D4 gene locus. *Proceedings of the National Academy of Sciences USA* 99: 309–314.

Dixson, A. F. (1998). *Primate Sexuality: Comparative Studies of the Prosimians, Monkeys, Apes, and Humans*. Oxford University Press, Oxford, UK.

Dixson, A. (2002). Sexual selection by cryptic female choice and the evolution of primate sexuality. *Evolutionary Anthropology* 1: 195–199.

Dixson, A. F., & Anderson, M. J. (2004). Effects of sexual selection upon sperm morphology and sexual skin morphology in primates. *International Journal of Primatology* 25: 1159–1171.

Domb, L. G., & Pagel, M. (2001). Sexual swellings advertise female quality in wild baboons. *Nature* 410: 204–206.

Domb, L. G., & Pagel, M. (2002). Evolutionary biology: Significance of primate sexual swellings. Reply. *Nature* 420: 143–143.

Doty, R. L., Ford, M., Preti, G., & Huggins, G. R. (1975). Changes in the intensity and pleasantness of human vaginal odors during the menstrual cycle. *Science* 190: 1316–1318.

Double, M., & Cockburn, A. (2000). Pre-dawn infidelity: Females control extra-pair mating in superb fairy-wrens. *Proceedings of the Royal Society of London* B 267: 465–470.

Drickamer, L. C., Gowaty, P. A., & Holmes, C. M. (2000). Free female mate choice in house mice affects reproductive success and offspring viability and performance. *Animal Behaviour* 59: 371–378.

Drost, J. B., & Lee, W. R. (1995). Biological basis of germline mutation: Comparisons of spontaneous germline mutation rates among *Drosophila*, mouse, and human. *Environmental and Molecular Mutagenesis* 25: 48–64.

Dufour, S. L., & Sauther, M. L. (2002). Comparative and evolutionary dimensions of the energetics of human pregnancy and lactation. *American Journal of Human Biology* 14: 584–602.

Dunbar, R. I. M. (1987). *Primate social systems.* Comstock, Ithaca, NY.

Dunson, D. B., Baird, D. D., Wilcox, A. J., & Weinberg, C. R. (1999). Day-specific probabilities of clinical pregnancy based on two studies with imperfect measures of ovulation. *Human Reproduction* 14: 1835–1839.

Dunson, D. B., Colombo, B., & Baird, D. D. (2002). Changes with age in the level and duration of fertility in the menstrual cycle. *Human Reproduction* 17: 1399–1403.

Dunson, D. B., Weinberg, C. R., Baird, D. D., Kesner, J. S., & Wilcox, A. J. (2001). Assessing human fertility using several markers of ovulation. *Statistics in Medicine* 20: 965–978.

Durante, K. M., Li, N. P., & Haselton, M. G. (2007). Ovulatory shifts in women's choice of dress: Naturalistic and experimental evidence. Paper presented at the annual meeting of the Human Behavior and Evolution Society, Williamsburg, VA.

Duvall, D., & Schuett, G. W. (1997). Straight-line movement and competitive mate searching in prairie rattlesnakes, *Crotalus viridis viridis. Animal Behaviour* 54: 329–334.

Eberhard, W. G. (1996). *Female Control: Sexual Selection by Cryptic Female Choice.* Princeton University Press, Princeton, NJ.

Egarter, C., Topcuoglu, A. M., Vogl, S., & Sator, M. (2002). Hormone replacement therapy with tibolone: Effects on sexual functioning in postmenopausal women. *Acta Obstetricia et Gynecologica Scandinavica* 81: 649–653.

Elbers, J. M. H., Asscheman, H., Seidell, J. C., & Gooren, L. J. G. (1999). Effects of sex steroid hormones on regional fat depots as assessed by magnetic resonance imaging in transsexuals. *American Journal of Physiology: Endocrinology and Metabolism* 276: E317–E325.

Ellegren, H., Gustafsson, L., & Sheldon, B. C. (1996). Sex ratio adjustment in relation to paternal attractiveness in a wild bird population. *Proceedings of the National Academy of Sciences USA* 93: 11723–11728.

Ellis, B. J. (1998). The partner-specific investment inventory: An evolutionary approach to individual differences in investment. *Journal of Personality* 66: 383–442.

Ellis, B. J., Bates, J. E., Dodge, K. A., Fergusson, D. M., Horwood, L. J., Pettit, G. S., & Woodward, L. (2003). Does father absence place daughters at special risk for early sexual activity and teenage pregnancy? *Child Development* 74: 801–821.

Ellis, B. J., & Garber, J. (2000). Psychosocial antecedents of variation in girls' pubertal timing: Maternal depression, stepfather presence, and marital and family stress. *Child Development* 71: 485–501.

Ellis, B. J., McFadyen-Ketchum, S., Dodge, K. A., Pettit, G. S., & Bates, J. E. (1999). Quality of early family relationships and individual differences in the timing of pubertal maturation in girls: A longitudinal test of an evolutionary model. *Journal of Personality and Social Psychology* 77: 387–401.

Ellis, H. (1903). Studies in the Psychology of Sex, vol. III: Analysis of the Sexual Impulse; Love and Pain; the Sexual Impulse in Women. F. A. Davis Co., Philadelphia, PA.

Ellis, H. (1922). *Little Essays of Love and Virtue.* A. & C. Black, London, UK.

Ellison, P. T. (1994). Extinction and descent. *Human Nature* 5: 155–165.

Ellison, P. T. (2001). *On fertile ground: A natural history of reproduction.* Harvard University Press, Cambridge, MA.

Ellison, P. T. (2003). Energetics and reproductive effort. *American Journal of Human Biology* 15: 342–351.

Ember, M., & Ember, C. R. (1979). Male-female bonding: A cross-species study of mammals and birds. *Behavior Science Research* 14: 37–56.

Emery, M. A., & Whitten, P. L. (2003). Size of sexual swellings reflects ovarian function in chimpanzees (*Pan troglodytes*). *Behavioral Ecology and Sociobiology* 54: 340–351.

Emlen, S. T, & Oring, L. W. (1977). Ecology, sexual selection, and the evolution of mating systems. *Science* 197: 215–223.

Engelhardt, A., Hodges, J. K., Niemitz, C., & Heistermann, M. (2005). Female sexual behavior, not sex skin swelling, reliably indicates the timing of the fertile phase in wild long-tailed macaques (*Macaca fascicularis*). *Hormones and Behavior* 47: 195–204.

Engelhardt, A., Pfiefer, J.-B., Heistermann, M., & Niemitz, C. (2004). Assessment of female reproductive status by male long-tailed macaques, *Macaca fascicularis*, under natural conditions. *Animal Behaviour* 67: 915–924.

Enlow, D. H. (1990). *Facial Growth,* 3rd ed. Harcourt Brace Jovanovich, Philadelphia, PA.

Enquist, M., Rosenberg, R. H., & Temrin, H. (1998). The logic of ménage à trois. *Proceedings of the Royal Society of London B* 265: 609–613.

Epel, E. S., McEwen, B., Seeman, T., Matthews, K., Castellazzo, G., Brownell, K. D., Bell, J., & Ickovics, J. R. (2000). Stress and body shape: Stress-induced cortisol secretion is consistently greater among women with central fat. *Psychosomatic Medicine* 62: 623–632.

Eshel, I., & Hamilton, W. D. (1984). Parent-offspring correlation in fitness under fluctuating selection. *Proceedings of the Royal Society of London* B 222: 1–24.

Etcoff, N. (1999). *Survival of the prettiest: The science of beauty.* New York: Doubleday.

Etkin, W. (1964). Types of social organization in birds and mammals. In *Social Behavior and Organization among Vertebrates* (ed. W. Etkin), pp. 256–298. University of Chicago Press, Chicago.

Ewen, J., Cassey, P., & Møller, A. P. (2004). Facultative primary sex ratio variation: A lack of evidence in birds? *Proceedings of the Royal Society of London* B 271: 1277–1282.

Eyre-Walker, A., & Keightley, P. D. (1999). High genomic mutation rates in hominids. *Nature* 397: 345–347.

Eyre-Walker, A., Woolfit, M., & Phelps, T. (2006). Distribution of fitness effects of new deleterious amino acid mutations in humans. *Genetics* 173, 891–900.

Falconer, D. S., & Mackay, T. F. C. (1996). *An Introduction to Quantitative Genetics*, 4th ed. Addison Wesley Longman, New York.

Feinberg, D. R., Jones, B. C., DeBruine, L. M., Moore, F. R., Law Smith, M. J., Cornwell, R. E., Tiddeman, B. P., Boothroyd, L. G., & Perrett, D. I. (2005). The voice and face of woman: One ornament that signals quality? *Evolution and Human Behavior* 26: 398–408.

Feinberg, D. R., Jones, B. C., Law Smith, M. J., Moore, F. R., DeBruine, L. M., Cornwell, R. E., Hillier, S. G., & Perrett, D. I. (2006). Menstrual cycle, trait estrogen level, and masculinity preferences in the human voice. *Hormones and Behavior* 49: 215–222.

Feinberg, D. R., Jones, B. C., Little, A. C., Burt, D. M., & Perrett, D. I. (2005). Manipulations of fundamental and formant frequencies influence the attractiveness of human male voices. *Animal Behaviour* 69: 561–568.

Ferin, M., Jewelewicz, R., & Warren, M. (1993). *The Menstrual Cycle: Physiology, Reproductive Disorders, and Infertility.* Oxford University Press, Oxford, UK.

Ferris, C. F., Snowden, C. T., King, J. A., Sullivan, J. M., Ziegler, T. E., Olson, D. P., Schultz-Darken, N. J., Tannenbaum, P. L., Ludwig, R., Wu, Z. J., Einspanier, A., Vaughan, J. T., & Duong, T. Q. (2004). Activation of neural pathways associated with sexual arousal in non-human primates. *Journal of Magnetic Resonance Imaging* 19: 168–175.

Fessler, D. M. T. (2001). Luteal phase immunosuppression and meat eating. *Rivista di Biologia—Biology Forum* 94: 403–426.

Fessler, D. M. (2002). Reproductive immunosuppression and diet. *Current Anthropology* 43: 19–61.

Fessler, D. M. (2003). No time to eat: An adaptationist account of periovulatory behavioral changes. *Quarterly Review of Biology* 78: 3–21.

Fessler, D. M. T., & Navarrete, C. D. (2003). Domain-specific variation in disgust sensitivity across the menstrual cycle. *Evolution and Human Behavior* 24: 406–417.

Fessler, D. M. T., Nettle, D., Afshar, Y., de Andrade Pinheiro, I., Bolyanatz, A., Mulder, M. B., Cravalho, M., Delagado, T., Gruzd, B., Corrcia, M. O., Khaltourina, D., Korotaycv, A., Marrow, J., Santiago de Souza, L., & Zbarauskaite, A. (2005). A cross-cultural investigation of the role of foot size in physical attractiveness. *Archives of Sexual Behavior* 34: 267–276.

Fink, B., Grammer, K., & Matts, P. J. (2006). Visible skin color distribution plays a role in the perception of age, attractiveness, and health in female faces. *Evolution and Human Behavior* 27: 433–442.

Fink, B., Grammer, K., & Thornhill, R. (2001). Human (*Homo sapiens*) facial attractiveness in relation to skin texture and color. *Journal of Comparative Psychology* 115: 92–99.

Fink, B., & Penton-Voak, I. S. (2002). Evolutionary psychology of facial attractiveness. *Current Directions in Psychological Science* 11: 154–158.

Firman, R. C., Simmons, L. W., Cummins, J. M., & Matson, P. L. (2003). Are body fluctuating asymmetry and the ratio of 2nd to 4th digit length reliable predictors of semen quality? *Human Reproduction* 18: 808–812.

Fisher, H. (2004). *Why we love: The nature and chemistry of romantic love*. Henry Holt, New York.

Fisher, H., Aron, A., & Brown, L. L. (2005). Romantic love: An fMRI study of a neural mechanism for mate choice. *Journal of Comparative Neurology* 493: 58–62.

Fisher, H. E. (1982). *The Sex Contract: The Evolution of Human Behavior*. Morrow, New York.

Fisher, H. S., Swaisgood, R. R., & Fitch-Snyder, H. (2003). Countermarking by male pygmy lorises (*Nycticebus pygmaeus*): Do females use odor cues to select mates with high competitive ability? *Behavioral Ecology and Sociobiology* 53: 123–130.

Fisher, M. L. (2004). Female intrasexual competition decreases female facial attractiveness. *Proceedings of the Royal Society of London B* (suppl.). DOI: 10.1098/rsbl.2004.0160. Online February 18, 2004.

Fisher, R. A. (1930). *The Genetical Theory of Natural Selection*. Oxford University Press, Oxford, UK.

Fishman, M. A., Stone, L., & Lotem, A. (2003). Fertility assurance through extrapair fertilizations and male paternity defense. *Journal of Theoretical Biology* 221: 103–114.

Flaxman, S. M., & Sherman, P. W. (2000). Morning sickness: A mechanism for protecting mother and embryo. *Quarterly Review of Biology* 75: 113–148.

Fleming, A. S., Corter, C., Stallings, J., & Steiner, M. (2002). Testosterone and prolactin are associated with emotional responses to infant cries in new fathers. *Hormones and Behavior* 42: 399–413.

Flinn, M. (1981). Uterine vs. agnatic kinship variability and associated cross-cousin marriage preferences: An evolutionary biological analysis. In *Natural Selection and Social Behavior* (eds. R. D. Alexander & D. W. Tinkle), pp. 439–475. Chiron, New York.

Flinn, M. V. (1987). Mate guarding in a Caribbean village. *Ethology and Sociobiology* 8: 1–28.

Flinn, M. V. (1988). Parent–offspring interactions in a Caribbean village: Daughter guarding. In *Human Reproductive Behavior: A Darwinian Perspective* (eds. L. Betzig, M. Borgerhoff Mulder, & P. Turke), pp. 189–200. Cambridge University Press, Cambridge, UK.

Floter, A., Nathorst-Boos, J., Carlstrom, K., & von Schoultz, B. (2002). Addition of testosterone to estrogen replacement therapy in oophorectomized women: Effects on sexuality and well-being. *Climacteric* 5: 357–365.

Foerster, K., Coulson, T., Sheldon, B. C., Pemberton, J. M., Clutton-Brock, T. H., & Kruuk, L. E. B. (2007). Sexually antagonistic genetic variation for fitness in red deer. *Nature* 447: 1107–1110.

Forbes, G. B. (1987). Human body composition: Growth, aging, nutrition, and activity. Springer-Verlag, New York.

Forstmeier, W., Kempenaers, B., Meyer, A., & Leisler, B. (2002). A novel song parameter correlates with extra-pair paternity and reflects male longevity. *Proceedings of the Royal Society of London B* 269: 1479–1485.

Fowler, K., & Partridge, L. (1989). A cost of mating in female fruitflies. *Nature* 338: 760–761.

Frank, R. H. (1988). *Passions Within Reason: The Strategic Role of the Emotions*. Norton, New York.

Franklin, R. D., & Kutteh, W. H. (1999). Characterization of immunoglobulins and cytokines in human cervical mucus: Influence of exogenous and endogenous hormones. *Journal of Reproductive Immunology*, 42: 93–106.

Frederick, D. A., & Haselton, M. G. (2007). Why is muscularity sexy? Tests of the fitness indicator hypothesis. *Personality and Social Psychology Bulletin* 33: 1167–1183.

Freed, L. A. (1987). The long-term pair bond of tropical house wrens: Advantage or constraint? *American Naturalist* 130: 507–525.

Freeman, A. L. J., & Wong, H. Y. (1995). The evolution of self-concealed ovulation in humans. *Ethology and Sociobiology* 16: 531–533.

Freeman, E. W., & Halbeich, U. (1998). Premenstrual syndrome. *Psychopharmacology Bulletin* 34: 291–295.

Freeman-Gallant, C. R., Meguerdichian, M., Wheelwright, N. T., & Sollecito, S. V. (2003). Social pairing and female mating fidelity predicted by restriction fragment length polymorphism similarity at the major histocompatibility complex in a songbird. *Molecular Ecology* 12: 3077–3083.

French, J. A., & Schaffner, C. A. (2000). Contextual influences on sociosexual behavior in monogamous primates. In *Reproduction in Context* (eds. K. Wallen & J. E. Schneider), pp. 325–353. MIT Press, Cambridge, MA.

Friberg, U., Lew, T. A., Byrne, P. G., & Rice, W. R. (2005). Assessing the potential for an ongoing arms race within and between the sexes: Selection and heritable variation. *Evolution* 59: 1540–1551.

Frisch, R. E. (1990). Body fat, menarche, fitness and fertility. In *Adipose Tissue and Reproduction* (ed. R. E. Frisch), pp. 1–26. Karger, Basel, Switzerland.

Frost, P. (1988). Human skin color: A possible relationship between its sexual dimorphism and its social perception. *Perspectives in Biology and Medicine* 32: 38–58.

Frost, P. (1994). Preference for darker faces in photographs at different phases of the menstrual cycle: Preliminary assessment of evidence for a hormonal relationship. *Perceptual and Motor Skills* 79: 507–514.

Fuller, R. C., & Houle, D. (2003). Inheritance of developmental instability. In *Developmental Instability: Causes and Consequences* (ed. M. Polak), pp. 157–186. Oxford University Press, Oxford, UK.

Fuller, R. C., Houle, D., & Travis, J. (2005). Sensory bias as an explanation for the evolution of mate preferences. *American Naturalist* 166: 437–446.

Furlow, F. B., Armijo–Pruett, T., Gangestad, S. W., & Thornhill, R. (1997). Fluctuating asymmetry and psychometric intelligence. *Proceedings of the Royal Society of London B* 264: 1–8.

Furlow, F. B., Gangestad, S. W., & Armijo–Prewitt, T. (1998). Developmental stability and human violence. *Proceedings of the Royal Society of London B* 265: 1–6.

Gallup, G. G., Jr. (1982). Permanent breast enlargement in human females: A sociobiological analysis. *Journal of Human Evolution* 11: 597–601.

Gallup, G. G., Jr., Burch, R. L., & Platek, S. M. (2002). Does semen have antidepressant properties? *Archives of Sexual Behavior* 31: 289–293.

Gangestad, S. W. (1993). Sexual selection and physical attractiveness: Implications for mating dynamics. *Human Nature* 4: 205–236.

Gangestad, S. W., Bennett, K. L., & Thornhill, R. (2001). A latent variable model of developmental instability in relation to men's number of sex partners. *Proceedings of the Royal Society of London B* 268: 1677–1684.

Gangestad, S. W., Garver-Apgar, C. E., Simpson, J. A., & Cousins, A. J. (2007). Changes in women's mate preferences across the ovulatory cycle. *Journal of Personality and Social Psychology* 92: 151–163.

Gangestad, S. W., Haselton, M. G., & Buss, D. M. (2006). Evolutionary foundations of cultural variation: Evoked culture and mate preferences. *Psychological Inquiry* 17: 75–95.

Gangestad, S. W., & Scheyd, G. J. (2005). The evolution of human physical attractiveness. *Annual Review of Anthropology* 34: 523–548.

Gangestad, S. W., & Simpson, J. A. (1990). Toward an evolutionary history of female sociosexual variation. *Journal of Personality* 58: 69–96.

Gangestad, S. W., & Simpson, J. A. (2000). The evolution of human mating: The role of trade-offs and strategic pluralism. *Behavioral and Brain Sciences* 23: 573–644.

Gangestad, S. W., & Simpson, J. A., eds. (2007). *The Evolution of Mind: Fundamental Issues and Controversies.* Guilford Press, New York.

Gangestad, S. W., Simpson, J. A., Cousins, A. J., Garver-Apgar, C. E., & Christensen, P. N. (2004). Women's preferences for male behavioral displays shift across the menstrual cycle. *Psychological Science* 15: 203–207.

Gangestad, S. W., & Thornhill, R. (1997a). Human sexual selection and developmental stability. In *Evolutionary Social Psychology* (eds. J. A. Simpson & D. T. Kenrick), pp. 169–195. Erlbaum, Mahwah, NJ.

Gangestad, S. W., & Thornhill, R. (1997b). The evolutionary psychology of extra-pair sex: The role of fluctuating asymmetry. *Evolution and Human Behavior* 18: 69–88.

Gangestad, S. W., & Thornhill, R. (1998). Menstrual cycle variation in women's preference for the scent of symmetrical men. *Proceeding of the Royal Society of London B* 265: 927–933.

Gangestad, S. W., & Thornhill, R. (1999). Individual differences in developmental precision and fluctuating asymmetry: A model and its implications. *Journal of Evolutionary Biology* 12: 402–416.

Gangestad, S. W., & Thornhill, R. (2003a). Facial masculinity and fluctuating asymmetry. *Evolution and Human Behavior* 24: 231–241.

Gangestad, S. W., & Thornhill, R. (2003b). Fluctuating asymmetry, developmental instability, and fitness: Toward model-based interpretation. In *Developmental Instability: Causes and Consequences* (ed. M. Polak, 62–80). Cambridge University Press, Cambridge, UK.

Gangestad, S. W., & Thornhill, R. (2007). The evolution of social inference processes: The importance of signaling theory. In *Evolutionary psychology and social cognition* (eds. J. P. Forgas, M. G. Haselton, & W. von Hippel), pp. 33–48. Psychology Press, New York.

Gangestad, S. W., Thornhill, R., & Garver, C. E. (2002). Changes in women's sexual interests and their partners' mate retention tactics across the menstrual cycle: Evidence for shifting conflicts of interest. *Proceedings of the Royal Society of London B* 269: 975–982.

Gangestad, S. W., Thornhill, R., & Garver-Apgar, C. E. (2005a). Women's sexual interests across the ovulatory cycle depend on primary partner fluctuating asymmetry. *Proceedings of the Royal Society of London B* 272: 2023–2027.

Gangestad, S. W., Thornhill, R., & Garver-Apgar, C. E. (2005b). Adaptations to ovulation. In *Handbook of Evolutionary Psychology* (ed. D. M. Buss), pp. 344–371. Wiley, New York.

Gangestad, S. W., Thornhill, R., & Garver-Apgar, C. E. (2007a). Fertility in the cycle predicts women's interest in sexual opportunism. Manuscript submitted for publication.

Gangestad, S. W., Thornhill, R., & Garver-Apgar, C. E. (2007b). Estrous effects on women's extra-pair sexual interest: No detected moderator effects of measures of relationship commitment. Manuscript in preparation.

Gangestad, S. W., Thornhill, R., Quinlan, R. J., & Flinn, M. V. (2007). Fluctuating asymmetry, attractiveness, and reproduction in a rural Caribbean village. Unpublished manuscript.

Gangestad, S. W., Thornhill, R., & Yeo, R. (1994). Facial attractiveness, developmental stability, and fluctuating asymmetry. *Ethology and Sociobiology* 15: 73–85.

Gangestad, S. W., & Yeo, R. A. (1997). Behavioral genetic variation, adaptation, and maladaptation: An evolutionary perspective. *Trends in Cognitive Science* 1: 103–108.

Gardner, M. P., Fowler, K., Barton, N. H, & Partridge, L. (2005). Genetic variation for total fitness in *Drosophila melanogaster:* Complex yet replicable patterns. *Genetics* 169: 1553–1571.

Garamszegi, L. Z., Eens, M., Hurtrez-Boussès, S., & Møller, A. P. (2005). Testosterone, testes size and mating success in birds: A comparative study. *Hormones and Behavior* 47: 389–409.

Garver-Apgar, C. E., Gangestad, S. W., Thornhill, R., Miller, R. D., & Olp, J. (2006). MHC alleles, sexually responsivity, and unfaithfulness in romantic couples. *Psychological Science* 17: 830–835.

Garver-Apgar, C. E., Cousins, A. J., Gangestad, S. W., & Thornhill, R. (2007). Intersexual conflict across women's ovulatory cycle. Manuscript in preparation.

Garver-Apgar, C. E., Gangestad, S. W., & Thornhill, R. (2007). Women's perceptions of men's sexual coerciveness change across the menstrual cycle. *Acta Psychologica Sinica* 39: 536–540.

Garver-Apgar, C. E., Gangestad, S. W., & Thornhill, R. (in press). Hormonal correlates of women's mid-cycle preference for the scent of symmetry. Evolution and Human Behavior.

Gaulin, S. J. C., & Boster, J. (1990). Dowry and female competition. *American Anthropologist* 92: 994–1005.

Gaulin, S. J. C., & Schlegel, A. (1980). Paternal confidence and paternal investment: A cross-cultural test of a sociobiological hypothesis. *Ethology and Sociobiology* 1: 301–309.

Geary, D. C. (1998). *Male, Female: The Evolution of Human Sex Differences.* American Psychological Association, Washington, D.C.

Geary, D. C. (2000). Evolution and proximate expression of human paternal investment. *Psychological Bulletin* 126: 55–77.

Geary, D. C., & Flinn, M. V. (2001). Evolution of human parental behavior and the human family. *Parenting* 1: 5–61.

Geary, T. W., & Reeves, J. J. (1992). Relative importance of vision and olfaction for detection of estrus by bulls. *Journal of Animal Sciences* 70: 2726–2731.

Geise, A. R., & Hedrick, P. W. (2003). Genetic variation and resistance to a bacterial infection in endangered Gila topminnow. *Animal Conservation* 6: 369–377.

Gesquiere, L. R., Wango, E. O., Alberts, S. C., & Altmann, J. (2007). Mechanisms of sexual selection: Sexual swellings and estrogen concentrations as fertility indicators and cues for male consort decisions in wild baboons. *Hormones and Behavior* 51: 114–125.

Getty, T. (2002). Signaling health versus parasites. *American Naturalist* 159: 363–371.

Getty, T. (2006). Sexually selected signals are not similar to sports handicaps. *Trends in Ecology and Evolution* 21: 83–88.

Ghiselin, M. T. (1969). *The Triumph of the Darwinian Method.* University of California Press, Berkeley, CA.

Gibson, R. M., & Langen, T. A. (1996). How do animals choose their mates? *Trends in Ecology and Evolution* 11: 468–470.

Gil, D., Graves, J. A., & Slater, P. J. B. (1999). Seasonal patterns of singing in the willow warbler: Evidence against the fertility announcement hypothesis. *Animal Behaviour* 58: 995–1000.

Gilbert, S. F. (2000). *Developmental Biology.* Sinauer, New York.

Gillespie, D. H. (1977). Natural selection for variances in offspring numbers: A new evolutionary principle. *American Naturalist* 11: 1010–1014.

Ginsberg, J. R., & Huck, U. W. (1989). Sperm competition in mammals. *Trends in Ecology and Evolution* 4: 74–79.

Girdwood, G. F. (1842). Theory of menstruation. *Lancet* 43: 825–830.

Girolami, L., & Bielert, C. (1987). Female perineal swelling and its effect on male sexual arousal: An apparent sexual releaser in the chacma baboon (*Papio ursinus*). *International Journal of Primatology* 8: 651–661.

Godin, J. G. J., Herdman, E. J. E., & Dugatkin, L. A. (2005). Social influences on female mate choice in the guppy, *Poecilia reticulata:* Generalized and repeatable trait-copying behavior. *Animal Behaviour* 69: 999–1005.

Goel, V., Grafman, J., Sadato, N., & Hallett, M. (1995). Modeling other minds. *Neuroreport* 6: 1741–1746.

Goetz, A. T., & Shackelford, T. K. (2006). Sexual coercion and forced in-pair copulation as sperm competition tactics in humans. *Human Nature* 17: 265–282.

Goetz, A. T., Shackelford, T. K., Weekes-Shackelford, V. A., Euler, H. A., Hoier, S., Schmitt, D. P., & LaMunyon, C. W. (2005). Mate retention, sperm displacement, and human sperm competition: A preliminary investigation of tactics to prevent and correct female infidelity. *Personality and Individual Differences* 38: 749–763.

Gomendio, M., Harcourt, A. H., & Roldán, R. S. (1998). Sperm competition in mammals. In *Sperm Competition and Sexual Selection* (eds. T. R. Birkhead & A. P. Møller), pp. 667–756. Academic Press, New York.

Gomendio, M., Martin-Coello, J., Crespo, C., Magana, C., & Roldan, E. R. S. (2006). Sperm competition enhances functional capacity of mammalian spermatozoa. *Proceedings of the National Academy of Sciences USA* 103: 15113–15117.

Goodwin, T. E., Eggert, M. S., House, S. J., Weddell, M. E., Schulte, B. A., & Rasmussen, L. E. L. (2006). Insect pheromones and precursors in female African elephant urine. *Journal of Chemical Ecology* 32: 1849–1853.

Gooren, L. (2005). Hormone treatment of the adult transsexual. *Hormone Research* 64: 31–36.

Goranson, N. C., Ebersole, J. P., & Brault, S. (2005). Resolving an adaptive conundrum: Reproduction of *Caenorhabditis elegans* is not sperm-limited when food is scarce. *Evolutionary Ecology Research* 7: 325–333.

Gordon, T. P., Gust, D. A., Busse, C. D., & Wilson, M. E. (1991). Hormones and sexual behavior associated with postconception perineal swelling in the sooty mangabey (*Cercocebus torquatus atys*). *International Journal of Primatology* 12: 585–597.

Gosling, L. M., & Roberts, S. C. (2001). Scent-marking by male mammals: Cheat-proof signals to competitors and mates. *Advances in the Study of Behaviour* 30: 169–217.

Gottschall, J. A., & Gottschall, T. A. (2003). Are per-incident rape-pregnancy rates higher than per-incident consensual-pregnancy rates? *Human Nature* 14: 1–20.

Gottschall, J. A., Berkey, R., Drown, C., Fleischner, M., Glotzbecker, M., Kernan, K., Magnan, T., Muse, K., Ogburn, C., Skeels, C., St. Joseph, S., Weeks, S., Welch, A., & Welch, E. (2005). The heroine with a thousand faces: Universal trends in the characterization of female folk tale protagonists. *Evolutionary Psychology* 3: 85–103.

Gould, S. J. (1987). Freudian slip. *Natural History* 94: 14–21.

Gould, S. J., & Lewontin, R. C. (1979). The spandrels of San Marco and the panglossian paradigm: A critique of the adaptationist program. *Proceedings of the Royal Society of London B* 205: 581–598.

Gowaty, P. A. (1996). Battles of the sexes and the origins of monogamy. In *Partnerships in Birds* (ed. J. M. Black), pp. 21–52. Oxford University Press, Oxford, UK.

Gowaty, P. A. (1997). Sexual dialectics, sexual selection, and variation in reproductive success. In *Feminism and Evolutionary Biology: Boundaries, Intersections and Frontiers* (ed. P. A. Gowaty), pp. 351–384. Chapman & Hall, New York.

Gowaty, P. A., Drickamer, L. C., & Schmid-Holmes, S. (2003). Male house mice produce fewer offspring with lower viability and poorer performance when mated with females they do not prefer. *Animal Behaviour* 65: 95–103.

Gower, D. B., & Ruperelia, B. A. (1993). Olfaction in humans with special reference to odorous 16-androstenes: Their occurrence, perception and possible social, psychological and sexual impact. *Journal of Endocrinology* 137: 167–187.

Grafen, A. (1990). Biological signals as handicaps. *Journal of Theoretical Biology* 144: 517–546.

Graham, J. H., Shimizu, K., Emlen, J. M., Freeman, D. C., & Merkel, J. (2003). Growth model and the expected distribution of fluctuating asymmetry. *Biological Journal of the Linnean Society* 80: 57–65.

Grammer, K. (1993). 5-"-androst-16-en-3"-on: A male pheromone? A brief report. *Ethology and Sociobiology* 14: 201–214.

Grammer, K., Fieder, M., & Filova, V. (1997). The communication paradox and possible solutions. In *New Aspects of Human Ethology* (eds. A. Schmitt, K. Atzwanger, K. Grammer, & K. Schäfer), pp. 91–120. Plenum Press, New York.

Grammer, K., Fink, B., Juette, A., Ronzal, G., & Thornhill, R. (2002). Female faces and bodies: N-dimensional feature space and attractiveness. In *Advances in Visual Cognition: Vol. 1. Facial Attractiveness: Evolutionary, Cognitive and Social Perspectives* (eds. G. Rhodes & L. Zebrowitz), pp. 91–125. Greenwood, Westport, CT.

Grammer, K., Fink, B., Møller, A. P., & Manning, J. T. (2005). Physical attractiveness and health: Comment on Weeden and Sabini. *Psychological Bulletin* 131: 658–661.

Grammer, K., Fink, B., Møller, A. P., & Thornhill, R. (2003). Darwinian aesthetics: Sexual selection and the biology of beauty. *Biological Reviews* 78: 385–407.

Grammer, K., Keki, V., Striebel, B., Atzmüller, M., & Fink, B. (2003). Bodies in motion: A window to the soul. In *Evolutionary Aesthetics* (eds. E. Voland & K. Grammer), pp. 295–324. Springer–Verlag, Heidelberg, Germany.

Grammer, K., Renninger, L., & Fischer, B. (2004). Disco clothing, female sexual motivation, and relationship status: Is she dressed to impress? *Journal of Sex Research* 41: 66–74.

Gravitt, P. E., Hildesheim, A., Herrero, R., Schiffman, M., Sherman, M. E., Bratti, M. C., Rodriguez, A. C., Morera, L. A., Cardenas, F., Bowman, F. P., Shah, K. V., Crowley-Nowick, P. A. (2003). Correlates of IL-10 and IL-12 concentrations in cervical secretions. *Journal of Clinical Immunology,* 23: 175–183.

Gray, E. M. (1997). Do red-winged blackbirds benefit genetically from seeking copulations with extra-pair males? *Animal Behaviour* 53: 625–639.

Gray, J. P., & Wolfe, L. D. (1983). Human female sexual cycles and the concealment of ovulation problem. *Journal of Social and Biological Structures* 6: 345–352.

Gray, P. B. (2003). Marriage, parenting, and testosterone variation among Kenyan Swahili men. *American Journal of Physical Anthropology* 122: 279–286.

Gray, P. B., Chapman, J. F., Burnham, T. C., McIntyre, M. H., Lipson, S. F., & Ellison, P. T. (2004). Human male pair bonding and testosterone. *Human Nature* 15: 119–131.

Gray, P. B, Kahlenberg, S. M., Barrett, E. S., Lipson, S. F., & Ellison, P. T. (2002). Marriage and fatherhood are associated with lower testosterone in males. *Evolution and Human Behavior* 23: 193–201.

Gray, P. B., Yang, C. F. J., & Pope, H. G. (2006). Fathers have lower salivary testosterone levels than unmarried men and married non-fathers in Beijing, China. *Proceedings of the Royal Society of London* B 273: 333–339.

Green, D. J., Peters, A., & Cockburn, A. (2002). Extra-pair paternity and mate-guarding behaviour in the brown thornbill. *Australian Journal of Zoology* 50: 565–580.

Greene, P. J. (1979). Promiscuity, paternity and culture. *American Ethnologist* 5: 151–159.

Greiling, H., & Buss, D. M. (2000). Women's sexual strategies: The hidden dimension of women's short-term mating. *Personality and Individual Differences* 28: 929–963.

Griffith, S. C. (2007). The evolution of infidelity in socially monogamous passerines: Neglected components of direct and indirect selection. *American Naturalist* 169: 274–281.

Griffith, S. C., Owens, I. P. F., & Thuman, K. A. (2002). Extra-pair paternity in birds: A review of interspecific variation and adaptive function. *Molecular Ecology* 11: 2195–2212.

Griggio, M., Matessi, G., & Pilastro, A. (2003). Male rock sparrow (*Petronia petronia*) nest defense correlates with female ornament size. *Ethology* 109: 659–669.

Griggio, M., Valera, F., Cassas, A. J., & Pilastro, A. (2005). Males prefer ornamental females: A field experiment of male choice in the rock sparrow. *Animal Behaviour* 69: 1243–1250.

Gross, J. J. (1998). The emerging field of emotion regulation: An integrative review. *Review of General Psychology* 2: 271–299.

Gross, J. J. (2002). Emotion regulation: Affective, cognitive and social consequences. *Psychophysiology*, 39: 281–291.

Gurven, M., Kaplan, H., & Gutierrez, M. (2006). How long does it take to become a proficient hunter? Implications for the evolution of extended development and long life span. *Journal of Human Evolution* 51: 454–470.

Gwynne, D. T. (1991). Sexual competition among females: What causes courtship role reversal? *Trends in Ecology and Evolution* 6: 118–121.

Gwynne, D. T. (2001). *Katydids and Bush-crickets: Reproductive Behavior and Evolution of the Tettigoniidae.* Cornell University Press, Ithaca, NY.

Hagelin, J. C., Jones, I. L., & Rasmussen, L. E. L. (2003). A tangerine-scented social odour in a monogamous seabird. *Proceedings of the Royal Society of London* B 270: 1323–1329.

Haig, D. (1991). Brood reduction and optimal parental investment when offspring differ in quality. *American Naturalist* 136: 550–556.

Haig, D. (1993). Genetic conflicts in human pregnancy. *Quarterly Review of Biology* 68: 495–532.

Hall, B. K. (2003). Descent with modification: The unity underlying homology and homoplasy as seen through an analysis of development and evolution. *Biological Reviews* 78: 409–433.

Hall, G. H., Noble, W. L., Lindow, S. W., & Masson, E. A. (2001). Long-term sexual co-habilitation offers no protection from hypertensive disease of pregnancy. *Human Reproduction* 16: 349–352.

Hall, M. L., & Magrath, R. D. (2000). Duetting and mate-guarding in Australian magpie-larks (*Grallina cyanoleuca*). *Behavioral Ecology and Sociobiology* 47: 180–187.

Hallgrímsson, B. (1998). Fluctuating asymmetry in the mammalian skeleton: Evolutionary and developmental implications. In *Evolutionary Genetics, Volume 30* (ed. M. K. Hecht et al.). Plenum, New York.

Halpern, M. (1992). Nasal chemical senses in reptiles: Structure and function. In *Hormones, Brain, and Behavior: Biology of the Reptilia. Vol. 18: Physiology E* (eds. C. Gans & D. Crews), pp. 423–524. University of Chicago Press, Chicago, IL.

Hamilton, W. D. (1980). Sex vs. no-sex vs. parasite. *Oikos* 35: 282–290.

Hamilton, W. D. (1982). Pathogens as causes of genetic diversity in their host populations. In *Population Biology of Infectious Diseases* (eds. R. M. Anderson & R. M. May pp. 122–143). Springer-Verlag, New York.

Hamilton, W. D. (1984). Significance of paternal investment by primates to the evolution of adult male–female associations. In *Primate Paternalism* (ed. D. M. Taub), pp. 309–335. Van Nostrand–Reinhold, New York.

Hamilton, W. D., & Zuk, M. (1982). Heritable true fitness and bright birds: A role for parasites. *Science* 218: 384–387.

Hammerstein, P., & Parker, G. A. (1987). Sexual selection: Games between the sexes. In *Sexual Selection: Testing the Alternatives* (eds. J. W. Bradbury & M. B. Andersson), pp. 119–142. Wiley, Chichester, UK.

Hansen, G. L. (1986). Extradyadic relations during courtship. *Journal of Sex Research* 22: 382–390.

Hansen, T. F., Alvarez-Castro, J. M., Carter, A. J. R., Hermisson, J., & Wagner, G. P. (2006). Evolution of genetic architecture under directional selection. *Evolution* 60: 1523–1536.

Harano, T., Yasui, Y., & Miyatake, T. (2006). Direct effects of polyandry on female fitness in *Callosobruchus chinensis*. *Animal Behaviour* 71: 539–548.

Harris, C. R. (2003). A review of sex differences in sexual jealousy, including self-report data, psychophysiological responses, interpersonal violence, and morbid jealousy. *Personality and Social Psychology Review* 7: 102–128.

Harshman, L. G., & Prout, T. (1994). Sperm displacement without sperm tranfer in *Drosophila melanogaster*. *Evolution* 48: 758–766.

Hart, C. W. M., Pilling, A. R., & Goodale, J. C. (1987). *The Tiwi of North Australia*, 3rd ed. Holt, Rinehart, & Winston, New York.

Hartung, J. (1985). Matrilineal inheritance: New theory and analysis. *Behavioral and Brain Sciences* 8: 661–670.

Haselton, M. G., & Gangestad, S. W. (2006). Conditional expression of women's desires and male mate retention efforts across the ovulatory cycle. *Hormones and Behavior* 49: 509–518.

Haselton, M., & Miller, G. F. (2006). Women's fertility across the cycle increases the short-term attractiveness of creative intelligence compared to wealth. *Human Nature*, 17: 50–73.

Haselton, M. G., Mortezaie, M., Pillsworth, E. G., Bleske-Recheck, A. M., & Frederick, D. A. (2007). Ovulatory shifts in human female ornamentation: Near ovulation, women dress to impress. *Hormones and Behavior* 51: 40–45.

Haselton, M. G., & Nettle, D. (2006). The paranoid optimist: An integrative evolutionary model of cognitive biases. *Personality and Social Psychology Review* 10: 47–66.

Hassebrauck, M. (2003). The effect of fertility risk on relationship scrutiny. *Evolution and Cognition* 9: 116–122.

Hasselquist, D. (1998). Polygyny in great reed warblers: A long-term study of factors contributing to male fitness. *Ecology* 79: 2376–2390.

Hasselquist, D., Bensch, S., & von Schantz, T. (1996). Correlation between male song repertoire, extrapair paternity and offspring survival in the great reed warbler. *Nature* 381: 229–232.

Hauser, M. D., Tsao, F., Garcia, P., & Spelke, E. S. (2003). Evolutionary foundations of number: Spontaneous representation of numerical magnitudes by cotton-top tamarins. *Proceedings of the Royal Society of London B* 270: 1441–1446.

Hausfater, G. (1975). *Dominance and Reproduction in Baboons* (*Papio cynocephalus*). Karger, Basel, Switzerland.

Havliček, J., Dvorakova, R., Bartos, L., & Flegr, J. (2006). Non-advertized does not mean concealed: Body odour change across the human menstrual cycle. *Ethology* 112: 81–90.

Havliček, J., Roberts, S. C., & Flegr, J. (2005). Women's preference for dominant male odour: Effects of menstrual cycle and relationship status. *Biology Letters*, 1: 256–259.

Hawkes, K. (1991). Showing off: Tests of an hypothesis about men's foraging goals. *Ethology and Sociobiology* 12: 29–54.

Hawkes, K. (2003). Grandmothers and the evolution of human longevity. *American Journal of Human Biology* 15: 380–400.

Hawkes, K. (2004). Mating, parenting, and the evolution of human pair bonds. In *Kinship and Behavior in Primates* (eds. B. Chapais & C. M. Berman), pp. 443–473. Oxford University Press, Oxford, UK.

Hawkes, K., & Bliege Bird, R. (2002). Showing off, handicap signaling, and the evolution of men's work. *Evolutionary Anthropology* 11: 58–67.

Hawkes, K., O' Connell, J. F., & Blurton Jones, N. G. (1991). Hunting patterns among the Hadza: Big game, common goals, foraging goals and the evolution of the human diet. *Philosophical Transactions of the Royal Society of London* B 334: 243–251.

Hawkes, K., O' Connell, J. F., & Blurton Jones, N. G. (2001). Hunting and nuclear families: Some lessons from the Hadza about men's work. *Current Anthropology* 42: 681–709.

Hebets, E. A., & Papaj, D. R. (2005). Complex signal function: Developing a framework of testable hypotheses. *Behavioral Ecology and Sociobiology* 57: 197–214.

Hedrick, P. W. (1998). Balancing selection and the MHC. *Genetica* 104: 207–214.

Hedrick, P. W. (2002). Pathogen resistance and genetic variation at MHC loci. *Evolution* 56: 1902–1908.

Hedrick, P. W., & Black, F. L. (1997a). HLA and mate selection: No evidence in South Amerindians. *American Journal of Human Genetics* 61: 505–511.

Hedrick, P. W., & Black, F. L. (1997b). Random mating and selection within families against homozygotes for HLA in South Amerindians. *Hereditas* 127: 51–58.

Hedricks, C., Piccinino, L. J., Udry, J. R., & Chimbira, T. H. (1987). Peak coital rate coincides with onset of luteinizing hormone surge. *Fertility and Sterility* 48: 234–238.

Hegedus, Z. L. (2000). The probable involvement of soluble and deposited melanins, their intermediates, and the reactive oxygen side-products in human diseases and aging. *Toxicology* 145: 85–101.

Heistermann, M., Ziegler, T., van Schaik, C. P., Launhardt, K., Winkler, P., & Hodges, J. K. (2001). Loss of oestrus, concealed ovulation and paternity confusion in free-ranging Hanuman langurs. *Proceedings of the Royal Society of London B* 268: 2445–2451.

Henderson, J. J. A., & Anglin, J. M. (2003). Facial attractiveness predicts longevity. *Evolution and Human Behavior* 24: 351–356.

Herman-Giddens, M. E., Slora, E. J., Wasserman, R. C., Bourdony, C. L., Bhapkar, M. V., Koch, G. G., & Hasemeier, C. M. (1997). Secondary sexual characteristics and menses in young girls seen in office practice: A study from the pediatric research in office settings network. *Pediatrics* 99: 505–517.

Hermans, E. P., Putnam, P., Baas, J. M., Koppeschaar, H. P., & van Honk, J. (2006). Single administration of testosterone reduces fear-potentiated startle. *Biological Psychiatry* 59: 872–874.

Heymann, E. W. (2003). Scent marking, paternal care, and sexual selection in Callitrichines. In *Sexual Selection and Reproductive Competition in Primates: New Perspectives and Directions* (ed. C. B. Jones), pp. 305–325. American Society of Primatologists, Norman, OK.

Hill, E. M. (1988). The menstrual cycle and components of human female sexual behavior. *Journal of Social and Biological Structures* 11: 443–455.

Hill, G. E. (2002). Red bird in a brown bag: The function and evolution of colorful plumage coloration. *Journal of Avian Biology* 31: 559–566.

Hill, K. (1982). Hunting and human evolution. *Journal of Human Evolution* 11: 521–544.

Hill, K. R., & Hurtado, A. M. (1996). *Ache Life History: The Ecology and Demography of a Forest People.* Aldine de Gruyter, New York.

Hirschfield, M. F., & Tinkle, D. W. (1975). Natural selection and the evolution of reproductive effort. *Proceedings of the National Academy of Sciences USA* 72: 2227–2231.

Hite, S. (1987). *Women and Love: A Cultural Revolution in Progress.* Grove Press, New York,.

Hohoff, C., Franzin, K., & Sachser, N. (2003). Female choice in a promiscuous wild guinea pig, the yellow-toothed cavy (*Galea musteloides*). *Behavioral Ecology and Sociobiology* 53: 341–349.

Hoi, H. (1997). Assessment of the quality of copulation partners in the monogamous bearded tit. *Animal Behaviour* 53: 277–286.

Holden, C. J., Sear, R., & Mace, R. (2003). Matriliny as daughter-biased investment. *Evolution and Human Behavior* 24: 99–112.

Holland, B., & Rice, W. R. (1998). Perspective: Chase-away sexual selection: Antagonistic seduction vs. resistance. *Evolution* 52: 1–7.

Holland, B., & Rice, W. R. (1999). Experimental removal of sexual selection reverses intrasexual antagonistic coevolution and removes a reproductive load. *Proceedings of the National Academy of Sciences USA* 96: 5083–5088.

Hönekopp, J., Bartholomé, T., & Jansen, G. (2004). Facial attractiveness, symmetry, and physical fitness in young women. *Human Nature* 15: 147–167.

Hoogland, J. L. (1995). *The Black-tailed Prairie Dog: Social Life of a Burrowing Mammal.* University of Chicago Press, Chicago.

Houde, A. E. (1997). *Sex, Color, and Mate Choice in Guppies.* Princeton University Press, Princeton, NJ.

Houle, D. (1992). Comparing evolvability and variability of traits. *Genetics* 130: 195–204.

Houle, D. (1998). How should we explain variation in the genetic variance in traits? *Genetica* 102/103: 241–253.

Houle, D., Morikawa, B., & Lynch, M. (1996). Comparing mutational variabilities. *Genetics* 143: 1467–1483.

Houtman, A. M. (1992). Female zebra finches choose extra-pair copulations with genetically attractive males. *Proceedings of the Royal Society of London* B 249: 3–6.

Hrdy, S. B. (1979). Infanticide among animals, a review, classification and examination of the implications for the reproductive strategies of females. *Ethology and Sociobiology* 1: 3–40.

Hrdy, S. B. (1981). *The Woman That Never Evolved.* Harvard University Press, Cambridge, MA.

Hrdy, S. B. (1997). Raising Darwin's consciousness: Female sexuality and the prehominid origins of patriarchy. *Human Nature* 8: 1–49.

Hrdy, S. B. (2000). The optimal number of fathers: Evolution, demography, and history in the shaping of female mate preferences. *Annals of the New York Academy of Sciences* 907: 75–96.

Hrdy, S. B., & Whitten, P. L. (1987). Patterning of sexual activity. In *Primate Societies* (eds. B. B. Smuts, D. L. Cheney, R. M. Seyfarth, R. W. Wrangham, & T. T. Struhsaker), pp. 370–384. University of Chicago Press, Chicago, IL.

Hua, C. (2001). *A Society Without Fathers or Husbands: The Na of China.* MIT Press, Cambridge, MA.

Hughes, K. A. (1995). The evolutionary genetics of male life-history characters in *Drosophila melanogaster. Evolution* 49: 521–537.

Hughes, S. M., Dispenza, F., & Gallup, G. G., Jr. (2004). Ratings of voice attractiveness predicts sexual behavior and body configuration. *Evolution and Human Behavior* 25: 295–304.

Hughes, S. M., Harrison, M. A., & Gallup, G. G., Jr. (2002). The sound of symmetry: Voice as a marker of developmental instability. *Evolution and Human Behavior* 23: 173–180.

Hummel, T., Gollisch, R., Wildt, G., & Kobal, G. (1991). Changes in olfactory perception during the menstrual cycle. *Experientia* 47: 712–715.

Hunter, M., & Davis, L. S. (1998). Female Adélie penguins acquire nest material from extrapair males after engaging in extrapair copulations. *Auk* 115: 526–528.

Hurst, J. L., & Rich, T. J. (1999). Scent marks as competitive signals of mate quality. In *Advances in Chemical Signals in Vertebrates* (eds. R. E. Johnston, D. Müller-Schwarze, & P. W. Sorensen), pp. 209–226. Kluwer Academic/Plenum, New York.

Ibáñez, L., Ong, K. K., Mongan, N., Jaaskelainen, J., Marcos, M. V., Hughes, I. A., de Zegher, F., & Dunger, D. B. (2003). Androgen receptor gene CAG repeat polymorphism in the development of ovarian hyperandrogenism. *Journal of Clinical Endocrinology and Metabolism* 88: 3333–3338.

Ihara, Y., Aoki, K., Tokumaga, K., Takahashi, K., & Juji, T. (2000). HLA and human mate choice: Tests on Japanese couples. *Anthropological Science* 108: 199–214.

Ishida, Y., Yahara, T., Kasuya, E., & Yamane, A. (2001). Female control of paternity during copulation: Inbreeding avoidance in feral cats. *Behaviour* 138: 235–250.

Jablonka, E., & Lamb, M. L. (2005). *Evolution in four dimensions.* MIT Press, Cambridge, MA.

Jackson, L. A. (1992). *Physical Appearance and Gender: Sociobiological and Sociocultural Perspectives.* State University of New York Press, Albany, NY.

Jacob, S., McClintock, M. K., Zelano, B., & Ober, C. (2002). Paternally inherited HLA alleles are associated with male choice of male odor. *Nature Genetics* 30: 175–179.

Jacobson, M. (1972). *Insect Sex Pheromones.* Academic Press, New York.

James, W. H. (1980). Implications of a time-dependent model of sexual intercourse within the menstrual cycle: A comment. *Journal of Biosocial Science* 12: 495–496.

Jameson, E. W. (1988). *Vertebrate Reproduction.* Wiley, New York.

Jankowiak, W. R., & Fischer, E. F. (1992). A cross-cultural perspective on romantic love. *Ethnology* 31: 148–155.

Jankowiak, W., Nell, M. D., & Buckmaster, A. (2002). Managing infidelity: A cross-cultural perspective. *Ethnology* 41: 85–101.

Jasieńska, G. (2003). Energy metabolism and the evolution of reproductive suppression in the human female. *Acta Biotheoretica* 51: 1–18.

Jasieńska, G., Lipson, S. F., Ellison, P. T., Thune, I., & Ziomkiewicz, A. (2006). Symmetrical women have higher potential fertility. *Evolution and Human Behavior* 27: 390–400.

Jasieńska, G., Ziomkiewicz, A., Ellison, P. T., Lipson, S. F., & Thune, I. (2004). Large breasts and narrow waists indicate high reproductive potential in women. *Proceedings of the Royal Society of London B* 271: 1213–1217.

Jawor, J. M., Gray, N., Beall, S. M., & Breitwisch, R. (2004). Multiple ornaments correlate with aspects of condition and behavior in female northern cardinals, *Cardinalis cardinalis*. *Animal Behaviour* 67: 875–882.

Jennions, M. D., Møller, A. P., & Petrie, M. (2001). Sexually selected traits and adult survival: A meta-analysis. *Quarterly Review of Biology* 76: 3–36.

Jennions, M. D., & Petrie, M. (2000). Why do females mate multiply? A review of the genetic benefits. *Biological Reviews* 75: 21–64.

Jeppsson, B. (1986). Mating by pregnant water voles (*Arvicola terrestris*): A strategy to counter infanticide by males. *Behavioral Ecology and Sociobiology* 19: 293–296.

Jöchle, W. (1973). Coitus-induced ovulation. *Contraception* 7: 523–564.

Johnsen, A., Andersson, S., Ornborg, J., & Lifjeld, J. T. (1998). Ultraviolet plumage ornamentation affects social mate choice and sperm competition in bluethroats (Aves: *Luscinia s. svecica*): A field experiment. *Proceedings of the Royal Society of London B* 265: 1313–1318.

Johnsen, A., Andersen, V., Sunding, C., & Lifjeld, J. T. (2000). Female bluethroats enhance offspring immunocompetence through extra-pair copulations. *Nature* 406: 296–299.

Johnsen, A., Lifjeld, J. T., Andersson, S., Ornborg, J., & Amundsen, T. (2001). Male characteristics and fertilisation success in male bluethroats. *Behaviour* 138: 1371–1390.

Johnson, K., Thornhill, R., Ligon, J. D., & Zuk, M. (1993). The direction of mothers' and daughters' preferences and the heritability of male ornaments in red jungle fowl (*Gallus-Gallus*). *Behavioral Ecology* 4: 254–259.

Johnson, K. L., & Tassinary, L. G. (2007). Compatibility of basic social perceptions determines perceived attractiveness. *Proceedings of the National Academy of Sciences USA* 104: 5246–5251.

Johnson, W., Gangestad, S. W., Segal, N. L., & Bouchard, T. J., Jr. (in press). Heritability of fluctuating asymmetry in a human twin sample: The effect of trait aggregation. *American Journal of Human Biology*.

Johnson, W., Segal, N. L., & Bouchard, T. J., Jr. (2007). Fluctuating asymmetry and general intelligence: No phenotypic or genetic association. *Intelligence*. doi: 10.1016/j.intell.2007.07.001.

Johnston, V. S., & Franklin, M. (1993). Is beauty in the eye of the beholder? *Ethology and Sociobiology* 14: 183–199.

Johnston, V. S., Hagel, R., Franklin, M., Fink, B., & Grammer, K. (2001). Male facial attractiveness: Evidence for hormone-mediated adaptive design. *Evolution and Human Behavior* 22: 251–267.

Johnstone, R. A. (1995). Honest advertisement of multiple qualities using multiple signals. *Journal of Theoretical Biology* 177: 87–94.

Johnstone, R. A. (1996). Multiple displays in animal communication: Back-up signals and multiple messages. *Philosophical Transactions of the Royal Society of London B*, 352: 329–338.

Jolly, A. (1972). *The Evolution of Primate Behavior.* Macmillan, New York.

Jones, B. C., Little, A. C., Penton-Voak, I., Tiddeman, B. P., Burt, D. M., & Perrett, D. I. (2001). Facial symmetry and judgments of apparent health: Support for a "good genes" explanation of the attractiveness–symmetry relationship. *Evolution and Human Behavior* 22: 417–429.

Jones, B. C., Little, A. C., Boothroyd, L., DeBruine, L. M., Feinberg, D. R., Law Smith, M. J., Cornwell, R. E., & Perrett, D. I. (2005). Commitment to relationships and preferences for femininity and apparent health in faces are strongest on days of the menstrual cycle when progesterone level is high. *Hormones and Behavior* 48: 283–290.

Jones, B. C., Perrett, D. I., Little, A. C., Boothroyd, L., Cornwell, R. E., Feinberg, D. R., Tiddeman, B. P., Whiten, S., Pitman, R. M., Hillier, S. G., Burt, D. M., Stirrat, M. R., Law Smith, M. J., & Moore, F. R. (2005). Menstrual cycle, pregnancy and oral contraceptive use alter attraction to apparent health in faces. *Proceedings of the Royal Society of London B*, 272: 347–354.

Jones, C. B. (2003). *Sexual Selection and Reproductive Competition in Primates: New Perspectives and Directions.* American Society of Primatologists, Norman, OK.

Jones, D. (1996). *Physical Attractiveness and the Theory of Sexual Selection: Results From Five Populations* (with a foreword by D. Symons). Museum of Anthropology, University of Michigan, Ann Arbor, MI.

Jones, D., & Hill, K. (1993). Criteria of facial attractiveness in five populations. *Human Nature* 4: 271–296.

Jones, R. E. (1978). *The Vertebrate Ovary: Comparative Biology and Evolution.* Plenum Press, New York.

Jones, R. E., Guillette, L. J., Summers, C. H., Tokarz, R. R., & Crews, D. (1983). The relationship among ovarian condition, steroid-hormones, and estrous behavior in *Anolis carolinensis*. *Journal of Experimental Zoology* 227: 145–154.

Kanazawa, S. (2001). Why father absence might precipitate early menarche: The role of polygyny. *Evolution and Human Behavior* 22: 329–334.

Kanazawa, S., & Vandermassen, G. (2006). Engineers have more sons, nurses have more daughters: An evolutionary psychological extension of Baron-Cohen's extreme male brain theory of autism. *Journal of Theoretical Biology* 233: 589–599.

Kanhai, R. C. J., Hage, J. J., van Diest, P. J., Bloemena, E., & Mulder, J. W. (2000). Short-term and long-term histologic effects of castration and estrogen treatment on breast tissue of 14 male-to-female transsexuals in comparison with two chemically castrated men. *American Journal of Surgical Pathology* 24: 74–80.

Kano, T. (1980). Social behavior of wild pygmy chimpanzees (*Pan paniscus*) of Wamba: A preliminary report. *Journal of Human Evolution* 9: 243–260.

Kaplan, H. (1997). The evolution of the human life course. In *Between Zeus and Salmon: The Biodemography of Longevity* (eds. K. Watcher & C. Finch), pp. 175–211. National Academy of Sciences, Washington, DC.

Kaplan, H., & Hill, K. (1985). Hunting ability and reproductive success among male Ache foragers: Preliminary tests. *Current Anthropology* 26: 131–133.

Kaplan, H., Hill, K., Lancaster, J., & Hurtado, A. M. (2000). A theory of human life history evolution: Diet, intelligence, and longevity. *Evolutionary Anthropology* 9: 156–185.

Kaplan, H., Lancaster, J., & Robson, A. J. (2003). Embodied capital and the evolutionary economics of the human life span. *Population and Development Review* 29: 152–182.

Kavaliers, M., & Colwell, D. D. (1995a). Odours of parasitized males induce aversive responses in female mice. *Animal Behavior* 50: 1161–1169.

Kavaliers, M., & Colwell, D. D. (1995b). Discrimination by female mice between the odours of parasitized males. *Proceedings of the Royal Society of London B* 261: 31–35.

Kavaliers, M., Choleris, E., Ågmo, A., Muglia, L. J., Ogawa, S., & Pfaff, D. W. (2005). Involvement of the oxytocin gene in the recognition and avoidance of parasitized males by female mice. *Animal Behaviour* 70: 693–702.

Keightley, P. D., Lercher, M. J., & Eyre-Walker, A. (2005). Evidence for widespread degradation of gene control regions in hominid genomes. *PLOS Biology* 3: 282–288.

Keller, M. C., & Miller, G. (2006). Resolving the paradox of common, harmful, heritable mental disorders: Which evolutionary models work best? *Behavioral and Brain Sciences* 29: 385–452.

Kelly, C. D., Godin, J. G. J., & Wright, J. M. (1999). Geographical variation in multiple paternity within natural populations of the guppy (*Poecilia reticulata*). *Proceedings of the Royal Society of London B* 266: 2403–2408.

Kenrick, D. T., Groth, G., Trost, M. R., & Sadalla, E. K. (1993). Integrating evolutionary and social exchange perspectives on relationships: Effects of gender, self-appraisal, and involvement level on mate selection criteria. *Journal of Personality and Social Psychology* 64: 951–969.

Kenrick, D. T., Sadalla, E. K., Groth, G., & Trost, M. R. (1990). Evolution, traits, and the stages of human courtship: Qualifying the parental investment model. *Journal of Personality* 58: 97–116.

Kinsey, A. C., Pomeroy, W. B., Martin, C. E., & Gebbard, P. H. (1953). *Sexual Behavior in the Human Female.* W. B. Saunders, Philadelphia, PA.

Kirchengast, S., & Gartner, M. (2002). Changes in fat distribution (WHR) and body weight across the menstrual cycle. *Collegium Antropologicum* 26: 47–57.

Kirchengast, S., Gruber, D., Sator, M., Knogler, W., & Huber, J. (1997). The fat distribution index: A new possibility to quantify sex-specific fat patterning in females. *Homo* 48: 285–295.

Kirchengast, S., & Huber, J. (2001a). Body composition characteristics and body fat distribution in lean women with polycystic ovary syndrome. *Human Reproduction* 16: 1255–1260.

Kirchengast, S., & Huber, J. (2001b). Fat distribution patterns in young amenorrheic females. *Human Nature* 12: 123–140.

Kirchengast, S., & Huber, J. (2004). Body composition characteristics and fat distribution patterns in young infertile women. *Fertility and Sterility* 81: 539–544.

Kirkpatrick, M. (1982). Sexual selection and the evolution of female choice. *Evolution* 36: 1–12.

Kjaer, K. W., Hansen, L., Eiberg, H., Utkus, A., Skovgaard, L. T., Leicht, P., Opitz, J. M., & Tommerup, M. (2005). A 72-year-old Danish puzzle resolved: Comparative analysis of phenotypes of families with different sized HOXD13 polyalanine expansions. *American Journal of Medical Genetics* 138A, 328–339.

Kleiman, D. G. (1977). Monogamy in mammals. *Quarterly Review of Biology* 52: 39–69.

Kleiman, D. G., & Mack, D. S. (1977). A peak in sexual activity during mid-pregnancy in the golden-lion tamarin, *Leontopithecus rosalia* (Primates: Callitrichidae). *Journal of Mammalogy* 58: 657–660.

Klusmann, D. (2002). Sexual motivation and the duration of partnership. *Archives of Sexual Behavior* 31: 275–287.

Klusmann, D. (2006). Sperm competition and female procurement of resources as explanations for a sex-specific time course in the sexual motivation of couples. *Human Nature* 17: 283–300.

Knight, T. M., Steets, J. A., Vamosi, J. C., Mazer, S. J., Burd, M., Campbell, D. R., Dudash, M. R., Johnston, M. O., Mitchell, R. J., & Ashman, T.-L. (2005). Pollen limitation of plant reproduction: Pattern and process. *Annual Review of Ecology, Evolution and Systematics* 36: 467–497.

Kodric-Brown, A., & Brown, J. H. (1984). Truth in advertising: The kinds of traits favored by sexual selection. *American Naturalist* 124: 309–323.

Kodric-Brown, A., & Brown, J. H. (1987). Anisogamy, sexual selection, and the evolution and maintenance of sex. *Evolutionary Ecology* 1: 95–105.

Kodric-Brown, A., & Nicoletto, P. F. (2005). Courtship behavior, swimming performance, and microhabitat use of Trinidadian guppies. *Environmental Biology of Fishes* 73: 299–307.

Koehler, N., Rhodes, G., & Simmons, L. W. (2002). Are human female preferences for symmetrical male faces enhanced when conception is likely? *Animal Behaviour* 64: 233–238.

Koehler, N., Rhodes, G., Simmons, L. W. & Zebrowitz, L. (2006). Do cyclic changes in women's face preferences target cues to long-term health? *Social Cognition* 24: 641–656.

Koehler, N., Simmons, L. W., Rhodes, G., & Peters, M. (2004). The relationship between sexual dimorphism in human faces and fluctuating asymmetry. *Proceedings of the Royal Society of London B*, 271: S233–S236.

Kokko, H. (1997). Evolutionarily stable strategies of age-dependent sexual advertisement. *Behavioral Ecology and Sociobiology* 41: 99–107.

Kokko, H., Brooks, R., Jennions, M. D., & Morley, J. (2003). The evolution of mate choice and mating biases. *Proceedings of the Royal Society of London B* 270: 653–664.

Kokko, H., Brooks, R., McNamara, J. M., & Houston, A. I. (2002). The sexual selection continuum. *Proceedings of the Royal Society of London B* 269: 1331–1340.

Kokko, H., & Jennions, M. (2002). It takes two to tango. *Trends in Evolution and Ecology* 18: 103–104.

Kokko, H., & Johnstone, R. A. (2002). Why is mutual mate choice not the norm? Operational sex ratios, sex roles and the evolution of sexually dimorphic and monomorphic signalling. *Philosophical Transactions of the Royal Society of London* B 357: 319–330.

Komdeur, J., Kraaijeveld-Smit, F., Kraaijeveld, K., & Edelaar, P. (1999). Explicit experimental evidence for the role of mate guarding in minimizing loss of paternity in the Seychelles warbler. *Proceedings of the Royal Society of London* B 266: 2075–2081.

Komers, P. E., Birgersson, B., & Ekvall, K. (1999). Timing of estrus in fallow deer is adjusted to the age of available mates. *American Naturalist* 153: 431–436.

Kondrashov, A. S. (2001). Sex and U. *Trends in Genetics* 17: 75–77.

Konner, M. (1982). *The tangled wing: Biological constraints on the human spirit.* New York: Holt, Rinehart, & Winston.

Koprowski, J. L. (2007). Alternative reproductive tactics and strategies of tree squirrels. In *Rodent Societies* (eds. J. Wolff & P. W. Sherman), pp. 52–79. University of Chicago Press, Chicago, IL.

Korpimäki, E., Lahti, K., May, C. A., Parkin, D. T., Powell, G. B., Tolonen, P., & Wetton, J. H. (1996). Copulatory behaviour and paternity determined by DNA fingerprinting in kestrels: Effects of cyclic food abundance. *Animal Behaviour* 51: 945–955.

Krasnoff, S. B., & Roelofs, W. L. (1988). Sex pheromone released as an aerosol by the moth, *Pyrrharctia isabella. Nature* 333: 263–265.

Krasnoff, S. B., & Roelofs, W. L. (1990). Evolutionary trends in the male phermone systems of arctiid moths: Evidence from studies of courtship in *Phragmatobia fuliginosa* and *Pyrrharctia isabella* (Lepidoptera: Arctiidae). *Zoological Journal of the Linnean Society* 99: 319–338.

Krasnoff, S. B., & Yager, D. D. (1988). Acoustic response to a pheromonal cue in the arctiid moth, *Pyrrharctia isabella. Physiological Entomology* 13: 433–440.

Krebs, E. A., Hunte, W., & Green, D. J. (2004). Plume variation, breeding performance and extra-pair copulations in the cattle egret. *Behaviour* 141: 479–499.

Krug, R., Pietrowsky, R., Fehm, H. L., & Born, J. (1994). Selective influence of the menstrual cycle on perception of stimuli with reproductive significance. *Psychosomatic Medicine* 56: 410–417.

Krug, R., Plihal, W., Fehm, H. L., & Born, J. (2000). Selective influence of the menstrual cycle on perception of stimuli with reproductive significance: An event-related potential study. *Psychophysiology* 37: 111–122.

Kruger, D. J. (2006). Male facial masculinity influences attributions of personality and reproductive strategy. *Personal Relationships* 13: 451–463.

Kruuk, L. E. B., Clutton-Brock, T. H., Slate, J., Pemberton, J. M., Brotherstone, S., & Guinness, F. E. (2000). Heritability of fitness in a wild mammal population. *Proceedings of the National Academy of Sciences USA* 97: 698–703.

Kruuk, L. E. B., Slate, J., Pemberton, J. M., & Clutton-Brock, T. H. (2003). Fluctuating asymmetry in a secondary sexual trait: No associations, individual stress, or inbreeding, and no heritability. *Journal of Evolutionary Biology* 16: 101–113.

Kumar, K. R., Archunan, G., Jeyaraman, R., & Narasimhan, S. (2000). Chemical characterization of bovine urine with special reference to oestrus. *Veterinary Research Communications* 24: 445–454.

Kuukasjärvi, S., Eriksson, C. J. P., Koskela, E., Mappes, T., Nissinen, K., & Rantala, M. J. (2004). Attractiveness of women's body odors over the menstrual cycle: The role of oral contraception and received sex. *Behavioral Ecology* 15: 579–584.

Laaksonen, M., Sarlio-Lahteenkorva, S., & Lahelma, E. (2004). Multiple dimensions of socioeconomic position and obesity among employees: The Helsinki Heath Study. *Obesity Research* 12: 1851–1858.

Lancaster, J. B. (1986). Human adolescence and reproduction: An evolutionary perspective. In *School-Age Pregnancy and Parenthood: Biosocial Dimensions* (eds. J. B. Lancaster & B. A. Hamburg), pp. 17–37. Aldine de Gruyter, New York.

Lancaster, J. B. (1997). The evolutionary history of human parental investment in relation to population growth and social stratification. In *Feminism and Evolutionary Biology* (ed. P. A. Gowaty), pp. 466–488. Chapman & Hall, New York.

Lancaster, J. B., & Kaplan, H. S. (in press). The endocrinology of the human adaptive complex. In *Endocrinology of Social Relationships* (eds. P. G. Gray & P. T. Ellison). Harvard University Press, Cambridge, MA.

Lancaster, J. B., & Lancaster, C. S. (1983). Parental investment: The hominid adaptation. In D. Ortner (Ed.), *Parental care in mammals* (pp. 347–387). New York: Plenum.

Landaeta-Hernandez, A. J., Yelich, J. V., Lemaster, J. W., Fields, M. J., Tran, T., Chase, C. C., Rae, D. O., & Chenoweth, P. J. (2002). Environmental, genetic and social factors affecting the expression of estrus in beef cows. *Theriogenology* 57: 1357–1370.

Lande, R. (1981). Models of speciation by sexual selection on polygenic traits. *Proceedings of the National Academy of Sciences USA* 78: 3721–3725.

Lange, I. G., Hartel, A., & Meyer, H. H. D. (2002). Evolution of oestrogen functions in vertebrates. *Journal of Steroid Biochemistry and Molecular Biology* 83: 219–226.

Langefors, A., Hasselquist, D., & von Schantz, T. (1998). Extra-pair fertilizations in the sedge warbler. *Journal of Avian Biology* 29: 134–144.

Langlois, J. H., Kalakanis, L., Rubenstein, A. J., Larson, A., Hallam, M., & Smoot, M. (2000). Maxims or myths of beauty? A meta-analytic and theoretical review. *Psychological Bulletin* 126: 390–423.

Langmore, N. E., Davies, N. B., Hatchwell, B. J., & Hartley, I. R. (1996). Female song attracts males in the alpine accentor *Prunella collaris*. *Proceedings of the Royal Society of London B* 263: 141–146.

Lank, D. B., Smith, C. M., Hanotte, O., Ohtonen, A., Bailey, S., & Burke, T. (2002). High frequency of polyandry in a lek mating system. *Behavioral Ecology* 13: 209–215.

Laska, M., & Freyer, D. (1997). Olfactory discrimination ability for aliphatic esters in squirrel monkeys and humans. *Chemical Senses* 22: 457–465.

Lassek, W., & Gaulin, S. J. (2006). Changes in body fat distribution in relation to parity in American women: A covert form of maternal depletion. *American Journal of Physical Anthropology* 131: 295–302.

Lassek, W. D., & Gaulin, S. J. C. (2008). Waist-hip ratio and cognitive ability: Is gluteofemoral fat a priveledged store for neurodevelopmental resources? *Evolution and Human Behavior* 29: 26–34.

Laumann, E. O., Gagnon, J. H., Michael, R. T., & Michaels, S. (1994). *The Social Organization of Sexuality.* University of Chicago Press, Chicago, IL.

Law Smith, M. J., Perrett, D. I., Jones, B. C., Cornwell, R. E., Moore, F. R., Feinberg, D. R., Boothroyd, L. G., Durrani, S. J., Stirrat, M. R., Whiten, S., Pitman, R. M., & Hillier, S. G. (2006). Facial appearance is a cue to estrogen levels in women. *Proceedings of the Royal Society of London B* 273: 135–140.

Lea, J., Halliday, T., & Dyson, M. (2000). Reproductive stage and history affect the phonotactic preferences of female midwife toads, *Alytes muletensis*. *Animal Behaviour* 60: 423–427.

Leach, E. (1988). The social anthropology of marriage and mating. In V. Reynolds & J. Kellett (Eds.), *Mating and marriage* (pp. 91–110). Oxford University Press, Oxford, UK.

Leamy, L., & Klingenberg, C. (2005). The genetics and evolution of fluctuating asymmetry. *Annual Review of Ecology, Evolution and Systematics* 36: 1–21.

LeBas, N. R., Hockman, L. R., & Ritchie, M. G. (2003) Nonlinear and correlational sexual selection on "honest" female ornamentation. *Proceedings of the Royal Society of London B* 270: 2159–2165.

Le Boeuf, B. J. (1967). Interindividual associations in dogs. *Behaviour* 29: 268–295.

Leinders-Zufall, T., Brennan, P., Widmayer, P., Chandramani, S. P., Maul-Pavicic, A., Jäger, M., Li, X., Breer, H., Zufall, F., & Boehm, T. (2004). MHC Class I peptides as chemosensory signals in the vomeronasal organ. *Science* 306: 1033–1037.

Lens, L., & Van Dongen, S. (1999). Evidence for organism-wide asymmetry in five bird species of a fragmented afrotropical forest. *Proceedings of the Royal Society of London* B 266: 1055–1060.

Leslie, A. M. (1987). Pretense and representation: The origins of theory of mind. *Psychological Review* 94: 412–426.

Leung, B., & Forbes, M. R. (1996). Fluctuating asymmetry in relation to stress and fitness: Effects of trait type as revealed by meta-analysis. *Ecoscience* 3: 400–413.

Leung, B., Forbes, M. R., & Houle, D. (2000). Fluctuating asymmetry as a bioindicator of stress: Comparing efficacy of analyses involving multiple traits. *American Naturalist* 155: 101–115.

Levin, R. J. (2002). The physiology of sexual arousal in the human female: A recreational and procreational synthesis. *Archives of Sexual Behavior, 31:* 405–411.

Lew, T. A., Morrow, E. H., & Rice, W. R. (2006). Standing genetic variance in female resistance to harm from males and its relationship to interlocus sexual conflict. *Evolution* 60: 97–105.

Lew, T. A., & Rice, W. R. (2005). Natural selection favours harmful male *Drosophila melanogaster* that reduce the survival of females. *Evolutionary Ecology Research* 7: 633–641.

Lieberman, D., Tooby, J., & Cosmides, L. (2003). Does morality have a biological basis? An empirical test of the factors governing moral sentiments relating to incest. *Proceedings of the Royal Society of London* B 270: 819–826.

Lieberman, D., Tooby, J., & Cosmides, L. (2007). The architecture of human kin detection. *Nature* 445: 727–731.

Ligon, J. D., Thornhill, R., Zuk, M., & Johnson, K. (1990). Male–male competition, ornamentation and the role of testosterone in sexual selection in red jungle fowl. *Animal Behaviour* 40: 367–373.

Ligon, J. D., & Zwartjes, P. W. (1995). Female red junglefowl choose to mate with multiple males. *Animal Behaviour* 49: 127–135.

Liley, N. R., & Stacey, N. E. (1983). Hormones, pheromones, and reproductive behavior in fish. *Fish Physiology* 9: 1–63.

Lim, H., & Greenfield, M. D. (2006). Female pheromonal chorusing in an arctiid moth, *Utetheisa ornatrix*. *Behavioral Ecology* doi: 10.1093/beheco/arl069

Linder, J. E., & Rice, W. R. (2005). Natural selection and genetic variation for female resistance to harm from males. *Journal of Evolutionary Biology* 18: 568–575.

Lipson, S. F., & Ellison, P. T. (1996). Comparison of salivary steroid profiles in naturally occurring conception and non-conception cycles. *Human Reproduction* 11: 2090–2096.

Little, A. C., Burt, D. M., Penton-Voak, I. S., & Perrett, D. I. (2001). Self-perceived attractiveness influences human female preferences for sexual dimorphism and symmetry in male faces. *Proceedings of the Royal Society of London, B* 268: 39–44.

Little, A. C., Jones, B. C., Burt, D. M., & Perrett, D. I. (2007). Preferences for symmetry in faces change across the menstrual cycle. *Biological Psychology* 76: 209–216.

Little, A. C., Jones, B. C., Penton-Voak, I. S., Burt, D. M., & Perrett, D. I. (2002). Partnership status and the temporal context of relationships influence human female preferences for sexual dimorphism in male face shape. *Proceedings of the Royal Society of London* B 269: 1095–1100.

Little, A. C., Jones, B. C., & Burriss, R. P. (2007). Preferences for masculinity in male bodies change across the menstrual cycle. *Hormones and Behavior* 51: 633–639.

Livshits, G., & Kobylianski, E. (1989). Study of genetic variance in the fluctuating asymmetry of anthropometric traits. *Annals of Human Biology* 16: 121–129.

Lloyd, E. A. (2005). *The Case of the Female Orgasm: Bias in the Science of Evolution.* Harvard University Press, Cambridge, MA.

Lohm, J., Grahn, M., Langefors, A., Andersen, O., Storset, A., & von Schantz, T. (2002). Experimental evidence for major histocompatibility complex-allele-specific resistance to a bacterial infection. *Proceedings of the Royal Society of London* B 269: 2029–2033.

Lombardi, J. (1998). *Comparative Vertebrate Reproduction.* Kluwer Academic, Boston, MA.

Lombardo, M. P., Thorpe, P. A., & Power, H. W. (1999). The beneficial sexually transmitted microbe hypothesis of avian copulation. *Behavioral Ecology* 10: 333–337.

Lovejoy, C. O. (1981). The origins of man. *Science* 211: 341–350.

Low, B. S. (1978). Environmental uncertainty and parental strategies of marsupials and placentals. *American Naturalist* 112: 197–213.

Low, B. S. (1989). Cross-cultural patterns in the training of children. *Journal of Comparative Psychology* 103: 311–319.

Low, B. S. (1990). Marriage systems and pathogen stress in human societies. *American Zoologist* 30: 325–339.

Low, B. S. (2001). *Why Sex Matters: A Darwinian Look at Human Behavior.* Princeton University Press, Princeton, NJ.

Low, B. S., Alexander, R. D., & Noonan, K. M. (1987). Human hips, breasts, buttocks: Is fat deceptive? *Ethology and Sociobiology* 4: 249–257.

Low, M. (2004). Female weight predicts the timing of forced copulation attempts in stitchbirds, *Notiomystis cincta. Animal Behaviour* 68: 637–644.

Low, M. (2005a). Female resistance and male force: Context and patterns of copulation in the New Zealand stitchbird *Notiomystis cincta. Journal of Avian Biology* 36: 436–448.

Low, M. (2005b). Factors influencing mate guarding and territory defence in the stitchbird (hihi) *Notiomystis cincta. New Zealand of Ecology* 29: 231–242.

Lubjuhn, T., Strohbach, S., Brun, J., Gerken, T., & Epplen, J. T. (1999). Extra-pair paternity in great tits (*Parus major*): A long-term study. *Behaviour* 136: 1157–1172.

Lumpkin, S. (1983). Female manipulation of male avoidance of cuckoldry in the ring dove. In *Social Behavior of Female Vertebrates* (ed. S. K. Wasser), pp. 91–112. Academic Press, New York.

Luque-Larena, J. J., López, P., & Gosálbez, J. (2003). Male dominance and female chemosensory preferences in the rock-dwelling snow vole. *Behaviour* 140: 665–681.

Luxen, M. F., & Buunk, B. P. (2006). Human intelligence, fluctuating asymmetry, and the peacock's tail: General intelligence (g) as an honest signal of fitness. *Personality and Individual Differences* 41: 897–902.

Lynch, K. S., Crews, D., Ryan, M. J., & Wilczynski, W. (2006). Hormonal state influence aspects of female mate choice in the Túngara Frog (*Physalaemus pustulosus*). *Hormones and Behavior* 49: 450–457.

Lynch, K. S., Rand, A. S., Ryan, M. J., & Wilczynski, W. (2005). Plasticity in female mate choice associated with changing reproductive states. *Animal Behaviour* 69: 689–699.

Lynch, M., Blanchard, J., Houle, D., Kibota, T., Schultz, S., Vassilieva, L., & Willis, J. (1999). Perspective: Spontaneous deleterious mutation. *Evolution* 53: 645–663.

Lynch, M., Latta, L., Hicks, J., & Giorgiana, M. (1998). Mutation, selection, and the maintenance of life-history variation in a natural population. *Evolution* 52: 727–733.

Mackintosh, J. A. (2001). The antimicrobial properties of melanocytes, melanosomes and melanin and the evolution of black skin. *Journal of Theoretical Biology* 211: 101–113.

Macrae, C. N., Alnwick, K. A., Milne, A. B., & Schloerscheidt, A. M. (2002). Person perception across the menstrual cycle: Hormonal influences on social-cognitive functioning. *Psychological Science* 13: 532–536.

Maestripieri, D., & Roney, J. R. (2005). Primate copulation calls and postcopulatory female choice. *Behavioral Ecology* 16: 106–113.

Maines, R. P. (1999). *The Technology of Orgasm: "Hysteria," the Vibrator and Women's Sexual Satisfaction.* The Johns Hopkins University Press, Baltimore, MD.

Manning, J. T. (1995). Fluctuating asymmetry and body weight in men and women: Implications for sexual selection. *Ethology and Sociobiology* 16: 145–153.

Manning, J. T. (2002). *Digit Ratio: A Pointer to Fertility, Behavior and Health.* Rutgers University Press, New Brunswick, NJ.

Manning, J. T., Barley, L., Walton, J., Lewis-Jones, D. I., Trivers, R. L., Singh, D., Thornhill, R., Rohde, P., Bereczkei, T., Henzi, P., Soler, M., & Szwed, A. (2000). The 2nd:4th digit ratio, sexual dimorphism, population differences, and reproductive success: Evidence for sexually antagonistic genes? *Evolution and Human Behavior* 21: 163–183.

Manning, J. T., Bundred, P. E., & Henzi, P. (2003). Melanin and HIV in sub-Saharan Africa. *Journal of Theoretical Biology* 223: 131–133.

Manning, J. T., Bundred, P. E., Newton, D. J., & Flanagan, B. F. (2003). The second- to fourth-digit ratio and variation in the androgen receptor gene. *Evolution and Human Behavior* 24: 399–405.

Manning, J. T., Koukourakis, K., & Brodie, D. A. (1997). Fluctuating asymmetry, metabolic rate and sexual selection in human males. *Evolution and Human Behavior* 18: 15–21.

Manning, J. T., Trivers, R. L., Singh, D., & Thornhill, R. (1999). The mystery of female beauty. *Nature* 399: 214–215.

Manning, J. T., Scutt, D., Whitehouse, G. H., & Leinster, S. J. (1997). Breast asymmetry and phenotypic quality in women. *Evolution and Human Behavior* 18: 223–236.

Manning, J. T., Scutt, D., Whitehouse, G. H., Leinster, S. J., & Walton, J. M. (1996). Asymmetry and the menstrual cycle in women. *Ethology and Sociobiology* 17: 129–143.

Manning, J. T., Scutt, D., Wilson, J., & Lewis-Jones, D. I. (1998). The ratio of the 2nd to the 4th digit length: A predictor of sperm numbers and concentrations of testosterone, luteinizing hormone and oestrogen. *Human Reproduction* 13: 3000–3004.

Manning, J. T., Scutt, D., & Lewis-Jones, D. I. (1998). Developmental stability, ejaculate size, and quality in men. *Evolution and Human Behavior* 19: 273–282.

Manning, J. T., & Wood, D. (1998). Fluctuating asymmetry and aggression in boys. *Human Nature* 9: 53–65.

Manson, W. C. (1986). Sexual cyclicity and concealed ovulation. *Journal of Human Evolution* 15: 21–30.

Marlowe, F. (1998). The nubility hypothesis: The human breast as an honest signal of residual reproductive value. *Human Nature* 9: 263–271.

Marlowe, F. (1999). Male care and mating effort among Hadza foragers. *Behavioral Ecology and Sociobiology* 46: 57–64.

Marlowe, F. (2001). Male contribution to diet and female reproductive success among foragers. *Current Anthropology* 42: 755–760.

Marlowe, F., Apicella, C., & Reed, D. (2005). Men's preferences for women's profile waist-to-hip ratio in two societies. *Evolution and Human Behavior* 26: 458–468.

Marlowe, F., & Wetsman, A. (2001). Preferred waist-to-hip ratio and ecology. *Personality and Individual Differences* 30: 481–489.

Marlowe, F. W. (2003a). A critical period for provisioning by Hadza men: Implications for pair bonding. *Evolution and Human Behavior* 24: 217–229.

Marlowe, F. W. (2003b). The mating system of foragers in the Standard Cross-Cultural Sample. *Cross-Cultural Research* 37: 282–306.

Marlowe, F. W. (2004). Is human ovulation concealed? Evidence from conception beliefs in a hunter–gatherer society. *Archives of Sexual Behavior* 33: 427–432.

Marlowe, F. W. (2005). Mate preferences among Hadza hunter-gatherers. *Human Nature* 15: 364–375.

Marshall, W. A., & Tanner, J. M. (1974). Puberty. In *Scientific Foundations of Pediatrics* (eds. J. A. Davis & J. Dobbing), Heineman, London, UK.

Martins, E. P. (2000). Adaptation and the comparative method. *Trends in Ecology and Evolution* 15: 296–299.

Mason, R. T. (1992). Reptilian pheromones. In *Hormones, Brain, and Behavior: Biology of the Reptilia*, Vol. 18: *Physiology E* (eds. C. Gans & D. Crews), pp. 114–228. University of Chicago Press, Chicago, IL.

Matsumoto-Oda, A. (1999). Female choice in the opportunistic mating of wild chimpanzees (*Pan troglodytes schweinfurthii*) at Mahale. *Behavioral Ecology and Sociobiology* 46: 258–266.

Maynard Smith, J. (1978). *The evolution of sex*. Cambridge University Press, Cambridge, UK.

Mazur, A., & Booth, A. (1998). Testosterone and dominance in men. *Behavioral and Brain Sciences* 21: 353–397.

Mazur, A., & Michalek, J. (1998). Marriage, divorce, and male testosterone. *Social Forces* 77: 315–330.

McCleery, R. H., Pettifor, R. A., Armbruster, P., Mayer, K., Sheldon, B. C., & Perrins, C. M. (2004). Components of variance underlying fitness in a natural population of the great tit *Parus major*. *American Naturalist* 164: E62–E72.

McClelland, E. E., Penn, D. J., & Potts, W. K. (2003). Major histocompatibility complex heterozygote superiority during co-infection. *Infection and Immunity* 71: 2079–2086.

McCracken, G. F., Burghardt, G. M., & Houts, S. E. (1999). Microsatellite markers and multiple paternity in the garter snake, *Thamnophis sirtalis*. *Molecular Ecology* 8: 1475–1479.

McDade, T. W. (2003). Life history theory and the immune system: Steps toward a human ecological immunology. *Yearbook of Physical Anthropology* 46: 100–125.

McFarland, R. (1997). Female primates: Fat or fit? In *The Evolving Female: A Life-history Perspective* (eds. M. E. Morbeck, A. Galloway, & A. L. Zihlman), pp. 163–178. Princeton University Press, Princeton, NJ.

McGraw, K. J. (2005). The antioxidant function of many animal pigments: Are there consistent health benefits of sexually selected colourants? *Animal Behaviour* 69: 757–764.

McIntyre, M. H., Chapman, J. F., Lipson, S. F., & Ellison, P. T. (2007). Index-to-ring finger length ratio (2D:4D) predicts levels of salivary estradiol, but not progesterone, over the menstrual cycle. *American Journal of Human Biology* 19: 434–436.

McIntyre, M. H., Gangestad, S. W., Gray, P. E., Chapman, J. F., Burnham, T. C., O'Rourke, M. T., & Thornhill, R. (2006). Romantic involvement often reduces men's testosterone levels—but not always: The moderating effect of extra-pair sexual interest. *Journal of Personality and Social Psychology* 91: 642–651.

Mendonca, M. T., & Crews, D. (1996). Effects of ovariectomy and estrogen replacement on attractivity and receptivity in the red-sided garter snake (*Thamnophis sirtalis parietalis*). *Journal of Comparative Physiology A: Sensory Neural and Behavioral Physiology* 178: 373–381.

Mendonca, M. T., Chernetsky, S. D., Nester, K. E., & Gardner, G. L. (1996). Effects of gonadal sex steroids on sexual behavior in the big brown bat, *Eptesicus fuscus*, upon arousal from hibernation. *Hormones and Behavior* 30: 153–161.

Meric, S., De Nicola, E., Iaccarino, M., Gallo, M., Di Gennaro, A., Morrone, G., Warnau, M., Belgiorno, V., & Pagano, G. (2005). Toxicity of leather tanning wastewater effluents in sea urchin early development and in marine microalgae. *Chemosphere* 61: 208–217.

Merilä, J., & Sheldon, B. C. (2000). Lifetime reproductive success and heritability in nature. *American Naturalist* 155: 301–310.

Mesnick, S. L. (1997). Sexual alliances: Evidence and evolutionary implications. In *Feminism and Evolutionary Biology: Boundaries, Intersections, and Frontiers* (ed. P. A. Gowaty), pp. 207–260. Chapman & Hall, New York.

Meuwissen, I., & Over, R. (1992). Sexual arousal across phases of the human menstrual cycle. *Archives of Sexual Behavior* 21: 101–119.

Michael, R. P., Bonsall, R. W., & Zumpe, D. (1976). Evidence for chemical communication in primates. *Vitamins and Hormones*, 34: 137–186.

Michael, R. P., & Zumpe, D. (1982). Influence of olfactory signals on the reproductive behaviour of social groups of rhesus monkeys (*Macaca mulatta*). *Journal of Endocrinology* 95: 189–205.

Michalek, K. G., & Winkler, H. (2001). Parental care and parentage in monogamous great spotted woodpeckers (*Picoides major*) and middle spotted woodpeckers (*Picoides medius*). *Behaviour* 138: 1259–1285.

Michl, G., Torok, T., Griffith, S. C., & Sheldon, B. C. (2002). Experimental analysis of sperm competition mechanisms in a wild bird population. *Proceedings of the National Academy of Sciences USA* 99: 5466–5470.

Milinski, M., Griffiths, S., Wegner, K. M., Reusch, T. B. H., Haas-Assenbaum, A., & Boehm, T. (2005). Mate choice decisions of stickleback females predictably modified by MHC peptide ligands. *Proceedings of the National Academy of Sciences USA* 102: 4414–4418.

Miller, E. M. (1995). Human breasts: Evolutionary origins as a deceptive signal of need for provisioning and temporary infertility. *Mankind Quarterly* 36: 135–150.

Miller, E. M. (1996). Concealed ovulation as a strategy for increasing per capita paternal investment. *Mankind Quarterly* 36: 297–333.

Miller, G. F. (2000). *The Mating Mind: How Sexual Choice Shaped the Evolution of Human Nature.* Anchor Books, New York.

Miller, G. F. (2003, June). A good sense of humor is a good genes indicator: Ovulatory cycle effects on the sexual attractiveness of male humor ability. Paper presented at the meeting of the Human Behavior and Evolution Society, Lincoln, NE.

Miller, G. F., & Penke, L. (2007). The evolution of human intelligence and the coefficient of additive genetic variance in brain size. *Intelligence* 35: 97–114.

Miller, G. F., Tybur, J., & Jordan, B. (2008). Ovulatory cycle effects on tip earnings by lap dancers: Economic evidence for human estrus? *Evolution and Human Behavior* 28: 375–381.

Mitton, J. B. (2000). *Selection in natural populations*. Oxford University Press, New York.

Miyata, H., Hayashida, H., Kuma, K., Mitsuyasu, K., & Yasunaga, T. (1987). Male-driven molecular evolution: A model and nucleotide sequence analysis. *Cold Spring Harbor Symposium in Quantitative Biology* 52: 863–867.

Mohle, U., Heistermann, M., Dittami, J., Reinberg, V., & Hodges, J. K. (2005). Patterns of anogenital swelling size and their endocrine correlates during ovulatory cycles and early pregnancy in free-ranging Barbary macaques (*Macaca sylvanus*) of Gibraltar. *American Journal of Primatology* 66: 351–368.

Møller, A. P. (1987). Mate guarding in the swallow Hirundo rustica: An experimental study. *Behavioral Ecology and Sociobiology* 21: 119–123.

Møller, A. P. (1990). Fluctuating asymmetry in male sexual ornaments may reliably reveal male quality. *Animal Behaviour* 40: 1185–1187.

Møller, A. P. (1991). Why mated songbirds sing so much: Mate guarding and male announcement of mate fertility status. *American Naturalist* 138: 994–1014.

Møller, A. P. (1997). Developmental stability and fitness: A review. *American Naturalist* 149: 916–932.

Møller, A. P. (1999). Asymmetry as a predictor of growth, fecundity and survival. *Ecology Letters* 2: 149–156.

Møller, A. P. (2000). Male paternal care, female reproductive success, and extrapair paternity. *Behavioral Ecology* 11: 161–168.

Møller, A. P. (2001). Sexual selection in the barn swallow. In *Model Systems in Behavioral Ecology: Integrating Empirical Approaches* (ed. L. A. Dugatkin), pp. 359–377. Princeton University Press, Princeton, N.J.

Møller, A. P. (2003). The evolution of monogamy: Mating relationships, paternal care and sexual selection. In *Monogamy: Mating Strategies and Partnerships in Birds, Humans and Other Mammals* (ed. U. H. Reichard & C. Boesch), pp. 29–41. Cambridge University Press, Cambridge, UK.

Møller, A. P. (2006). A review of developmental instability, parasitism and disease. *Infection, Genetics and Evolution* 6: 133–140.

Møller, A. P., & Cuervo, J.J. 2000. The evolution of paternity and paternal care in birds. *Behavioral Ecology II:* 472–485.

Møller, A. P. & Alatalo, R. (1999). Good genes effects in sexual selection. *Proceedings of the Royal Society of London B* 266: 85–91.

Møller, A. P., Christe, P., & Lux, E. (1999). Parasitism, host immune function, and sexual selection. *Quarterly Review of Biology* 74: 3–20.

Møller, A. P., & Cuervo, J. J. (2000). The evolution of paternity and paternal care in birds. *Behavioral Ecology* II: 472–485.

Møller, A. P., & Cuervo, J. J. (2003). Asymmetry, size and sexual selection: Factors affecting heterogeneity in relationships between asymmetry and sexual selection. In *Developmental Instability: Causes and Consequences* (ed. M. Polak), pp. 262–278. Oxford University Press, Oxford, U.K.

Møller, A. P., & Cuervo, J. J. (2004). Sexual selection, germline mutation rate and sperm competition. *BMC Evolutionary Biology* 3: 1–11.

Møller, A. P., Gangestad, S. W., & Thornhill, R. (1999). Nonlinearity and the importance of fluctuating asymmetry as a predictor of fitness. *Oikos* 86: 366–368.

Møller, A. P., Garamszegi, L. Z., Gil, D., Hurtrez-Boussès, S., & Eens, M. (2005). Correlated evolution of male and female testosterone profiles in birds and its consequences. *Behavioral Ecology and Sociobiology* 58: 534–544.

Møller, A. P., & Jennions, M. D. (2001). How important are direct benefits of sexual selection? *Naturwissenschaften* 88: 401–415.

Møller, A. P., & Legendre, S. (2001). Allee effect, sexual selection and demographic stochasticity. *Oikos* 92: 27–34.

Møller, A. P., & Pomiankowski, A. (1993). Why have birds got multiple sexual ornaments? *Behavioral Ecology and Sociobiology* 32: 167–176.

Møller, A. P., Saino, N., Taramino, G., Galeotti, P., & Ferrario, S. (1998). Paternity and sexual signaling: Effects of a secondary sexual character and song on paternity in the barn swallow. *American Naturalist* 151: 236–242.

Møller, A. P., Soler, M., & Thornhill, R. (1995). Breast asymmetry, sexual selection and human reproductive success. *Ethology and Sociobiology* 16: 207–219.

Møller, A. P., & Swaddle, J. P. (1997). *Asymmetry, Developmental Stability, and Evolution.* Oxford University Press, Oxford, UK.

Møller, A. P., & Thornhill, R. (1997). A meta-analysis of the heritability of developmental stability. *Journal of Evolutionary Biology* 10: 1–16.

Møller, A. P., & Thornhill, R. (1998a). Male parental care, differential parental investment by females, and sexual selection. *Animal Behaviour* 55: 1507–1515.

Møller, A. P., & Thornhill, R. (1998b). Bilateral symmetry and sexual selection: A meta-analysis. *American Naturalist* 151: 174–192.

Møller, A. P., Thornhill, R., & Gangestad, S. W. (2005). Direct and indirect tests for publication bias: Asymmetry and sexual selection. *Animal Behaviour* 70: 497–506.

Monfort, S. L., Wasser, S. K., Mashburn, K. L., Burke, M., Brewer, B. A., & Creel, S. R. (1997). Steroid metabolism and validation of noninvasive endocrine monitoring in the African wild dog (*Lycaon pictus*). *Zoo Biology* 16: 533–548.

Montagna, W. (1985a). The evolution of human skin: Preface. *Journal of Human Evolution* 14: 1–2.

Montagna, W. (1985b). The evolution of human skin. *Journal of Human Evolution* 14: 3–22.

Montgomerie, R., & Bullock, H. (1999, June). Fluctuating asymmetry and the human female orgasm. Paper presented at the annual meetings of the Human Behavior and Evolution Society, Salt Lake City, UT.

Montgomerie, R., & Thornhill, R. (1989). Fertility advertisement in birds: A means of inciting male–male competition? *Ethology* 81: 209–220.

Moore, A. J., Gowaty, P. A., & Moore, P. J. (2003). Females avoid manipulative males and live longer. *Journal of Evolutionary Biology* 16: 523–530.

Morgan, H. L. (1877). *Ancient society.* Henry Holt, New York.

Morris, D. (1967). *The Naked Ape: A Zoological Study of the Human Animal.* McGraw-Hill, New York.

Morris, D. J., & van Aarde, R. J. (1985). Sexual behavior of the female porcupine, *Hystrix africaeaustralis. Hormones and Behavior* 19: 400–412.

Morris, M. (1993). Telling tails explain the discrepancy in sexual partner reports. *Nature* 365: 437–440.

Morrison, S. F., Keogh, J. S., & Scott, I. A. W. (2002). Molecular determination of paternity in a natural population of the multiply mating polygynous lizard *Eulamprus heatwolei. Molecular Ecology* 11: 535–545.

Mougeot, F. (2000). Territorial intrusions and copulation patterns in red kites, Milvus milvus, in relation to breeding density. *Animal Behaviour* 39: 633–642.

Mougeot, F. S. (2004). Breeding density, cuckoldry risk and copulation behaviour during the fertile period in raptors: A comparative analysis. *Animal Behaviour* 67: 1067–1076.

Mougeot, F., Arroyo, B. E., & Bretagnolle, V. (2001). Decoy presentations as a means to manipulate the risk of extrapair copulation: an experimental study in a semicolonial raptor, the Montagu's harrier (*Circus pygargus*). *Behavioral Ecology* 12: 1–7.

Mozuraitis, R., Buda, V., Liblikas, I., Unelius, C. R., & Borg-Karlson, A. K. (2002). Parthenogenesis, calling behavior, and insect-released volatiles of leafminer moth, *Phyllonorycter emberizaepenella*. *Journal of Chemical Ecology* 28: 1191–1208.

Mueller, U., & Mazur, A. (1997). Facial dominance in *Homo sapiens* as honest signaling of male quality. *Behavioral Ecology*, 8: 569–579.

Müller, J. K., & Eggert, A.-K. (1989). Paternity assurance by "helpful" males: Adaptations to sperm competition in burying beetles. *Behavioral Ecology and Sociobiology* 24: 245–249.

Muller, M. N., Kahlenberg, S.M. & Wrangham, R.W. (in press), male aggression against females and sexual coercion in chimpanzees. In *Sexual Coercion in Primates: An Evolutionary Perspective on Male Aggression Against Females*. Harvard University Press, Cambridge, MA.

Muller, M. N., Thompson, M. E., & Wrangham, R. W. (2006). Male chimpanzees prefer mating with older females. *Current Biology* 16: 2234–2238.

Muma, K. E., & Weatherhead, P. J. (1991). Plumage variation and dominance in captive female red-winged blackbirds. *Canadian Journal of Zoology* 69: 49–54.

Murdock, G. P., & White, D. R. (1969). Standard Cross-Cultural Sample. *Ethnology* 9: 329–369.

Nachman, M. W., & Crowell, S. L. (2000). Estimation of the mutation rate per nucleotide in humans. *Genetics* 156: 297–304.

Nation, J. L. (2002). *Insect Physiology and Biochemistry*. CRC Press, Boca Raton, FL.

Neel, J. V., & Weiss, K. M. (1975). The genetic structure of a tribal population, the Yanamama Indians: 12. Biodemographic studies. *American Journal of Physical Anthropology* 42: 25–52.

Neff, B. D., & Pitcher, T. E. (2005). Genetic quality and sexual selection: An integrated framework of good genes and compatible genes. *Molecular Ecology* 14: 19–38.

Nelson, R. J. (1995). *An Introduction to Behavioral Endocrinology*. Sinauer, Sunderland, MA.

Nelson, R. J. (2000). *An Introduction to Behavioral Endocrinology*, 2nd ed. Sinauer, Sunderland, MA.

Neudorf, D. L., Stutchbury, B. J. M.,, & Piper, W. H. (1997). Covert extraterritorial behavior of female hooded warblers. *Behavioral Ecology* 8: 595–600.

Nieschlag, E., Kramer, U., & Nieschlag, S. (2003). Androgens shorten the longevity of women: Sopranos last longer. *Experimental and Clinical Endocrinology and Diabetes* 111: 230–231.

Novotny, M. V., Ma, W., Zidek, L., & Daer, E. (1999). Recent biological insights into puberty acceleration, estrus induction and puberty delay in the house mouse. In *Advances in Chemical Signals in Vertebrates* (eds. R. E. Johnston, D. Müller-Schwarze, & P. W. Sorenson), pp. 99–116. Kluwer Academic/Plenum Press, New York.

Nunes, S., Fite, J. E., & French, J. A. (2000). Variation in steroid hormones associated with infant care behaviour and experience in male marmosets (*Callithrix kuhlii*). *Animal Behaviour* 60: 1–9.

Nunes, S., Fite, J. E., Patera, K. J., & French, J. A. (2001). Interactions among paternal behavior, steroid hormones, and parental experience in male marmosets (*Callithrix kuhlii*). *Hormones and Behavior* 39: 70–82.

Nunn, C. L. (1999). The evolution of exaggerated sexual swellings in primates and the graded-signal hypothesis. *Animal Behaviour* 58: 229–246.

Nunn, C. L. (2003). Behavioural defenses against sexually transmitted diseases in primates. *Animal Behaviour* 66: 37–48.

Nunn, C. L., van Schaik, C. P., & Zinner, D. (2001). Do exaggerated sexual swellings function in female mating competition in primates? A comparative test of the reliable indicator hypothesis. *Behavioral Ecology* 12: 646–654.

Ober, C., Hyslop, T., Elias, S., Weitkamp, L. R., & Hauck, W. W. (1998). Human leukocyte antigen matching and fetal loss: Results from a 10-year prospective study. *Human Reproduction* 13: 33–38.

Ober, C., Weitkamp, L. R., Cox, N., Dytch, H., Kostyu, D., & Elias, S. (1997). HLA and mate choice in humans. *American Journal of Human Genetics* 61: 497–504.

O'Connell, H. E., Sanjeevan, K. V., & Hutson, J. H. (2005). Anatomy of the clitoris. *Journal of Urology* 174: 1189–1195.

Okami, P. (2004). True, new, and important: An introduction to the special issue. *Journal of Sex Research* 41: 2–4.

Oliver, M. B., & Hyde, J. S. (1993). Gender differences in sexuality: A meta-analysis. *Psychological Bulletin* 114: 29–51.

Oliver-Rodriguez, J. C., Guan, Z., & Johnston, V. S. (1999). Gender differences in late positive components evoked by human faces. *Psychophysiology* 36: 176–185.

Olsson, M., Madsen, T., Nordby, J., Wapstra, E., Ujvari, B., & Wittsell, H. (2003). Major histocompatibility genes and mate choice in sand lizards. *Proceedings of the Royal Society of London* B 270: S254–S256.

O'Neill, A. C., Fedigan, L. M., & Zeigler, T. C. (2004). Ovarian cycle phase and same-sex mating behavior in Japanese macaque females. *American Journal of Primatology* 63: 25–31.

Orians, G. H. (1969). On the evolution of mating systems in birds and mammals. *American Naturalist* 103: 589–603.

*Orteiza, N., Linder, J. E., & Rice, W. R. (2005). Sexy sons from re-mating do not recoup the direct costs of harmful male interactions in the *Drosophila melanogaster* laboratory model system. *Journal of Evolutionary Biology* 18: 1315–1323.

Osorio-Beristain, M., & Drummond, H. (1998). Non-aggressive mate guarding by the blue-footed booby: A balance of female and male control. *Behavioral Ecology and Sociobiology* 43: 407–415.

Otte, D. (1974). Effects and functions in the evolution of signaling systems. *Annual Review of Ecology and Systematics* 5: 385–417.

Otter, K., Ratcliffe, L., Michaud, D., & Boag, P. T. (1998). Do female black-capped chickadees prefer high-ranking males as extra-pair partners? *Behavioral Ecology and Sociobiology* 43: 25–36.

Otter, K. A., Stewart, I. R. K., McGregor, P. K., Terry, A. M. R., Dabelsteen, T., & Burke, T. (2001). Extra-pair paternity among Great Tits Parus major following manipulation of male signals. *Journal of Avian Biology* 32: 338–342.

Packer, C. (1979). Inter-troop transfer and inbreeding avoidance in *Papio anubis*. *Animal Behaviour* 27: 1–36.

Pagel, M. (1994). Evolution of conspicuous estrous advertisement in old-world monkeys. *Animal Behaviour* 47: 1333–1341.

Pagel, M. (1997). Desperately seeking father: A theory of parent-offspring resemblance. *Animal Behaviour* 53: 973–981.

Pagel, M., & Meade, A. (2006). Bayesian analysis of correlated evolution of discrete characters by reversible-jump Markov chain Monte Carlo. *American Naturalist* 167: 808–825.

Palmer, A. R. (1999). Detecting publication bias in meta-analysis: A case study of fluctuating asymmetry and sexual selection. *American Naturalist* 154: 220–233.

Palmer, A. R. (2000). Quasireplication and the contract of error: Lessons from sex ratios, heritabilities and fluctuating asymmetry. *Annual Review of Ecology and Systematics* 31: 441–480.

Palombit, R. A. (1999). Infanticide and the evolution of pair bonds in nonhuman primates. *Evolutionary Anthropology* 7: 117–129.

Parker, G. A. (1970). Sperm competition and its evolutionary consequences in the insects. *Biological Reviews* 45: 525–567.

Parker, G. A. (1979a). Sexual selection and sexual conflict. In *Sexual Selection and Reproductive Competition in Insects* (eds. M. Blum & N. Blum), pp. 123–166. Academic Press, New York.

Parker, G. A. (1979b). Sperm competition and its evolutionary consequences in the insects. *Biological Reviews* 45: 525–568.

Parker, G. A. (1983). Mate quality and mating decisions. In *Mate Choice* (ed. P. P. G. Bateson), pp. 141–166. Cambridge University Press, Cambridge, MA.

Parker, G. A. (1984). Sperm competition and the evolution of animal mating strategies. In *Sperm Competition and the Evolution of Animal Mating Strategies* (ed. R. L. Smith), pp. 1–60. Academic Press, New York.

Parker, G. A., Baker, R. R., & Smith, V. G. F. (1972). The origin and evolution of gamete dimorphism and the male-female phenomenon. *Journal of Theoretical Biology* 36: 529–533.

Parker, G. A., & Pearson, R. G. (1976). A possible origin and adaptive significance of the mounting behaviour shown by some female mammals in oestrus. *Journal of History* 10: 241–245.

Parker, T. H. (2003). Genetic benefits of mate choice separated from differential maternal investment in red junglefowl (*Gallus gallus*). *Evolution* 57: 2157–2165.

Parker, T. H., & Ligon, J. D. (2003). Female mating preferences in red junglefowl: A meta-analysis. *Ethology, Ecology, and Evolution* 15: 63–72.

Parsons, P. A. (1992). Fluctuating asymmetry: A biological monitor of environmental and genomic stress. *Heredity* 68: 361–364.

Pärt, T., & Qvarnström, A. (1997). Badge size in collared flycatchers predicts outcome of male competition over territories. *Animal Behaviour* 54: 893–899.

Partridge, L. (1980). Mate choice increases a component of offspring fitness in fruit flies. *Nature* 283: 290–291.

Partridge, L. (1983). Non-random mating and offspring fitness. In *MateChoice* (ed. P. Bateson), pp. 227–256. Cambridge University Press, New York.

Patrat, C., Auer, J., Fauque, P., Leandri, R. L., Jouannet, P., & Serres, C. (2006). Zona pellucida fertilized human oocytes induces a voltage-dependent calcium influx and the acrosome reaction in spermatozoa, but cannot be penetrated by sperm. *BMC Developmental Biology* 6: 59.

Pause, B. M., Sojka, B., Krauel, K., Fehmwolfsdorf, G., & Ferstl, R. (1996). Olfactory information-processing during the course of the menstrual cycle. *Biological Psychology* 44: 31–54.

Pawlowski, B. (1999a). Loss of oestrus and concealed ovulation in human evolution: The case against the sexual selection hypothesis. *Current Anthropology* 40: 257–275.

Pawlowski, B. (1999b). Permanent breasts as a side effect of subcutaneous fat tissue increase in human evolution. *Homo* 50: 149–162.

Pawlowski, B., & Dunbar, R. I. M. (2005). Waist-to-hip ratio versus body mass index as predictors of fitness in women. *Human Nature* 16: 164–177.

Pawlowski, B., & Grabarczyk, M. (2003). Center of body mass and the evolution of female body shape. *American Journal of Human Biology* 15: 144–150.

Pawlowski, B., & Jasienska, G. (2005). Women's preferences for sexual dimorphism in height depend on menstrual cycle phase and expected duration of relationship. *Biological Psychology* 70: 38–43.

Pazol, K. (2003). Mating in the Kakamega forest blue monkeys (*Cercopithecus mitis*): Does female sexual behavior function to manipulate paternity assessment? *Behaviour* 140: 473–499.

Pearse, D. E., Janzen, F. J., & Avise, J. C. (2002). Multiple paternity, sperm storage, and reproductive success of female and male painted turtles (*Chrysemys picta*) in nature. *Behavioral Ecology and Social Biology* 51: 164–171.

Pedersen, M. C., Dunn, P. O., & Whittingham, L. A. (2006). Extraterritorial forays are related to a male ornamental trait in the common yellowthroat. *Animal Behaviour* 72: 479–486.

Pedersen, S. B., Kristensen, K., Hermann, P. A., Katzenellenbogen, J. A., & Richelsen, B. (2004). Estrogen controls lipolysis by up-regulating alpha 2A-adrenergic receptors directly in human adipose tissue through the estrogen receptor alpha: Implications for the female fat distribution. *Journal of Clinical Endocrinology and Metabolism* 89: 1869–1878.

Penn, D. J., & Potts, W. K. (1999). The evolution of mating preferences and major histocompatibility complex genes. *American Naturalist* 153: 145–164.

Penn, D. J., Damjanovich, K., & Potts, W. (2002). MHC heterozygosity confers a selective advantage against multiple-strains infections. *Proceedings of the National Academy of Sciences USA* 99: 11260–11264.

Penton-Voak, I. S., & Chen, J. Y. (2004). High salivary testosterone is linked to masculine male facial appearance in humans. *Evolution and Human Behavior* 25: 229–241.

Penton-Voak, I. S., Jacobson, A., & Trivers, R. (2004). Populational differences in attractiveness judgements of male and female faces: Comparing British and Jamaican samples. *Evolution and Human Behavior* 25: 355–370.

Penton-Voak, I. S., Little, A. C., Jones, B. C., Burt, D. M., Tiddeman, B. P., & Perrett, D. I. (2003). Female condition influences preferences for sexual dimorphism in faces of male humans (*Homo sapiens*). *Journal of Comparative Psychology* 117: 264–271.

Penton-Voak, I. S., & Perrett, D. I. (2000). Female preference for male faces changes cyclically: Further evidence. *Evolution and Human Behavior* 21: 39–48.

Penton-Voak, I. S., & Perrett, D. I. (2001). Male facial attractiveness: Perceived personality and shifting female preferences for male traits across the menstrual cycle. *Advances in the Study of Behavior* 30: 219–259.

Penton-Voak, I. S., Perrett, D. I., Castles, D. L., Kobayashi, T., Burt, D. M., Murray, L. K., & Minamisawa, R. (1999). Female preference for male faces changes cyclically. *Nature* 399: 741–742.

Perrett, D. I., Lee, K. J., Penton-Voak, I., Rowland, D., Yoskikawa, S., Burt, D. M., Henzi, S. P., Castles, D. L., & Akamatsu, S. (1998). Effects of sexual dimorphism on facial attractiveness. *Nature* 394: 884–887.

Perrett, D. I., May, K. A., & Yoshikawa, S. (1994). Facial shape and judgments of female attractiveness. *Nature* 368: 239–242.

Petralia, S. M., & Gallup, G. G., Jr. (2002). Effects of a sexual assault scenario on handgrip strength across the menstrual cycle. *Evolution and Human Behavior* 23: 3–10.

Petrie, M. (1992). Copulation frequency in birds: Why do females copulate more than once with the same male? *Animal Behaviour* 44: 790–792.

Petrie, M., Doums, C., & Møller, A. P. (1998). The degree of extra-pair paternity increases with genetic variability. *Proceedings of the National Academy of Sciences USA* 95: 9390–9395.

Petrie, M., & Kempenaers, B. (1998). Extra-pair paternity in birds: Explaining variation between species and populations. *Trends in Ecology and Evolution* 13: 52–58.

Petrie, M., & Roberts, G. (2007). Sexual selection and the evolution of evolvability. *Heredity* 98: 198–205.

Pickard, A. R., Holt, W. V., Green, D. I., Cano, M., & Abaigar, T. (2003). Endocrine correlates of sexual behavior in the Mohor gazelle (*Gazella dama mhorr*). *Hormones and Behavior* 44: 303–310.

Pierce, E. P., & Lifjeld, J. T. (1998). High paternity without paternity-assurance behavior in the purple sandpiper, a species with high paternal investment. *Auk* 115: 602–612.

Pilastro, A., Evans, J. P., Sartorelli, S., & Bisazza, A. (2002). Male phenotype predicts insemination success in guppies. *Proceedings of the Royal Society of London B* 269: 1325–1330.

Pillsworth, E. G., & Haselton, M. G. (2006). Male sexual attractiveness predicts differential ovulatory shifts in female extra-pair attraction and male mate retention. *Evolution and Human Behavior* 27: 247–258.

Pillsworth, E. G., Haselton, M. G., & Buss, D. M. (2004). Ovulatory shifts in female sexual desire. *Journal of Sex Research* 41: 55–65.

Pipitone, R. N., & Gallup, G. G., Jr. (2007). Voice attractiveness varies across the menstrual cycle. Paper presented at the annual meeting of the Human Behavior and Evolution Society, Williamsburg, VA.

Pischedda, A., & Chippendale, A. K. (2006). Inrtalocus sexual conflict diminishes the benefits of sexual selection. *PLOS Biology* 4: 2099–2103.

Pitcher, T. E., Neff, B. D., Rodd, F. H., & Rowe, L. (2003). Multiple mating and sequential mate choice in guppies: Females trade up. *Proceedings of the Royal Society of London B* 270: 1623–1629.

Pizzari, T. (2003). Food, vigilance, and sperm: The role of male direct benefits in the evolution of female preference in a polygynous bird. *Behavioral Ecology* 47: 593–601.

Pizzari, T., Cornwallis, C. K., Levlie, H., Jakobsson, S., & Birkhead, T. R. (2003). Sophisticated sperm allocation in male fowl. *Nature* 426: 70–74.

Platek, S. M., Burch, R. L., Panyavin, I. S., Wasserman, B. H., & Gallup, G. G., Jr. (2002). Reaction to children's faces: Resemblance affects males more than females. *Evolution and Human Behavior* 23: 159–166.

Platek, S. M., Critton, S. R., Burch, R. L., Frederick, D. A., Myers, T. E., & Gallup, G. G., Jr. (2003). How much paternal resemblance is enough? Sex differences in hypothetical investment decisions, but not in the detection of resemblance. *Evolution and Human Behavior* 24: 81–87.

Platek, S. M., Keenan, J. P., & Mohamed, F. B. (2005). Sex differences in the neural correlates of child facial resemblance: An event-related fMRI study. *Neuroimage* 25: 1336–1344.

Platek, S. M., Raines, D. M., Gallup, G. G., Mohamed, F. B., Thomson, J. W., Myers, T. E., Panyayin, I. S., Levin, S. L., Davis, J. A., Fonteyn, L. C. M., & Arigo, D. R. (2004). Reactions to children's faces: Males are more affected by resemblance than females are, and so are their brains. *Evolution and Human Behavior* 25: 394–405.

Plavkan, J. M., & van Schaik, C. P. (1999). Intrasexual competition and canine dimorphism in anthropoid primates. *American Journal of Physical Anthropology* 87: 461–477.

Poiani, A. (2006). Complexity of seminal fluid: A review. *Behavioral Ecology and Sociobiology* 60: 289–310.

Polak, M., & Stillabower, E. M. (2004). The relationship between genotype, developmental instability, and mating performance: Disentangling the epigenetic causes. *Proceedings of the Royal Society of London* B 271: 1815–1821.

Pomiankowski, A. (1987). The costs of choice in sexual selection. *Journal of Theoretical Biology* 128: 195–218.

Pomiankowski, A., & Møller, A. P. (1995). A resolution of the lek paradox. *Proceedings of the Royal Society of London* B 260: 21–29.

Pond, C. M. (1978). Morphological aspects of the ecological significance of fat distribution in wild vertebrates. *Annual Review of Ecology and Systematics* 9: 519–570.

Pond, C. M. (1981). Storage. In *Physiological Ecology: An Evolutionary Approach to Resource Use* (eds. C. R. Townsend & P. Calow), pp. 190–219. Blackwell Scientific, Oxford, UK.

Poole, J. H. (1989). Mate guarding, reproductive success and female choice in African elephants. *Animal Behaviour* 37: 842–849.

Poole, J. H. (1999). Signals and assessment in African elephants: Evidence from playback experiments. *Animal Behavior* 58: 185–193.

Poran, N. S. (1994). Cycle attractivity of human female odors. *Advances in the Biosciences* 93: 555–560.

Potts, W. K., Manning, C. J., & Wakeland, E. K. (1991). Mating patterns in seminatural populations of mice influenced by MHC genotype. *Nature* 352: 619–621.

Pozo, J., & Argente, J. (2003). Ascertainment and treatment of delayed puberty. *Hormone Research* 60: 35–48.

Pradhan, R. R, Englehardt, A., van Schaik, C. P., & Maestripieri, D. (2006). The evolution of female copulation calls in primates: A review and a new model. *Behavioral Ecology and Sociobiology* 59: 333–343.

Preston, B. T., Stevenson, I. R., & Wilson, K. (2003). Soay rams target reproductive activity towards promiscuous females' optimal insemination period. *Proceedings of the Royal Society of London* B 270: 2073–2078.

Preti, G., Wysocki, C. J., Barnhart, K. T., Sondheimer, S. J., & Seyden, J. J. (2003). Male axillary extracts contain pheromones that affect pulsatile secretion of luteinizing hormone and mood in women participants. *Biology of Reproduction* 68: 2107–2116.

Prokosch, M. D., Yeo, R. A., & Miller, G. F. (2005). Intelligence tests with higher g-loadings show higher correlations with body symmetry: Evidence for a general fitness factor mediated by developmental stability. *Intelligence* 33: 203–213.

Provost, M. P., Quinsey, V. L., & Troje, N. F. (2007). Differences in gait across the menstrual cycle and their attractiveness to men. *Archives of Sexual Behavior* doi: 10.1007/s10508–007-9219-7.

Pruett-Jones, S. (1992). Independent versus non-independent mate choice: Do females copy each other? *American Naturalist* 140: 1000–1009.

Puts, D. A. (2005). Mating context and menstrual phase affect women's preferences for male voice pitch. *Evolution and Human Behavior* 26: 388–397.

Puts, D. A. (2006a). The case of the female orgasm: Bias in the science of evolution. *Archives of Sexual Behavior* 35: 103–108.

Puts, D. A. (2006b). Cyclic variation in women's preferences for masculine traits: Potential hormonal causes. *Human Nature* 17: 114–127.

Puts, D. A., & Dawood, K. (2006). The evolution of female orgasm: Adaptation or by-product? *Twin Research and Human Genetics* 9: 467–472.

Puts, D. A., Gaulin, S. J. C., & Verdolini, K. (2006). Dominance and the evolution of sexual dimorphism in human voice pitch. *Evolution and Human Behavior* 27: 283–296.

Queller, D. C. (1997). Why do females care more than males? *Proceedings of the Royal Society of London* B 264: 1555–1557.

Quinlan, R. J., & Quinlan, M. B. (2008). Human lactation, pair-bonds and alloparents: A cross-cultural analysis. *Human Nature* 19: 87–102.

Quinlan, R. J., Quinlan, M. B., & Flinn, M. V. (2003). Parental investment and age of weaning in a Caribbean village. *Evolution and Human Behavior* 24: 1–16.

Quinsey, V. L., & Lalumiere, M. L. (1995). Evolutionary perspectives on sexual offending. *Sexual Abuse: A Journal of Research and Treatment* 7: 301–315.

Quinsey, V. L., Rice, M. E., Harris, G. T., & Reid, K. S. (1993). The phylogenetic and ontogenetic development of sexual age preferences in males: Conceptual and measurement issues. In *The Juvenile Sex Offender* (eds. H. E. Barbaree, W. L. Marshall, & S. M. Hudson), pp. 143–163. Guilford Press, New York.

Qvarnström, A. (1999). Different reproductive tactics in male collared flycatchers signalled by size of secondary sexual character. *Proceedings of the Royal Society of London* B 266: 2089–2093.

Radford, S. L., Croft, D. B., & Moss, G. L. (1998). Mate choice in female red-necked pademelons, *Thylogale thetis* (Marsupialia: Macropodidae). *Ethology* 104: 217–231.

Ragoobirsingh, D., Morrison, E. Y. S., Johnson, P., & Lewis-Fuller, E. (2004). Obesity in the Caribbean: The Jamaican experience. *Diabetes, Obesity and Metabolism* 6: 23–27.

Rahman, Q., Wilson, G. D., & Abrahams, S. (2004). Developmental instability is associated with neurocognitive performance in heterosexual and homosexual men but not in women. *Behavioral Neuroscience* 118: 243–247.

Rajanarayanan, S., & Archunan, G. (2004). Occurrence of flehmen in male buffaloes (*Bubalus bubalis*) with special reference to estrus. *Theriogenology* 61: 861–866.

Rako, S. (2000). Testosterone supplemental therapy after hysterectomy with or without concomitant oophorectomy: Estrogen alone is not enough. *Journal of Women's Health and Gender-based Medicine* 9: 917–923.

Ramarao, N., Gray-Owen, S. D., & Meyer, T. F. (2000). Helicobacter pylori induces but survives the extracellular release of oxygen radicals from professional phagocytes using its catalase activity. *Molecular Microbiology* 38: 103–113.

Rantala, M. J., Eriksson, C. J. P., Vainikka, A., & Kortet, R. (2006). Male steroid hormones and female preference for male body odor. *Evolution and Human Behavior* 27: 259–260.

Rasmussen, L. E. L. (1999). Evolution of chemical signals in the Asian elephant, *Elephas maximus*: Behavioral and ecological influences. *Journal of Biosciences* 24: 241–251.

Rasmussen, L. E. L. (2001). Source and cyclic release pattern of (Z)-7-dodecenyl acetate, the pre-ovulatory pheromone of the female Asian elephant. *Chemical Senses* 26: 611–623.

Rasmussen, L. E. L.. Krishnamurthy, V., & Sukamar, R. (2005). Behavioural and chemical confirmation of the preovulatory pheromone, (Z)-7-dodecenyl acetate in wild Asian elephant: Its relationship to musth. *Behaviour* 142: 351–396.

Rasmussen, L. E. L., Lee, T. D., Roelofs, W. L., Zang, A., & Daves, G. D. (1996). Insect pheromone in elephants. *Nature* 379: 684.

Rasmussen, L. E. L., & Murru, F. L. (1992). Long-term studies of serum concentrations of reproductively related steroid-hormones in individual captive carcharinids. *Australian Journal of Marine and Freshwater Research* 43: 273–281.

Rätti, O., Hovi, M., Lundberg, A., Telegstrom, H., & Alatalo, R. V. (1995). Extra-pair paternity and male characteristics in the pied flycatcher. *Behavioral Ecology and Sociobiology* 37: 419–425.

Real, L. A. (1990). Search theory and mate choice: I. Models of single-sex discrimination. *American Naturalist* 136: 376–405.

Reeder, D. M. (2003). The potential for cryptic female choice in primates: Behavioral, anatomical, and physiological considerations. In *Sexual Selection and Reproductive Competition in Primates: New Perspectives and Directions* (ed. C. B. Jones), pp. 255–303. American Society of Primatologists, Norman, OK.

Reeve, H. K., & Sherman, P. W. (1993). Adaptation and the goals of evolutionary research. *Quarterly Review of Biology* 68: 1–32.

Reeve, H. K., & Sherman, P. W. (2001). Optimality and phylogeny: A critique of current thought. In *Adaptationism and Optimality* (eds. S. Hecht & E. Sober), pp. 64–113. Cambridge University Press, Cambridge, UK.

Regan, P. C. (1996). Rhythms of desire: The association between menstrual cycle phases and female sexual desire. *Canadian Journal of Human Sexuality* 5: 145–156.

Reichard, U. H., & Boesch, C., eds. (2003). *Monogamy: Mating Strategies and Partnerships in Birds, Humans and Other Mammals.* Cambridge University Press, Cambridge, UK.

Reichert, K. E., Heistermann, M., Hodges, J. K., Boesch, C., & Hohmann, G. (2002). What females tell males about their reproductive status: Are morphological and behavioral cues reliable signals of ovulation in bonobos (*Pan paniscus*)? *Ethology* 108: 583–600.

Reynolds, E. L. (1951) The distribution of subcutaneous fat in childhood and adolescence. *Monographs of the Society for Research in Child Development*, Inc., Volume 55, Serial No. 53. No. 2. Fayerweather Hall, Northwestern University, Evanston, Ill.

Rhodes, G. (2006). The evolutionary psychology of facial beauty. *Annual Review of Psychology* 57: 199–226.

Rhodes, G., Chan, J., Zebrowitz, L. A., & Simmons, L. W. (2003). Does sexual dimorphism in human faces signal health? *Proceedings of the Royal Society of London B,* 270: S93–S95.

Rhodes, G., Simmons, L. W., & Peters, M. (2005). Attractiveness and sexual behavior: Does attractiveness enhance mating success? *Evolution and Human Behavior* 26: 186–201.

Rice, W. R. (1996). Sexually antagonistic male adaptation triggered by experimental arrest of female evolution. *Nature* 381: 232–234.

Rice, W. R., & Chippendale, A. K. (2001). Intersexual ontogenetic conflict. *Journal of Evolutionary Biology* 14: 685–693.

Rice, W. R., & Holland, B. (1997). The enemies within: Intragenomic conflict, interlocus contest evolution (ICE), and the intraspecific Red Queen. *Behavioral Ecology and Sociobiology* 41: 1–10.

Richardson, D. S., & Burke, T. (1999). Extra-pair paternity in relation to male age in Bullock's orioles. *Molecular Ecology* 8: 2115–2126.

Riley, H. T., Bryant, D. M., Carter, P. E., & Parkin, D. T. (1995). Extrapair fertilizations and paternity defense in house martins, *Delichon urbica*. *Animal Behaviour* 49: 495–509.

Ringo, J. (1996). Sexual receptivity in insects. *Annual Review of Entomology* 41: 473–494.

Rikowski, A., & Grammer, K. (1999). Human body odour, symmetry and attractiveness. *Proceedings of the Royal Society of London B* 266: 869–874.

Rissman, E. F. (1991). Evidence that neural aromatization of androgen regulates the expression of sexual behavior in female musk shrews. *Journal of Neuroendocrinology* 3: 441–448.

Roberts, C. W., Walker, W., & Alexander, J. (2001). Sex-associated hormones and immunity to protozoan parasites. *Clinical Microbiology Reviews* 14: 476–488.

Roberts, J. A., & Uetz, G. W. (2005). Information content of female chemical signals in the wolf spider, *Schizocosa ocreata:* Male discrimination of reproductive state and receptivity. *Animal Behaviour* 70: 217–223.

Roberts, S. C., Havlicek, J., Flegr, J., Hruskova, M., Little, A. C., Jones, B. C., Perrett, D. I., & Petrie, M. (2004). Female facial attractiveness increases during the fertile phase of the menstrual cycle. *Proceedings of the Royal Society of London B* 271: S270–S272.

Robertson, B. C., Degnan, S. M., Kikkawa, J., & Moritz, C. C. (2001). Genetic monogamy in the absence of paternity guards: The Capricorn silvereye, *Zosterops lateralis chlorocephalus*, on Heron Island. *Behavioral Ecology* 12: 666–673.

Robertson, S. A. (2005). Seminal plasma and male factor signalling in the female reproductive tract. *Cell and Tissue Research* 322: 43–52.

Robertson, S. A., Bromfield, J. J., & Tremellen, K. P. (2003). Seminal "priming" for protection from preeclampsia: A unifying hypothesis. *Journal of Reproductive Immunology* 59: 253–265.

Robertson, S. A., & Sharkey, D. J. (2001). The role of semen in induction of maternal immune tolerance to pregnancy. *Seminars in Immunology* 13: 243–254.

Robillard, P., Chaline, J., Chaout, G., & Hulsey, T. C. (2003). Preeclampsia/eclampsia and the evolution of the human brain. *Current Anthropology* 44: 130–134.

Robson, A. J., & Kaplan, H. S. (2003). The evolution of human life expectancy and intelligence in hunter-gatherer economies. *American Economic Review* 93: 150–169.

Rodríguez-Gironés, M. A., & Enquist, M. (2001). The evolution of female sexuality. *Animal Behaviour* 61: 695–704.

Røed, K. H., Holand, ø., Mysterud, A., Tverdal, A., Kumpula, J., & Niemenen, M. (2007). Male phenotypic quality influences offspring sex ratio in a polygynous ungulate. *Proceedings of the Royal Society of London B* 274: 727–733.

Roff, D., & Reale, D. (2004). The quantitative genetics of fluctuating asymmetry: A comparison of two models. *Evolution* 58: 47–58.

Rohwer, S., & Rohwer, F. C. (1978). Status signalling in Harris sparrows: Experimental deceptions achieved. *Animal Behaviour* 26: 1012–1022.

Rolland, C., MacDonald, D. W., de Fraipont, M., & Berdoy, M. (2004). Free female choice in house mice: Leave best for last. *Behaviour* 140: 1371–1388.

Romano, J. E. (1997). Effect of service and oxytocin on estrus duration in goats. *Small Ruminant Research* 23: 213–216.

Roney, J. (2005). An alternative explanation for menstrual phase effects on women's psychology and behavior. Paper presented at the annual meeting of the Human Behavior and Evolution Society, Austin, TX.

Roney, J. R., Hansen, K. N., Durante, K. M., & Maestripieri, D. (2006). Reading men's faces: Women's mate attractiveness judgments track men's testosterone and interest in infants. *Proceedings of the Royal Society of London* B 273: 2169–2175.

Roney, J. R., Mahler, S. V., & Maestripieri, D. (2003). Behavioral and hormonal responses of men to brief interactions with women. *Evolution and Human Behavior* 24: 365–375.

Roney, J. R., & Simmons, Z. L (2008). Women's estradiol predicts preference for facial cues of men's testosterone. *Hormones and Behavior* 53: 14–19.

Rossi, A. B., & Vergnanini, A. L. (2000). Cellulite: A review. *Journal of the European Academy of Dermatology and Venereology* 14: 251–262.

Roulin, A. (2004). Proximate basis of the covariation between a melanin-based female ornament and offspring quality. *Oecologia* 140: 668–675.

Roulin, A., Jungi, T. W., Pfister, H., & Dijkstra, C. (2000). Female barn owls (*Tyto alba*) advertise good genes. *Proceedings of the Royal Society of London* B 267: 937–941.

Rowe, L., & Houle, D. (1996). The lek paradox and the capture of genetic variance by condition-dependent traits. *Proceedings of the Royal Society of London* B 263: 1415–1421.

Rucas, S. (2004). Female intersexual social behaviors among the Tsimané of Bolivia. Ph.D. dissertation, Department of Anthropology, University of New Mexico, Albuquerque, NM.

Rucas, S. L., Kaplan, H. S., Winking, J., Gurven, M., Gangestad, S., & Crespo, M. (2006). Female intrasexual competition and reputational effects on attractiveness among the Tsimané of Bolivia. *Evolution and Human Behavior* 27: 40–52.

Saino, N., Leoni, B., & Romano, M. (2006). Human digit ratios depend on birth order and sex of older siblings and predict maternal fecundity. *Behavioral Ecology and Sociobiology* 60: 34–45.

Saino, N., & Møller, A. P. (1995). Testosterone correlates of mate guarding, singing and aggressive behavior in male barn swallows, *Hirundo rustica*. *Animal Behaviour* 49: 465–472.

Saino, N., Primmer, C. R., Ellegren, H., & Møller, A. P. (1999). Breeding synchrony and paternity in the barn swallow (*Hirundo rustica*). *Behavioral Ecology and Sociobiology* 45: 211–218.

Saino, N., Romano, M., & Innocenti, P. (2006). Length of index and ring fingers differentially influence sexual attractiveness of men's and women's hands. *Behavioral Ecology and Sociobiology* 60: 447–454.

Salmon, W. C. (1984). *Scientific Explanation and the Causal Structure of the World.* Princeton University Press, Princeton, NJ.

Sangupta, M., & Karmakar, B. (2007). Genetics of anthropometric asymmetry in an endogamous Indian population: Vaidyas. *American Journal of Human Biology* 19: 399–408.

Santoro, N., Crawford, S. L., Allsworth, J. E., Gold, E. B., Greendale, G. A., Korenman, S., Lasley, B. L., McConnell, D., McGaffigan, P., Midgely, R., Schocken, M., Sowers, M., & Weiss, G. (2003). Assessing menstrual cycles with urinary hormone assays. *American Journal of Physiology: Endocrinology and Metabolism* 284: E521–E530.

Santos, P. S. C., Schinemann, J. A., Gabardo, J., & Bicalho, M. D. (2005). New evidence that the MHC influences odor perception in humans: A study with 58 southern Brazilian students. *Hormones and Behavior* 47: 384–388.

Savage-Rumbaugh, E. S., & Wilkerson, B. J. (1978). Socio-sexual behavior in *Pan paniscus* and *Pan troglodytes*: A comparative study. *Journal of Human Evolution* 24: 327–344.

Savic, I., Berglund, H., & Lindström, P. (2005). Brain response to putative pheromones in homosexual men. *Proceedings of the National Academy of Sciences USA* 102: 7356–7361.

Say, L., Devillard, S., Natoli, E., & Pontier, D. (2002). The mating system of feral cats (*Felis catus* L.) in a sub-Antarctic environment. *Polar Biology* 25: 838–842.

Say, L., Pontier, D., & Natoli, E. (2001). Influence of oestrus synchronization on male reproductive success in the domestic cat (*Felis catus* L.). *Proceedings of the Royal Society of London B* 268: 1049–1053.

Sax, A., Hoi, H., & Birkhead, T. R. (1998). Copulation and sperm use by female bearded tits, *Panurus biarmicus*. *Animal Behaviour* 56: 1199–1204.

Scarbrough, P. S., & Johnston, V. S. (2005). Individual differences in women's facial preferences as a function of digit ratio and mental rotation ability. *Evolution and Human Behavior* 26: 509–526.

Schaefer, K., Fink, B., Grammer, K., Mitteroecker, P., Gunz, P., & Bookstein, F. L. (2006). Female appearance: Facial and bodily attractiveness as shape. *Psychology Science* 48: 187–204.

Scheib, J. E. (2001). Context-specific mate choice criteria: Women's trade-offs in the contexts of long-term and extra-pair mateships. *Personal Relationships* 8: 371–389.

Scheib, J. E., Gangestad, S. W., & Thornhill, R. (1999). Facial attractiveness, symmetry and cues of good genes. *Proceedings of the Royal Society of London B* 266: 1913–1917.

Scheyd, G. J., Garver-Apgar, C. E., & Gangestad, S. W. (2007). Physical attractiveness: Signals of phenotypic quality and beyond. In *Foundations of Evolutionary Psychology* (eds. C. Crawford & D. Krebs), pp. 239–259. Erlbaum, Hillsdale, NJ.

Schiml, P. A., Wersinger, S. R., & Rissman, E. F. (2000). Behavioral activation of the female neuroendocrine axis. In *Reproduction in Context: Social and Environmental Influences on Reproductive Physiology and Behavior* eds. K. Wallen & J. E. Schneider pp. 445–472. MIT Press, Cambridge, MA.

Schroder, I. (1993). Concealed ovulation and clandestine copulation: A female contribution to human evolution. *Ethology and Sociobiology* 14: 381–389.

Schroeder, E. T., Singh, A., Bhasin, S., Storer, T. W., Azen, C., Davidson, T., Martinez, C., Sinha-Hikim, I., Jaque, S. V., Terk, M., & Sattler, F. R. (2003). Effects of an oral androgen on muscle and metabolism in older, community-dwelling men. *American Journal of Physiology: Endocrinology and Metabolism* 284: E120–E128.

Schuett, G. W., & Duvall, D. (1996). Head lifting by female copperheads, Agkistrodon contortrix, during courtship: Potential mate choice. *Animal Behaviour* 51: 367–373.

Schulte, B. A., & Rasmussen, L. E. L. (1999). Signal receiver interplay in the communication of male condition by Asian elephants. *Animal Behaviour* 57: 1265–1274.

Schultheiss, O. C., Campbell, K. L., & McClelland, D. C. (1999). Implicit power motivation moderates men's testosterone responses to real and imagined dominance success. *Hormones and Behavior* 36: 234–241.

Schultheiss, O. C., Dargel, A., & Rohde, W. (2003). Implicit motives and gonadal steroid hormones: Effects of menstrual cycle phase, oral contraceptive use, and relationship status. *Hormones and Behavior* 43: 293–301.

Schultheiss, O. C., & Rohde, W. (2002). Implicit power motivation predicts men's testosterone changes and implicit learning in a contest situation. *Hormones and Behavior* 41: 195–202.

Schuster, J. P., & Schaub, G. A. (2001). Experimental chagas disease: The influence of sex and psychoneuroimmunological factors. *Parasitology Research* 87: 994–1000.

Schwagmeyer, P. L., & Parker, G. A. (1987). Queuing for mates in thirteen-lined ground squirrels. *Animal Behaviour* 35: 1015–1025.

Schwagmeyer, P. L., & Parker, G. A. (1990). Male mate choice as predicted by sperm competition in thirteen-lined ground squirrels. *Nature* 348: 62–64.

Scott, A. P., & Vermierssen, E. L. M. (1994) Production of conjugated steroids by teleost gonads and their role as phermones. *Perspective in Comparative Endocrinology*, Symposium Volume of National Research Council of Canada, pp. 645–654.

Scutt, D., & Manning, J. T. (1996). Symmetry and ovulation in women. *Human Reproduction* 11: 2477–2480.

Searcy, W. A., & Nowicki, S. (2005). *The Evolution of Animal Communication: Reliability and Deception in Signaling Systems.* Princeton University Press, Princeton, NJ.

Sellen, D. W., & Smay, D. B. (2001). Relationship between subsistence and age at weaning in "pre-industrial" societies. *Human Nature* 12: 47–87.

Semple, S., & McComb, K. (2000). Perception of female reproductive state from vocal cues in a mammal species. *Proceedings of the Royal Society of London B* 267: 707–712.

Service, R. (1998). New role for estrogen in cancer? *Science* 279: 1631–1632.

Setchell, J. M., Charpentier, M., & Wickings, E. J. (2005). Mate guarding and paternity in mandrills: Factors influencing alpha male monopoly. *Animal Behaviour* 70: 1105–1120.

Setchell, J. M., Charpentier, M. J. E., Bedjabaga, I. B., Reed, P., Wickings, E. J., & Knapp, L. A. (2006). Secondary sexual characters and female quality in primates. *Behavioral Ecology and Sociobiology* 61: 305–315.

Setchell, J. M., & Kappeler, P. M. (2003). Selection in relation to sex in primates. *Advances in the Study of Behavior* 33: 87–173.

Setchell, J. M., & Wickings, E.J. (2004). Sexual swellings in mandrills (*Mandrillus sphinx*): A test of the reliable indicator hypothesis. *Behavioral Ecology* 15: 438–445.

Sgro, C. M., & Hoffman, A. A. (1998). Effects of temperature extremes on genetic variances for life history traits in *Drosophila melanogaster* as determined from parent-offspring comparisons. *Journal of Evolutionary Biology* 11: 1–20.

Shackelford, T. K., Goetz, A. T., Guta, F. E., & Schmitt, D. P. (2006). Mate guarding and frequent in-pair copulation in humans: Concurrent or compensatory anti-cuckoldry tactics? *Human Nature* 17: 239–252.

Shackelford, T. K., LeBlanc, G. J., Weekes-Shakelford, V. A., Bleske-Rechek, A. L., Euler, H. A., & Hoier, S. (2002). Psychological adaptation to human sperm competition. *Evolution of Human Behavior* 23: 123–138.

Shackelford, T. K., & Pound, N., eds. (2006). *Sperm Competition in Humans: Classic and Contemporary Readings.* Springer, New York.

Shackelford, T. K., Weekes-Shackelford, V. A., LeBlanc, G. J., Bleske, A. L., Euler, H. A., & Hoier, S. (2000). Female coital orgasm and male attractiveness. *Human Nature* 11: 299–306.

Shahnoor, N., & Jones, C. B. (2003). A brief history of the study of sexual selection and reproduction competition in primatology. In *Sexual Selection and Reproductive Competition in Primates: New Perspectives and Directions* (ed. C. B. Jones), pp. 1–43. American Society of Primatologists, Norman, OK.

Shalit, W. (1999). *A Return to Modesty: Discovering the Lost Virtue.* Free Press, New York.

Sheldon, B. C., Davidson, P., & Lindgren, G. (1999). Mate replacement in experimentally widowed collared flycatchers (*Ficedula albicollis*): Determinants and outcomes. *Behavioral Ecology and Sociobiology* 46: 141–148.

Sheldon, B. C., & Ellegren, H. (1999). Sexual selection resulting from extrapair paternity in collared flycatchers. *Animal Behaviour* 57: 285–298.

Sheldon, B. C., Merila, L., Qvarnström, A., Gustafson, L., & Ellegren, H. (1997). Paternal genetic contribution to offspring condition predicted by size of male secondary sexual character. *Proceedings of the Royal Society of London B* 264: 297–302.

Sheldon, M. S., Cooper, M. L., Geary, D. C., Hoard, M., & DeSoto, M. C. (2006). Fertility cycle patterns in motives for sexual behavior. *Personality and Social Psychology Bulletin* 32: 1659–1673.

Sherman, P. W., & Neff, B. D. (2003). Father knows best. *Nature* 425: 136–137.

Shine, R., Phillips, B., Waye, H., LeMaster, M., & Mason, R. T. (2003). Chemosensory cues allow courting male garter snakes to assess body length and body condition of potential mates. *Behavioral Ecology and Sociobiology* 54: 162–166.

Siefferman, L., & Hill, G. E. (2005). Evidence for sexual selection on structural plumage coloration in female eastern bluebirds (*Sialia sialis*). *Evolution* 59: 1819–1828.

Silk, J. (2001). Book review of *Tree of Origin: What Primate Behavior Can Tell Us about Human Social Evolution* (ed. Frans B. M. de Waal). *Evolution and Human Behavior* 22: 443–448.

Silk, J., Alberts, S. C., & Altmann, J. (2003). Social bonds of female baboons enhance infant survival. *Science* 302: 1231–1234.

Sillén-Tullberg, B., & Møller, A. P. (1993). The relationship between concealed ovulation and mating systems in anthropoid primates: A phylogenetic analysis. *American Naturalist* 141: 1–25.

Simmons, L. W. (2001). *Sperm Competition and its Evolutionary Consequences in the Insects.* Princeton University Press, Princeton, NJ.

Simmons, L. W. (2005). The evolution of polyandry: Sperm competition, sperm selection, and offspring viability. *Annual Review of Ecology, Evolution and Systematics* 36: 125–146.

Simmons, L. W., Firman, R. C., Rhodes, G., & Peters, M. (2004). Human sperm competition: Testis size, sperm production and rates of extrapair copulation. *Animal Behavior* 68: 297–302.

Simpson, J. A., Gangestad, S. W., Christensen, P. N., & Leck, K. (1999). Fluctuating asymmetry, sociosexuality, and intrasexual competitive tactics. *Journal of Personality and Social Psychology* 76: 159–172.

Singh, B. B., Berman, B. M., Simpson, R. L., & Annechild, A. (1998). Incidence of premenstrual syndrome and remedy usage: A national probability sample study. *Alternative Therapies in Health and Medicine* 4: 75–79.

Singh, D. (1993). Adaptive significance of female physical attractiveness: Role of waist-to-hip ratio. *Journal of Personality and Social Psychology* 65: 293–307.

Singh, D. (1995). Female health, attractiveness and desirability for relationships: Role of breast asymmetry and waist-to-hip ratio. *Ethology and Sociobiology* 16: 465–481.

Singh, D. (2002a). Waist-to-hip ratio changes across the menstrual cycle. Paper presented at the annual meeting of the Human Behavior and Evolution Society, London, England.

Singh, D. (2002b). Female mate value at a glance: Relationship of waist-to-hip ratio to health, fecundity and attractiveness. *Neuroendocrinology Letters* 23 (Suppl. 4), 81–91.

Singh, D., & Bronstad, P. M. (2001). Female body odour is a potential cue to ovulation. *Proceedings of the Royal Society of London B* 268: 797–801.

Singh, D., & Randall, P. K. (2007). Beauty is in the eye of the surgeon: Waist-hip ratio (WHR) and women's attractiveness. *Personality and Individual Differences* 34: 329–340.

Singh, D., Renn, P., & Singh, A. (2007). Did the perils of abdominal obesity affect the depiction of female beauty in the sixteeth to eighteenth century British literature? Exploring the health and beauty link. *Proceedings of the Royal Society of London B* 274: 891–894.

Skau, P. A., & Folstad, I. (2003). Do bacterial infections cause reduced ejaculate quality? A meta-analysis of antibiotic treatment of male infertility. *Behavioral Ecology* 14: 40–47.

Slob, A. K., Bax, C. M., Hop, W. C. J., Rowland, D. L., & ten Bosch, J. J. V. W. (1996). Sexual arous-ability and the menstrual cycle. *Psychoneuroendocrinology* 21: 545–558.

Small, M. (1993). *Female Choices: Sexual Behavior of Female Primates.* Cornell University Press, Ithaca, NY.

Small, M. F. (1996). "Revealed" ovulation in humans? *Journal of Human Evolution* 30: 483–488.

Smith, E. A. (2002). *Why do good hunters have higher reproductive success?* Paper presented at the meetings of the American Anthropological Association, New Orleans, LA.

Smith, M. J. (1992). Evidence from the estrous cycle for male-induced ovulation in Bettongia-penicillata (*Marsupialia*). *Journal of Reproduction and Fertility* 95: 283–289.

Smith, N. G. C., & Eyre-Walker, A. (2002). Adaptive protein evolution in *Drosophila. Nature* 415: 1022–1024.

Smith, R. L. (1984a). Human sperm competition. In *Sperm Competition and the Evolution of Animal Mating Systems* (ed. R. L. Smith), pp. 601–660. Academic Press, London, UK.

Smith, R. L., ed. (1984b). *Sperm Competition and the Evolution of Animal Mating Systems.* Academic Press, London, UK.

Smock, T., Albeck, D., & Stark, P. (1998). A peptidergic basis for sexual behavior in mammals. *Progress in Brain Research* 119: 467–481.

Smuts, B. (1992). Male aggression against women: An evolutionary perspective. *Human Nature* 3: 1–44.

Smuts, B. B. (1985). *Sex and Friendship in Baboons.* Aldine de Gruyter, New York.

Smuts, B. B., & Smuts, R. W. (1993). Male aggression and sexual coercion of females in nonhuman primates and other mammals: Evidence and theoretical implications. *Advances in the Study of Behavior* 22: 1–63.

Soler, C., Núñez, M., Gutiérrez, R., Núñez, J., Medina, P., Sancho, M., Álvarez, J., & Núñez, A. (2003). Facial attractiveness in men provides clues to semen quality. *Evolution and Human Behavior* 24: 199–207.

Soler, J. J., Cuervo, J. J., Møller, A. P., & de Lope, F. (1998). Nest building is a sexually selected behavior in the barn swallow. *Animal Behaviour* 56: 1435–1442.

Soltis, J. (2002). Do primate females gain nonprocreative benefits by mating with multiple males? Theoretical and empirical considerations. *Evolutionary Anthropology* 11: 187–197.

Sommer, S. (2005). Major histocompatibility complex and mate choice in a monogamous rodent. *Behavioral Ecology and Sociobiology* 58: 181–189.

Spritzer, M. D., Meikle, D. B., & Solomon, N. G. (2005). Female choice based on male spatial ability and aggressiveness among meadow voles. *Animal Behaviour* 69: 1121–1130.

Spuhler, J. N. (1979). Continuities and discontinuities in anthropoid-hominid behavioral evolution: Bipedal locomotion and sexual receptivity. In *Evolutionary Biology and Human Social Behavior: An Anthropological Perspective* (eds. N. A. Chagnon & W. Irons), pp. 454–461. Duxbury Press, North Scituate, MA.

Stacey, P. B. (1982). Female promiscuity and male reproductive success in social birds and mammals. *American Naturalist* 120: 51–64.

Stallmann, R. R., & Froehlich, J. W. (2000). Primate sexual swellings as coevolved signal systems. *Primates* 41: 1–16.

Stanback, M., Richardson, D. S., Boix-Hinzen, C., & Mendelsohn, J. (2002). Genetic monogamy in Monteiro's hornbill, *Tockus monteiri*. *Animal Behaviour* 63: 787–793.

Stanford, C. B., Wallis, J., Pongo, E., & Goodall, J. (1994). Hunting decisions in wild chimpanzees. *Behaviour* 131: 1–18.

Stanger, K. F., Coffman, B. S., & Izard, M. K. (1995). Reproduction in coquerels dwarf lemur (*Mirza coquereli*). *American Journal of Primatology* 36: 223–237.

Steklis, H. D., & Whiteman, C. H. (1989). Lost of estrus in human evolution: Too many answers, too few questions. *Ethology and Sociobiology* 10: 417–434.

Stephens, D. W., & Krebs, J. R. (1986). *Foraging Theory*. Princeton University Press, Princeton, NJ.

Steverink, D. W. B., Soede, N. M., Groenland, G. J. R., van Schle, F. W., Noordhuizen, J. P. T. M., & Kemp, B. (1999). Duration of estrus in relation to reproduction results in pigs on commercial farms. *Journal of Animal Science* 77: 801–809.

Stewart, A. D., Morrow, E. H., & Rice, W. R. (2005). Assessing putative interlocus sexual conflict in *Drosophila melanogaster* using experimental evolution. *Proceedings of the Royal Society of London* B 272: 2029–2035.

Stige, L. C., Stagsvold, T., & Vøllestad, L. A. (2005). Individual fluctuating asymmetry in pied flycatchers (*Ficedula hypoleuca*) persists across moults, but is not heritable and not related to fitness. *Evolutionary Ecology Research* 7: 381–406.

Stockley, P. (2002). Sperm competition risk and male genital anatomy: Comparative evidence for reduced duration of female sexual receptivity in primates with penile spines. *Evolutionary Ecology* 16: 123–137.

Storey, A. E., Walsh, C. J., Quinton, R. L., & Wynne-Edwards, K. E. (2000). Hormonal correlates of paternal responsiveness in new and expectant fathers. *Evolution and Human Behavior* 21: 79–95.

Strassmann, B. I. (1981). Sexual selection, paternal care, and concealed ovulation in humans. *Ethology and Sociobiology* 2: 31–40.

Strassmann, B. I. (1996a). Menstrual cycle visits by Dogon women: A hormonal test distinguishes deceit from honest signaling. *Behavioral Ecology* 7: 304–315.

Strassmann, B. I. (1996b). The evolution of endometrial cycles and menstruation. *Quarterly Review of Biology* 71: 181–220.

Strassmann, B. I. (1999). Comments. *Current Anthropology* 40: 268.

Streeter, S. A., & McBurney, D. H. (2003). Waist-hip ratio and attractiveness: New evidence and a critique of "a critical test." *Evolution and Human Behavior* 24: 88–98.

Stumpf, R. M., & Boesch, C. (2005). Does promiscuous mating preclude female choice? Female sexual strategies in chimpanzees (*Pan troglodytes verus*) of the Taï National Park, Côte d'Ivoire. *Behavioral Ecology and Sociobiology* 57: 511–524.

Stutchbury, B. J. M., Piper, W. H., Neudorf, D. L., Tarof, S. A., Rhymer, J. M., Fuller, G., & Fleischer, R. C. (1997). Correlates of extra-pair fertilization success in hooded warblers. *Behavioral Ecology and Sociobiology* 40: 119–126.

Sugiyama, L. S. (2004). Is beauty in the context-sensitive adaptations of the beholder? Shiwar use of waist-to-hip ratio in assessments of female mate value. *Evolution and Human Behavior* 25: 51–62.

Swaddle, J. P., & Reierson, G. W. (2002). Testosterone increases perceived dominance but not attractiveness of human males. *Proceedings of the Royal Society of London B,* 269: 2285–2289.

Swenson, R. O. (1997). Sex-role reversal in the tidewater goby, *Eucyclogobius newberryi. Environmental Biology of Fishes* 50: 27–40.

Symons, D. (1979). *The Evolution of Human Sexuality.* Oxford University Press, Oxford, UK.

Symons, D. (1982). Another woman that never existed. *Quarterly Review of Biology* 57: 297–300.

Symons, D. (1987). An evolutionary approach: Can Darwin's view of life shed light on human sexuality? In *Theories of Human Sexuality* (eds. J. H. Greer & W. T. O'Donahue), pp. 91–125. Plenum Press, New York.

Symons, D. (1992). On the use and misuse of Darwinism in the study of human behavior. In *The Adapted Mind* (eds. J. Barkow, L. Cosmides, & J. Tooby), pp. 137–159. Oxford University Press, Oxford, UK.

Symons, D. (1995). Beauty is in the adaptations of the beholder: The evolutionary psychology of human female sexual attractiveness. In *Sexual Nature/Sexual Culture* (eds. P. R. Abramson & S. D. Pinkerton), pp. 80–118. University of Chicago Press, Chicago, IL.

Szalay, F. S., & Costello, R. K. (1991). Evolution of permanent estrus displays in hominids. *Journal of Evolution* 20: 439–464.

Sztatecsny, M., Jehle, R., Burke, T., & Hoedl, W. (2006). Female polyandry under male harassment: The case of the common toad (*Bufo bufo*). *Journal of Zoology* 270: 517–522.

Takahashi, L. K. (1990). Hormonal regulation of sociosexual behavior in female mammals. *Neuroscience and Biobehavioral Reviews* 14: 403–413.

Takahata, Y., Ihobe, H., & Idani, G. (1996). Comparative copulations of chimpanzees and bonobos: Do females exhibit proceptivity or receptivity? In *Great Ape Societies* (eds. W. G. McGrew, L. F. Marchant, & T. Nishida), pp. 146–155. Cambridge University Press, Cambridge, UK.

Tarín, J. J., & Gómez-Piquer, V. (2002). Do women have a hidden heat period? *Human Reproduction* 17: 2243–2248.

Tarof, S. A., Dunn, P. O., & Whittingham, L. A. (2005). Dual functions of a melanin-based ornament in the common yellowthroat. *Proceedings of the Royal Society of London B* 272: 1121–1127.

Taylor, P. D., & Williams, G. C. (1982). The lek paradox is not resolved. *Theoretical Population Biology* 22: 392–409.

Tchernof, A., Poehlman, E. T., & Despes, J. P. (2000). Body fat distribution, the menopause transition, and hormone replacement therapy. *Diabetes and Metabolism* 26: 12–20.

Temple, J. L., Schneider, J. E., Scott, D. K., Korutz, A., & Rissman, E. F. (2002). Mating behavior is controlled by acute changes in metabolic fuels. *American Journal of Physiology: Regulatory Integrative and Comparative Physiology* 282: R782–R790.

Tenaza, R. R. (1989). Female sexual swellings in the Asian colobine *Simias concolor*. *American Journal of Primatology* 17: 81–86.

Thibaut, J. W., & Kelley, H. H. (1959). *The social psychology of groups*. Wiley, New York.

Thiessen, D. (1994). Environmental tracking by females: Sexual lability. *Human Nature* 5: 167–202.

Thoma, R. J., Yeo, R. A., Gangestad, S., Halgren, E., David, J., Paulson, K. M., & Lewine, J. D. (2006). Developmental instability and the neural dynamics of the speed-intelligence relationship. *NeuroImage* 32: 1456–1464.

Thoma, R. J., Yeo, R. A., Gangestad, S. W., Halgren, E., Sanchez, N. M., & Lewine, J. D. (2005). Cortical volume and developmental instability are independent predictors of general intellectual ability. *Intelligence* 33: 27–38.

Thoma, R. J., Yeo, R. A., Gangestad, S. W., Lewine, J., & Davis, J. (2002). Fluctuating asymmetry and the human brain. *Laterality* 7: 45–58.

Thompson, A. P. (1983). Extramarital sex: A review of the research literature. *Journal of Sex Research* 19: 1–22.

Thompson, K. V., & Monfort, S. L. (1999). Synchronization of oestrous cycles in sable antelope. *Animal Reproduction Science* 57: 185–197.

Thornhill, N. W. (1991). An evolutionary analysis of rules regulating human inbreeding and marriage. *Behavioral and Brain Sciences* 14: 247–260.

Thornhill, N. W., & Thornhill, R. (1987). Evolutionary theory and rules of mating and marriage. In *Sociobiology and Psychology: Ideas, Issues and Applications* (eds. C. Crawford, M. Smith, & D. Krebs), pp. 373–400. Erlbaum, Hillsdale, NJ.

Thornhill, R. (1976). Sexual selection and paternal investment in insects. *American Naturalist* 110: 153–163.

Thornhill, R. (1979). Male and female sexual selection and the evolution of mating systems in insects. In *Sexual Selection and Reproductive Competition in Insects* (eds. M. S. Blum & N. A. Blum), pp. 81–132. Academic Press, New York.

Thornhill, R. (1983). Cryptic female choice and its implications in the scorpionfly *Harpobittacus nigriceps*. *American Naturalist* 122: 765–788.

Thornhill, R. (1984a). Alternative hypotheses for traits believed to have evolved by sperm competition. In *Sperm Competition and the Evolution of Animal Mating Systems* (ed. R. Smith), pp. 151–178. Academic Press, New York.

Thornhill, R. (1984b). Alternative female choice tactics in the scorpionfly *Hylobittacus apicalis* (Mecoptera) and their implications. *American Zoologist* 24: 367–383.

Thornhill, R. (1988). The jungle fowl hen's cackle incites male competition. *Verhalten Duertzch Zoologie Gestation* 81: 145–154.

Thornhill, R. (1990). The study of adaptation. In *Interpretation and Explanation in the Study of Behavior* (eds. M. Bekoff & D. Jamieson), Vol. 2, pp. 31–62. Westview Press, Boulder, CO.

Thornhill, R. (1992a). Female preference for the pheromone of males with low fluctuating asymmetry in the Japanese scorpionfly (*Panorpa japonica*: Mecoptera). *Behavioral Ecology* 3: 277–283.

Thornhill, R. (1992b). Fluctuating asymmetry and the mating system of the Japanese scorpionfly, *Panorpa japonica*. *Animal Behaviour* 44: 867–879.

Thornhill, R. (1997). The concept of an evolved adaptation. In *Characterizing Human Psychological Adaptations* (ed. M. Daly), pp. 4–13. Wiley, London, UK.

Thornhill, R. (2006). Foreword: Human sperm competition and woman's dual sexuality. In *Sperm Competition in Humans: Classic and Contemporary Readings* (eds. T. K. Shackelford & N. Pound), pp. v–xvii. Kluwer Academic Press, New York.

Thornhill, R., & Alcock, J. (1983). *The Evolution of Insect Mating Systems*. Harvard University Press, Cambridge, MA.

Thornhill, R., & Furlow, F. B. (1998). Stress and human behavior: Attractiveness, women's sexual development, post-partum depression, and baby's cry. In *Advances in the Study of Behavior* (eds. P. J. B. Slater, M. Milinski & A. Møller), pp. 319–369. Academic Press, New York.

Thornhill, R., & Gangestad, S. W. (1993). Human facial beauty: Averageness, symmetry and parasite resistance. *Human Nature* 4: 237–269.

Thornhill, R., & Gangestad, S. W. (1994). Human fluctuating asymmetry and human sexual behavior. *Psychological Science* 5: 297–302.

Thornhill, R., & Gangestad, S. W. (1996). Human female orgasm: Adaptation or phylogenetic holdover? *Animal Behaviour* 52: 853–855.

Thornhill, R., & Gangestad, S. W. (1999a). Facial attractiveness. *Trends in Cognitive Science* 3: 452–460.

Thornhill, R., & Gangestad, S. W. (1999b). The scent of symmetry: A human sex pheromone that signals fitness? *Evolution and Human Behavior* 20: 175–201.

Thornhill, R., & Gangestad, S. W. (2003a). Evolutionary theory led to evidence for a male sex pheromone that signals symmetry. *Psychological Inquiry* 14: 318–325.

Thornhill, R., & Gangestad, S. W. (2003b). Do women have evolved adaptation for extra-pair copulation? In *Evolutionary Aesthetics* (eds. E. Voland & K. Grammer), pp. 341–368. Springer-Verlag, Heidelberg, Germany.

Thornhill, R., & Gangestad, S. W. (2006). Facial sexual dimorphism, developmental stability and parasitic infections in men and women. *Evolution and Human Behavior* 27: 131–144.

Thornhill, R., Gangestad, S. W., & Comer, R. (1995). Human female orgasm and mate fluctuating asymmetry. *Animal Behaviour* 50: 1601–1615.

Thornhill, R., Gangestad, S. W., Miller, R., Scheyd, G., Knight, J., & Franklin, M. (2003). MHC, symmetry, and body scent attractiveness in men and women. *Behavioral Ecology* 14: 668–678.

Thornhill, R., & Grammer, K. (1999). The body and face of woman: One ornament that signals quality? *Evolution and Human Behavior* 20: 105–120.

Thornhill, R., & Møller, A. P. (1997). Developmental stability, disease and medicine. *Biological Reviews* 72: 497–528.

Thornhill, R., Møller, A. P., & Gangestad, S. W. (1999). The biological significance of fluctuating asymmetry and sexual selection: A reply to Palmer. *American Naturalist* 154: 234–241.

Thornhill, R., & Palmer, C. T. (2000). *A Natural History of Rape: Biological Bases of Sexual Coercion*. MIT Press, Cambridge, MA.

Thornhill, R., & Thornhill, N. W. (1983). Human rape: An evolutionary analysis. *Ethology and Sociobiology* 4: 137–173.

Thornton, J. W. (2001). Evolution of vertebrate steroid receptors from an ancestral estrogen receptor by ligand exploitation and serial genome expansions. *Proceedings of the National Academy of Sciences USA* 98: 5671–5676.

Thornton, J. W., Need, E., & Crews, D. (2003). Resurrecting the ancestral steroid receptor: Ancient origin of estrogen signaling. *Science* 301: 1714–1717.

Thurz, M. R., Thomas, H. C., Greenwood, B. M., & Hill, A. V. S. (1997). Heterozygote advantage for HLA class-II type in hepatitis B virus infection. *Nature Genetics* 17: 11–12.

Thusius, K. J., Peterson, K. A., Dunn, P. O., & Whittingham, L. A. (2001). Male mask size is correlated with mating success in the common yellowthroat. *Animal Behaviour* 62: 435–446.

Tinbergen, N. (1963). On aims and methods of ethology. *Zeitschrift fur Tierpsychologie* 20: 410–433.

Tobias, J., & Seddon, N. (2000). Territoriality as a paternity guard in the European robin, *Erithacus rubecula*. *Animal Behaviour* 60: 165–173.

Tooby, J. (1982). Pathogens, polymorphism, and the evolution of sex. *Journal of Theoretical Biology* 97: 557–576.

Török, J., Michl, G., Garamszegi, L. Z., & Barna, J. (2003). Repeated inseminations required for natural fertility in a wild bird population. *Proceedings of the Royal Society of London* B 270: 641–647.

Tovee, M. J., Maisey, D. S., Emery, J. L., & Cornelissen, P. L. (1999). Visual cues to female physical attractiveness. *Proceedings of the Royal Society of London* B 266: 211–218.

Townsend, J. M. (1998). *What Women Want—What Men Want: Why the Sexes Still See Love and Commitment So Differently*. Oxford University Press, New York, N.Y.

Tricas, T. C., Maruska, K. P., & Rasmussen, L. E. L. (2000). Annual cycles of steroid hormone production, gonad development, and reproductive behavior in the Atlantic stingray. *General and Comparative Endocrinology* 118: 209–225.

Trivers, R. L. (1972). Parental investment and sexual selection. In *Sexual Selection and the Descent of Man, 1881–1971* (ed. B. Campbell), pp. 136–179. Aldine, Chicago, IL.

Trivers, R. L. (1974). Parent-offspring conflict. *American Zoologist* 14: 249–264.

Trivers, R. L. (1985). *Social Evolution*. Benjamin/Cummings, Menlo Park, CA.

Troisi, A., & Carosi, M. (1998). Female orgasm rate increases with male dominance in Japanese macaques. *Animal Behaviour* 56: 1261–1266.

Tryjanowski, P., & Hromada, M. (2005). Do males of the great grey shrike, *Lanius excubitor*, trade food for extrapair copulations? *Animal Behaviour* 69: 529–533.

Turke, P. W. (1984). Effects of ovulatory concealment and synchrony on protohominid mating systems and parental roles. *Ethology and Sociobiology* 5: 33–44.

Turelli, M., & Barton, N. H. (2004). Polygenic variation maintained by balancing selection: Pleiotropy, sex-dependent allelic effects, and G x E interactions. *Genetics* 166: 1053–1079.

Uy, J. A. C., Patricelli, G. L., & Borgia, G. (2001). Complex mate searching in the satin bowerbird *Ptilonorhynchus violaceus*. *American Naturalist* 158: 530–542.

Valera, F., Hoi, H., & Kristin, A. (2003). Male shrikes punish unfaithful females. *Behavioral Ecology* 14: 403–408.

van Anders, S. M., Hamilton, L. D., Schmidt, N., & Watson, N. V. (2007). Associations between testosterone secretion and sexual activity in women. *Hormones and Behavior* 51: 477–482.

Van de Crommacker, J., Richardson, D. S., Groothuis, T. G. G., Eising, C. M., Dekker, A. L., & Komdeur, J. (2004). Testosterone, cuckoldry risk and extra-pair opportunities in the Seychelles warbler. *Proceedings of the Royal Society of London* B 271: 1023–1031.

Van Dongen, S. (2000). The heritability of fluctuating asymmetry: A Bayesian hierarchical model. *Annals Zoologica Fennici* 37: 15–23.

Van Dongen, S. (2006). Fluctuating asymmetry and developmental instability in evolutionary biology: Past, present and future. *Journal of Evolutionary Biology* 19: 1727–1743.

Van Dongen, S., Talloen, W., & Lens, L. (2005). High variation in developmental instability under non-normal developmental error: A Bayesian perspective. *Journal of Theoretical Biology* 236: 263–275.

VanGoozen, S. H. M., Frijda, N. H., Wiegant, V. M., Endert, E., & VanderPoll, N. E. (1996). The premenstrual phase and reactions to aversive events: A study of hormonal influences on emotionality. *Psychoneuroendocrinology* 21: 479–497.

van Honk, J., Peper, J. S., & Schutter, D. J. L. G. (2005). Testosterone reduces unconscious fear but not consciously experienced anxiety: Implications for the disorders of fear and anxiety. *Biological Psychiatry* 58: 218–225.

van Honk, J., Tuiten, A., van den Hout, M., Koppeschaar, H., Thijssen, J., de Haan, E., & Verbaten, R. (2000). Conscious and preconscious selective attention to social threat: Different neuroen-docrine response patterns. *Psychoneuroendocrinology* 25: 577–591.

van Honk, J., Tuiten, A., Verbaten, R., van den Hout, M., Koppeschaar, H., Thijssen, J., & de Haan, E. (1999). Correlations among salivary testosterone, mood, and selective attention to threat in humans. *Hormones and Behavior* 36: 17–24.

van Schaik, C. P. (2000). Infanticide by male primates: The sexual selection hypothesis revis-ited. In *Infanticide by Males and Its Implications* (eds. C. P. Van Schaik & C. Janson), pp. 27–60. Cambridge University Press, Cambridge, UK.

van Schaik, C. P., & Dunbar, R. I. M. (1990). The evolution of monogamy in large primates: A new hypothesis and some crucial tests. *Behaviour* 115: 30–62.

van Schaik, C. P., Hodges, J. K., & Nunn, C. L. (2000). Paternity confusion and the ovarian cycles of female primates. In *Infanticide by Males and Its Implications* (eds. C. P. J. C. van Schaik), pp. 361–387. Cambridge University Press, Cambridge, MA.

van Schaik, C. P., Prodham, G. R., & Van Noordwijk, M. A. (2004). Mating conflict in primates: Infanticide, sexual harassment and female sexuality. In *Sexual Selection in Primates: New and Comparative Perspectives* (eds. P. Kappler & C. van Schaik), pp. 131–150. Cambridge University Press, Cambridge, UK.

Van Valen, L. (1962). A study of fluctuating asymmetry. *Evolution* 10: 139–146.

Van Valen, L. (1973). A new evolutionary law. *Evolutionary Theory* 1: 1–30.

VanVliet, J. H., & VanEerdenburg, F. J. C. M. (1996). Sexual activities and oestrus detection in lactating Holstein cows. *Applied Animal Behaviour Science* 50: 57–69.

Vaughn, J. T., Macmillan, K. L., Anderson, G. A., & D'Occhio, M. J. (2003). Effects of mating behav-iour and the ovarian follicular state of female alpacas on conception. *Australian Veterinary Journal* 81: 86–90.

von Schantz, T., Bensch, S., Grahn, M., Hasselquist, D., & Wittzell, H. (1999). Good genes, oxida-tive stress and condition-dependent sexual signals. *Proceedings of the Royal Society of London* B 266: 1–12.

Wagner, R. H. (1992). The pursuit of extra-pair copulations by female razorbills: How do females benefit? *Behavioral Ecology and Sociobiology* 29: 455–464.

Waitt, C., Gerald, M. S., Little, A. C., & Kraiselburd, E. (2006). Selective attention toward female secondary sexual color in male rhesus monkeys. *American Journal of Primatology* 68: 738–744.

Waitt, C. Little, A. C., Wolfensohn, S., Honess, P. Brown, A. P., Buchanan-Smith, H. M., & Perrett, D. I. (2003). Proceedings of the Royal Society of London B 270: 144–146.

Wakano, J. Y., & Ihara, Y. (2005). Evolution of male parental care and female multiple mating: Game-theoretical and two-locus diploid models. *American Naturalist* 166: E32–E44.

Walker, A. E. (1997). *The Menstrual Cycle.* Routledge, New York.

Wallen, K. (2000). Risky business: Social context and hormonal modulation of primate sex drive. In *Reproduction in Context* (eds. K. Wallen & J. E. Schneider), pp. 289–323. MIT Press, Cambridge, MA.

Wallis, J., & Goodall, J. (1993). Anogenital swelling in pregnant chimpanzees of Gombe National Park. *American Journal of Primatology* 31: 89–98.

Wallner, S. J., Luschnigg, N., Schnedl, W. J., Lahousen, T., Sudi, K., Crailsheim, K., Møller, R., Tafeit, E., & Horejsi, R. (2004). Body fat distribution of overweight females with a history of weight cycling. *International Journal of Obesity* 28: 1143–1148.

Watson, P. J. (1986). Transmission of a female sex phermone thwarted by males in the spider *Linyphia litigosa* (Linyphiidae). *Science* 233: 219–221.

Watson, P. J., & Thornhill, R. (1994). Fluctuating asymmetry and sexual selection. *Trends in Ecology and Evolution* 9: 21–25.

Waynforth, D. (1998). Fluctuating asymmetry and human male life history, traits in rural Belize. *Proceedings of the Royal Society of London B* 265: 1497–1501.

Weatherhead, P. J., & Robertson, R. J. (1979). Offspring quality and the polygyny threshold: The sex son hypothesis. *American Naturalist* 113: 201–208.

Wedekind, C., & Füri, S. (1997). Body odor preference in men and women: Do they aim for specific MHC combinations or simply heterozygosity? *Proceedings of the Royal Society of London B* 264: 1471–1479.

Wedekind, C., Seebeck, T., Bettens, F., & Paepke, A. J. (1995). MHC-dependent mate preferences in humans. *Proceedings of the Royal Society of London, B* 260: 245–249.

Wedell, N., Gage, M. J. G., & Parker, G. A. (2002). Sperm competition, male prudence and sperm-limited females. *Trends in Ecology and Evolution* 17: 313–320.

Wedell, N., Kvarnemo, C., Lessells, C. K. M., & Tregenza, T. (2006). Sexual conflict and life histories. *Animal Behaviour* 71: 999–1011.

Wegner, K. M., Reusch, T. B. H., & Kalbe, M. (2003). Multiple parasites are driving major histocompatibility complex polymorphism in the wild. *Journal of Evolutionary Biology* 16: 224–232.

Weeden, J., & Sabini, J. (2005). Physical attractiveness and health in Western societies: A review. *Psychological Bulletin* 131: 635–653.

Weingrill, T., Lycett, J. E., Barrett, L., Hill, R. A., & Henzi, S. P. (2003). Male consortship behaviour in chacma baboons: The role of demographic factors and female conceptive probabilities. *Behaviour* 140: 405–427.

Weisfeld, G. (1997). Puberty rites as clues to the nature of human adolescence. *Cross-cultural Research* 31: 27–54.

Weisfeld, G. E. (1999). *Evolutionary Principles of Human Adolescence.* Basic Books, New York.

Weisfeld, G. E., & Woodward, L. (2004). Current evolutionary perspectives on adolescent romantic relations and sexuality. *Journal of the American Academy of Child and Adolescent Psychiatry* 43: 11–19.

Weiss, S. L. (2002). Reproductive signals of female lizards: Pattern of trait expression and male response. *Ethology* 108: 793–813.

Weiss, S. L. (2006). Female-specific color is a signal of quality in the striped plateau lizard (*Sceloporus virgatus*). *Behavioral Ecology* 17: 726–732.

Welling, L. L. M., Jones, B. C., DeBruine, L. M., Conway, C. A., Law Smith, M. J., Little, A. C., Feinberg, D. R., Sharp, M. A., & Al-Dujaili, E. A. S. (2007). Raised salivary testosterone in women is associated with increased attraction to masculine faces. *Hormones and Behavior* 52: 156–161.

West-Eberhard, M. J. (1984). Sexual selection, competitive communication and species-specific signals in insects. In *Insect Communication* (ed. T. Lewis), pp. 283–324. Academic Press, New York.

West-Eberhard, M. J. (2003). *Developmental Plasticity and Evolution.* Oxford University Press, New York.

Westermarck, E. (1929). *Marriage.* New York: Jonathan Cape & Harrison Smith.

Westneat, D. F., & Sherman, P. W. (1997). Density and extra-pair fertilizations in birds: A comparative analysis. *Behavioral Ecology and Sociobiology* 41: 205–215.

Weston, E. M., Friday, A. E., & Liò, P. (2007). Biometric evidence that sexual selection has shaped the hominin face. *PLOS One* 2: e710.

White, L. M., Hosack, D. A., Warren, R. J., & Fayrerhosken, R. A. (1995). Influence of mating on duration of estrus in captive white-tailed deer. *Journal of Mammalogy* 76: 1159–1163.

Whiteman, E. A., & Cote, I. M. (2004). Monogamy in marine fishes. *Biological Reviews* 79: 351–375.

Whitlock, M. C., & Fowler, K. (1997). The instability of studies of instability. *Journal of Evolutionary Biology* 10: 63–67.

Whittier, J. M., & Tokarz, R. R. (1992). Physiological regulation of sexual behavior in female reptiles. In *Hormones, Brain, and Behavior: Biology of the Reptilia,* Vol. 18: *Physiology E* (eds. C. Gans & D. Crews), pp. 24–69. University of Chicago Press, Chicago, IL.

Wikipedia, The Free Encyclopedia. (5/18/08.) Estrous Cycle. Http://en.wikipedia.org/wiki/Estrous_cycle.

Wilcox, A. J., Baird, D. D., Dunson, D. B., McConnaughey, D. R., Kesner, J. S., & Weinberg, C. R. (2004). On the frequency of intercourse around ovulation: Evidence for biological influences. *Human Reproduction* 19: 1539–1543.

Wilcox, A. J., Dunson, D., & Baird, D. D. (2000). The timing of the "fertile window" in the menstrual cycle: Day-specific estimates from a prospective study. *British Medical Journal* 321: 1259–1262.

Wilcox, A. J., Dunson, D. B., Weinberg, C. R., Trussell, J., & Baird, D. D. (2001). Likelihood of conception with a single act of intercourse: Providing benchmark rates for assessment of post-coital contraceptives. *Contraception* 63: 211–215.

Wilcox, A. J., Weinberg, C. R., & Baird, D. D. (1995). Timing of sexual intercourse in relation to ovulation: Effects on the probability of conception, survival of the pregnancy, and sex of the baby. *New England Journal of Medicine* 333: 1517–1521.

Williams, G. C. (1966). *Adaptation and Natural Selection: A Critique of Some Current Evolutionary Thought*. Princeton University Press, Princeton, NJ.

Williams, G. C. (1992). *Natural Selection: Domains, Levels and Challenges*. Oxford University Press, Oxford, UK.

Williams, R. J., & Lenington, S. (1993). Factors modulating preferences of female house mice for males differing in T-complex genotype: Role of T-complex genotype, genetic background, and estrous condition of females. *Behavior Genetics* 23: 51–58.

Willson, W. F., & Burley, N. (1983). *Mate Choice in Plants*. Princeton University Press, Princeton, NJ.

Wilson, E. O. (1975). *Sociobiology: The New Synthesis*. Harvard University Press, Cambridge, MA.

Wilson, M., & Daly, M. (1992). The man who mistook his wife for a chattel. In *The Adapted Mind: Evolutionary Psychology and the Generation of Culture* (eds. J. H. Barkow, L. Cosmides, & J. Tooby), pp. 289–326. Oxford University Press, Oxford, UK.

Wilson, M., & Mesnick, S. L. (1997). An empirical test of the bodyguard hypothesis. In *Feminism and Evolutionary Biology* (ed. P. A. Gowaty), pp. 505–511. Chapman & Hall, New York.

Winking, J., Kaplan, H. S., Gurven, M., & Rucas, S. (2007). Why do men marry and why do they stray? *Proceedings of the Royal Society of London B* 274: 1643–1649.

Wolf, L. L. (1975). Prostitution behavior in a tropical hummingbird. *Condor* 77: 140–144.

Wolfe, K., & Schulman, S. (1984). Male response to "strange" females as a function of female reproductive value among chimpanzees. *American Naturalist* 123: 163–174.

Wolff, J. O. (1998). Breeding strategies, mate choice, and reproductive success in American bison. *Oikos* 83: 529–544.

Wolff, J. O., & Macdonald, D. W. (2004). Promiscuous females protect their offspring. *Trends in Ecology and Evolution* 19: 127–134.

Wood, B., & Constantino, P. (2004). Human origins: Life at the top of the tree. In *Assembling the Tree of Life* (eds. J. Cracraft & M. J. Donoghue), pp. 517–535. Oxford University Press, New York.

Woodward, K., & Richards, M. H. (2005). The parental investment model and minimum mate choice criteria in humans. *Behavioral Ecology* 16: 57–61.

Wrangham, R. W. (1993). The evolution of sexuality in chimpanzees and bonobos. *Human Nature: An Interdisciplinary Biosocial Perspective* 4: 47–79.

Wrogemann, D., & Zimmermann, E. (2001). Aspects of reproduction in the eastern rufous mouse lemur (*Microcebus rufus*) and their implications for captive management. *Zoo Biology* 20: 157–167.

Wysocki, D., & Halupka, K. (2004). The frequency and timing of courtship and copulation in blackbirds, *Turdus merula*, reflect sperm competition and sexual conflict. *Behaviour* 141: 501–512.

Yamazaki, K., Beauchamp, G. K., Curran, M., Baird, J., & Boyse, E. A. (2000). Parent-progeny recognition as a function of MHC odortype identity. *Proceedings of the National Academy of Sciences USA* 97: 10500–10502.

Yasui, Y. (1998). The "genetic benefits" of female multiple mating reconsidered. *Trends in Ecology and Evolution* 13: 246–250.

Yeo, R., Gangestad, S. W., Thoma, R. J., Shaw, P. K., & Repa, K. (1997). Developmental instability and cerebral lateralization. *Neuropsychology* 11: 552–561.

Yeo, R. A., Gangestad, S. W., Edgar, C., & Thoma, R. J. (1999). The evolutionary–genetic underpinnings of schizophrenia: The developmental instability model. *Schizophrenia Research* 39: 197–202.

Yeo, R. A., Hill, D., Campbell, R., Vigil, J., & Brooks, W. M. (2000). Developmental instability and working memory ability in children: A magnetic resonance spectroscopy investigation. *Developmental Neuropsychology* 17: 143–159.

Yezerinac, S. M., & Weatherhead, P. J. (1997). Reproductive synchrony and extra-pair mating strategy in a socially monogamous bird, *Dendroica petechia*. *Animal Behaviour* 54: 1393–1403.

Young, L. J., Nag, P. K., & Crews, D. (1995). Regulation of estrogen-receptor and progesterone-receptor messenger-ribonucleic-acid by estrogen in the brain of the whiptail lizard (*Cnemidophorus uniparens*). *Journal of Neuroendocrinology* 7: 119–125.

Yu, D. E., & Shepard, G. H. (1998). Is beauty in the eye of the beholder? *Nature* 396: 321–322.

Yund, P. O. (2000). How severe is sperm limitation in natural populations of marine free-spawners? *Trends in Ecology and Evolution* 15: 10–13.

Zahavi, A. (1975). Mate selection: A selection for a handicap. *Journal of Theoretical Biology* 53: 205–214.

Zahavi, A. (1977). Cost of honesty (Further remarks on handicap principle). *Journal of Theoretical Biology* 67: 603–605.

Zahavi, A., & Zahavi, A. (1997). *The Handicap Principle: A Missing Piece of Darwin's Puzzle*. Oxford University Press, New York.

Zebrowitz, L. A., & Rhodes, G. (2004). Sensitivity to "bad genes" and the anomalous face over-generalization effect: Cue validity, cue utilization, and accuracy in judging intelligence and health. *Journal of Nonverbal Behavior* 28: 167–185.

Zeh, J. A., & Zeh, D. W. (1996). The evolution of polyandry: I. Intragenomic conflict and genetic incompatibility. *Proceedings of the Royal Society of London B* 263: 1711–1717.

Zeh, J. A., & Zeh, D. W. (1997). The evolution of polyandry: II. Post-copulatory defenses against genetic incompatibility. *Proceedings of the Royal Society of London B* 264: 69–75.

Zeh, J. A., & Zeh, D. W. (2001). Reproductive mode and the genetic benefits of polyandry. *Animal Behavior* 61: 1051–1063.

Zinner, D., & Deschner, T. (2000). Sexual swellings in female hamadryas baboons after male take-overs: "Deceptive" swellings as a possible female counter-strategy against infantcide. *American Journal of Primatology* 52: 157–168.

Zinner, D., Alberts, S. C., Nunn, C. L., & Altmann, J. (2002). Significance of primate sexual swelling. *Nature* 420: 142.

Zinner, D. P., Nunn, C. L., van Schaik, C. P., & Kappeler, P. M. (2004). Sexual selection and exaggerated sexual swellings of female primates. In *Sexual Selection in Primates: New and Comparative Perspectives* (eds. P. Kappeler & C. van Schaik), pp. 71–89. Cambridge University Press, Cambridge, UK.

Zuk, M., Thornhill, R., Ligon, J. D., Johnson, K., Austad, S., Ligon, S. H., Thornhill, N. W., & Costin, C. (1990). The role of male ornaments and courtship behavior in female mate choice of red jungle fowl. *American Naturalist* 136: 459–473.

Zuk, M., Thornhill, R., Ligon, D. J., & Johnson, K. (1990). Parasites and mate choice in red jungle fowl. *American Zoologist* 30: 235–244.

INDEX